Grass
Its Production and Utilizati

Grass
Its Production and Utilization

Third Edition

Edited by
Alan Hopkins
Institute of Grassland and Environmental Research, North Wyke, Okehampton, Devon, UK

Published for
the British Grassland Society
by
Blackwell Science Ltd

Blackwell
Science

Editorial Offices:
Osney Mead, Oxford OX2 0EL
25 John Street, London WC1N 2BL
23 Ainslie Place, Edinburgh EH3 6AJ
350 Main Street, Malden
MA 02148 5018, USA
54 University Street, Carlton
Victoria 3053, Australia
10, rue Casimir Delavigne
75006 Paris, France

Other Editorial Offices:

Blackwell Wissenschafts-Verlag GmbH
Kurfürstendamm 57
10707 Berlin, Germany

Blackwell Science KK
MG Kodenmacho Building
7–10 Kodenmacho Nihombashi
Chuo-ku, Tokyo 104, Japan

Iowa State University Press
A Blackwell Science Company
2121 S. State Avenue
Ames, Iowa 50014-8300, USA

First published 1980, reprinted 1982
Second edition 1989, reprinted 1994
Third edition published 2000
Reprinted 2001, with corrections

Set in 10 on 12.5 pt Times
by Best-set Typesetter Ltd., Hong Kong
Printed and bound in Great Britain by
MPG Books Ltd,
Bodmin, Cornwall

DISTRIBUTORS

Marston Book Services Ltd
PO Box 269
Abingdon
Oxon OX14 4YN
(*Orders:* Tel: 01235 465500
Fax: 01235 465555)

USA
Blackwell Science, Inc.
Commerce Place
350 Main Street
Malden, MA 02148 5018
(*Orders:* Tel: 800 759 6102
781 388 8250
Fax: 781 388 8255)

Canada
Login Brothers Book Company
324 Saulteaux Crescent
Winnipeg, Manitoba R3J 3T2
(*Orders:* Tel: 204 837-2987)
Fax: 204 837-3116)

Australia
Blackwell Science Pty Ltd
54 University Street
Carlton, Victoria 3053
(*Orders:* Tel: 03 9347 0300
Fax: 03 9347 5001)

A catalogue record for this title
is available from the British Library

ISBN 0-632-05017-9

Library of Congress
Cataloging-in-Publication Data

Grass : its production and utilization / edited
by Alan Hopkins. – 3rd ed.
p. cm.
Includes bibliographical references.
ISBN 0-632-05017-9 (hb)
1. Grasses. 2. Range management.
3. Rangelands. 4. Grasses – Economic
aspects. I. Hopkins, Alan. II. British
Grassland Society.

SB197.G7 2000
633.2 – dc21

99-059353

For further information on
Blackwell Science, visit our website:
www.blackwell-science.com

Contents

List of Contributors

DAVID E. BEEVER, Department of Agriculture, University of Reading, Earley Gate, Reading, RG6 2AT, UK.

DAVID F. CHAPMAN, Department of Animal Production, Institute of Land and Food Resources, University of Melbourne, Parkville, Victoria 3052, Australia.

GEORGE E.J. FISHER, Kemira Agro UK Ltd, Ince, Chester, CH2 4LB, UK.

MARGARET GILL, Natural Resources International, Chatham Maritime, Chatham, Kent, ME4 4TB, UK.

DAVID H. HIDES, c/o. Institute of Grassland and Environmental Research, Plas Gogerddan, Aberystwyth, SY23 3EB, UK.

ALAN HOPKINS, Institute of Grassland and Environmental Research, North Wyke, Okehampton, Devon, EX20 2SB, UK.

RICHARD G. JEFFERSON, English Nature, Northminster House, Peterborough, Cambridgeshire, PE1 1UD, UK.

RAYMOND JONES, Institute of Grassland and Environmental Research, Plas Gogerddan, Aberystwyth, SY23 3EB, UK.

GRAHAM C. LEWIS, Institute of Grassland and Environmental Research, North Wyke, Okehampton, Devon, EX20 2SB, UK.

ATHOLE H. MARSHALL, Institute of Grassland and Environmental Research, Plas Gogerddan, Aberystwyth, SY23 3EB, UK.

C. SINCLAIR MAYNE, Agricultural Research Institute of Northern Ireland, Hillsborough, Co Down, BT26 6DR, UK

JOHN P. McINERNEY OBE, Agricultural Economics Unit, University of Exeter, Lafrowda House, St German's Rd., Exeter, EX4 6TL, UK.

ROGER J. MERRY, Institute of Grassland and Environmental Research, Plas Gogerddan, Aberystwyth, SY23 3EB, UK.

NICK W. OFFER, SAC, Auchincruive, Ayr, KA6 5HW, UK.

BRIAN F. PAIN, Institute of Grassland and Environmental Research, North Wyke, Okehampton, Devon, EX20 2SB, UK.

ANTHONY J. PARSONS, AgResearch Grasslands, Private Bag 11008, Palmerston North, New Zealand.

HEATHER J. ROBERTSON, English Nature, Northminster House, Peterborough, Cambridgeshire, PE1 1UD, UK.

ANDREW J. ROOK, Institute of Grassland and Environmental Research, North Wyke, Okehampton, Devon, EX20 2SB, UK.

ROGER D. SHELDRICK OBE, c/o Institute of Grassland and Environmental Research, North Wyke, Okehampton, Devon, EX20 2SB, UK.

MICHAEL K. THEODOROU, Institute of Grassland and Environmental Research, Plas Gogerddan, Aberystwyth, SY23 3EB, UK.

DANNY THOROGOOD, Institute of Grassland and Environmental Research, Plas Gogerddan, Aberystwyth, SY23 3EB, UK.

IAIN A. WRIGHT, Macaulay Land Use Research Institute, Craigiebuckler, Aberdeen, AB15 8QH, UK.

DAVID YOUNIE, SAC, Ferguson Building, Craibstone Estate, Bucksburn, Aberdeen, AB21 9YA, UK.

Foreword

by C.K. Mackie, President of the British Grassland Society, 1998–1999

This, the third edition of *Grass: its Production and Utilization*, not only updates the previous versions but also widens the range of topics covered, reflecting the changes that have taken place in land use and the agri-food industry over the last ten years.

The British Grassland Society asked the new editor, Alan Hopkins, to widen the scope and expand the team of contributors enlisted by his predecessor Professor W. Holmes. The areas covered include herbage seed production, organic farming, farm wastes, amenity grass and nature conservation.

Many developments from research in recent years have considerably enhanced our knowledge of efficient grassland use and led to the introduction of new technologies. This is particularly evident within the sections on grazing management and forage conservation.

Grass: its Production and Utilization is recognized as an authoritative textbook on grassland science and husbandry and is essential reading for students, researchers, advisers and forward-looking farmers.

Preface to the third edition

The first edition of this book, under the editorship of Professor W. Holmes, appeared in 1980, and a second edition with revisions and additional material followed in 1989.

In 1996 the Council of the British Grassland Society invited me to organize and prepare a third edition, with a brief that it should be a book for the twenty-first century, wider in its scope than the previous editions, and reflecting the impact of grassland research and policy developments since the 1980s. The book maintains its emphasis on the technical aspects of production and utilization under agricultural management, but it has been expanded to include areas not covered in previous editions, or that had been dealt with only superficially. It is intended to meet the requirements of students in a range of disciplines concerned with agriculture and rural studies, as well as reference work for farmers, scientists and other professionals.

The content of this book is the responsibility of the authors and editor. The contributions of many colleagues are also acknowledged, and in some instances text and figures from the previous editions of this book have been included. In particular we thank Dr M.J. Robson and Professors J.C. Murdoch and W. Holmes for permission to incorporate material from their previous contributions.

Alan Hopkins
Institute of Grassland and Environmental Research
North Wyke
Okehampton
Devon EX20 2SB
UK

Chapter 1
Introduction

A. Hopkins

1.1 Objectives

The term 'grassland' refers to a plant community in which grasses (*Graminaea*) are usually the dominant species, with forbs (herbaceous dicotyledonous species, including legumes) present in variable amounts, but trees and shrubs absent or only minor constituents. In the British Isles, grassland is a major agricultural resource, supporting both intensive and extensive systems of ruminant livestock production. It is also important for livestock agriculture in many other European countries, and in temperate and tropical regions elsewhere in the world (Breymeyer 1990). Grassland also has important non-agricultural roles for amenity, recreation and wildlife conservation. These aspects, together with the environmental aspects of grassland farming, have received increased recognition and scientific study, particularly since the 1980s.

It is the aim of this book to provide a wide-ranging text on the current state of knowledge on the production, feeding value and utilization of grassland in a temperate climate, and its role in the wider environment. The information assembled by the authors has been updated and increased in scope compared with the previous editions of this book, reflecting scientific advances and policy changes affecting the management of grassland. The authors have drawn largely, although not exclusively, on British and European examples, but the principles can be applied to temperate grasslands in other areas of the world. The book is intended to meet the requirements of students in a range of disciplines associated with agriculture and rural studies. It is also intended to serve as a reference work for farmers and others having a professional interest in agriculture and countryside management, and to provide a broad outline of the various aspects of grassland science for more specialist scientists. The reader requiring more detailed sources of information on the topics covered will find a comprehensive reference list at the end of each chapter.

1.2 The development and distribution of grasslands

There are few areas of natural grassland in the British Isles. The climax vegetation of most humid temperate regions is forest, and almost all of the grazing lands are the result of past human activities. Forest clearance for timber, cultivation of clearings for crops, and the control of the regeneration by cattle and sheep, or by cutting, have formed the majority of our pastures. The enormous economic value of sheep for wool, as the principal source of fibre for clothing, and their ability to forage during winter probably account for the prominence of sheep in the early pastoral scene and the conversion of woodland grazings into pastures and meadows (Green 1990). These influences have resulted in large areas of the British Isles developing, and remaining, as grassland. Events that occurred over many centuries have also been repeated in other temperate regions. In New Zealand, for example, a considerable cattle and sheep industry was developed by forest clearance to grassland during a period of about 80 years (Levy 1955). Whether in the British Isles or elsewhere, the driving force in the development of grasslands has been population growth and the increased demand for the products of ruminant livestock, combined with physical constraints on the land which restrict the opportunities for other, more profitable, forms of agriculture.

In the UK, based on information obtained from agricultural census returns and sample surveys of farms, the grassland area can be divided into temporary grass, permanent grass and rough grazing, and the major species whose presence and abundance are relevant to the present argument can be examined (Lazenby 1981, Green 1990). The UK's annual agricultural census defines temporary grassland as that sown within the previous five years. Almost all grassland in this category has been sown to ryegrasses (*Lolium perenne* or *L. multiflorum*), either alone or with a legume. Other cultivated grasses, such as cocksfoot (*Dactylis glomerata*), timothy (*Phleum pratense*) and meadow fescue (*Festuca pratensis*), are minor sward constituents or restricted to special situations, e.g. on drought-prone sites or for winter hardiness. Permanent grassland includes older sown swards consisting of a mix of sown species and volunteer grasses, e.g. meadow grasses (*Poa* spp.), bent (*Agrostis* spp.) and Yorkshire fog (*Holcus lanatus*), that invade the ageing sward, through to semi-natural swards, often recognized as having important landscape or nature conservation values. The term 'rough grazing' refers mainly to unenclosed land, predominantly in the cool wet uplands of the north and west. The short growing season and leached acid soils give rise to vegetation of low forage value, including the potentially injurious bracken fern (*Pteridium aquilinum*), and shrubby species such as heather and bilberry (*Calluna vulgaris* and *Vaccinium* spp.). However, grasses usually predominate, notably the purple moor-grass (*Molinia caerulea*), hair-grasses (*Deschampsia* spp.), mat-grass (*Nardus stricta*) and sheep's fescue (*Festuca ovina*). Rough grazings in lowland areas include salt marsh, heaths and enclosed areas affected by slopes or obstructions that limit their use.

Table 1.1 shows the areas of Great Britain designated as temporary grass, permanent grass or rough grazing at intervals since 1875, when compilation of official statistics began. Changes in the total area of agricultural land as a result of urbanization and changes in the definition of the categories prevent precise comparison, but throughout this period the area of permanent and temporary grassland on the lowlands has exceeded the area of arable crops.

The distribution of grassland and its contribution as a proportion of total agricultural land is not uniform, but reflects the combination of rainfall, temperature, soil and topography (Fig. 1.1). The highest levels of grass production,

Table 1.1 Changes in land use in Great Britain from 1875 to 1997 expressed in M ha.

Land use	1875	1938*	1944*	1980	1997
Total crops	5.57	3.36	5.55	4.89	4.93
Temporary grass	1.76	1.44	1.71	1.69	1.22
Permanent grass	5.39	7.01	4.37	4.64	4.64
Rough grazing	—	6.50	6.69	6.12	5.69

* The years 1938 and 1944 were years of minimal and maximal tillage, respectively.

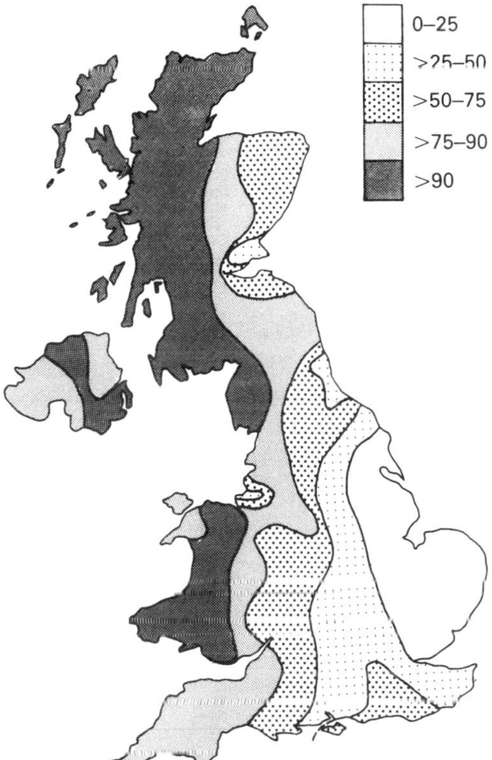

Fig. 1.1 Percentage of agricultural land under grass and rough grazing in the UK (from agricultural census data 1995).

and the livestock it supports, are achieved where rainfall is well distributed, soils have a good water-holding capacity and grass growth is not limited by extremes of environment. The maritime climate of western Britain meets these requirements well (Lazenby & Down 1982, Green 1990). The oceanic influence decreases from west to east, giving lower rainfall and colder winters, as well as drier summers which allow grain to ripen, and harvesting and cultivation operations to be completed in a timely manner. Also, in the east, flatter land with drier summers is better suited to large-scale cultivation and the growing and harvesting of arable crops, while in the west, slopes and soil conditions often limit the extent of land that can be cultivated in any one year. However, the longer growing season allows high levels of grass production to be sustained from permanent swards. On steeper sites and on upland pastures, grazing by sheep and beef cattle extend the area for food production on to marginal land where crops for human consumption could not be grown.

1.3 Classification of grasslands – ecological

The National Vegetation Classification (NVC) (Rodwell 1991, 1992) provides a systematic classification of British vegetation, and replaces the earlier non-systematic classifications of Tansley (1939) and Ratcliffe (1977). (A description of grasslands in Ireland is given by O'Sullivan, 1982.) The main NVC grassland and mire communities that occur on agriculturally utilized grassland are summarized in Tables 1.2 and 1.3 (see also Crofts & Jefferson 1994). This diversity of types, including sub-divisions of the main communities, reflects the complexity of regional geology, climate, hydrology and management. Four main NVC groups of grassland are recognized, based on soil type: mesotrophic, calcicolous, calcifugous and fen meadow, as defined below.

Table 1.2 National Vegetation Classification of grassland communities in Great Britain: lowland neutral (mesotrophic) grasslands of low botanical interest.

MG1	*Arrhenatherum elatius* grassland. Unmanaged coarse grassland on neutral soils, e.g. road verges and neglected agricultural habitats
MG6	*Lolium perenne – Cynosurus cristatus*. The major permanent pasture type of lowlands, often brought about by the action of fertilizers, drainage, and herbicides on other MG types, or by the deterioration of MG7
MG7	*Lolium perenne* – reseeded grassland. The major sown grassland type in Britain
MG9	*Holcus lanatus – Deschampsia cespitosa* and
MG10	*Holcus lanatus – Juncus effesus* pastures. Characteristic of permanently moist soils in lowland areas, often where *D. cespitosa* or *Juncus* have invaded MG6 or MG7 under poor drainage conditions

Table 1.3 National Vegetation Classification of grassland communities in Great Britain: lowland mesotrophic (MG), calcareous (CG), acid (U) and mire (M) communities of high botanical interest.

Mesotrophic grasslands	
MG2	*Filipendula ulmaria – Arrhenatherum elatius*: northern tall herb grassland on Carboniferous limestone
MG3	*Anthoxanthum odoratum – Geranium sylvaticum*: northern hay meadows
MG4	*Alopecurus pratensis – Sanguisorba officinalis*: flood meadows
MG5	*Cynosurus cristatus – Centaurea nigra*: lowland hay meadow and pasture (widely scattered throughout the British lowlands)
MG8	*Cynosurus cristatus – Caltha palustris*: flood pasture including former water meadows
MG11	*Festuca rubra – Agrostis stolonifera – Potentilla anserina*: inundation grassland
MG13	*Agrostis stolonifera – Alopecurus geniculatus*: inundation grassland, silver meadows
Calcareous grasslands	
CG1	*Festuca ovina – Carlina vulgaris*: warm temperate limestone grassland
CG2	*Festuca ovina – Helictotrichon pratensis*: species-rich chalk grassland
CG3	*Bromopsis erectus* grassland
CG4	*Brachypodium pinnatum* grassland
CG5	*Bromopsis erectus – Brachypodium pinnatum* grassland
CG6	*Helictotrichon pubescens* grassland
CG7	*Festuca ovina – Pilosella officinarum – Thymus polytrichus* grassland
CG8	*Sesleria caerulea – Scabiosa columbaria*: magnesian limestone grassland
CG9	*Sesleria caerulea – Galium sterneri*: Carboniferous limestone grassland
CG10	*Festuca ovina – Agrostis capillaris – Thymus polytrichus* grassland
Acid grasslands	
U1	*Festuca ovina – Agrostis capillaris – Rumex acetosella* grassland
U2	*Deschampsia flexuosa* grassland
U3	*Agrostis curtistii* grassland
U4	*Festuca ovina – Agrostis capillaris – Galium saxatile* grassland
U5	*Nardus stricta – Galium saxatile* grassland
U6	*Juncus squarrosus – Festuca ovina* grassland
	(U4 – U6 are principally communities of upland areas)
Mires	
M22	*Juncus subnodilosus – Cirsium palustre* fen meadow
M23	*Juncus effusus/acutiflorus – Galium palustre* rush pasture
M24	*Molinia caerulea – Cirsium dissectum* fen meadow
M25	*Molinia caerulea – Potentilla erecta* mire
M26	*Molinia caerulea – Crepis paludosa* mire

(a) Mesotrophic

Grasslands of circum-neutral clay and loam soils, including semi-natural communities (MG1–MG5, MG8–MG13) and anthropogenic grasslands created by fertilizers, lime and reseeding (MG6–MG7).

(b) Calcicolous

Grasslands on nutrient-poor soils derived from calcareous parent materials (CG1–CG10).

(c) Calcifugous (acid)
Grasslands on nutrient-poor acid substrates (U1–U6).

(d) Fen meadow
Wet grassland or mire communities developed over acid or circum-neutral peat, or wet mineral soils (M22–M26).

The NVC is a valuable tool in relation to the description and management of grassland of nature conservation value (considered in greater detail in Chapter 11). However, it is of limited value for understanding the ecology and management of intensively managed grasslands, the majority of which fall within either the MG6 *Lolium perenne* – *Cynosurus cristatus* grassland type, or MG7 *Lolium perenne* ley grasslands (Table 1.3).

1.4 Classification of grasslands – agricultural

In contrast to ecologists, agronomists have placed emphasis upon a limited number of economically important species in description and classification, mainly grasses, sown legumes and common weeds. A simple method of botanical analysis for assessing agricultural value was used for the first grassland surveys in the 1930s, and provided the basis of grassland mapping of England and Wales (Davies 1941). Lowland swards were classified into one of seven main pasture types (Table 1.4). The proportion of perennial ryegrass was the basis of the classification; e.g. first-grade was defined as >30% perennial ryegrass, plus white clover and other productive grasses, and second-grade as 15–30% ryegrass and a greater abundance of *Agrostis*. These two grades accounted for less

Table 1.4 Agricultural classification of grassland swards and major rough grazings (source: Davies 1941).

A	*Lowland types*
	First-grade ryegrass pasture
	Second-grade ryegrass pasture
	Agrostis – ryegrass pasture
	Agrostis pasture
	Agrostis with rushes and sedge
B	*Fescue pasture types*
	Mountain and hill fescue pasture
	Downland fescue pasture
C	*Rough and hill grazing*
	Nardus – fescue moorland
	Molinia and Molinia – Nardus moorland
	Cotton grass and deer grass moors
	Heather moor and fell
	Bilberry moor

Table 1.5 The average botanical composition of swards on enclosed grassland in England and Wales, classified by age (source: Hopkins & Davies 1994).

Species	Sward age			
	1–4 years	5–8 years	9–20 years	Over 20 years
Lolium spp.	64	47	38	24
Other sown grasses*	6	7	7	5
Sown legumes	6	5	5	4
Poa spp.	10	12	11	9
Agrostis spp.	7	17	22	27
Holcus lanatus	2	5	7	10
Festuca rubra	<1	1	2	7
Other grasses, etc.	<1	1	2	7
Dicotyledonous spp.	4	5	6	7

* Mainly *Dactylis glomerata* and *Phleum pratense*.

than 8% of lowland permanent pasture in the 1930s. Surveys carried out in later years (Forbes *et al.* 1980, Green 1982, Peel & Green 1984) continued to emphasize the percentage contribution of perennial ryegrass in relation to sward age, management and output. By the 1980s, perennial ryegrass had become so abundant (Table 1.5) that about 70% of over-20-year-old swards (and most of the younger sown grass) would have fitted the descriptions of first- or second-grade pasture.

1.5 Leys and permanent pastures

The impetus for the agricultural improvement of grassland in Britain during the second half of the twentieth century began with the ploughing-up campaign of the 1939–1945 wartime emergency. This was greatly influenced by the doctrine of ley farming (Stapledon & Davies 1942) and had been preceded by a long period of agricultural depression.

The essentials of ley farming were the rotation of crops with grass in a regular sequence, with temporary grass, sometimes referred to as a ley, normally lasting from 2 to 5 years. This provided a balanced system of farming with many technical and managerial advantages, including the sustainable management of soil fertility and some control of pests, weeds and diseases. In subsequent decades the support for ley farming waned as the availability of herbicides and cheap sources of fertilizers reduced the need for crop rotations. The high capital investment needed to support both livestock production and arable cropping also contributed to greater on-farm specialization.

These factors led to a resurgence of interest in permanent, or at least long-term, swards. Improved knowledge of grassland management has been a major factor in this trend, in part a legacy of the ley farming doctrine that grass should be regarded as a crop to be exploited by agronomic treatment. One area where ley farming is still considered to be an integral part of agricultural production is in organic farming (see Chapter 14) where, without the use of artificial fertilizers or agro-chemicals, grass/legume leys provide the residual fertility for subsequent arable crops.

1.6 Grassland production

Grass as a crop receiving and transforming solar energy into the products of photosynthesis has great potential for the production of biomass. The factors that determine how grass grows are described in Chapter 3, and the role of management inputs, soil nutrient status, environment and sward composition in determining levels of herbage production are the subject of Chapter 4. On temperate grasslands the sward covers the ground almost completely, so that the leaf area index (total leaf area in relation to ground area) is typically from 2 to 6, and light energy is received throughout the year. In these conditions, provided temperatures are not below 5°C and soil moisture is not limiting, some sward growth can be maintained throughout the year, although in practice there is a marked seasonal variation in production. In the British Isles, growing days range from 200 to 250 days in the hilly areas and summer-drought-prone areas of eastern England, to over 300 days in some lowland western regions (Fig. 1.2). Although maximal yields of about 25 t dry matter (DM) ha^{-1} could be produced from perennial ryegrass in the best grass-growing areas of Britain or New Zealand (Cooper 1970), and even greater theoretical potentials have been calculated (Leafe 1988), the very high inputs of nitrogen (N) required would be both uneconomic and environmentally damaging. In practice, annual herbage production from newly sown perennial ryegrass swards receiving about 250 kg fertilizer N ha^{-1} has been shown to vary between 10 and 18 t DM ha^{-1} (under a 6-weekly cutting regime) in seasons with good growing conditions, the higher values being associated with exceptionally good grass-growing sites (Hopkins *et al.* 1995). Production levels from grass/white clover swards, without inputs of N fertilizer, are similar to those of grass swards receiving up to 200 kg N ha^{-1} (Davies & Hopkins 1996).

1.7 Grass as a source of feed for livestock

In temperate latitudes where annual herbage production has a marked seasonal variation, with maximum production in late spring and early summer, efficient use of the grass crop depends on harvesting seasonal surplus production for conserved feed for winter rations. Forage was traditionally conserved as hay, made in late summer, and hay is still regarded as an important feed, particularly for young stock and horses. However, the widespread adoption

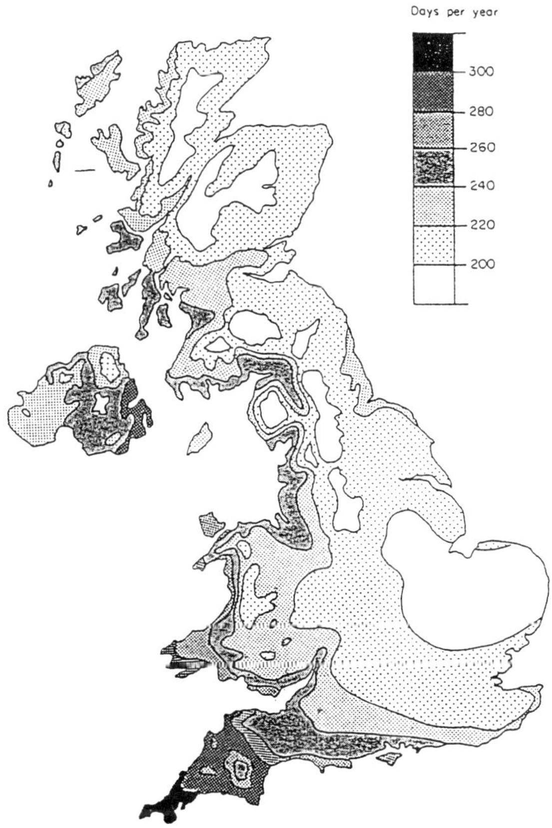

Fig. 1.2 Grass growing days – soil temperature adjusted for drought and altitude. (Reproduced from *Grass and Forage Science*, **36** (4) p. 253.)

of silage has been one of the major changes in grassland management practices in the UK since the 1970s, allowing a huge increase in the total amount of conserved forage. There have been considerable advances in silage-making technology during this period, including the development of additives to improve fermentation and of big-bale systems. These aspects are described in Chapter 8.

Whether supplied as conserved feed or grazed in situ, the feeding value of any forage depends on its ability to supply nutrients to the animal, and it has three main components: voluntary intake, nutrient content and the ability of the animal to absorb and utilize the nutrients. This subject is described in some detail in Chapter 7. Feed intake is affected by foraging and grazing behaviour, the subject of Chapter 9, and the management needed to control feed availability and nutritional quality that determine animal responses is described in Chapter 10.

1.8 Grass utilization

Except in the case of swards managed for amenity purposes, grass needs to be harvested either by mowing or by grazing animals, and subsequently utilized by animals, directly or after storage, before it is converted into useful products. Levels of utilized output are commonly expressed in terms of energy, e.g. GJ ha^{-1} of utilized metabolizable energy (UME). A high proportion of UME is used to meet the maintenance requirements of livestock, and UME output from grass shows large variations between farms, even for similar swards and under similar inputs of fertilizer N (Forbes *et al.* 1980). Losses occur under grazing, particularly under poor drainage conditions when poaching occurs, as well as in the harvesting, conservation and feeding out of the crop as hay or silage. In studies carried out on lowland commercial farms, on average only 67% of the herbage production was utilized by livestock (Peel *et al.* 1988).

1.9 Grassland production and environmental objectives

Increased production from grassland has been stimulated by government and EC policies that sought to address problems of food shortages and low levels of farm productivity, and through the availability of research findings and technological innovations. For example in the UK, between the mid-1940s and the mid-1980s, the output of beef and lamb doubled and that of milk trebled. This agricultural success was achieved at a considerable environmental cost, including the loss of many wildlife habitats and landscapes (Hopkins & Hopkins 1994). Surplus production of many agricultural commodities since the 1980s has required policy makers to impose limits on production, and to address the needs of the wider countryside and encourage low-input production systems and alternative land-use activities. There is also a growing demand for amenity grassland for sport and informal recreation, which is the subject of Chapter 12.

In many areas, such as designated Environmentally Sensitive Areas (Sheldrick 1997), farmers have been encouraged to adopt or maintain traditional practices through financially supported management agreements designed to deliver environmental benefits, including nature protection or public access. Research that supports grassland production has also become focused on the integration of production and environmental goals, including management for biodiversity, on technologies that improve nutrient management and limit pollution, and on low-input systems such as organic farming. These issues are considered in Chapters 11, 13 and 14, respectively.

1.10 References

Breymeyer A.I. (ed) (1990) *Ecosystems of the World 17A. Managed Grasslands – Regional Studies*. Elsevier, Amsterdam.

Cooper J.P. (1970) Potential production and energy conversion in temperate and tropical grasses. *Herbage Abstracts* **40**, 1–15.

Crofts A. & Jefferson R.G. (1994) *The Lowland Grassland Management Handbook*. English Nature/The Wildlife Trusts, Peterborough.

Davies D.A. & Hopkins A. (1996) Production benefits of legumes in grassland. In Younie D. (ed) *Legumes in Sustainable Farming Systems*, 234–46. British Grassland Society Occasional Symposium, No. 30. BGS, Reading.

Davies W. (1941) The grassland map of England and Wales – explanatory notes. *Agriculture* **48**, 112–21.

Forbes T.J., Dibb C., Green J.O., Hopkins A. & Peel S. (1980) *Factors Affecting the Productivity of Permanent Grassland – a National Farm Study*, 140 pp. GRI/ADAS Permanent Pasture Group, Hurley.

Green J.O. (1982) *A Sample Survey of Grassland in England and Wales, 1970–72*, 39 pp. Grassland Research Institute, Hurley.

Green J.O. (1990) The distribution and management of grassland in the British Isles. In Breymeyer A.I. (ed) *Ecosystems of the World 17A. Managed Grasslands – Regional Studies*, 15–35. Elsevier, Amsterdam.

Hopkins A. & Davies R.R. (1994) Changing grassland utilization in the United Kingdom and its implications for pollen production and hay fever. *Grana* **33**, 71–5.

Hopkins A. & Hopkins J.J. (1994) UK grasslands now: agricultural production and nature conservation. In Peel S. & Haggar R.J. (eds) *Grassland Management and Nature Conservation*, 10–19. British Grassland Society Occasional Symposium, No. 28. BGS, Reading.

Hopkins A., Murray P.J., Bowling P.J., Rook A.J. & Johnson J. (1995) Productivity and nitrogen uptake of ageing and newly sown swards of perennial ryegrass (*Lolium perenne* L.) at different sites and with different nitrogen fertilizer treatments. *European Journal of Agronomy* **4**, 65–75.

Lazenby A. (1981) British grasslands: past, present and future. *Grass and Forage Science* **36**, 243–66.

Lazenby A. & Down K.L. (1982) Realizing the potential of British grasslands: some problems and possibilities. *Applied Geography* **2**, 171–88.

Leafe E.L. (1988) Introduction – the history of improved grasslands. In Jones M.B. & Lazenby A. (eds) *The Grass Crop*, 1–23. Chapman & Hall, London.

Levy E.B. (1955) *Grasslands of New Zealand*. 2nd edn. Government Printer, Wellington.

O'Sullivan A.M. (1982) The lowland grasslands of Ireland. *Journal of Life Sciences, Royal Dublin Society* **3**, 131–42.

Peel S. & Green J.O. (1984) Sward composition and output on grassland farms. *Grass and Forage Science* **39**, 107–10.

Peel S., Matkin E.A. & Huckle C.A. (1988) Herbage growth and utilized output from grassland on dairy farms in southwest England: case studies from five farms, 1982 and 1983. II. Herbage utilization. *Grass and Forage Science* **43**, 71–8.

Ratcliffe D.A. (ed) (1977) *A Nature Conservation Review. Vols 1 & 2*. Cambridge University Press, Cambridge.

Rodwell J.S. (ed) (1991) *British Plant Communities 2. Mires and Heaths*. Cambridge University Press, Cambridge.

Rodwell J.S. (ed) (1992) *British Plant Communities. 3. Grassland and Montane Communities*. Cambridge University Press, Cambridge.

Sheldrick R.D. (ed) (1997) *Grassland Management in Environmentally Sensitive Areas.* British Grassland Society Occasional Symposium, No. 32. BGS, Reading.
Stapledon R.G. & Davies W. (1942) *Ley Farming*. Penguin, Harmondsworth.
Tansley A.G. (1939) *The British Islands and their Vegetation*. Cambridge University Press, Cambridge.

Chapter 2
Sward Establishment and Renovation

R.D. Sheldrick

2.1 Introduction

The composition of a newly sown sward is largely a reflection of the mixture of species and varieties that was sown, together with such weed species that have gained entry during the establishment phase. As the sward ages, its evolving composition becomes more dependent on the various climatic and environmental influences at that site, and particularly on the management imposed. Vegetative expansion of successful sward components, and germination and establishment from seed (either from the soil seed bank, or seed dispersal and seed rain from flowering species, particularly in hay meadows) ensures that gaps are filled with those species or ecotypes that can best survive in the particular circumstances. Such a succession from the sown species (predominantly perennial ryegrass, *Lolium perenne*, in the British Isles) to unsown (and usually less preferred) grass genera such as *Poa*, *Agrostis*, *Holcus* and *Festuca*, together with an increase in the range of dicotyledonous (broad-leaved) species, has been described as sward deterioration (Charles & Haggar 1979). This is arguably a rather pejorative term for an entirely natural ecological progression towards a species composition adapted to the environment. At moderate levels of fertilizer application, such old swards may be capable of good herbage production responses (Hopkins *et al.* 1990) and highly economic levels of output (Elliot *et al.* 1974, 1978). Also, by ameliorating some of the conditions of management or sward environment, an increase in the proportion of ryegrass and other highly valued species may be possible, and grassland surveys have shown that this has occurred on many farms (Hopkins *et al.* 1988). There are also a number of management options for renovating swards, e.g. by introducing additional seeds, improvement of soil drainage, or using selective herbicides to modify the botanical composition.

Nevertheless, there are situations when it is necessary to establish a new sward, either following arable cropping, or to replace an existing sward which is considered to have an inappropriate composition for its intended use. In Great Britain, for instance, some 0.2 million hectares of grassland are sown each year. Successful management in establishment and post-establishment phases is an important foundation for the subsequent productivity of the sward. This chapter discusses the management options for improving the productivity of grassland by sowing or sward renovation, the features of herbage species, varieties, and seeds mixtures, and the technical aspects of seedbed preparation and sward establishment.

2.2 Options for sward improvement and reseeding

When a new grass ley has to be established following arable cropping, conventional cultivation for seedbed preparation will be carried out, in most cases, before sowing the new sward. The same will be true if an entirely different forage species is to be sown, such as red clover (*Trifolium pratense*) or lucerne (*Medicago sativa*). In particular circumstances, herbicide treatment of the former vegetation followed by direct drilling may be more appropriate and will preserve the load-bearing properties of the soil.

However, when seeding grass immediately after grass with the intention of replacing a sward which is considered to have deteriorated too far, as indicated by a loss of production or herbage quality, it is essential to identify those ecological or management factors that have led to the decline and correct them first. There is strong evidence that reseeding will increase yields only temporarily over those of a similarly managed permanent sward. The productivity gain is due partly to the release of mineral nutrients by cultivation; later, at sub-optimal rates of N input, yields will soon fall back and the extra herbage generated may be little more than will compensate for that lost during the reseeding process (Hopkins *et al.* 1990). The composition of the permanent sward is a reflection of its management and of various environmental factors (Davies 1960), and 'attention to the management, soil nutrient status and drainage conditions favourable to sown species is probably more appropriate than repeated reseeding into an environment to which other species are better adapted' (Hopkins 1986).

Faced with a decline in sward production, a range of options of increasing cost, speed and related severity of intervention are available. From the outset, attention must be given to the major factors of drainage, lime status, soil levels of phosphorus and potassium, and grazing and cutting management. Close attention to sward height guidelines when grazing, and avoidance of over-long cutting intervals should reduce sites for possible colonization by weeds, either grasses or broad-leaved species. Such measures, together with an improvement in the N nutrition of the sward, may improve its competitive position and increase the proportion of preferred agricultural grasses such as ryegrasses, cocksfoot (*Dactylis glomerata*) and timothy (*Phleum pratense*) to a point where sward productivity and quality are acceptable (Smith & Allcock 1985). However, whatever additional action is contemplated, the major factors mentioned earlier must be corrected. If the sward is open, new seeds of a preferred species may be broadcast with fertilizer and harrowed in, or trodden in by mob-stocked sheep. Alternatively, seeds can be suspended in slurry and applied by tanker (e.g. Jones & Roberts 1989), a technique particularly favoured for thinning swards of Italian ryegrass (*Lolium multiflorum*).

A more costly option is to use a specialized drill for slot-seeding or strip-seeding new seed into an existing sward. Such drills cut through the turf and place the seeds (of grass or clover species) in contact with mineral soil to provide good conditions for germination. To increase the chances of success, vigorous regrowth by the old sward can be checked by herbicide or management to reduce competition during the establishment of the newly introduced components.

Finally, the most costly option is to plough, cultivate and sow. This is costly not only in terms of the mechanical operations, but also in terms of production lost during the establishment and early growth of the new sward. When carried out efficiently it can be the quickest method and the most reliable in achieving a complete alteration of sward composition. All of the options are considered in more detail later in this chapter.

2.3 Species for sowing

Plant breeders collect, manipulate and select grass and clover species to improve productivity, seasonality of growth, resistance to disease or environmental stress, and other desirable characteristics. Before new cultivars can be sold within the EC, they must be entered on a National List: after 2 years on the National List of a specific country they can then be entered on the EC Common Catalogue and sold throughout the Community. Comprehensive tests are carried out before new varieties can be entered on the UK National list (Weddell *et al.* 1997). A valuable spin-off from such testing is the compilation of recommended lists by the National Institute of Agricultural Botany (NIAB) for England and Wales, the Scottish Agricultural College (SAC) for Scotland and the Department of Agriculture for Northern Ireland (DANI).

2.3.1 Grasses

Perennial ryegrass

Perennial ryegrass is the most widely sown grass in British agriculture, and in the 19th and early 20th centuries when grassland productivity was generally very low, high-grade fattening pastures were ranked according to their content of this species (Lazenby 1981). It shows rapid establishment from seed and strong tillering to produce a dense sward that withstands grazing, and responds well to fertile conditions and inputs of N. When regularly utilized in medium- to high-input strategies, the yield of DM and the nutritive quality are usually higher than for other species. Significant advances have been made in breeding for yield and persistence (Camlin 1997). There are numerous recommended varieties, classified according to heading date as early, intermediate or late. Heading date is defined by the emergence of the ear in at least half of the fertile, reproductive tillers in the crop. In central England, heading dates vary from about 7 May to 15 June, and in Scotland from 17 May to 28 June. A particularly cold spring may delay the start of heading by early varieties by 5–7 days, but subsequent crop development catches up quickly. Recommended varieties for the UK, together with their main characteristics, are listed in the annual Recommended Lists of the National Institute of Agricultural Botany (e.g. NIAB 1998) and the variety testing organizations in Scotland and Northern Ireland (SAC 1998, Gilliland 1998, respectively).

Perennial ryegrass does not thrive under very dry conditions or on infertile soils, when it rapidly becomes stemmy and poorly tillered. In countries with the severe winter weather associated with continental climates (e.g. central Europe and parts of North America), the crop may not survive and other grasses must be grown which show greater winter-hardiness.

Tetraploid varieties of perennial ryegrass have been bred which tend to be slightly higher yielding, with higher sugar and water contents, and higher digestibility. Their seed size is nearly double that of the normal diploid varieties. With higher sugar levels and digestibility, the tetraploids have good intake characteristics (Hageman *et al.* 1993) and their development is regarded as a major advance in grass breeding (Camlin 1997). However, they have fewer, larger tillers than diploids, leading to more open swards and possible weed ingress. Nevertheless, the lower tiller density can also give greater compatibility with white clover. For the same reasons, they are generally less suited to swards that may be left for a hay cut.

Italian ryegrass

Italian ryegrass is the second most widely sown grass species in the UK and establishes vigorously, even from autumn sowings in October in southern England, to give a sward with a 2-year life. It is most productive in the first year after sowing, producing heavy crops with numerous upright fertile tillers, and responding well in intensive systems with high inputs of nitrogen. Spring growth starts rather earlier than in perennial ryegrass, so after an early grazing, large crops of good quality can be cut for silage. The second cut must be taken no longer than 4 weeks after the first, as heading is very rapid in the second crop and quality falls accordingly (Corrall *et al.* 1979). Autumn growths may be grazed to lessen winter damage because, except in the seedling stage, the crop is less winter-hardy than perennial ryegrass. Recommended varieties are listed by NIAB, DANI and SAC as for perennial ryegrass.

Westerwolds ryegrass (*L. multiflorum* var. Westerwoldicum or *L. westerwoldicum*) establishes very rapidly and is an annual species used for catch-cropping, or in pioneer cropping mixtures for heaps of mining waste and similar situations.

Hybrid ryegrass

Hybrid ryegrasses (*Lolium perenne* × *multiflorum*) have been bred in an attempt to combine the productivity of Italian ryegrass with the greater longevity of the perennial forms. Most varieties are tetraploid and resemble the Italian parent more than the perennial, and the earlier released varieties from New Zealand (e.g. Grasslands Manawa) suffered from winter-hardiness problems in the UK. Unfortunately, despite breeders' improvements to overcome these limitations, they have been widely regarded only as alternatives to Italian ryegrass, with which their performance is compared in short-term evaluations by the variety testing organizations (Camlin 1997). They have a place in medium-term leys for intensive production.

Other sown grass species

Timothy (*Phleum pratense*) has very small seeds and is slow to establish, best yields often being obtained in the second year after sowing. It is winter hardy, and is used in leys in cooler areas of Britain and the countries of northern Europe where ryegrasses would not survive. It is also well adapted to heavy, wetter soil types. It flowers late and has a lower digestibility than ryegrass at the same stage, but is a useful component in hay mixtures due to its palatability. It combines well with less vigorous species such as meadow fescue (*Festuca pratensis*), and with white clover (*Trifolium repens*).

Cocksfoot (*Dactylis glomerata*) is comparatively drought-tolerant and is used on sandy textured soils in dry areas. It is slow to establish, but grows rapidly and develops a conspicuous coarse, tufted habit if swards are under-stocked. It has a lower digestibility and is less palatable than ryegrass. Its use in the UK has declined in recent decades, despite breeders' attempts to create varieties with improved acceptability to livestock, but it is widely used in many parts of continental Europe, being better adapted than ryegrasses to the greater climatic extremes.

Meadow fescue (*Festuca pratensis*) grows best in wet soils, but establishes slowly and lacks persistency in intensive, high N systems. It was originally used in mixtures with timothy and white clover because of its compatibility with the legume, but the open swards allowed weed invasion. Tall fescue (*Festuca arundinacea*) grows well in hot, dry or moist conditions and is very winter hardy. While being productive and responsive to N fertilizer, it is slow to establish and of poor digestibility compared with ryegrass. It is popular in regions with a continental type of climate, e.g. parts of North America (Barnes *et al.* 1995). Red fescue (*Festuca rubra*) is a constituent of old pastures and a species for amenity sowings, but it produces acceptable yields on infertile soils and is sometimes used in mixtures for hill reseeding where the main agricultural grasses would be at the limits of their range. Many varieties have been bred for sports turf and amenity situations because of its winter greenness and durability (see Chapter 10).

Several varieties of brome (*Bromus* spp.) have been released. Prairie grass (*Bromus willdenowii*) cv. Grasslands Matua was released in New Zealand (Rumball 1974) and has since been introduced elsewhere. It shows drought tolerance on sandy soils and good autumn growth, but may be susceptible to frost damage in northern Britain. Its open, upright habit makes it suited to cutting or rotational grazing, but also allows weed invasion.

2.3.2 Legumes

Nitrogen fixation by legumes is the driving force behind low-input systems of grassland farming, particularly where economic returns cannot justify reliance on nitrogen fertilizers, or where their use is prohibited, e.g. in organic farming (see Chapter 14). In the mixed sward, legumes contribute to the crop yield and also provide N for the benefit of the associated grasses. White clover is by far the most important legume in British agriculture and in many other temperate areas, notably New Zealand. Red clover and lucerne play lesser roles (although lucerne is also very important in terms of world agriculture), whilst sainfoin (*Onobrychis viciifolia*) and lotus (trefoils) (*Lotus corniculatus* and *L. pedunculatus*) have only minor specialized roles in the UK (Sheldrick *et al.* 1995).

White clover

White clover is a perennial plant with small seeds that are best broadcast on the soil surface (see below). It is always grown in association with grass. Its creeping stolons extend leaves into the sward canopy. The tap-rooted seedling plant eventually dies, leaving many independently rooted daughter plants developed from the nodes on the stolons.

In the UK, classification of varieties in the Recommended Lists is based on leaf size. Small-leaved varieties (e.g. S184, Gwenda) have thin, highly branched stolons, and are productive and persistent in short swards under sheep grazing. Medium-leaved varieties are typical general-purpose varieties (e.g. Menna, Donna). Large-leaved varieties (e.g. Siwan, Alice) have thick, less-branched stolons; they can withstand N inputs better than other varieties, and are suited to dairy systems as they will also withstand grazing integrated with cutting. They are generally regarded as less persistent. All types are very plastic, however, and leaf size is no longer an absolute guide to persistency and best use. Many instances have occurred where a small-leaved type has flourished under intensive use, although there are fewer instances of the reverse. Although often regarded principally as a component of the grazed sward, cutting a crop for silage in early summer before the grass tillers have elongated sufficiently to over-top the clover's leaves will enhance the clover content of the sward (Sheldrick *et al.* 1993).

Red clover

In the British Isles, outside the requirements of organic farming systems, red clover is a very minor crop, although it is widely used in other countries of western and northern Europe. It may be drilled as a pure stand, or more usually with a companion grass (commonly an intermediate, possibly tetraploid, ryegrass, but timothy or meadow fescue are alternatives), and forms a 2-year crop for silage with some autumn grazing (Frame 1990). It can also successfully be strip-seeded into an existing grass base (Haggar & Koch 1983). Red clover has declined in use because of the trend to longer-term swards, its sensitivity to increased cutting frequency (Sheldrick *et al.* 1986) and the susceptibility of popular tetraploid varieties to necrotic mosaic virus, which developed in the 1980s. Red clover is also used in some amenity wild flower sowings.

Lucerne

Lucerne is a perennial plant, forming very deep tap roots in soils alkaline to depth (e.g. over chalk or limestone), and is a specialist cutting crop. It is best sown without a companion grass, allowing use of weed-control chemicals when it is dormant in winter. The deep rooting habit renders it one of the most drought-resistant of forages. Crops last generally 3–4 years in Britain. Its cultivation is a highly specialized business in warmer climates (Barnes *et al.* 1995).

2.3.3 Seeds mixtures

In the past, complex mixtures of grass species, legumes and herbs were sown on the supposition that at least some of the species would establish and thrive to provide a sward (Lazenby 1981). Modern herbage varieties have been bred to critical standards to display particular attributes of growth pattern, quality features or tolerance of pest, disease or environmental conditions. To sow such varieties in broadly based mixtures fails to exploit the potential of these attributes. The precise conditions for which the mixture is intended should be established, and choices made on this basis. Information on varieties is now compiled in databases, and advisers can access this data to provide tailor-made mixtures. Combining ability is particularly important when choosing grass and white clover varieties for sowing in association (Evans *et al.* 1989).

For swards intended for cutting, grass varieties must be chosen with similar heading dates, so the sward can be harvested at optimum yield and digestibility. Mixing a small range of varieties that exhibit resistance to the major pest and disease problems likely to arise may provide some insurance against epidemic attack. For swards intended solely for grazing, there may be greater latitude in choice based on heading date. The former perception that late-flowering varieties were the more persistent under grazing may not now be true. Recommended lists of perennial ryegrass varieties show good persistency in all groupings. Before choosing any seeds mixture, the conditions the sward must meet should be defined, and then the lists of recommended varieties (e.g. NIAB 1998, SAC 1998) for the particular geographical area should be consulted for up-to-date information on the best varieties of each herbage species that meet those conditions. For a discussion of the possible synergistic benefits of mixtures compared with sowing single varieties, see Ingram (1997).

However, it is right not to exaggerate the importance of the choice of seeds mixtures for longer-term swards. Obviously, for a two-year sward of Italian ryegrass, or red clover and ryegrass, the sown components will exert a big influence, but environment, and above all management, exert an over-riding influence on the composition of the older sward (Snaydon 1979). So having chosen the appropriate seeds mixture for the conditions, it is then even more important to correct the management. Examples of several types of seeds mixtures are given in Frame (1992) and Turner (1997).

2.4 Sward establishment

2.4.1 Seedbed preparation

If the choice of varieties to sow is to have a worthwhile influence on the future sward, optimum seedbed conditions for germination and seedling growth must be provided, together with every means to prevent weeds gaining a foothold, or attack by pests and diseases.

Before starting, the field drainage should be examined, and soil samples taken to establish the pH at the surface and at plough depth, together with the P and K status and possible trace element deficiencies. The objective of the cultivation is to create a firm, fine and level seedbed. This will usually include ploughing, cultivating or discing, harrowing and rolling. The former vegetation should be well buried, together with the fungal diseases and pests associated with it. The seedbed must be firm and fine to ensure moisture can reach the seeds by capillary action through the fine pore spaces, and level to ensure that the drilling machinery can ride smoothly over the surface and place the seed to the correct depth consistently. A crumbling roller with a levelling board can be a valuable tool. After ploughing, operations must be carefully timed to ensure a crumbly soil structure is obtained, without smearing, before there has been excessive drying out. Lime should be spread to correct soil acidity, and mineral fertilizers or manures should be applied and cultivated into the seedbed. A target soil pH for the seedbed of a grass-only sward is 5.8–6.0, and above 6.0 for sowings of grass–white clover. Detailed fertilizer requirements for a range of soil nutrient situations are given in MAFF (1994), and typical elemental NPK requirements for average conditions are shown in Table 2.1. In organic farming systems, seedbed nutrients can come from permitted sources, slurry or farmyard manure (see Chapter 14). No nitrogen should be incorporated in the seedbed for grass–white clover sowings.

Table 2.1 Average seedbed requirements for mineral elements.

Element	Rate ($kg\,ha^{-1}$)
N	60
P	25
K	40

Final cultivation to provide a surface tilth must ensure sub-surface clods or old sward debris are not brought to the top, and final consolidation is often best done with a Cambridge or ring roller. Loose, puffy or hollow seedbeds must be avoided, but over-consolidation at the surface leading to capping and run-off is equally serious.

2.4.2 Time of sowing

Grass seeds can germinate throughout the year in most of Britain and similar areas with a temperate maritime climate, except during cold spells in winter or dry spells in summer. However, to fit patterns of farming, it is usual to sow new swards either in spring (March–May in Britain) or in the late summer (August–early September), whilst in upland areas there is usually less flexibility in sowing dates, July–early August being suitable in most years. Spring sowings provide fresh grazing later the same season, perhaps for weaned lambs; late summer sowings allow up to two cuts of silage from the old sward or a crop of winter barley to be harvested first, and will provide a fresh sward the following spring. Grass swards sown in spring will experience improving conditions of light and temperature and should tiller and develop rapidly. Many dicotyledonous weeds can be controlled by topping, and will soon be eliminated by competition from the sward, although annual meadowgrass (*Poa annua*) can be a problem in open swards. Swards sown in the autumn experience declining day-length and temperature during their establishment, and are at risk if there is severe autumn weather. While they are less likely to suffer from weed invasion, except by chickweed (*Stellaria media*), attack by frit fly (*Oscinella frit*) can be a serious problem (see Chapter 6).

Whenever a legume is included, sowing must be completed before mid-August in southern Britain, and up to a fortnight earlier further north.

2.4.3 Seed rates

Increasing the seed rate is not a reliable strategy to compensate for an inadequately prepared seedbed, or failure to take measures to control pests (Ellis *et al.* 1990). It might be supposed that there is an optimum number of seeds to sow per unit area, and the actual seed rate is a product of this and the individual seed weight. However, species with larger seeds and thus greater seed reserves generally have better emergence rates. Therefore, fewer seeds need to be sown to gain a satisfactory seedling population. The range of typical seed-rates for ryegrasses is shown in Table 2.2.

Other grass species that may be included in mixtures for particular situations in Britain, and are often more widely used in western Europe, are timothy (very small-seeded), cocksfoot, meadow fescue and tall fescue (for characteristics, see above). Inclusion in grass seeds mixtures can range from 2 kg ha^{-1} of the fine-seeded timothy to 6 kg ha^{-1} of larger-seeded species. For pure sowings, timothy may be sown at 12–15 ha^{-1}, meadow fescue and cocksfoot at 25 ha^{-1}, and tall fescue at up to 40 ha^{-1}. Further details of grass seedling morphology and the physiological changes during germination and emergence appear in Chapter 3.

When white clover is to be included in the mixture, the grass seed-rate must be lowered to avoid early competition with the clover seedlings. Techniques to separate the grass and clover components spatially may also be employed (see below, Section 2.4.4). For swards being sown in fertile, lowland conditions where the clover component is not just a nutritional benefit, but is intended, through its N-fixing ability, to drive the productivity of the sward, the clover seed should comprise 25% by weight of the total. Typical combinations would be 12 kg of perennial ryegrass seed and 4 kg of white clover seed per hectare, or 15 kg of ryegrass with 5 kg of clover (Sheldrick *et al.* 1995). In organic farming systems, particularly when following a sequence of cereal crops, a mixture with more grass (up to 25 kg ha^{-1}) and less clover (3–4 kg ha^{-1}) may allow less weed ingress and still provide satisfactory clover establishment.

The ryegrass varieties should be chosen for their particular attributes necessary for the site, as well as their compatibility with clover. The clover varieties should also reflect the intended use of the sward. As timothy establishes slowly, 2–4 kg of timothy per hectare could be added to a mixture without detriment to the establishment of the clover.

Red clover swards intended for cutting are sown with 12–15 kg ha^{-1} of clover seed together with 4–5 kg of an intermediate perennial ryegrass, so that this component is not too mature and prejudicial to the overall quality of the herbage when the first cut is taken in May (Sheldrick *et al.* 1986). Italian ryegrass was the traditional companion for red clover, but modern bred varieties are now too aggressive for good clover contents when used in

Table 2.2 Seed rates for ryegrass species and varieties.

Species/variety	Seed rate (kg ha^{-1})
Westerwolds ryegrass	35–45
Italian ryegrass	30–35
Perennial ryegrass	25–30*

* The lower part of this range is appropriate for diploid mixtures, the upper for mixtures with the larger-seeded tetraploids or hybrids.

intensive systems. In organic systems, Italian ryegrass may prove acceptable. The more persistent late-flowering types of red clover are included in amenity sowings, usually at 2–4 kg ha^{-1} in complex mixtures with other broad-leaved species.

2.4.4 Drilling

Historically, grass seeds were sown broadcast from a grass barrow, a V-shaped wooden trough, 12 feet (3.7 m) long and centred over a wheel and pair of handles rather like a wheel barrow. The drive from the wheel rotated small brushes that broadcast the seeds through apertures of adjustable size (Watson & More 1949). Light harrowing and rolling followed. This had the advantage of achieving low inter-seedling competition, with the sward rapidly developing a uniform cover, although such shallow sowings ran a risk of drying out. Seed-rates up to 50% greater than those in Table 2.2 were often used.

Modern specialist grass drills have their coulters set closer (8–10 cm) than cereal drills (15–20 cm), and to reduce intra-row competition, it is normal for half the seed to be drilled in one direction, and the other half at an angle to the first. Even so, there is more space between the drills for weed growth than if the seeds were broadcast. The advantage of the drill is that it places seed at a uniform depth. Optimum depths range from 1.0 cm for small-seeded grasses such as timothy, to 2.5–3.0 cm for larger seeded ryegrasses. Drilling should be carried out slowly to achieve uniform sowing depth.

Grass–white clover swards are best sown in a two-stage operation by drilling the grass component at its optimum depth, ring rolling, and then broadcasting the white clover seed by passing over the field a second time with the coulters of the drill in the raised position. White clover seeds are very small and many will fail to emerge after germination if sown too deeply. The field should then be ring-rolled at right angles to the previous rolling (Sheldrick *et al.* 1995). If grass and clover are already mixed on purchase, they should be broadcast as above on to a ring-rolled surface, then lightly harrowed and rolled heavily again. Inoculation of white clover seeds with *Rhizobia* is not normally necessary on lowland sites, but may be advantageous on peaty, upland soils.

Red clover seeds are much larger than those of white clover, and may be drilled. However, to avoid early inter-specific competition, the grass companion should be drilled first, and the red clover drilled separately at an angle. Inoculation is generally not necessary.

2.4.5 Under-sowing

To raise farm productivity during the period while a spring sown sward is establishing, the grass may be undersown in a spring cereal. An early maturing, stiff-strawed, upright variety must be chosen, and sown at three-quarters the normal rates of seed and fertilizer. The grass seed mixture is then sown immediately afterwards. If the cereal crop lodges it can smother the grass, and this is particularly serious if it is a grass–clover sward. A better option may be to mix the cereal with forage peas or vetches and remove the cover-crop in good time as whole-crop silage. In such a case, a later maturing cereal variety can be used to improve the digestibility when cut (Frame 1992).

2.4.6 Direct drilling

Direct drilling sounds an attractive option to establish a new sward, since it cuts out the time spent on ploughing and cultivation, particularly if the site is steep or stony. The technique has been extensively reviewed by Naylor *et al.* (1983). However, experiments at several sites in Belgium suggest that conventional soil cultivation is more reliable (Vliegher *et al.* 1986). Direct drilling can be used where grass follows grass, or where grass follows a previous arable crop. If the old sward or the weed cover in the arable situation is desiccated with herbicide (e.g. glyphosate), direct drilling may be carried out 10–14 days later. However, difficulties can arise through (i) the old sward or perennial deep-rooted weeds not being completely killed, (ii) toxic agents being released by the decaying sward or trash from former cereal crops, (iii) a compacted or rutted soil surface, or (iv) failure of the drill to place the new seed in contact with mineral soil on account of its riding up on the old matt or surface trash. The interval between destruction of the old sward and drilling is critical for slug control (Welty *et al.* 1981). Insect pests can also move from the dying sward to attack the new seedlings, so pest control becomes especially important (Ellis *et al.* 1990).

Cutting or grazing down the sward beforehand not only allows the herbicide spray to penetrate thoroughly, but also minimizes the quantity of material decaying and releasing toxins, and makes the drilling more reliable. However, there must be sufficient leaf area to accept the spray. If compaction is severe, or if the soil surface is badly deformed by poaching or rutting, cultivation may be the best option.

2.5 Post-sowing management

Competition between emerging grass seedlings is extreme, but by the start of the first full harvest year a minimum of 5000–7000 tillers per m^2 would indicate successful establishment (Frame 1992). In grass–white clover swards, a count of 120–150 white clover seedlings per m^2 in the first half of March in the year after sowing would also indicate satisfactory establishment (ITEB 1987). Similarly, a count of over 150 red clover seedlings per m^2 in the late autumn after sowing would indicate a good red clover crop that would yield well through the two ensuing years (Sheldrick *et al.* 1995). The same density of lucerne seedlings in the spring after sowing would be regarded as optimal (Lane 1998).

Whatever the method of sowing or its timing, the aim of post-sowing management for a grass sward must be to achieve a densely tillered, leafy sward, eliminating pests and the weeds that manage to colonize. Pest and weed control methods are discussed in Chapter 6. The best management to adopt is light periodic grazing, ideally using sheep, but young cattle are also suitable, as lightweight stock are less likely to poach the as yet unconsolidated soil surface. Sward surface height should not be allowed to exceed 8–10 cm, and it should be grazed down to 4–6 cm. Grazing should be avoided until there is sufficient root anchorage and stock removed immediately in wet conditions, a situation likely to be more usual with late summer sowings as autumn conditions set in. Red clover must be cut or grazed down particularly close at the end of the season, to avoid lush, leafy herbage remaining through the winter and attracting attack by clover rot (*Sclerotinia trifoliorum*). The beneficial effects of sheep grazing during sward establishment have long been recognized as encouraging tillering and providing gentle surface consolidation, which may reduce the slug population (Ferguson *et al.* 1988). Gentle consolidation not only builds a sound surface that will withstand the weight of larger cattle and dairy cows at a later date, but it also improves the root anchorage of the grass seedlings and reduces the risk of seedling death through frost heave of the soil in the subsequent winter.

Weeds that are unpalatable to the stock and remain after a period of grazing should be controlled by topping. This action may be sufficient to control them, and will avoid the use of herbicides.

2.6 Sward renovation

As mentioned in Section 2.2, there are alternatives to carrying out a full reseeding operation. Such options may be less expensive, but will not change the composition of the sward instantly. Referring to these options as 'renovation' implies that the sward still has some good features and does not require complete replacement.

2.6.1 Correcting management practices

Pressures in modern farming may mean that poor practices grow unchecked, and swards decline in their content of preferred species and their productivity. Poor drainage needs to be corrected, e.g. by moling, soil samples need to be taken and tested for acidity and nutrient status (MAFF 1994), and any weed or pest problems should be identified and corrected. Practices that create bare patches in the sward should be eliminated as far as possible, examples being over-grazing in spring, under-grazing in summer, excessive intervals between silage cuts, poaching by cattle in wet periods and wheeling damage by heavily laden silage trailers or slurry tankers.

Correction of these poor management practices, together with a review of the nitrogen nutrition, may be sufficient to halt the decline. If it is not intended to raise nitrogen inputs above approximately 200 kg N ha^{-1} per annum, the possibility of adopting a low nitrogen input strategy should be considered, with white clover introduced to provide a nitrogen source. If clover is to be sown, either by broadcasting or strip-seeding (see below), efforts should be made to control all perennial broad-leaved weeds (e.g. docks (*Rumex* spp.), thistles (*Cirsium* spp.), nettles (*Urtica dioica*)) before this is done. The choice and 'window of opportunity' for the subsequent use of herbicides that will not risk eliminating the clover is very limited.

2.6.2 Broadcasting seed

For a variety of reasons a sward may have bare patches or a very low tiller density. Stock may have been out-wintered and fed in a field, or control measures may have been taken against a serious weed infestation (e.g. docks). In such situations grass or clover seed can be introduced by broadcasting, together with fertilizer. Chain harrowing or light discing will prepare the surface for the seed. Clover seeds must be mixed with a small quantity of lime first, in order to coat them and prevent sedimenting in the hopper when mixed with the much larger prills of fertilizer. If applied through a spinner, checks must be made on how far the much lighter seed component is thrown, and bout-widths adjusted accordingly. A further light harrowing and ring rolling will incorporate the broadcast seed. Alternatively, the seed can be trodden in by 'mob-stocked' sheep.

2.6.3 Slurry seeding and seed dispersal by livestock

Slurry seeding is a special form of broadcasting where the seeds are suspended in slurry and spread through a tanker. The technique has been used to renovate ageing stands of Italian ryegrass (Jones & Roberts 1989), where it was noted that the slurry reduced the population of annual meadowgrass. It has also been used to introduce white clover. In New Zealand hill country, a long-established technique has been to graze sheep on mature flowering white clover and then turn them into a paddock without clover. The seeds pass through the sheep and become distributed in the new paddock (Suckling 1952).

2.6.4 Strip seeding

Specialist drills of British, European or New Zealand design are available to open the turf and sow fresh seeds into mineral soil (Frame 1992). When introducing new grass varieties, vigorous upright cultivars should be selected, such as Italian, hybrid or tetraploid ryegrasses. The method is particularly valuable for introducing white clover, as the stolons can spread from the original strip to colonize the neighbouring sward. Many strip-seeders have a band spray facility to allow grass suppression either side of the strip to reduce competition for the new seedlings. Swards destined for strip seeding should be grazed or cut very low before treatment, to check immediate regrowth. If cut, it is possible to spray overall with a low-dose contact herbicide, which will control bottom grasses such as *Poa annua* or *Agrostis stolonifera*, but not affect the preferred species such as ryegrass, as they will have few leaves left to intercept the spray.

On light soils or in drier areas, strip seeding must be carried out in spring or late summer. On moisture-retentive soils, or in wetter areas, the option to strip-seed in May or June following a silage cut provides the advantage of warm soil and long day length to accelerate the establishment of the new seedlings, as well as providing extra production from the sward.

Some strip-seeders also have a fertilizer hopper, and the placement of potassium, and particularly of soluble phosphate, in close proximity to the developing seedling is a great advantage (Culleton *et al.* 1988).

As the old sward remains adjacent to the new seedlings, there is a reservoir of pests and disease inoculum at hand. Measures to check pests and diseases are very important in this situation (Ellis *et al.* 1990).

It must be recognized that the competition experienced by the new seedlings in a strip-seeding, or indeed a broadcast renovation, will be extreme, both above and below ground. Measures must be taken to give access to light for the establishing seedlings by delaying the regrowth of the old sward, and then managing it appropriately. The availability of moisture, the provision of nutrients and the control of pests and diseases are all essential to a successful outcome. Once the new seedlings are established, management of the sward should follow that outlined earlier for the newly sown sward.

Sward renovation can do little if there are problems of soil compaction, or if the surface is severely deformed by poaching or wheel rutting from machinery. In such cases, ploughing, cultivation and full reseeding may be the most satisfactory course.

2.7 Sward establishment in special situations

Surplus food production in the European Union has resulted in policies to reduce agricultural production, including the adoption of 'set-aside' land. The establishment of a green cover crop can be achieved by natural regeneration, by sowing grass or a wild-bird foraging crop or, in designated Environmentally Sensitive Areas, by sowing arable set-aside to a botanically diverse mixture of grasses and wild flower seed (Mackie & Wightman 1997).

Allowing reversion of arable land by natural regeneration of species from the buried seed-bank or quick colonizers can appear unsightly, and the vegetation bears little resemblance to the semi-natural vegetation of earlier times.

Although a cheap option initially, it may be expensive in the long run, with possible soil erosion, leaching of N and multiplication of weed problems. Sowing a grass sward will control erosion, take up N from the soil and reduce weed ingress. Seeds mixtures must be inexpensive, and the species chosen must grow slowly to avoid the cost of repeated toppings.

Opportunities for creating botanically diverse grassland on previously species-poor (but agriculturally productive) grassland, and on former arable land, have been encouraged by agri-environmental schemes in the UK and other EU countries (see Chapter 11). Regeneration of species-rich grassland is often difficult, as high levels of P are adsorbed on the clay colloids in previously intensively managed sites and these militate against the successful establishment of the most valued components of a species-rich sward. Non-intensive management of old grassland can be supported by payments linked to environmental benefits, particularly in Environmentally Sensitive Areas (Sheldrick 1997). On land that was previously intensively managed, attractive wild flower meadows may be created by sowing less competitive grasses (e.g. *Agrostis* and *Festuca* species) with clovers and other dicotyledonous species, e.g. ox-eye daisy (*Leucanthemum vulgare*), ribwort plantain (*Plantago lanceolata*), yarrow (*Achillea millefolium*) and self-heal (*Prunella vulgaris*). Grass and wild flower seeds mixtures for these situations and other non-agricultural sites are provided by several commercial seed companies. Full lists of recommended species for various soils can be found in Wells *et al.* (1989).

2.8 References

Barnes R.F., Miller D.D. & Nelson C.J. (1995) *Forages*. 5th edn. Iowa State University Press, Ames, IA.

Camlin M.S. (1997) Grasses. In Weddell J.R. (ed) *Seeds of Progress*, 2–14. British Grassland Society Occasional Symposium, No. 31. BGS, Reading.

Charles A.H. & Haggar R.J. (eds) (1979) *Changes in Sward Composition and Productivity*, 253 pp. British Grassland Society Occasional Symposium, No. 10. BGS, Hurley.

Corrall A.J., Lavender R.H. & Terry C.P. (1979) *Grass Species and Varieties*. Grassland Research Institute, Hurley, Technical Report No. 26, Tables 13–16.

Culleton N., McCarthy V., Tunney H. & Murphy W.E. (1988) Aspects of direct drilling of grass seed and seedling growth. *Proceedings of the 12th General Meeting of the European Grassland Federation, Dublin*, 204–8.

Davies W. (1960) *The Grass Crop*. 2nd edn. Spon, London.

Elliott J.G., Oswald A.K., Allen G.P. & Haggar R.J. (1974) The effect of fertilizer and grazing on the botanical composition and output of an *Agrostis/Festuca* sward. *Journal of the British Grassland Society* **29**, 29–35.

Elliott J.G., Dale R.J. & Barnes F. (1978) The performance of beef animals on a permanent pasture. *Journal of the British Grassland Society* **33**, 41–8.

Ellis S.A., Clements R.O. & Bale J.S. (1990) A comparison of the effects of sward improvement on invertebrates, sward establishment and herbage yield. *Annals of Applied Biology* **116**, 343–56.

Evans D.R., Hill J., Williams T.A. & Rhodes I. (1989) Co-existence and the productivity of white clover – perennial ryegrass mixtures. *Theoretical and Applied Genetics* **77**, 65–70.

Ferguson C.M., Barratt B.I.P. & Jones P.A. (1988) Control of the grey field slug (*Deroceras reticulatum* Muller) by stock management prior to direct-drilled pasture establishment. *Journal of Agricultural Science, Cambridge* **111**, 443–9.

Frame J. (1990) The role of red clover in United Kingdom pastures. *Outlook on Agriculture* **19**, 49–55.

Frame J. (1992) *Improved Grassland Management*, 24–30. Farming Press, Ipswich.

Gilliland T.J. (1998) Grass and clover – recommended varieties for Northern Ireland. Department of Agriculture for Northern Ireland, Belfast.

Hageman I.W., Lantinga E.A., Schlepers H. & Neuteboom J.H. (1993) Herbage intake, digestibility characteristics and milk production of a diploid and two tetraploid cultivars of perennial ryegrass. *Proceedings of XVII International Grassland Congress, Palmerston North, 1993*, 459–60.

Haggar R.J. & Koch D.W. (1983) Slot-seeding investigations. 3. The productivity of slot-seeded red clover compared with all-grass swards receiving nitrogen. *Grass and Forage Science* **38**, 45–53.

Hopkins A. (1986) Botanical composition of permanent grassland in England and Wales, in relation to soil, environment and management factors. *Grass and Forage Science* **41**, 237–46.

Hopkins A., Wainwright J., Murray P.J., Bowling P.J. & Webb M. (1988) 1986 survey of upland grassland in England and Wales: changes in age structure and botanical composition since 1970–72 in relation to grassland management and physical features. *Grass and Forage Science* **43**, 185–98.

Hopkins A., Gilbey J., Dibb C., Bowling P.J. & Murray P.J. (1990) Response of permanent and re-seeded grassland to fertilizer nitrogen. 1. Herbage production and herbage quality. *Grass and Forage Science* **45**, 43–55.

Ingram J. (1997) Some theoretical aspects of mixtures. In Weddell J.R. (ed) *Seeds of Progress*, 176–88. British Grassland Society Occasional Symposium, No. 31. BGS, Reading.

ITEB (L'Institut Technique de l'Elevage Bovin) (1987) *Des patures riches en trefle blanc*, 32. ITEB/EDE de Bretagne, Brittany.

Jones E.L. & Roberts J.E. (1989) Sward maintenance of *Lolium multiflorum* by slurry seeding. *Grass and Forage Science* **44**, 27–30.

Lane G.P.F. (1998) A novel technique for establishing lucerne (*Medicago sativa*) by undersowing in forage maize. In Lane G.P.F. and Wilkinson J.M. (eds) *Alternative Forages for Ruminants*, 21–34. Chalcombe Publications, Canterbury.

Lazenby A. (1981) British grasslands: past, present and future. *Grass and Forage Science* **36**, 243–66.

Mackie C.K. & Wightman P.S. (1997) Set-aside and alternative uses of grass. In Weddell J.R. (ed) *Seeds of Progress*, 46–55. British Grassland Society Occasional Symposium, No. 31. BGS, Reading.

MAFF (Ministry of Agriculture Fisheries and Food) (1994) *Fertilizer Recommendations for Agricultural and Horticultural Crops.* Reference Book 209, 6th edn. HMSO, London.

Naylor R.E.L., Marshall A.H. & Matthews S (1983) Seed establishment in directly drilled sowings. *Herbage Abstracts* **53**, 73–91.

NIAB (National Institute of Agricultural Botany) (1998) *Recommended List of Grasses and Herbage Legumes 1996/97.* NIAB, Cambridge.

Rumball W. (1974) 'Grasslands Matua' prairie grass. *New Zealand Journal of Experimental Agriculture* **2**, 1–5.

SAC (Scottish Agricultural College) (1998) *Recommended Grass and Clover Varieties for Scotland*. Technical Note T.435. SAC, Edinburgh.

Sheldrick R.D. (ed) (1997) *Grassland Management in Environmentally Sensitive Areas.* British Grassland Society Occasional Symposium, No. 32. 294 pp. BGS, Reading.

Sheldrick R.D., Lavender R.H. & Tewson V.J. (1986) The effects of frequency of defoliation, date of first cut and heading date of a perennial ryegrass companion on the

yield, quality and persistence of diploid and tetraploid broad red clover. *Grass and Forage Science* **41**, 137–49.
Sheldrick R.D., Lavender R.H. & Martyn T.M. (1993) Management options for increasing white clover contents of swards without re-sowing. *Grass and Forage Science* **48**, 223–30.
Sheldrick R., Newman G. & Roberts D. (1995) *Legumes for Milk and Meat*, 2nd edn. 15–74. Chalcombe Publications, Canterbury.
Smith A. & Allcock P.J. (1985) The influence of species diversity on sward yield and quality. *Journal of Applied Ecology* **22**, 185–98.
Snaydon R.W. (1979) Selecting the most suitable species and cultivars. In Charles A.H. & Haggar R.J. (eds) *Changes in Sward Composition and Productivity*, 179–89. British Grassland Society Occasional Symposium, No. 10. BGS, Hurley.
Suckling F.E.T. (1952) Dissemination of white clover (*Trifolium repens L.*) by sheep. *New Zealand Journal of Science and Technology* **33A**, 64–77.
Turner R.W. (1997) Grass mixtures – an industry perspective. In Weddell J.R. (ed) *Seeds of Progress*, 189–99. British Grassland Society Occasional Symposium, No. 31. BGS, Reading.
Vliegher A. de, Carlier L. & Ketels T. (1986) Problemes relatifs au semis des prairies sans labour. II. Essais au cours de la période 1982–84. *Revue de l'Agriculture* **39**, 319–29.
Watson J.A.S. & More J.A. (1949) *Agriculture. The Science and Practice of British Farming*, 9th edn. Oliver and Boyd, Edinburgh.
Weddell J.R., Gilliland T.J. & McVittie J. (1997) Evaluation procedures: past, present and future. In Weddell J.R. (ed) *Seeds of Progress*, 220–25. British Grassland Society Occasional Symposium, No. 31. BGS, Reading.
Wells T.C.E., Cox R. & Frost A. (1989) The establishment and management of wildflower meadows. *Focus on Nature Conservation*, No. 21. Nature Conservancy Council, Peterborough.
Welty L.E., Anderson R.L., Delaney R.H. & Hensleigh P.F. (1981) Glyphosate timing effects on establishment of sod-seeded legumes and grasses. *Agronomy Journal* **73**, 813–17.

Chapter 3
The Principles of Pasture Growth and Utilization

A.J. Parsons and D.F. Chapman

This chapter is divided broadly in two: the first part (Sections 3.1–3.4) deals with the structure and form of the major forage grasses and the legume white clover *(Trifolium repens)*, and concentrates on the origin of new plant parts. It gives a précis of the factors affecting the production of leaves and branches (tillers/stolons) and some background on photosynthesis. The second part (Sections 3.5–3.16) deals with the principles of growth and utilization of grasses and legumes, based on the flow of energy and organic matter from photosynthesis to animal intake, at both the whole-paddock and the plant or bite scale. It proposes a theory for the balance of grasses and legumes in mixed pastures, and so a basis for understanding what controls their contribution to the diet and the cycling of nitrogen. As ecologists have long recognized, interactions between even just two species can be very complex (May 1974, 1981). Agricultural familiarity with grass and clover does not alter this fact.

Those concerned immediately with understanding how to maximize the 'utilization' of grass may wish to begin at Section 3.5, but we must recognize that there are limitations to enhancing grass growth and its utilization that are imposed by plant structure. Many plant breeding programmes have focused on manipulating the form of the plant, notably with respect to leaf size, for example in legumes, and the capacity for branching (as tillers or stolons). Recognizing how grass and legume plants grow, and how an increase in dry matter is expressed in plant parts, is valuable in qualifying our expectations of the performance of grassland. It has been said 'He who make two blades of grass grow where one grew before is the benefactor of mankind . . . and he who obscurely works to find the laws of such growth is the greater benefactor of mankind' (Henry Augustus Rowland, 1848–1901). Reading both parts of this chapter should qualify whether these are realistic, or even necessary steps, to optimizing the growth, utilization and sustainability of grasslands. From Section 3.17 onwards, we suggest how our understanding provides a basis for practical grassland management.

3.1 The grass plant

3.1.1 The seed

Grass seeds develop from a single fertilized ovule within the ovary of a flower (floret). As well as the ovary, with its twin stigmas, each floret contains three stamens and two very small scales (lodicules), all enclosed within two large bracts (palea and lemma) (Fig. 3.la). A number of florets make up a spikelet, and a number of spikelets the inflorescence (Fig. 3.lc). The great bulk of the

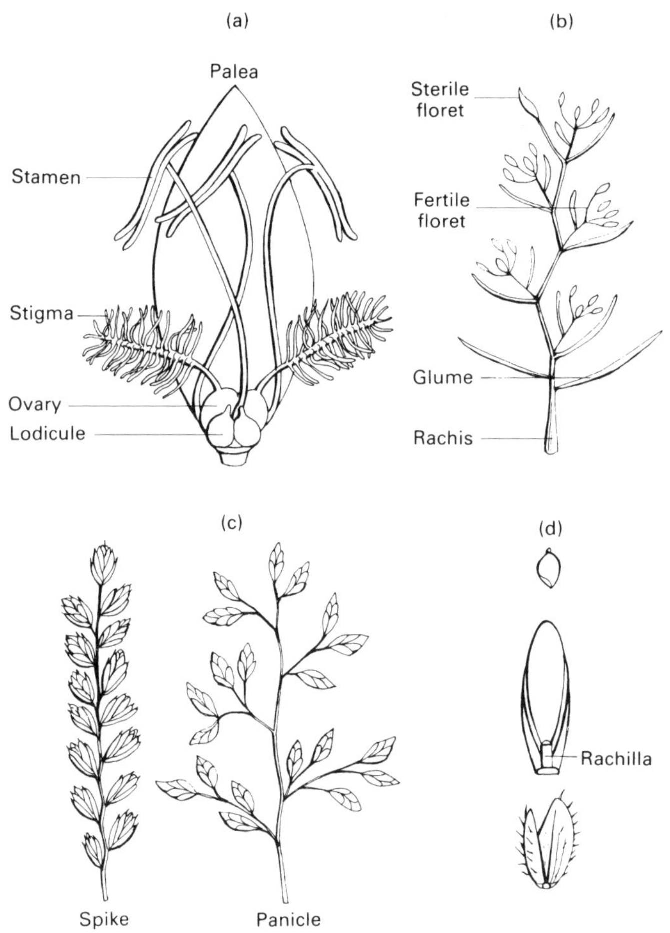

Fig. 3.1 (a) Floret with lemma removed to expose reproductive organs. (b) Diagrammatic spikelet showing two basal glumes, five fertile florets and one sterile (terminal) floret. (c) Types of inflorescence: spike and panicle. (d) Types of 'seed'. From the top: naked grain of *Phleum pratense*, enclosed grain of *Lolium perenne* and an entire spikelet of *Holcus lanatus* (after Robson *et al.* 1988).

seed is starchy endosperm used to sustain the embryo during germination and early seedling growth. The embryo consists of a primary shoot (plumule) within a protective sheath (coleoptile) and a primary root (radicle) also within a sheath (coleorhiza). These are attached by a short mesocotyl to a flat shield-like structure (scutellum) abutting the endosperm (Fig. 3.2a). As the seed matures, its

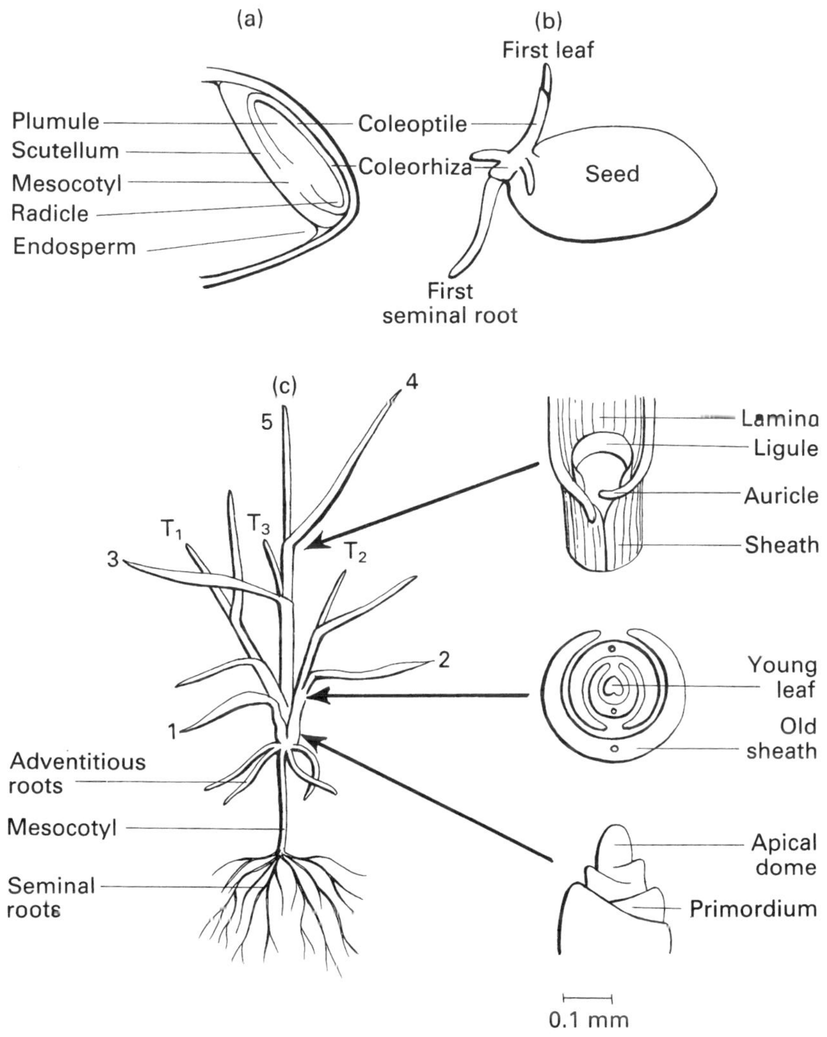

Fig. 3.2 (a) Diagrammatic cross section of the embryo region of a seed. (b) Germinating seed. (c) Young plant with five leaves on the main stem, four of them fully expanded and three subtending daughter tillers (T_1, T_2 and T_3). Also, from the top: the junction of lamina and sheath, a cross section of the pseudo-stem (sometimes sheaths are folded rather than rolled) and the vegetative main stem apex (after Robson *et al.* 1988).

coat fuses with the ovary wall to form a dry, one-seeded, indehiscent fruit known in the *Gramineae* as a grain. When sown, the grain may be naked, as in timothy *(Phleum pratense)*, or enclosed by a persistent lemma and palea as in perennial ryegrass *(Lolium perenne)*. In some agriculturally less important grasses, the 'seed' may be the entire spikelet (Fig. 3.1d).

3.1.2 Germination and seedling emergence

The germination of a viable seed is dependent largely on an adequate supply of water and oxygen, and a favourable temperature. The first stage in germination involves the passive uptake of water and takes about 12–24 h. Following this, cells in the aleurone layer surrounding the endosperm begin secreting enzymes which break the starch down into sugars. These are absorbed by the scutellum, and pass on through the mesocotyl to the embryo. There they supply the energy and the raw materials necessary for the growth of the plumule and radicle; when these break through the seed coat, germination is complete (Fig. 3.2b).

The radicle grows downwards through the coleorhiza with one or more secondary roots arising at its base. These develop, through branching, into a highly efficient seminal root system on which the young seedling relies initially for its supply of water and inorganic nutrients. As the radicle grows downwards, the young shoot or plumule is carried towards the soil surface by the elongation of the mesocotyl. This elongation is a response to darkness which ceases when the tip of the coleoptile emerges into light, thereby positioning the shoot apex just below the soil surface (Fig. 3.2c).

3.1.3 Shoot structure and leaf turnover

The shoot apex consists of a meristematic apical dome (Fig. 3.2c, and see diagrams by Jewiss, 1993) that continually lays down 'primordia' from which leaves develop in alternating order along its opposing flanks. The point of attachment of each leaf to the shoot apex is called a node, and the stem tissues separating one node from the next are 'internodes'.

Each grass leaf develops as a blade (or 'lamina') connected at an angle to a generally shorter sheath, the junction being marked by a membranous ligule and, in some species, two claw-like auricles. To achieve this structure, as a leaf primordium differentiates, cell division and expansion remain restricted to a meristem in the leaf base, which is divided in two by a band of parenchyma cells from which the ligule ultimately forms. The upper portion of the meristem is associated with lamina growth, the lower with that of the sheath. This restriction of leaf growth to a basal meristem means that leaves are extruded linearly, which is typical of monocotyledonous plants, and that the tip of a leaf is always older than the base.

Each new leaf grows up within the encircling bases of older leaves (the first leaf being within the coleoptile), so that what appears to be a 'stem' is, in many cases, a collection of sheaths rolled or folded one inside the other, with the oldest sheath on the outside and young expanding leaves in the centre (Fig. 3.2c). In grasses such as *L. perenne*, the 'true stem' (which comprises the apical meristem and the accumulating nodes and internodes) is located at the base of this 'pseudo-stem', concealed entirely by encircling sheaths, and remains short (<1 cm) and at ground level as long as the shoot remains vegetative. Note that this structure ensures that in many grasses all the growing tissues are close to ground level and so can escape grazing.

Apart from producing its own leaves, the true stem may branch to form 'tillers'. Every time the apical meristem produces a new leaf, it also produces an axillary meristem which is located on the opposite side of the internode and so in the axil of the previous leaf (see Jewiss 1993). Hence there is one at every node. Each of these is a potential site for a new tiller. When an axillary bud becomes active, its apex produces its own leaves and secondary tillers can form from the buds in the axils of these. This provides the *potential* for an exponential increase in tiller, and so in leaf, production, but there are limits to the number (and size) of structures that can be sustained per unit ground area (see Section 3.1.5).

In tussock-forming grasses, the leaves of tillers generally grow upward within the encircling sheaths of the subtending leaves to emerge at the ligule, whereas in stoloniferous or rhizomatous grasses, such as rough-stalked meadow grass *(Poa trivialis)* or couch *(Elytrigia repens)*, the tiller more often breaks through the base of the encircling sheath. The angle of attachment of tillers greatly alters how erect or prostrate the plant can become. In stoloniferous and rhizomatous grasses, the vegetative stem internodes can elongate to create a more creeping habit.

In seedlings, the number of live leaves on the main axis increases until there are five or six. Thereafter, the number of live leaves per tiller falls to a number that remains constant throughout the life of the plant, e.g. in *L. perenne* at 3.0 live leaves per tiller. *Leaves continue to be produced, but each time a new leaf appears, the oldest leaf on that tiller dies* (Hunt 1965). Thus the rate of leaf appearance, the number of live leaves and the longevity of leaves are closely interrelated (see Section 3.3.1). Moreover, for a given energy and carbon supply, the length of time a leaf expands will determine the size of leaves (and so tillers) produced. Species differ substantially in the outcome of these components of leaf turnover, and it has been proposed this can be used to characterize their morphology (Lemaire & Chapman 1996). For example, under the same growing conditions, *L. perenne* maintains three leaves per tiller, with leaves appearing every 11 days, and so leaves live some 33 days; tall fescue *(Festuca arundinacea)* maintains 2.5 leaves per tiller, leaves appear every 22 days, and so leaves live some 57 days (Lemaire 1988, Robson *et al.* 1988). Because each leaf expands for longer, leaves (and tillers) are larger in *F. arundinacea* than in *L. perenne*, but note (see Section 3.1.5) that this means there is likely to be fewer of them per unit ground area.

Although the shoot initially relies on the seminal or tap root, later adventitious roots arise from buds on the nodes at the base of the main shoot apex and that of daughter tillers. Given the profusion of tillering, this soon establishes a dense mat of roots in many grasses. Because of their anatomical connections (all being attached at a node), individual roots live little longer than the tiller to which they are attached, and so few roots survive one year (see Jewiss 1972, also Garwood 1967 in Jones & Lazenby 1990: p. 146).

3.1.4 Branching (site filling) and fragmentation

In perennial ryegrass growing from seed, the first tiller usually emerges from the axil of the first leaf on the main shoot as soon as this leaf and its successor are fully expanded. Subsequent tillers on this axis generally arise in acropetal succession, initially keeping pace with leaf production.

During growth from seed and free from competition, the early pattern of tillering in many north temperate species tends towards that shown in Fig. 3.3, where virtually all sites are filled (Robson *et al.* 1988). However, as a plant grows larger and denser, and adjacent tillers and plants come to 'compete' for limiting resources, the pattern of tillering diverges markedly from the exponential and many sites remain unfilled (Simon & Lemaire 1987). In mature grass communities typically <10% of potential tiller sites bear tillers. The degree of site filling (Neuteboom & Lantinga 1989) is a better indication of the success of branching than is indicated by counts of the number of tillers per plant. This is because, as time passes, the older internodes of tillers (and indeed whole older tillers) die and the plants fragment. Thus, the identity of the original plant, the product of a single seed, is lost. The community now comprises a population of plant fragments with a wide frequency distribution of tiller numbers per 'plant' (Brock & Thomas 1991). The mean number of tillers per 'plant' under realistic field managements is, e.g. in the case of *L. perenne*, typically four to five. Of these, fewer than 15% have more than seven tillers and some 60% have less than five tillers (Brock & Thomas 1991). The fragments may remain close together, but can be considered independent in that there is no organic connection between them. The proportion of plant fragments in different classes of plant complexity (topology) varies between sheep- and cattle-grazed swards, and also shows marked seasonal variation (Brock *et al.* 1996, Hume & Brock 1997), but the significance of the topologies of fragments, and the marked seasonal changes in these to growth, persistence and utilization has yet to be resolved.

There are surprisingly few theories to explain whether it is important to have a large number of tillers (branches) per plant, or even whether it is valuable to have a large number of tillers per unit ground area (as opposed

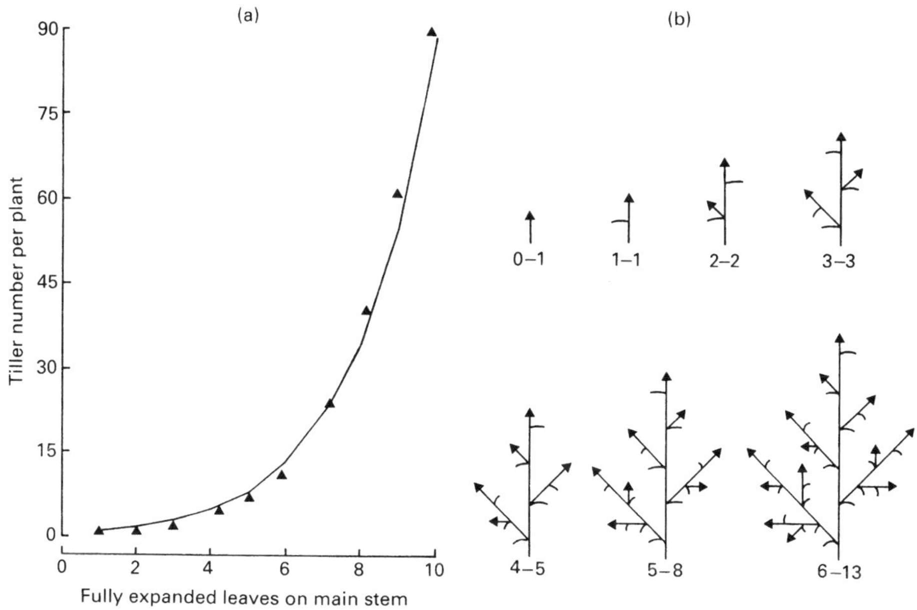

Fig. 3.3 Tiller production in young, spaced plants of *Lolium perenne* growing in favourable conditions (Robson 1982b). (a) Increase in tiller number per leaf interval on the main stem (▲), approximates to the Fibonacci series (____): each number is the sum of its two predecessors (1, 1, 2, 3, 5, 8, 13, 21, etc.). (b) Generally a tiller (▲) appears in a leaf axil once that leaf's successor is fully expanded. The numbers under each diagrammatic plant (e.g. 6–13) refer to fully expanded leaves on the main stem (6) and the total tiller number per plant (13) (after Robson *et al.* 1988).

to fewer larger ones). Provided they have the capacity to grow prostrately, some species (e.g. *Paspalum*) can persist with few very large tillers (leaves up to 1 cm in width) in communities comprising very large numbers of very small tillers (e.g. of red fescue; leaves <1 mm in width), suggesting that either strategy is successful, but there is a well-established relationship between the numbers and size of units of any one species that can be sustained per unit of ground.

3.1.5 Allometric constraints to the tiller population

There is a strong and intuitive link between the number and size of leaves/tillers that are sustained in grass populations, and indeed across a vast range of plant species (Kays & Harper 1974, Harper 1977, Westoby 1984, Matthew *et al.* 1995, 1996) (Fig. 3.4). In plant biology, this was long referred to as the –3/2 boundary 'rule', reflecting the slope of the relationship between the logarithm of unit (e.g. tiller) mass and log population density in closed communities. More

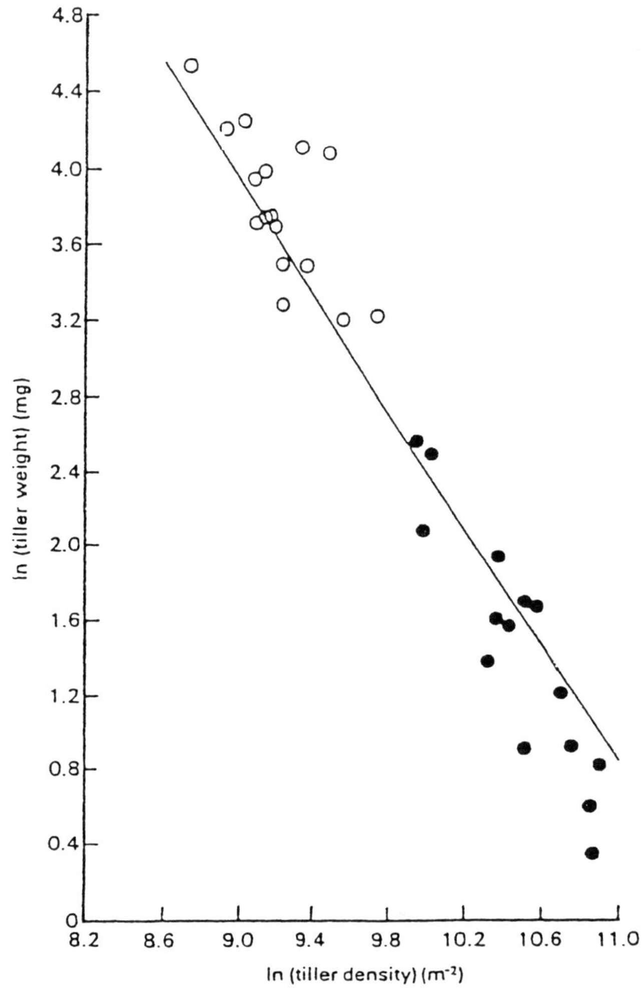

Fig. 3.4 Relationship between ln (tiller weight) and ln (tiller density) in perennial ryegrass/perennial ryegrass dominant swards. ○, cut swards, 30 June–17 November. ●, grazed swards, 15 July–4 September. Slope of line = –3/2. (Reproduced by permission from Davies 1988.)

recent studies have proposed that the relationship between numbers and size reflects the same kind of allometry as is seen in animals, so that population density scales as mass raised to the power of –3/4 (West *et al.* 1997, Enquist *et al.* 1998). A single grass genotype can demonstrate a wide range of combinations of numbers and size of tillers, e.g. *L. perenne* typically will have 10 000 large tillers m^{-2} under infrequent defoliation, but substantially greater numbers (and smaller size) tillers under frequent defoliation (Table 3.1). However, there are limits to this plasticity. Different grass species, under the same management, adopt markedly different combinations of numbers and size of units. At a leaf

Table 3.1 The effect of some cutting and grazing managements on sward structure and the expression of reproductive development in a perennial ryegrass sward in June (Johnson & Parsons 1985b).

	No. of tillers m^{-2}	% of reproductive tillers	Weight of elongated stem ($g\,DM\,m^{-2}$)	Stem length (cm)	LAI during spring
Cutting					
Uncut until 7 June	8330	74	548.0	—	—
4-weekly cuts until 7 June	12097	69	388.2	—	—
Continuous grazing					
Sward surface height (cm)					
3	43464	14	44.2	1.3	1.6
6	33765	31	105.5	3.6	2.3
9	20132	47	201.7	7.1	3.8
12	14311	59	333.0	9.2	4.6

area index (LAI is leaf area per unit ground area) of 3.0, *F. arundinacea*, for example, generates 4000–6000 large tillers m^{-2} (Mazzanti *et al.* 1994), while *L. perenne* develops 10000 15000 much smaller tillers m^{-2}. Moreover, the same sward growth rates can be achieved by markedly different combinations of plant size and density. These mechanisms have been shown to account for similar yields from contrasting managements that generate quite different sward structures, illustrating great buffering or compensating capacity in temperate pasture types (Chapman & Clark 1984, and see Section 3.14), but there are limits to this capacity, as described in some detail by Chapman & Lemaire (1993).

Although in most grass (and indeed forage legume) canopies less than 10% of the potential sites for the production of tillers (branches) are filled, it is salutary to calculate just how many sites *need* to be filled to sustain the tiller population. Whether it is a high population or a low one, to sustain tiller numbers requires simply that each tiller, in its lifetime, produces just one surviving daughter tiller, and so provides a one for one replacement. If a vegetative grass tiller survives for 1 year (for tiller life histories, see Jewiss 1972) it can produce some 17 leaves and so 17 new tiller sites. Hence, to sustain any tiller population requires only that 1 in 17 (*c.* 6%) are filled. However, this is certainly not to say that tiller site filling and vegetative 'recruitment' is never a limitation. The capacity for a rapid increase in tiller (and so leaf) population per unit ground area is essential for the establishment and vegetative dispersal of a genotype not only from seed, but thereafter following any major trauma to population numbers, e.g. winter mortality. In many cases the capacity to increase site filling rapidly is paramount.

However, we must not overlook the role of reproductive dispersal (by seed) in the abundance and yield of grasses. In closed communities, competition from established grass plants ensures that the germination of seeds plays only a small role in the recruitment of new plants (and so tillers) to the community (Crawley 1990), as well as in the recruitment or invasion of herbs. However, disturbance, e.g. by treading (poaching/pugging) or by rabbits or moles, can lead to gaps in a community where recruitment from seed is a major component of grass population density, and is often a major source of change in the diversity, including weed content, of species in the pasture (Burke & Grime 1996).

3.1.6 Reproductive development – events at the apex

Flowering can have a major effect on many aspects of the physiology, growth and utilization of the grass crop (Section 3.15), but here we concentrate on changes in the structure and fate of the apex – a subject extensively researched in the 1950s and 1960s.

Most north temperate perennial grasses must pass through winter conditions of low temperatures and/or short day lengths if they are to flower, although such a requirement may be less marked or entirely absent in biennials and annuals, respectively. Among the perennials, *P. pratense* is the only one in which no winter requirement has been identified. Satisfaction of a winter requirement causes no visible change in the shoot apex. Confirmation that it has occurred can be gained only retrospectively by observing the development of the shoot. The critical day length varies with species and cultivar, ranging from 8 to 13h in the ryegrasses and from 15 to 16h in timothy (Cooper 1951, Ryle & Langer 1963). Even after the shoot apex has made the transition to reproductive growth, long days are still necessary for normal inflorescence development (Langer & Ryle 1958), with the timing of ear emergence again depending on genotype and environment. Early flowering ryegrasses, for example, are particularly sensitive to temperatures in March and April, late flowering types to those in April and May.

The first sign of reproductive development is an acceleration of leaf primordia production and a lengthening of the shoot apex. Soon afterwards, bud primordia develop in the axils of the older leaf primordia, giving the apex a characteristic 'double ridge' appearance (Jewiss 1993). These developments at the shoot apex, together representing 'ear initiation', occur very early in the progression to flowering, e.g. in early February in *L. perenne* varieties that 'flower' or 'head' in late May. Development starts about one-third of the way down the apex, spreading both towards the tip and the base, and in many species the terminal meristem itself commits to reproductive development. This altered differentiation of primordia means that leaf production eventually ceases on that tiller. So in many grass species, tillers are in fact annuals. Perennation for the plant depends on producing new vegetative tillers during spring.

The changes at the apex are subsequently accompanied by a wave of cell division and expansion in the younger stem internodes which elongates the true stem and so lifts the developing inflorescence up through the sheath tube. This culminates in the emergence of the ear from the sheath of the flag leaf, the last leaf produced. The ear may be a spike, as in *L. perenne*, with each bud primordium developing into a spikelet bearing 3–10 closely packed florets, or it may be a panicle, as in *F. arundinacea*, with each bud developing into a primary branch, or branches, terminated by a spikelet (Fig. 3.1c).

3.2 The white clover plant

Despite many obvious differences in appearance, there are many similarities between the growth and development of legumes such as white clover and that of many grasses (for details and field cases, see Brock & Thomas 1991, Brock *et al.* 1996).

Following emergence from its very small seed, white clover initially develops a short vertical primary stem, carrying a rosette of trifoliate leaves and bearing a tap root. However, branches (stolons) soon arise in the axils of leaves on the primary stem, and these grow out horizontally, close to the soil surface. As in grasses, the meristematic apex (growing point) of each stolon lays down primordia from which a succession of leaves develop, each one comprising a trifoliate lamina (three leaflets) attached by a petiole to its node. The distinction between petiole and leaflets is evident at a very early stage. Unlike grass leaves, which extend in a linear fashion from a basal meristem, clover leaflets expand in two dimensions, which is typical of dicotyledonous plants; cell division and expansion take place uniformly across the surface of each of the leaflets, which remain folded until expansion is almost complete. Carlson (1966) has defined ten stages (scored 0.1–1.0) in the externally visible development of clover leaves (Fig. 3.5a). Petiole extension comes from a meristem positioned just below the leaflets and which moves up with them; it starts early in the development of the leaf, and is rapid. Final petiole length depends greatly on the light environment within the sward, with a high degree of overlap between supposedly short- versus long-petioled varieties. When a sward grows without interruption (whether from seed or after a harvest), and the light conditions worsen at the base of the increasingly dense canopy, successive leaves are borne on progressively longer petioles. Thus, each leaf in turn is positioned in high light close to the top of the canopy by the time its leaflets are unfolded (Dennis & Woledge 1985).

On an unshaded and undefoliated stolon, growing in otherwise favourable conditions, some ten or more live leaves may accumulate (Fig. 3.5b), but this situation rarely obtains in the field, where the stolons (like grass tillers) each bear only two or three fully expanded leaves, even when these escape grazing (Chapman 1983, 1986).

In clover, each node bears two root primordia, the lower one of which may form a root if it makes contact with moist soil (Thomas 1987a). The presence of roots at a node has important implications for the fate of that node and all branches, etc., attached to it (see below).

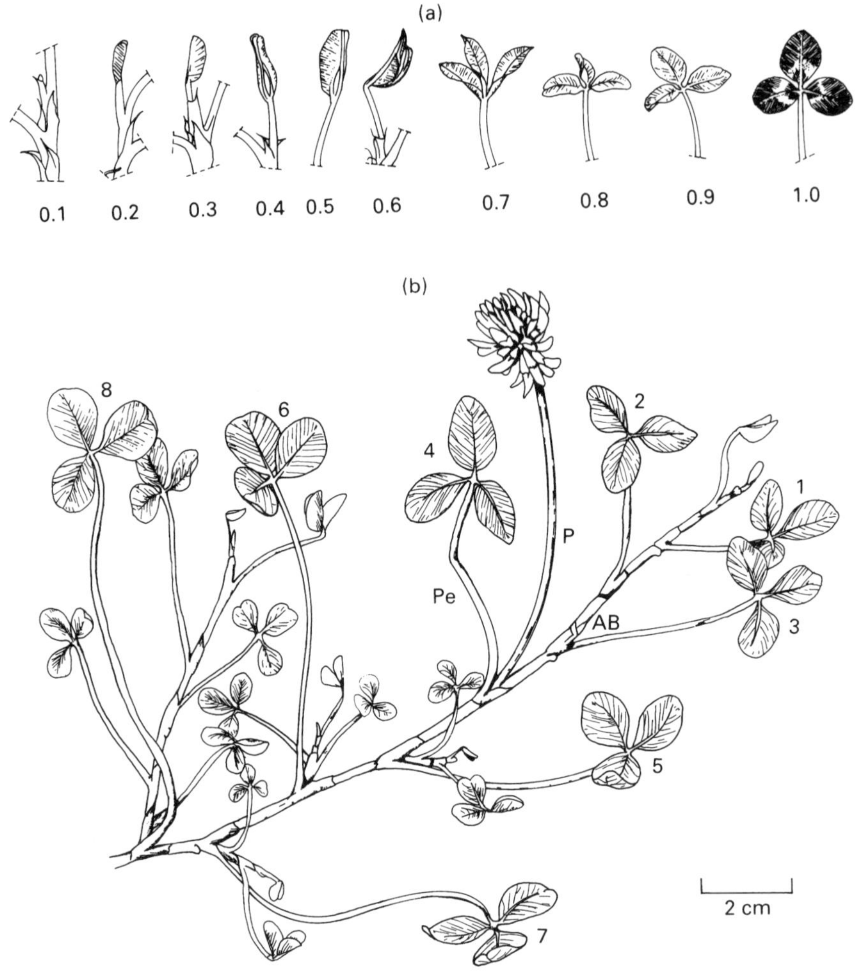

Fig. 3.5 (a) Ten stages in the development of a leaf of *Ladino* clover (from Carlson 1966, redrawn by Thomas 1987a). (b) An unshaded and undefoliated parent stolon of white clover bearing eight fully expanded leaves (numbered 1–8, youngest to oldest). Note: the axillary bud (AB) in the axil of leaf 3; the flower on its peduncle (P) in the axil of leaf 4, itself on its petiole (Pe); young daughter stolons in the axils of leaves 5–7; an older daughter in the axil of leaf 8 (after Thomas 1987b).

3.2.1 Branching and fragmentation

As in grasses, each node bears a single axillary bud which may develop into a daughter branch (stolon), offering the *potential* for an exponential increase in branch and leaf numbers. On unshaded stolons, growing free from competition and in favourable conditions, nearly all potential branch sites will develop branches (Fig. 3.5b), but in the field, the overall proportion of branch site filling is again closer to 10% (Chapman 1983, Hay *et al.* 1991, Newton *et al.* 1992). In clover, as in grass, the same general constraints on the number and size of units in a monoculture community apply (see Section 3.1.5). This limits the proportion of sites that can sustain branches. Of course, in a mixed community those 'boundary' or allometry rules apply, possibly assymetrically, to the *mixed* population (Enquist *et al.* 1998). The response to management of a clover monoculture, in the numbers and size of units, also parallels that of grass. Also, just as in grasses, it follows that each stolon axis needs, on average, to produce only one *surviving* daughter stolon in its lifetime (and so an overall degree of site filling of some 6–10%) for any given size population to be maintained. This requires that enough branches must be produced for at least one to survive all losses – including grazing. Some detail on work on the control of axillary bud outgrowth (site filling) in clover and grasses, and the subsequent fate of branches, is described in Section 3.2.2.

In clover, the primary (rosette) stem and tap root of white clover typically survive as little as 18 months (Hollowell 1966, Caradus 1990). When they die, the plant becomes fragmented and the daughter stolons, which may themselves have branched by then, become independent. The fragmentation of plants is more obvious than in many grasses simply because greater distances are involved, although one complicating factor is that at times, e.g. in winter, up to 90% of stolons become buried (Hay 1983, Hay *et al.* 1987). The size and topology of the fragments is similar to that in *L. perenne* and *F. arundinacea* (Brock & Thomas 1991, Brock *et al.* 1996, Hume & Brock 1997). As each daughter stolon grows forward at its apex, so the oldest portion dies back from one rooted node to the next. In this way, the fragments of the original plant migrate and disperse through the sward. There are marked seasonal changes in the frequency distributions of topology (e.g. 'branching order') of these migrating units (Brock *et al.* 1988, Hay *et al.* 1989a,b, 1990, Fothergill & Davies 1993, Fothergill *et al.* 1996). In summer and autumn the units are typically more complex (stolons themselves bear daughter stolons), whereas in spring there is an increase in the proportion of small unbranched units (cf., one tiller per 'plant' in grasses), but whether this is the cause or consequence of stress is uncertain. The possession of roots at a node greatly delays the senescence of that part of the plant (Sackville-Hamilton & Harper 1989), and so the distribution of roots along a stolon can have a profound impact on the topology of the sub-units that form, by altering when and where fragmentation occurs (Thomas 1987a). As in grasses, the implications of plant complexity for survival is strongly debated (see Brock 1988, Hay *et al.* 1988, Brock & Kim 1994, Brock & Caradus 1995, Brock & Hay 1995).

3.2.2 Control of axillary bud development

Concern over the low and patchy distribution of clover in pastures has led to much work on the factors controlling axillary bud development in clover (Newton *et al.* 1992, Hay & Newton 1996, Newton & Hay 1996). The work on bud development in clover is now beginning to be paralleled by equivalent work in grasses. Clearly the form of the plants (one creeping and stoloniferous, the other more upright) represents two alternative strategies for survival. The horizontal elongated stolons of clover, like those of some grasses, enable clover rapidly to invade any gaps or otherwise favourable patches that appear in a canopy, but even then, the sustained migration of clover means that its growing points soon leave even those gaps (Cain *et al.* 1995, Marriott *et al.* 1997a,b). Complex behavioural analyses are required to weigh up the benefits of these alternative strategies (Lerdau 1972, Caraco & Kelly 1991, Hutchings & de Kroon 1994, Oborny 1994).

Successful vegetative recruitment in white clover (i.e. the establishment of a new branch and therefore of a new growing point) requires that an axillary bud should pass through a number of stages. First, the bud must remain vegetative and not become reproductive; second, the vegetative bud must be viable; third, the viable bud must be stimulated to start outgrowth into a branch; fourth, the young (incipient) branch must mature to be self-sufficient and able to survive physiological separation from the parent stolon. A range of internal and external factors determine the fate of a bud through these stages.

In relation to flowering, some axillary buds become reproductive during spring (Fig. 3.5b). White clover is a long-day plant, and requires a critical day length of greater than 13.5–15.0 h, depending on the variety, before it will flower. About half the buds that experience the appropriate sequence of environmental conditions subsequently flower. In grasses, the inflorescence is terminal and prevents further production of leaves by that particular shoot axis, which subsequently dies. In clover, the flowers are axillary – 'in place of' branches (see Thomas 1987b). Thus, the main shoot apex continues to produce leaves and survives.

Those buds that remain vegetative may lose viability at a very early stage of development (within the apical dome) in response to external factors such as defoliation (i.e. removal of leaves from older nodes) (Hay & Newton 1996, Newton & Hay 1996). Buds also lose viability as they age (Newton *et al.* 1992). The outgrowth of viable buds has been shown to be affected by internal factors (Newton *et al.* 1992), such as the proximity of a nodal root, as well as a range of external factors. Rooting at stolon nodes is important to the success of stolon branches (Chapman 1983, Newton & Hay 1994). Although if root growth at a particular node is suppressed the ability of the associated bud to start outgrowth is not impaired, the subsequent successful outgrowth of that stolon branch may be. Both grasses and clover exploit the benefits of clonal growth, e.g. showing a capacity to distribute sugars and water between all connected units and all nodes (e.g. Chapman *et al.* 1991). However, there are clear priorities in the way resources are distributed. The relationship between a stolon branch and the roots at its subtending node appears particularly close. A high proportion of the carbon fixed by the young stolon goes to those particular roots (Chapman *et al.* 1992) which in turn may supply the stolon with water and nutrients, particularly

N and P (Chapman & Hay 1993, Lotscher & Hay 1996). The external environment of the bud also influences whether a bud starts to grow out into a branch. Particular emphasis has been placed on the quality of light experienced by the bud itself (e.g. Thompson & Harper 1988). However, it is now clear that for grasses (Murphy & Briske 1994), as well as white clover (Robin *et al.* 1994), it is the light quality (ratio of red to far-red light) experienced by the developing leaf subtending a bud that is the critical factor. Many other external factors have been shown to influence outgrowth, including mineral nutrient supply, light quantity, defoliation and soil moisture as well as defoliation (Newton & Hay 1996). In the early stages after a bud has started to grow out, the young branch receives nutrients from the parent stolon. This 'maternal support' may be the key to the high survival rate of incipient branches compared with the vulnerability of young seedlings, which would be the equivalent stage in plants recruiting from seed. Eventually, the new branch attains a positive carbon status and develops its own roots; in this condition it is potentially physiologically independent from the parent and able to survive fragmentation. The risks of death at this stage are broadly the same as for mature branches, with a frequent cause of death being damage or removal by grazing (Chapman *et al.* 1994).

In summary, it is clear that the recruitment of new branches (growing points) does not depend upon a simple relationship between an axillary bud and its immediate environment, but rather is the outcome of a complex of past and present events. In the field about 50% of axillary buds are deemed viable, that is can be 'persuaded' to grow out under the appropriate conditions (Newton *et al.* 1992), although in reality only 25% of all axillary buds actually grow out to some extent (Chapman 1983, Newton *et al.* 1992). Of these, about one-half (10% of the total buds) form a branch (Hay *et al.* 1991). These numbers are very similar in clover and in ryegrass (P.C.D. Newton, personal communication, 1998). It should also be noted that the majority of the 'active' buds are found between nodes three and eight along a stolon (Chapman 1983, Newton *et al.* 1992). It seems likely that this 'window of opportunity' may in part be due to the maturity and vigour of the axillary buds, although this cannot be a complete explanation. Although there are conditions under which death of incipient branches may be high (Soussana *et al.* 1995), in common with other clonal species (de Kroon & Kwant 1991, Murphy & Briske 1994), it is the birth rate of clover branches (the outgrowth stage) rather than their death rate which appears to be most limiting to branch numbers, implying little futile branching.

3.3 Factors affecting leaf growth

3.3.1 Grass

A comprehensive account of the factors affecting leaf growth and tillering would take several chapters, and detailed introductions can be found in Kemp

(1981), Schnyder (1986), Robson *et al.* (1988) and Lemaire & Chapman (1996) (see also the series of papers Volenec & Nelson 1981, 1982, 1983, 1984a,b, Nelson 1982). Here we concentrate on grasses, and provide an overview of the aspects of leaf and tiller production that is necessary to appreciate the limitations that the growth of a plant as a series of units imposes on growth and utilization (the flows of matter and energy) in the crop as a whole. We emphasize that to understand the response of the crop to external factors, it is valuable to consider not just how the factor affects the extension of a leaf, but also the rate of appearance of leaves, the number of units (tillers) actually producing those leaves, and how the demand for resources for that crop-scale increase in leaf growth can be met.

Leaf growth involves cell division; the subsequent expansion of the new cells (which requires 'water pressure') and the deposition of materials, e.g. cellulose, which increases the rigidity and mass of the cell walls. This latter process accounts for most of the accumulation of dry matter by the crop. Each of these processes, as well as the rate at which new leaves appear, can be affected in different ways by factors such as temperature, light, water, nutrients (notably nitrogen) and (see later) by defoliation management. However, in many major forage grasses the set inter-relationship between leaf appearance rate, final leaf size, the number of leaves growing and the number of live leaves on each tiller restricts the possible outcomes (Fig. 3.6). Moreover, the width of a leaf in grasses is partly predetermined. It reflects the basal circumference of the apex at the time the leaf was in the primordial stage, and so the previous overall size of that tiller. We will now take these factors one at a time (a rare instance in nature).

Temperature

Temperature is one of the major environmental determinants of leaf growth. Increasing temperature over the range 5–25°C increases the rate of appearance, as well as the rate of extension of leaves. In mid-summer, new leaves appear on ryegrass tillers every 7–10 days, but at five times that interval at the lower temperatures in mid-winter (see Davies 1977 in Jones & Lazenby 1990: p. 138). Temperature also affects the duration of extension and so the final size of leaves (Fig. 3.6). In general 'high temperature' leaves extend more rapidly, for a shorter period, but to a greater final length than 'low temperature' leaves; they tend to be longer in relation to their width, thinner, and have proportionately more lamina relative to sheath (Robson *et al.* 1988). Because the rate of appearance of leaves is increased as temperature increases, so too is the rate of production of sites for new tillers, so that although temperature has less direct affect on site filling (Mitchell 1953), and high temperatures may decrease site filling, the number of tillers producing leaves at the faster rate is increased at higher temperatures.

Leaf growth increases with temperature up to an optimum in the region of 20–25°C for most north temperate grasses, with the night temperature optimum

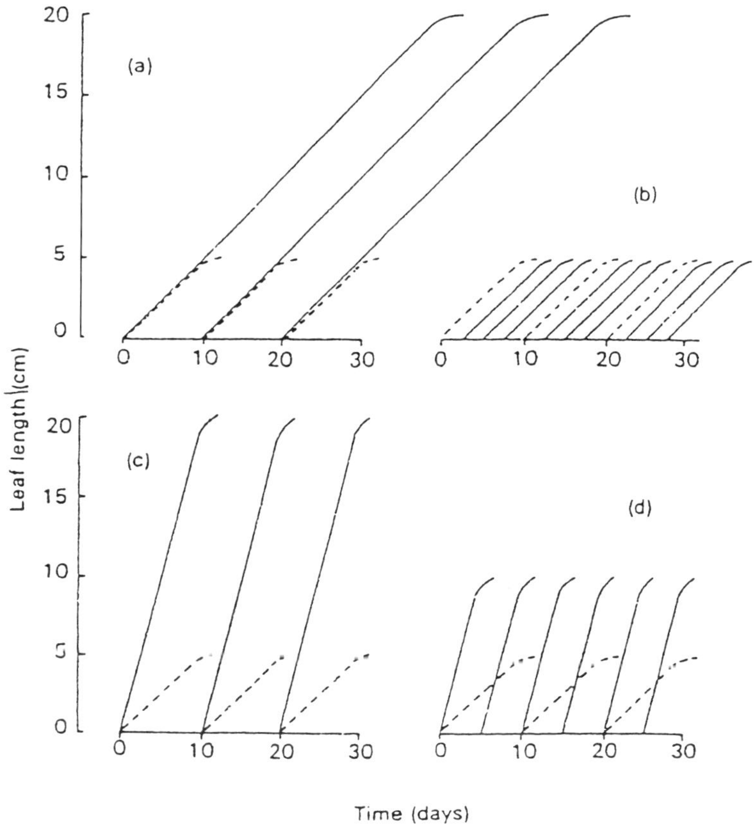

Fig. 3.6 Shoots of *Festuca arundinacea* produce four times as much leaf at 25°C (—) as at 10°C (- - -). Three ways in which this could, in theory, be achieved are: (a) each leaf grows for 4 times as long to 4 times the final length; (b) successive leaves appear 4 times as frequently, but grow at the same rate, for the same time, to the same final length; (c) each leaf grows at 4 times the rate to 4 times the final length. In practice, (d) the plants tend to both a greater rate of leaf appearance (2×) and a greater rate of leaf extension (4×), but with the duration of extension halved to give leaves of twice the final length, and with no change in the number of growing leaves. (Reproduced by permission, Robson *et al.* 1988.)

equal to or slightly lower than that of the day (Evans *et al.* 1964, Robson 1972, 1973c). Sustained periods at higher temperatures than this decrease leaf growth. Although the response to temperature is not linear, it has proved of great practical value to use an accumulator of temperature over time ('day degrees') to monitor the development of grass (and cereal) crops and for timing the first applications of fertilizer (see Chapter 4). Day degrees are typically accumulated only when above an assumed threshold, e.g. of 6°C.

Different parts of a grass leaf may be at widely different temperatures, the illuminated leaf blade being as much as 20–30°C warmer than the base (Peacock

1975a). The key temperature for leaf growth is that of the meristematic zone near the shoot apex (Peacock 1975a,b). The rate of extension of leaves can respond to a temperature change within minutes, but the response of leaf extension to temperature is not fixed. Leaves extend faster in spring than at the same temperature in autumn (see Section 3.15) (Peacock 1975c).

Light

Light (total energy) is clearly of importance in providing resources, through photosynthesis, for the creation of new cells, their increases in dry matter and so for leaf expansion on each and every tiller (see references to Volenec & Nelson above). During even short periods of low-light intensity, photosynthesis and hence the supply of raw materials for growth are reduced, but leaf expansion is less affected than might be supposed. The temporary storage of intermediate metabolites in sheath bases, and in the growing leaves themselves, provides a buffer against short-term fluctuations in the current supply of photosynthetic products (Gordon *et al.* 1977). Moreover, during more sustained shade, more photosynthate is retained by the shoot at the expense of root, and the new leaves produced are thinner ('shade' leaves) than those produced in bright light ('sun' leaves). As a result, leaf area is maximized per unit weight of plant tissue, offsetting to some extent the effects of reduced carbon assimilation per unit of that leaf area. Total light energy increases both leaf appearance rate and site filling (Mitchell 1953). Within plant communities, shade from other plants affects not just the amount of light, but also the light quality. Shade by other plants lowers the ratio of red to far-red wavelengths. Low R–FR ratios reduce site filling (see Section 3.2.2).

Water

Water is of profound importance in leaf growth, and the rate of leaf extension is a sensitive indicator of water stress. A shortage of water limits leaf extension even before stomata close to reduce water loss. Water stress results from an imbalance between the evaporative demand on the plant and the ability of the plant to meet that demand from the soil. During a dry period, as soil water deficit builds up, leaf expansion is restricted first during daylight hours when evaporative demand is highest, and only later at night. Cell expansion (which requires water pressure) is more sensitive to water stress than is cell division. Thus, unexpanded cells may accumulate during a dry period and then expand rapidly when rain comes, offsetting, in part, some of the effects of drought (e.g. Clark *et al.* 1999). Water stress severely reduces leaf appearance, and so tiller site production, and hence tiller numbers per unit of ground area (Barker *et al.* 1985).

Nitrogen

Nitrogen (N) plays a major role in leaf growth via its involvement in cell division and as a primary component of enzymes for all living systems and processes (see Volenec & Nelson above). However, N deficiency slows leaf expansion more than it does photosynthesis, which leads to some accumulation of sugars in the leaves. Nitrogen is such an effective stimulant to grass growth because it increases both leaf production and photosynthesis directly. Nitrogen (the avoidance of N deficiency) greatly increases the rate of leaf extension, but is only loosely related to the rate of leaf appearance (Lemaire 1988, Gastal *et al.* 1992). The increased rate of extension of leaves on existing tillers in turn stimulates greater light capture and so photosynthesis. Because the rate of leaf appearance is little affected, the rate of production of new tiller sites changes little, but there is a substantial increase in the degree of site filling (Lemaire & Chapman 1996), and so there can be an increase in the number of tillers producing leaves at the increased rate. Of course, faster growth means a more rapid approach to high LAI, the mutual shading of leaves, and the ceiling yield (albeit a greater one). In terms of population growth, N will increase the speed at which the sward builds up tiller numbers, and so reduce the time taken to reach the 'boundary' in Fig. 3.4. Nitrogen will therefore substantially increase the number of tillers present in a sward if it is sustained at a given low LAI, but it will accelerate the decline in tiller numbers if the faster growth is allowed to lead sooner to a high LAI, and a fall in site-filling, and later self-thinning, occurs.

3.3.2 Clover

Clover shows similar overall changes in leaf growth in response to external factors as does grass (although largely avoiding those due to N deficiency). Leaves appear on *T. repens* clover stolons at a similar rate to *L. perenne* at the same intermediate temperatures (10–25°C), and under similar defoliation managements (Parsons *et al.* 1991a). As in grass, the rate of extension (now two-dimensional) is slowed substantially by water stress and low temperature. It is widely considered that low temperatures slow leaf growth in clover more than in grasses (Mitchell & Lucanus 1960). However, rates are similar at, e.g., 10°C, but lower at temperatures around 2–5°C, implying a difference in temperature threshold for growth (e.g. Chapman *et al.* 1983). At low temperatures, such as over winter, clover produces small leaves of lower specific leaf area (SLA) on short petioles. With high temperatures or low light intensities, large, thin typically 'shade' leaves are produced on elongated petioles. Overall, clover leaves typically have higher SLA (300 $cm^2 g^{-1}$ DW) than in N-fertilized (250 $cm^2 g^{-1}$ DW) or N-deficient (200 $cm^2 g^{-1}$ DW) grass (Parsons *et al.* 1991a). Clover leaves, like grass leaves, are very sensitive to water deficiencies. Pulvinal cells at the base of leaflets respond by folding each trifoliate leaflet in half. This response reduces C inputs, and clover leaf-growth appears more adversely affected by drought, but it conserves water, and recovery of leaf growth from drought can be faster than in many grasses, although the duration of the drought is clearly critical.

The rate of appearance of clover leaves (and so potential sites for daughter stolons) increases substantially with increasing temperature (up to 25–30°C), while site filling in clover declines to some extent with temperatures above 10°C (Bienhart 1963, Hoglund & Williams 1984). The net effect, as in grasses, is that branching increases with temperature and there is a fairly broad optimum temperature range for branching (Hart 1987). Site filling is reduced by an inadequate supply of inorganic nutrients, particularly phosphorus (P) and potassium (K). There has been considerable work in clover on the role of light, notably of perceived shade (red:far-red ratio) on bud outgrowth and so site filling (see Section 3.2.2). As in grasses, water stress severely limits leaf appearance and site filling, and so restricts branching.

3.4 Some background on photosynthesis

Tissue production and ultimately crop yield depend on photosynthesis, the process by which solar radiation, intercepted by green leaves, provides the energy to convert CO_2 and water into simple sugars. Over 90% of the plant dry weight consists of organic compounds derived from those simple sugars. This is why 'carbon' (-fixation, -balance) is often used as a currency for changes in dry matter (DM) and energy. (Note that DM is *c.* 40–44% C, so DM = C × 100/44.)

Green leaves, like all living tissues, also respire; they oxidize substrates to produce energy necessary for metabolism with the release of CO_2 and water, essentially the reverse of photosynthesis. In well-illuminated grass leaves, the photosynthetic uptake of CO_2 vastly exceeds their respiratory efflux, but at progressively lower light intensities the gap narrows until, at the 'compensation point', net photosynthesis (the excess of photosynthesis over respiration) is zero. At even lower light intensities, respiration exceeds photosynthesis and the leaf is in negative carbon balance. The resulting relationship between 'net' photosynthesis and increments in light intensity is a classic one of 'diminishing returns' (e.g. a non-rectangular hyperbola) in that each successive increment in light intensity brings a progressively smaller return in carbon fixation. The net photosynthesis of individual leaves saturates (i.e. it reaches a plateau at a value representing the maximum capacity of which that particular leaf is capable) at relatively low light intensities (e.g. 150–200 $J\,m^2\,s^{-1}$ PAR, measured perpendicular to the leaf surface (Prioul 1971)). However, grass leaves in a canopy are not held at 90° to the light. Because the response of leaves to light is curved, it follows that a given amount of solar radiation will cause more carbon fixation if it is spread over a large leaf area (at low mean intensity) rather than limited to a small leaf area (at high intensity). Photosynthesis in a grass canopy of high LAI can remain unsaturated even at light intensities of 500–600 $J\,m^2\,s^{-1}$ PAR (photosynthetically active radiation) (see review by Woledge & Parsons 1986). At low LAI, prostrate leaves are beneficial to light interception and so to photosynthesis, but at high LAI erect leaves are better as light is distributed more effectively at lower mean intensity (Rhodes 1969, Robson 1973b).

Temperature

This has many direct effects on photosynthesis and respiration, as well as indirect effects via its role in the rate of leaf production. The light-saturated rate of net photosynthesis of ryegrass leaves increases almost three-fold between 5 and 25°C (Woledge & Dennis 1982), although vapour pressure deficit as well as water stress, which often accompany high temperatures, can completely counter that effect in the field.

There are two developmental features of grass leaves that greatly modify the photosynthetic capacity of the grass crop. These are age and leaf development in shade, which both reduce leaf and canopy photosynthesis.

Age

As leaves age, their photosynthetic capacity declines, starting soon after full expansion and well before any visible sign of senescence, with the older tip declining before the younger base. In general, although leaves continue to make a positive contribution to photosynthesis up to the time of their programmed senescence, the decline is slowed by low temperatures and low light, and these leaves live longer unless shade is extreme (Woledge & Jewiss 1969, Woledge 1971). The rate of photosynthesis which a leaf exhibits in low light intensities declines more slowly with age than its light-saturated rate (Woledge 1972). Since many leaves in a dense grass crop, especially the older ones, are poorly illuminated, the adverse effect on photosynthesis is less than might otherwise be expected. Managements have relatively little effect on the leaf age structure of grass canopies (see Parsons *et al.* 1988a) so there are few benefits to one management versus another on this account.

Shade

Development in shade has been shown to have a profound effect on leaf photosynthetic capacity. Leaves developed in low light have a poorer photosynthetic capacity to function at high light intensities, although they also have lower rates of respiration (Treharne & Eagles 1969, Prioul 1971, Woledge 1971, 1973), and lower stomatal and residual conductance (Prioul 1971, Prioul *et al.* 1975, Woledge 1977, Woledge & Dennis 1982). Leaves expanded in low light are thinner (Woledge 1971); they may have fewer stomata (Wilson & Cooper 1969) and fewer mesophyll cells per unit leaf area, fewer and smaller chloroplasts, and a reduced activity of RuBP carboxylase/oxygenase (Prioul *et al.* 1975, 1980a).

This is an issue because, in a vegetative sward, successive new leaves expand from the base of the canopy, and so as the canopy becomes increasingly dense, new leaves develop a poor capacity for photosynthesis in the high light into which at least a portion of each leaf ultimately emerges (Woledge 1973, Woledge

& Leafe 1976). Leaves that are grown artificially free from shade continually develop high rates of photosynthesis. It has been proposed that there is some capacity for 're-adaptation' of photosynthesis in portions of the leaf (Prioul *et al.* 1980b), but studies on whole leaves show limited capacity for re-adaptation (Woledge 1971). The decrease in leaf net photosynthesis due to the effects of development in shade may lead to a shortfall of some 30% in C fixation in the vegetative grass canopy, and this detracts significantly from dry matter production (Robson 1973b). However, this restriction is largely overcome during reproductive development (see Section 3.15). Clearly the maintenance of a low LAI, such as by continuous grazing, avoids this problem and sustains a high efficiency in successive leaves, but of course, photosynthesis is reduced then by the sustained low LAI.

Nitrogen

N has a profound effect on the photosynthesis of individual leaves, as much of the organic N in leaves is in the form of carboxylases, the major photosynthetic enzymes. N deficiency substantially impairs leaf photosynthetic capacity as the organic N content of leaves falls towards 3% (Woledge & Pearse 1985). Since this happens quite frequently in the field, even in heavily fertilized crops, photosynthesis will often be limited to some extent. The major impact of N deficiency on photosynthesis, however, is by decreasing the rate of leaf expansion and so light interception.

In the case of clover, the same general principles apply. Differences arise in that its leaf canopy is not only more prostrate, but the leaves actually track the movement of the sun (see Thomas 1987a). Clover canopies are therefore more effective in light interception at low LAI, but saturate at a lower LAI and lower light intensities than grass ones. Clover leaves clearly avoid the impact of nitrogen deficiency, but most important, leaves avoid the adverse effects of development in shade (see Section 3.12).

3.5 The overall trade-off between growth and utilization

The most notable feature of growth and utilization in the grass crop is that the major harvestable component is the leaves – the very tissues that are essential for the photosynthesis that sustains the growth and dry matter production of the pasture. The way the crop is harvested, and in particular the management of leaf area, has a profound effect on the amount grown as well as on yield. However, because of the rapid turnover of leaves in the sward, the crop must be harvested repeatedly or a large proportion of the tissue produced will die unharvested and be lost. Leaves, or portions of leaves, that are not harvested within the lifespan of a leaf (e.g. some 33 days for *L. perenne*; 57 days for *F. arundinacea*) wither, lose nutritive value, and in many cases fall from the canopy and become unavailable. To optimize the utilization of grass, it is necessary to strike a balance between the amount of leaf area that remains in the sward during a sequence of defoliations, regrowths to provide photosynthesis for growth, and the amount of leaf actually harvested to achieve a yield.

The effects of the amount of leaf material that is sustained long-term in the pasture on the balance of the major physiological processes involved in growth and utilization are shown in Fig. 3.7 (see Parsons *et al.* 1983b). We can regard these figures, for now, as mean values, averaged over time and/or across patches/paddocks on a farm. In this sense, the same *general* principles apply whether the total area is harvested by continuous grazing, rotational grazing or cutting (see Parsons *et al.* 1988b).

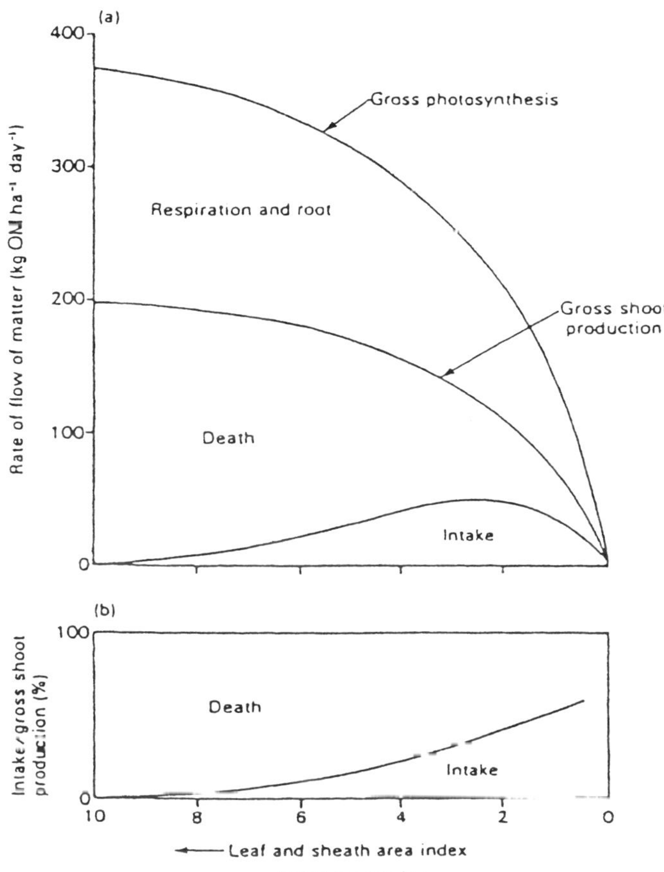

Fig. 3.7 The effect of the intensity of grazing (the average leaf area index at which the sward is sustained) on the balance of the major physiological components of growth and utilization (after Parsons *et al.* 1983b).

In swards maintained on average at a high LAI, canopy gross photosynthesis is close to a maximum, and the maximum amount of new tissue is produced. The 'losses' of material due to respiration are also close to a maximum, although the proportion of gross uptake lost in respiration (*c.* 45%) is little altered by the LAI sustained (Robson 1973b). However, to maintain the sward at a high LAI only a small proportion of the leaves produced can be harvested. As a result, most of the leaf tissue dies unharvested. At a sustained high LAI (left side of graph in Fig. 3.7), gross shoot production may be close to 30 t OM ha^{-1}, yet of this only 4 t OM ha^{-1} may be harvested and 26 t OM contributes to the soil organic matter turnover and respiration. Such a situation is not uncommon, and on some farms in the UK levels of utilized metabolizable energy (UME) output from grass of just 40 GJ ha^{-1} (equivalent to approximately 4 t OM ha^{-1}) have been recorded on high-quality soils receiving large inputs of fertilizer nitrogen (Peel & Matkin 1982). Clearly, to increase utilization the intensity of defoliation must be increased.

In swards maintained at a lower LAI (right side of Fig. 3.7), a far greater proportion of the leaf is harvested and a smaller proportion lost. Not all light is intercepted, and canopy photosynthesis and shoot growth are reduced, but the increase in the *proportion* of gross tissue production harvested more than outweighs the decrease in the *amount* of grass grown, and the amount harvested is actually increased. Indeed, the maximum yield (here 10 t ha^{-1}) is achieved, not because this management leads to maximum light interception and photosynthesis, but because it achieves the best balance between photosynthesis, gross tissue production, yield (intake) and death. At very high intensities of defoliation, where only a very small leaf area is sustained (extreme right, Fig. 3.7), all components of production and utilization are reduced, and the sward is clearly overgrazed.

It is important to note that if we define the efficiency of utilization as the proportion of all grass grown (gross production, net primary production) that is harvested, then maximum yield per ha is achieved at a point when the efficiency of harvest is close to just 50%, so as much plant material dies and contributes to soil organic matter as is harvested.

The principles of grass growth and utilization described here have been derived using simple dynamic mechanistic models (e.g. Johnson & Parsons 1985a, b, Woodward 1998) to rationalize field-based measurements of mass flux per unit ground area (e.g. Parsons *et al.* 1983a,b, King *et al.* 1984) and of the appearance, turnover and fate of individual leaves and tillers (e.g. Grant *et al.* 1983, Parsons *et al.* 1991a,b, Mazzanti & Lemaire 1994).

3.6 Different managements – or a question of scale?

A great deal of effort has gone into establishing the principles of growth and utilization under different managements, and so developing a theory for optimizing these, but for many years this fuelled controversy, rather than resolved it, as to the relative merits of alternative managements such as continuous versus rotational grazing (see McMeekan 1960, Raymond 1981). However, the differences between managements are in principle more imagined than real. It has long been recognized that even continuous grazing is actually rotational grazing at the bite scale (see Clark *et al.* 1984 and references therein). At the *bite* scale, or *plant* scale, all managements involve a succession of discrete defoliations; that is the plant suffers a sequence of defoliations separated by variable periods of uninterrupted regrowth. Even under continuous grazing defoliation is never a continuous process.

The same theory can therefore be used to identify and explain the optimal defoliation strategy for all managements, although this has only been possible recently following the development of fine-scale spatial models of the grazing process (e.g. Schwinning & Parsons 1999). We go through this theory below, and we see how the managements differ largely in the way they create and control spatial heterogeneity within and between paddocks (Parsons & Chapman 1998).

3.7 Regrowth – and the ceiling to yield

First we must consider how leaf area, photosynthesis and the losses of dry matter change as the crop regrows uninterrupted from a single severe defoliation, with a view to identifying an optimum timing of harvest.

When grass swards are allowed to regrow uninterrupted, they do not endlessly accumulate dry matter. Soon after the leaves have come to intercept all incident light, the pasture will reach a ceiling yield. These principles were established in seminal studies, e.g. by McCree & Troughton (1966a, b) and by Robson (1973a, b), and are illustrated in Fig. 3.8.

When defoliation removes virtually all leaf tissue, rates of photosynthesis are substantially reduced and respiration rates may initially exceed the uptake of carbon in photosynthesis on a 24-h basis. The sward will consequently make a net loss in weight, but new leaf tissue can still be produced from 'reserves'. These may be sugars, stored in many cases in sheath bases (Pollock & Jones 1979, Volenec 1986), or remobilized structural material and proteins (see Davidson & Milthorpe 1966a, b, and review by Richards 1993). In many north temperate grasses these come almost entirely from the stubble, and far less than had been supposed from the root. The root suffers, however, in that its supply of current assimilates is reduced.

The period over which new tissue growth can be supported from 'reserves' is very limited, and in any case the consumption of these resources represents a net *loss* of weight and so is not 'growth' in the common sense. Any subsequent accumulation of dry matter must depend on the capacity of the crop to re-establish a leaf area, and to restore inputs from photosynthesis.

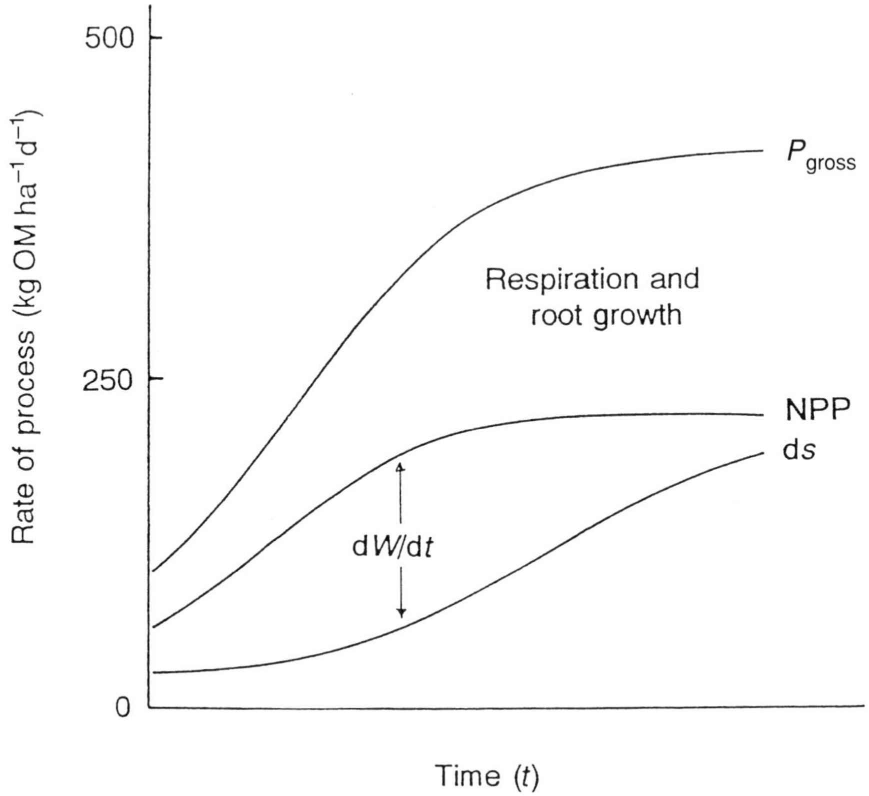

Fig. 3.8 Changes in the rates of gross canopy photosynthesis (P_{gross}), gross shoot tissue production (P_{net} or NPP) and tissue death (d*s*) with time during the growth of a grass sward from a low to a ceiling LAI.

There may be an overriding limitation to regrowth if a large proportion of meristems has been removed, although in agriculturally improved grasses, in vegetative growth, very few meristems are lost. Given that these same meristems may have been supporting, prior to defoliation, far higher rates of tissue production than now, it is unlikely that growth is limited by meristem numbers, but rather that this is the result of a reduced supply of carbon from photosynthesis.

As the carbon invested in new leaves enables the crop to intercept an increasing proportion of the available light, canopy 'gross' photosynthesis (P_{gross}) increases. However, 'gross' photosynthesis reaches a maximum rate soon after all the available light is intercepted. Thereafter, gross canopy photosynthesis may even decline, as the photosynthetic capacity of the leaves of some grass species declines as successive new leaves develop in shade (see Section 3.4) (Woledge 1977, 1978).

During regrowth, the rate of respiration of the pasture also increases. Some 25% of gross photosynthesis is consumed in the synthesis of new tissues (this is a relatively fixed biochemical stoichiometry; see the review of Thornley & Johnson 1990). The efficiency depends on the biochemical composition of the tissue created, but on average, for every 75 units of carbon in the new tissue produced, a further 25 units are respired (Penning de Vries 1972). There is also an increase in respiration associated with the 'maintenance' (e.g. active processes and protein turnover) of the increasing mass of live tissue (McCree 1970, 1974). A 'typical' rate would be equivalent to a loss of dry weight of about a 1.5% day^{-1}. In improved grassland species, respiration rarely exceeds 40–45% of 'gross' photosynthesis – hence it is divided roughly equally between 'synthesis' and 'maintenance' (Robson 1973b).

However, during regrowth there is also an increase in the rate of senescence of leaf (per unit ground area), 'd*s*' in Fig. 3.8. This increase in the rate of senescence largely reflects the increase in the size of leaves currently involved in the turnover of tissue in the sward (as opposed to an increased 'density-dependent' mortality). During regrowth, changes in the rate of loss by senescence inevitably lag behind changes in the rate of gross photosynthesis and gross tissue production as, at any point in time, the size of leaves currently dying reflects the size of leaves produced previously. Once the rate of gross tissue production has reached a maximum, the rate of senescence 'catches up' and comes to equal the rate of gross production (Fig. 3.8) and the rate of net accumulation of live tissue (the instantaneous growth rate, dW/dt) declines to zero.

It is important to stress that the mechanism by which cool temperate grassland species, typified by *L. perenne,* achieve a ceiling yield is not that which has sometimes been proposed (e.g. Donald 1961). Mature, shaded tissue does *not* build up inexorably until its respiratory demands equal gross photosynthesis, leaving no carbon from which new tissue can be synthesized. Rather, leaf death intervenes, ensuring that there is always sufficient carbon to sustain leaf production at a high rate. The ceiling crop is not static therefore, but dynamic. The ceiling yield simply means there is no further *net* accumulation of dry matter.

The balance of these processes accounts for the classic sigmoid growth curve in dry weight, *W*, seen in numerous agronomic studies of crops growing from seed or after a harvest (Fig. 3.9a) and the changes in the instantaneous growth rate, dW/dt (Fig. 3.9b).

3.8 The optimal timing of harvest – marginal value theorem

The principles for understanding the optimal timing of harvest owe much to the marginal value theorem, which is a central tenet of foraging theory in ecology where it is used to describe the optimal time for an 'animal' or herd to move on from one 'patch' to the next (Charnov 1976, Stephens & Krebs 1986). In the context of grassland management, the concept explains when a farmer should

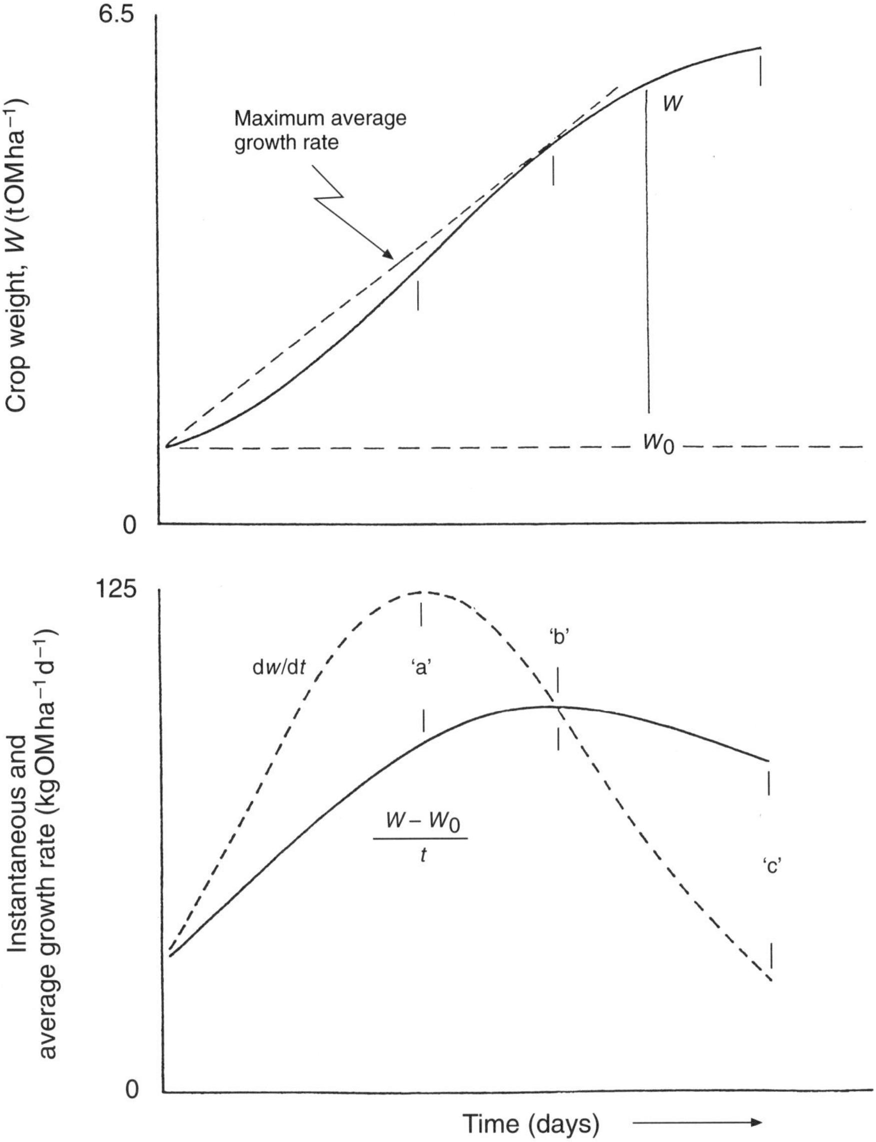

Fig. 3.9 Corresponding changes in the weight of the crop (W), the instantaneous growth rate (dW/dt) and the average growth rate ($(W - W_0)/t$). See text for an explanation of the effects of harvesting the sward at times a, b and c and for an explanation of the maximum average growth rate (after Parsons & Penning 1988).

harvest the material accumulated in the current patch or paddock, so terminating that period of regrowth, and move on to the next paddock or patch.

In Fig. 3.9, the crop is shown regrowing from a low initial weight, W_0, and increasing in weight, W, over time. Assuming the sward is defoliated back to its initial weight, then harvesting at time 'c' (see Fig. 3.9) would clearly achieve close to the maximum yield from a single period of regrowth ($W - W_0$ is greatest). However, in this case the instantaneous growth rate (dW/dt, the slope of the line W at any point in time) would have been decreasing for some time, and would currently be close to zero. Harvesting the sward at time 'a' (Fig. 3.9) would mean interrupting the growth of the sward at the time when the instantaneous growth rate was at a maximum, although the amount removed at harvest would be reduced. However, maximum utilization is achieved if the sward is harvested when the average growth rate, $(W - W_0)/t$, is at a maximum (Parsons & Penning 1988, and see Morley 1968), as this takes account not only of the material that has accumulated, but also of the total time that has elapsed since the start of regrowth. The average growth rate can be envisaged as a line drawn from the initial (residual) sward weight, W_0, to any point on the growth curve of W. The maximum average growth rate, the steepest possible slope, is when the line is a tangent to the curve, as in Fig. 3.9 at 'b', and this defines the optimum time to harvest the sward, the best duration for regrowth.

Following defoliation to a low LAI, the average growth rate initially increases (see solid line in Fig. 3.9b), but thereafter changes relatively gradually as the period of regrowth is extended. Note that the *maximum average growth rate* is achieved (at time 'b', Fig. 3.9b) after the time of the maximum instantaneous growth rate, but before the time of the ceiling yield. It corresponds to the time when the current instantaneous growth rate first falls below the long-term average growth rate (hence the 'marginal value' of continuing this period of regrowth is now declining), and so mathematically it is when the plots of dW/dt and $(W - W_0)/t$ intersect.

3.9 Effects of 'residual sward state' on regrowth

In Sections 3.7 and 3.8, we considered how grass communities regrow from a *low* initial (residual) mass, height or leaf area to a ceiling yield. That analysis emphasized the importance to utilization of understanding changes not only in the instantaneous growth rate, but also in the average growth rate. In practice, however, not all paddocks (or patches in a paddock) are defoliated consistently and uniformly to such a small residual state, and we have not yet considered whether this is in any case desirable. Operational constraints (e.g. in moving animals) can mean defoliation can give rise to all manner of residual paddock states, and animals biting potentially at random can mean there is spatial variance (heterogeneity) within the residual state (some patches are grazed short, others are left virtually untouched). Next, we must consider how the residual sward state affects regrowth and the optimum timing for harvest.

The effect of more lenient defoliation, and so greater residual states, on regrowth and its components, can be seen in Fig. 3.10. Following severe defoliation (e.g. line 1 in Fig. 3.10a) the rate of canopy net photosynthesis (gross tissue

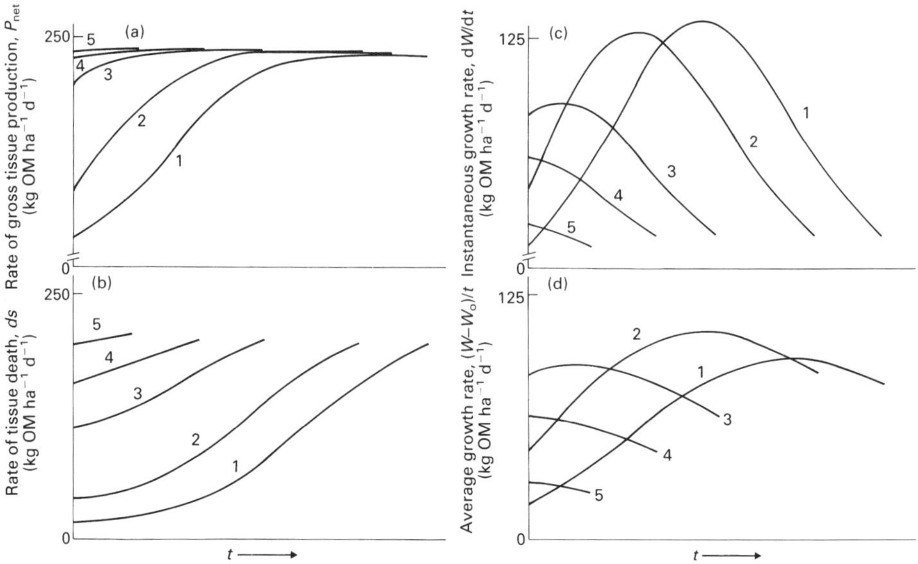

Fig. 3.10 The effects of the severity of intermittent defoliation on (a) the rate of gross tissue production (P_{net}), (b) the rate of tissue death (ds), (c) the instantaneous growth rate (dW/dt) and (d) the average growth rate ($(W - W_0)/t$), as the duration of regrowth is extended over time, t. The LAIs to which the swards had been cut, and from which they are regrowing, were 0.5, 1.1, 3.4, 5.3 and 6.8 for lines 1 to 5, respectively (after Parsons *et al.* 1988b).

production) is low initially, reflecting the low initial leaf area, and it takes a long time before the maximum rate of net photosynthesis is regained. Following more lenient defoliation (lines 2 and 3) the rate of net photosynthesis is greater initially, reflecting the greater residual leaf area, and the maximum rate of photosynthesis is regained sooner. However, the severity of defoliation also affects the loss of matter in leaf turnover and death (Fig. 3.10b). Following severe defoliation, death rates are low initially, there is a long delay before the rate of death equals the rate of net photosynthesis (gross tissue production) and there is no further net gain in weight. Following more lenient defoliation, not only is the rate of photosynthesis higher initially, but so too is the rate of tissue death, and maximum death rates are achieved sooner. The effects of the severity of defoliation on the instantaneous growth rate, and on the average growth rate, are shown in Fig. 3.10c and d, respectively. Following lenient defoliation, the maximum average growth rate is achieved sooner, after a shorter duration of regrowth (e.g. line 2 compared with 1, or 3 with 2). Moreover, after the more lenient defoliations (lines 4 and 5), average growth rates are lower overall and may actually decline throughout regrowth.

We can summarize how the initial conditions for regrowth (residual sward state) affect growth rate, as in Fig. 3.11. Note that, although regrowth from severe defoliation follows a classic 'logistic' growth form (and so a 'humped

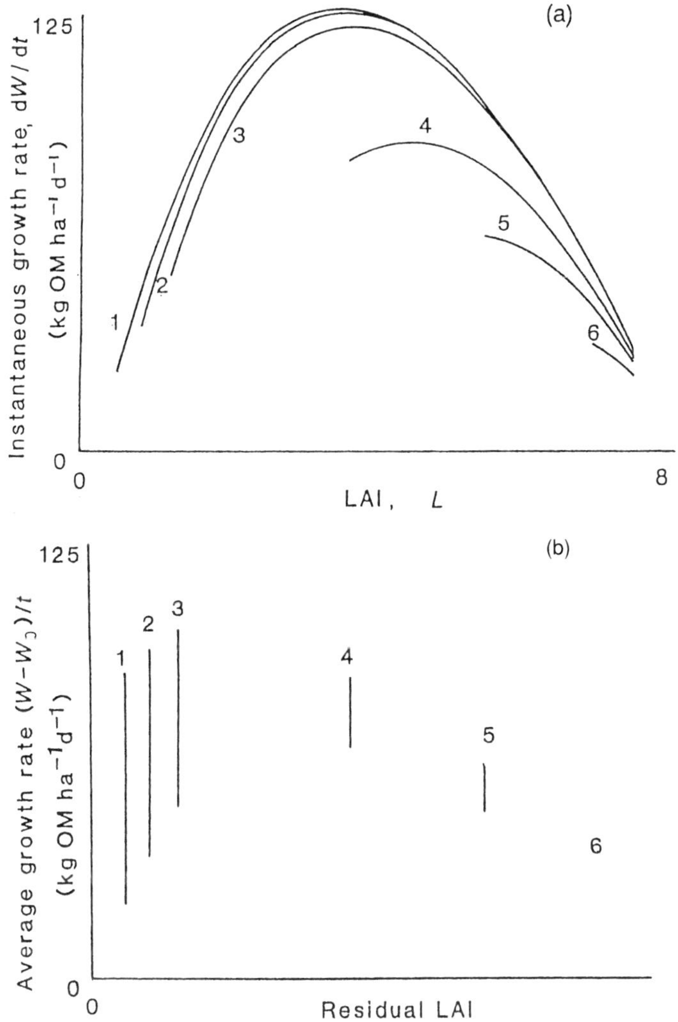

Fig. 3.11 The effect of residual sward state (six examples as in Fig. 3.10) on the growth rate of pastures. (a) Following lenient defoliation (high residual sward state as in '4'), growth rates are lower initially than in a severely defoliated (as in '1') sward passing up through that same LAI. Leniently defoliated swards do not re-trace the corresponding section of the growth curve seen under severe defoliation. (b) Residual sward state alone, however, provides a poor indication of yield (average growth rate) as swards may show a wide range of growth rates depending on how long the sward is allowed to regrow (after Parsons *et al.* 1988b).

shape pattern in dW/dt), the regrowth of a sward from other defoliation intensities does not simply retrace the corresponding sections of this growth curve. Following more lenient defoliation, the growth rate per unit LAI can be lower initially, due the higher initial rate of senescence (the leaves dying are larger) in the leniently defoliated crop.

To derive a theory for optimizing utilization for all possible initial (residual) sward states, we must first provide a growth curve that takes account of the effects of the residual state on net growth, and so takes account of the factors affecting regrowth. We could use simulation models (notably compartmental ones) that take account of the age structure of the leaves in the community, but few practical pasture growth models incorporate this complexity. However, the same phenomena can be simulated simply using a modified logistic (Schwinning & Parsons 1999) with growth rate d*w*/d*t* represented by

$$\frac{dW}{dt} = \mu\left(1 - \frac{W_0}{W_{max}}\right)W\left(1 - \frac{W}{W_{max}}\right)$$

This, combined with parameters to introduce variations in the shape of the growth curve (e.g. by Cacho 1993), provides a simple but responsive basis for describing growth under a wide range of circumstances.

3.10 Optimal solutions for residual sward state and the duration of regrowth

We have established how growth rate changes following any one given starting pasture/patch state, and have identified a theoretical basis for the optimum timing of harvest – the time when the average growth rate is maximum – in each case. We can therefore plot out for all starting (residual) patch states how the initial patch state affects the theoretical maximum average growth rate that is achieved in each case, and the duration of regrowth – the number of days – necessary to achieve this (see Fig. 3.12).

The solid lines in Fig. 3.12 show how the greatest overall yield (the global optimum solution) is achieved using severe defoliation (to around 5 cm) and when this is associated with relatively long defoliation intervals. More lenient defoliation (greater starting state) demands more frequent defoliation in order to achieve the maximum growth rate, but even then, this 'local' optimum may be less than the global optimum seen under infrequent severe defoliation. This is all in keeping with the consensus that has emerged from extensive cutting and grazing trials, despite the initial controversy surrounding their interpretation (see re-analyses by Parsons 1988, Parsons *et al.* 1988b and references therein).

However, these optimal solutions represent the yields that would be achieved if areas (field or bite scale) were defoliated uniformly and instantaneously – that is if *all* areas were cut perfectly to the same starting point and allowed to regrow precisely for the allotted optimum time, but can this be done under grazing?

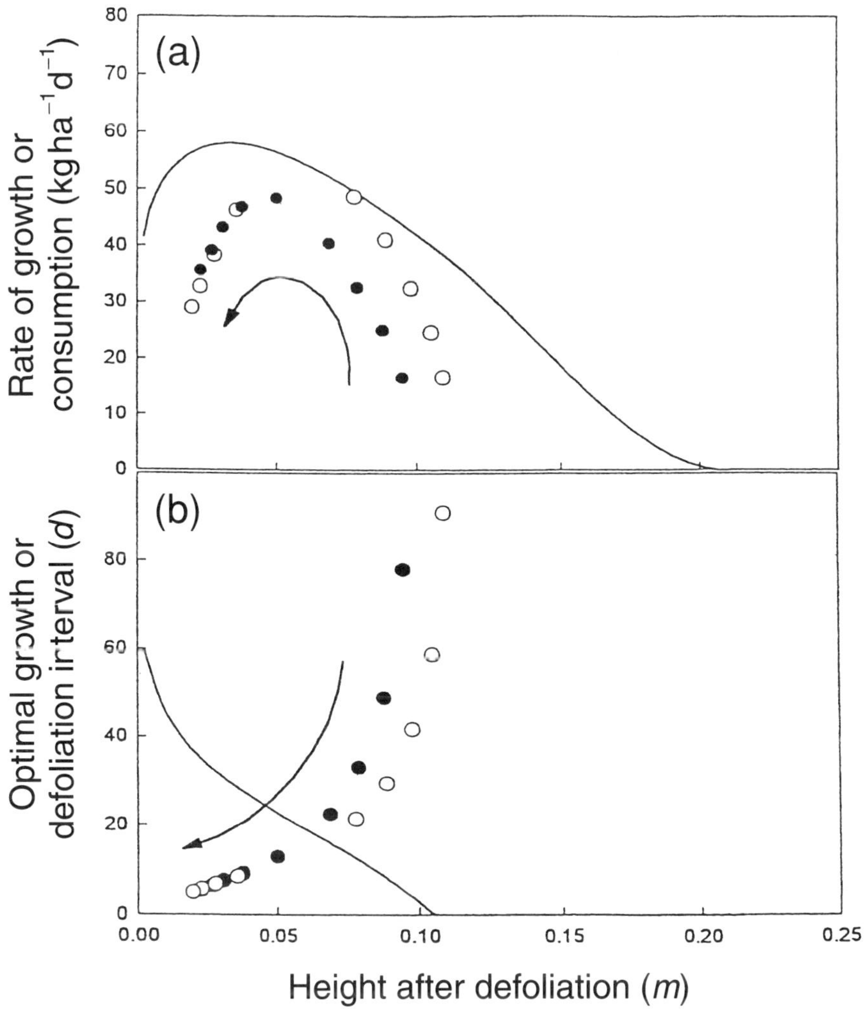

Fig. 3.12 The effect of residual sward state on (a) the maximum yield (maximum average growth rate) that may be achieved, and (b) the defoliation interval required to achieve this as predicted by a model that seeks optimal solutions for all combinations of the residual sward state ('severity') and the timing of harvest ('frequency') in a series of defoliations in all cases. The solid lines show the optimal solutions. The open circles show the combinations of residual states, defoliation intervals and yields that emerge in patches when animals are assumed to graze patches in sequence (deterministically), and the solid circles are when animals graze patches at random. Arrows show the direction of increasing stock density. See text for explanation (from Schwinning & Parsons 1999).

3.11 Operational constraints to utilization under grazing

A fascinating insight into the constraints for harvesting grass emerges when we consider how effective animals may be at realizing these optima, given the way animals graze, and notably how stock density affects this. Even if animals were to graze all the bite-sized areas in the field in strict sequence (and so not missing some, and grazing others twice), low stock densities mean that patches are grazed either too leniently and/or too infrequently to achieve maximum yield (see open circles in Fig. 3.12). High stock densities mean the opposite. There is one stock density that could achieve the local optimum, but even this may not be the global optimum. Moreover, when animals do not graze the patches in strict sequence, but at random (closed circles in Fig. 3.12), this creates variance in both the residual patch states, which would be apparent as spatial heterogeneity, as well as in defoliation intervals. This means that even controlling stock density will not achieve the optimum yields.

In short, Fig. 3.12 exposes some of the major constraints to harvesting grass with animals. Animals are constrained in that, unlike with cutting, they cannot reliably remove all leaves down to the desired residual state in all patches in single bites (this gives variance in residual patch state). Moreover, the time they take attempting to harvest patches constrains just how many patches they can harvest in a given time (this gives inappropriate defoliation intervals as well as variance in the defoliation interval). The mechanical constraints to harvesting grass bite by bite are described in Chapter 9 (see also Penning 1986, Penning *et al.* 1995, and reviews by Spalinger & Hobbs 1992, Parsons *et al.* 1994a and Ungar 1996).

3.12 The functioning of clover, relative to grass, in mixed swards

The most substantial difference between grasses and clover is in their response to mineral nitrogen. As a legume, clover clearly has the advantage over grass in terms of nitrogen inputs (via N fixation via nitrogenase) at low soil mineral-N supply. Clover maintains some 5% N in the leaves, and so maintains high photosynthetic capacity, and meristems function effectively even when soil mineral-N supply is very low. Under equivalent conditions grasses may be severely N-deficient, with leaf N contents of <3%. This said, even in regions of sustained low mineral-N supply (as occurs in New Zealand), clover remains at <20% of the mixture (Caradus *et al.* 1996, Chapman *et al.* 1996), suggesting other intrinsic differences in the growth efficiency of these species.

At higher mineral-N availability (or rate of supply), *both* clover and grass species are capable of increasing N inputs by increasing N uptake (i.e. via nitrate reductase). Indeed, the yield of clover monocultures growing dependent on mineral N (e.g. at sustained high mineral N supply) greatly exceeds that of clover growing at low mineral N supply and dependent on N fixation (Ryle *et al.* 1979, Arnott 1984) – mineral N is not a poison to clover. However, because grass growth is so severely depressed at low mineral N, grasses are stimulated more by N than is clover, and so in mixed swards clover becomes disadvantaged *relative* to grass as mineral-N supply increases.

It is not totally resolved why grasses at high mineral-N availability generally outyield (and so potentially outcompete) clover at high mineral-N. Many clover genotypes retain between 15 and 70% of total N inputs from fixation when mineral-N supply increases in the short term (as in a urine patch). The yield of clover monocultures growing dependent on N fixation is only 60% that of clover monocultures growing dependent on mineral N (Arnott 1984). This implies some increased costs of N fixation (relative to uptake). In clover dependent on N fixation, the cost has been estimated as some 24% of gross photosynthesis (Ryle *et al.* 1979). The retention of N fixation could represent one disadvantage for clover, compared with grass, at times when both are exposed to short-term high mineral-N, but the retention would be beneficial in the longer term, as any loss of nodules would prejudice performance when mineral N subsequently declined. We must also not overlook the fact that the relative costs of uptake and fixation are not fixed. The costs of N uptake can vary depending on the electrical balance of the ions (and thus the form of N) involved (Haynes & Goh 1978).

It was long considered that one of the main deficiencies of clover was that it could be 'shaded out' by grass, particularly when high levels of N fertilizer were applied. However, it has been shown that this is *not* the case in any simple sense (e.g. see Davidson & Robson 1985, Dennis & Woledge 1985, Davidson *et al.* 1986, Woledge 1988). In clover, petiole extension ensures that a greater proportion of clover leaf area is disposed in the upper layers of the canopy than in grass, and so mature clover leaves experience high light. Moreover, clover leaves escape the reduction in photosynthetic capacity experienced by grass leaves when these develop in shade in dense canopies. In both species, leaves developed in shade have poor capacity to photosynthesize in high light. However, unlike in grasses, clover leaves do not complete the development of photosynthetic capacity until the leaflets have been carried up, by the extending petiole, into high light near the top of the canopy (i.e. Carlson scale, 0.7–0.8). Therefore, even in dense canopies, clover leaves sustain high photosynthetic capacity to function in high light. In this respect, petiole elongation in clover acts in the same way as elongation of the flowering stem in grasses, but there may be costs associated with this strategy (Parsons *et al.* 1991a, b). In grasses, the ratio of carbon invested into lamina *versus* non-lamina in each phytomer (tissues produced in one leaf appearance interval) is fixed, in *L. perenne* at close to 0.7, regardless of unit size. In white clover the ratio declines (e.g. from 0.7 to 0.45) as larger units are produced.

Of course, having a higher proportion of its leaves in the upper layers of the canopy makes clover more liable to defoliation even under cutting (passive selection). In addition, sheep, and to a surprising extent cattle and goats, have been shown to have a strong (albeit partial) preference for a mixed diet. Sheep select a diet comprising some 70% clover when given free access to both species (Parsons *et al.* 1994b), and this has a strong influence on grass *vs.* legume growth (Thornley *et al.* 1995, Schwinning & Parsons 1996a). Genotypes bred to have large leaves (and long petioles) have been shown to have greater yields but poorer persistence than smaller leaf genotypes, particularly under frequent or severe defoliation (e.g. Caradus *et al.* 1997).

Clover does, however, appear more vulnerable than grass during winter. Not only are leaf growth and petiole extension reduced more in clover than in grass at low temperatures (e.g. of 2–5°C) at this time of year, but grass may also overtop the clover. Perhaps most importantly, a greater proportion of leaf is lost from clover over winter than is the case with grass. As a result, the clover content (amount and proportion) of the sward is generally lower leaving the winter each year than entering it. So, although under a wide range of managements, both infrequent cutting and even close grazing by sheep, clover can be seen to increase in proportion throughout the growing season (Parsons *et al.* 1991a), this annual setback to growth in winter clearly helps constrain long-term clover content. However, it is notable that even in countries such as New Zealand, with warmer climates than are found in most of the British Isles, clover content is still typically low. More complex consideration is necessary to explain this.

3.13 Grass–legume balance – a problem and a theory

There are a number of features of the grass/legume association that are a major concern for those relying on legume-based grassland. Agronomic studies have long been at pains to explain the following observations: (i) clover *can* persist long-term in mixtures with grass (so there is no sustained advantage to one species), although (ii) clover is present typically in small proportions – some 2–20% even in countries (e.g. NZ) which rely on legume-based grassland (Caradus *et al.* 1996); (iii) the clover content of swards is patchy (Cain *et al.* 1995, Edwards *et al.* 1996a), and (iv) the patches 'move around' (Cain *et al.* 1995). Most importantly, (v) there is long-term uncertainty in clover content – clover abundance fluctuates considerably over 'good' and 'bad' clover years, with crashes (see Fothergill *et al.* 1996) – although the very few long-term studies of clover content under consistent managements (see Steele & Shannon 1982, Rickard & McBride 1986) reveal sustained cycles of clover, with a period (cycle length) of approximately 4 years (Schwinning & Parsons 1996c).

Many insights into the complexity of interactions between two or more populations, and the scope for the coexistence of species, have come from studies in theoretical ecology (see Begon *et al.* 1990). Those theories account for how complex patterns of abundance may arise spatially and over time. In broad terms, the capacity of species to coexist (rather than for one to competitively exclude the other) requires some form of trade-off, so that one species does well in some circumstances (places, times or conditions) and the other species under other circumstances. The trade-off may be relative differences in population growth (Volterra 1926, Lotka 1932), resource use (Tilman 1982) or spatial invasion and extinctions (Nee & May 1992).

In the context of grass/clover there have been many valuable theories as to what attributes of clover, and its response to management, might cause the balance and uncertainty in clover content. Previous theories involve physiological interactions between clover and grass, e.g. through nitrogen cycling (Harper 1977), or problems of clover population dynamics – density-dependence rules for stolon births and deaths – and dispersal (Cain *et al.* 1995). A recent series of studies combines these approaches and provides one possible explanation for all five phenomena listed above (Thornley *et al.* 1995, Schwinning & Parsons 1996a–c).

The theory is based first on how grass and clover may interact through the nitrogen cycle, and the implications of their contrasting response to the supply of mineral N. During periods (or in places) of low mineral-N supply from the soil, it is argued that clover will have the growth advantage over grass owing to its capacity for N fixation, and so it should increase in abundance in the mixture. However, an increase in clover content will potentially increase the supply of mineral N, either as N-rich clover leaves and nodules turn over and die, or as animals eating clover deposit N-rich urine and dung, albeit in patches, around the field. However, if soil mineral-N supply increases during the period of clover dominance, this will progressively pass the relative advantage back to grass (see Section 3.12). During a period of grass dominance, clover content (and so N fixation) may now decline. With reduced N inputs from N fixation, the grass may run down the available mineral N supply, aided by net losses of N, e.g. by leaching, and so the relative advantage may return to clover. The 'cycle' potentially repeats itself. The two species may generate between them a mineral-N content in the soil that balances their relative advantages.

Although the interaction between plant species is widely assumed to be 'competition', an analysis suggests that an interaction like the one suggested above has more in keeping with 'predator–prey' associations, as the grass 'exploits' the N fixed by clover. These types of interactions are well recognized as being prone to generating population cycles. In Fig. 3.13a, the nature of this proposed interaction can be seen from a plot of the 'isoclines' (see Begon *et al.* 1990 – the combinations of grass population and clover population that give zero net growth in grass, and in clover, respectively). Whether legumes actually cycle in abundance would depend on the scale and presence of time delays in the transfers of N, but the theory does explain how grass and legumes can coexist, with or without cycles. It also proposes how grass/legume mixtures may not only respond to, but also have the capacity to self-regulate, the availability of N in soil and its associated losses to the environment via changes in the balance of the grasses and legumes.

However, this explanation alone cannot explain the long-term *field-scale* cycling in clover abundance. One can envisage the interaction above taking place in each of a myriad small patches in a field. In this theory, the random

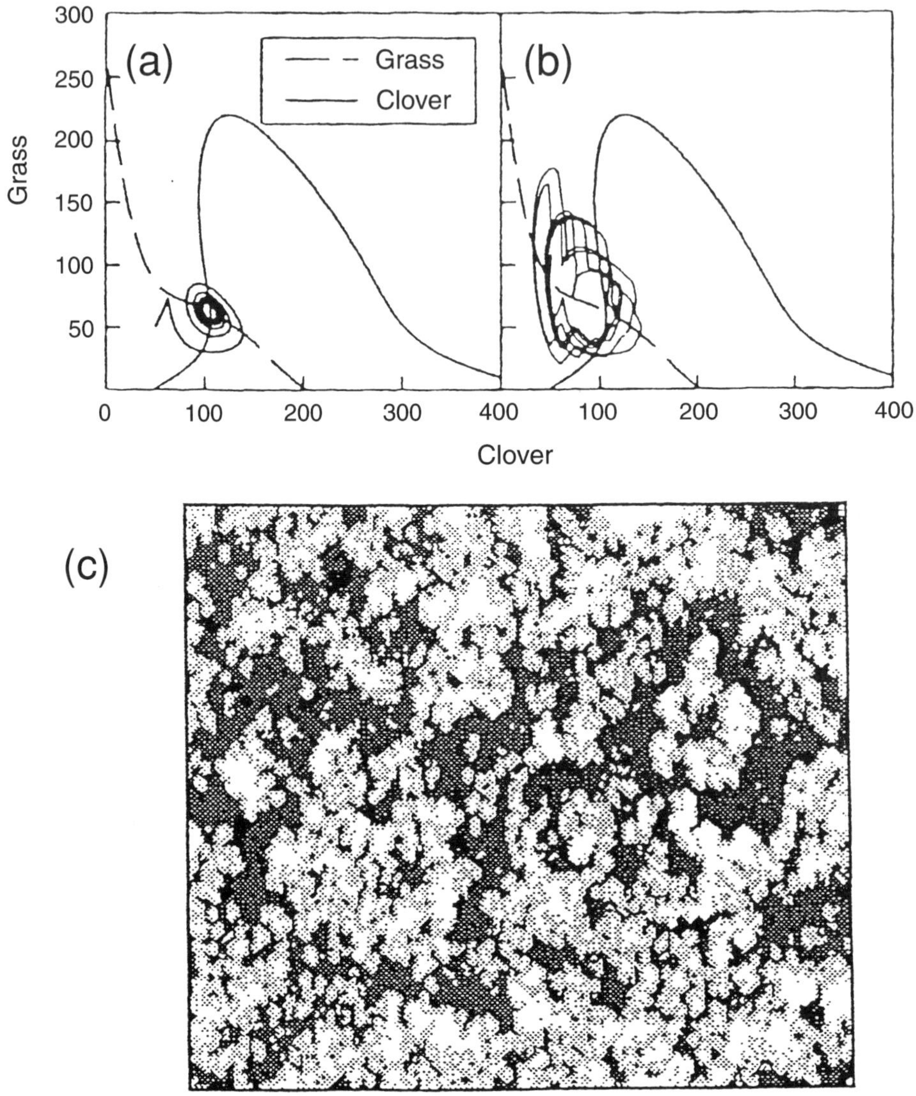

Fig. 3.13 Zero net growth isoclines for grass and clover show the combinations of grass mass and clover mass in a selectively grazed mixture that led to no change in grass mass (---) or no change in clover mass (—). The point of intersection of these is therefore the combinations of grass and clover that can be sustained (at equilibrium) in the mixture. Populations of grass and clover may cycle around this equilibrium showing damped oscillations (a), but the cycles may be made more marked and sustained by the repeated random deposition of urine (b). From Schwinning & Parsons (1996a). In (c), a spatial pasture model predicts the spontaneous formation of legume patches. White, legume-dominant areas; light grey, grass-dominant grass/clover areas (high soil N); dark grey, grass only areas (high soil N); black, grass-only areas (low soil N). From Schwinning & Parsons (1996b).

deposition of urine would greatly perturb the interaction between grass and clover, *locally*, and it is well recognized that such a perturbation will enhance or sustain the tendency for local populations to cycle (May 1974, and see Fig. 3.13b), but the randomness of urine deposition alone, sooner or later, will also ensure that the changes in clover content between patches move out of phase. This will cause the fluctuations at the *field scale* to become damped. However, field-scale cycles in clover content can re-emerge when we consider spatial dispersal – notably how effectively the clover in any clover-dominant patches manages to spread locally in the mixture (Schwinning & Parsons 1996b). Patches of high (and low) clover content could arise spontaneously (see Fig. 3.13c) from random events such as urine deposition. Any local cycle between grass and legume advantage should then be evident as a travelling wave (cf. a 'fairy ring') of high clover content dispersing from the high clover content patches. Hence, the patches move around, although subsequent urine deposits and plant extinctions would soon perturb this spreading pattern. In principle, the existence of patches of clover, and the local dispersal of clover from these, creates sufficient 'memory' of the fate of clover from one year to the next to regenerate field-scale population cycles in clover content (Schwinning & Parsons 1996b).

A model theory of this nature is extremely difficult to validate, although it contains very few and well-established basic assumptions. However, the predictions closely mimic the long-term oscillations in clover content seen in the very few long-term (>25 years) field studies (Steele & Shannon 1982, Rickard & McBride 1986), and notably the approximately 4-year period of clover fluctuations. Certainly, clover may fluctuate in direct response to other external factors, e.g. diseases and pests, and large changes in management (which would be detected best in a clover monoculture), but what is explained here is the possibility of an intrinsic long-term uncertainty in clover content. Randomly timed impacts of disease or severe winter mortality would again perturb the intrinsic cycle and create more sustained field-scale uncertainty in clover content (Schwinning & Parsons 1996c).

As a consequence of the complexity of the interaction between grass and clover, some seemingly straightforward propositions for increasing legume content, by modifying clover performance, can be shown to produce paradoxical results (Chapman *et al.* 1996). Some improvements in clover performance can be shown to actually decrease clover content – this is the 'paradox of enrichment' in predator–prey parlance (see Begon *et al.* 1990).

3.14 Modifying factors – adaptation to defoliation

Superimposed on the general principles of plant and community structure, and the general theory for maximizing utilization, are changes in the way plants grow in response to defoliation that can in some cases ameliorate, and others exacerbate, the impacts of defoliation on the growth and utilization of grass.

A previously uncut grass or clover plant will lose the bulk of its photosynthetic surface when harvested to a low residual *height* and so find itself in negative carbon balance, where the plant loses weight. As we have seen, the plants must re-establish a photosynthetic surface from existing, or new, shoot meristems, or go into terminal decline. It is intuitive that many grass species adapt to repeated exposure to defoliation by disposing a far greater proportion of their leaf area close to the ground, where it is physically more unlikely to be cut or grazed (see Chapter 9). This is achieved either by reducing leaf and tiller size (while sustaining LAI by increasing tiller number) and/or by adopting a more prostrate habit. Hence, although some grass species adapt their physiology to 'tolerate' defoliation (e.g. by having good strategies for storing substrates and meristems), others adapt to 'avoid' defoliation (Briske 1996) so that the assimilation of carbon is constrained by a low LAI, but continues without such traumatic interruption.

Grassland managements, through the frequency of defoliation (cf. duration of regrowth) and the severity of defoliation (closeness to ground as well as leaf area removal), therefore have a marked effect on the structure of the grass sward. For example, under sustained close continuous grazing by sheep to maintain a low mean height (e.g. 3 cm) as well as a low LAI (e.g. 2.0), many grass species develop a dense sward comprising a very large number of very small tillers – albeit each growing at a relatively low rate. Numbers fluctuate seasonally, but may reach some 30 000–50 000 tillers m^{-2} in mid-summer (Grant *et al.* 1983, Parsons *et al.* 1983a), falling steeply but still to sustain *c.* 20 000 m^{-2} over winter, even in cool temperate regions. By contrast, swards maintained at a high mean height (e.g. 9 cm) in an attempt to maintain a greater LAI (e.g. of 4.0) develop far fewer (<10 000 m^{-2}) and much larger tillers. Note that at times when the greater LAI is sustained, the photosynthesis and gross production of herbage is substantially (nearly two times) greater in this sward, despite the low tiller population, and growth rates per tiller will be some eight times greater. So again, the low meristem numbers are not a limiting factor here.

However, the capacity for adapting sward structure can prove deleterious, notably when there are changes in management. Although a structure appropriate to sustained close continuous grazing develops gradually, this structure can be lost rapidly. Dense swards allowed to grow to a high LAI, such as for a single heavy conservation cut, show a rapid and substantial loss of tiller numbers (e.g. Parsons *et al.* 1984). This self-thinning (Harper 1977, and see Ong 1978) is part of the 'boundary rule'/allometric constraint, in that as tillers get larger in a closed canopy, numbers must fall. It is widely recognized as beneficial to the growth of the remaining tillers, but it can have serious consequences for production if the same sward is then returned, with a now inappropriate structure, to close and more frequent continuous grazing. By the same principle, swards maintained at a large height, in an attempt to maintain a high leaf area, can actually decline in leaf area within that height. Hence the sward develops longer-term a structure that is not valuable even for sustaining a high standing biomass, and is extremely inappropriate when the vegetation is defoliated more severely, as a very low LAI remains. This is the major limiting factor, but under these circumstances, the low meristem numbers per unit ground area may indeed become a limitation to expressing growth once leaf area is restored, and most serious, the initially small number of large meristems constrains the crop to regenerating its inappropriate structure.

This deterioration in structure of swards maintained at large standing height has long been recognized (Hunt & Brougham 1967), and is as much the basis for dismissing any notion that 'frequent lenient' defoliation gives optimum production (as was once proposed, see Brougham 1958, Davidson & Donald 1958) as is the realization that sustained high leaf area does not give rise to maximum net growth rates in any case (as in Figs 3.7 and 3.12; and see Hunt 1965, Smetham 1975).

Management also affects the expression of reproductive development (see Table 3.1 and Johnson & Parsons 1985b) and the consequences to growth and utilization (see Sections 3.15 and 3.16, below). Because flowering in so many temperate grass species requires that the plant experiences a period of cold, and because the stimulus induced by cold is not substantially transferred between parts of the plant, flowering in spring is, in general, restricted to those tillers that were present throughout the winter. Defoliation and other management variables in the subsequent spring therefore have very little effect on the *number* of tillers that flower. They do, however, have a big effect on the *proportion* of tillers that flower, but largely by affecting the number of new and therefore vegetative tillers produced during spring. Maintaining a low mean height or LAI in spring greatly increases the number of new vegetative tillers produced at this time. Moreover, as the apices of the reproductive tillers are elevated into the grazed horizon, they are immediately removed, leading to a marked reduction in the amount of elongated stem tissue present (Table 3.1). Under lenient management, a large proportion of the overwintered tillers flower the following spring, leading to a marked accumulation of elongated stems in the crop.

3.15 Modifying factors – seasonal changes in physiology

Also superimposed on the general principles of grass growth and utilization are seasonal changes in the physiology of the crop. Many of these are associated with the progress toward reproductive development, and flowering, in spring. These affect the duration over which the crop can continue to accumulate dry matter, and so the ceiling yield achieved (Parsons & Robson 1980, 1981a, b,

1982). They can also increase growth in spring (and so alter the seasonal pattern of production) and total annual production, but can exacerbate some of the deleterious changes in sward structure and the implications of this for the decline in nutritive value of the forage.

Many temperate perennial grasses show a marked seasonal pattern of accumulation of 'reserve' carbohydrates, typically fructans (Pollock & Jones 1979). These accumulate throughout summer, autumn and winter, but are 'mobilized' in early spring when they may alleviate some of the restrictions to growth imposed by the low leaf area and light energy receipt at this time. They also play a role in frost tolerance, and their consumption in growth too early in spring can prejudice survival (Cooper 1964).

Coincident with the mobilization of 'reserve' carbohydrates, there is a change in the response of leaf extension to temperature (Peacock 1975c, Parsons & Robson 1980). Leaves extend faster at a given temperature in spring than at the same temperature in autumn. Conversely, this makes it possible for plants to achieve the *same* given rate of leaf area expansion in early spring even though, in southern England for example, mean soil surface temperatures are 8°C lower than at a time of equivalent light energy receipt in autumn.

Moreover, during spring, individual grass leaves now sustain a high photosynthetic capacity by avoiding the adverse effects on leaf photosynthesis of development in shade. Elongation of the internodes on flowering stems elevates young leaves to the top of the leaf canopy, where they develop and photosynthesize in high light. This effect has been shown to be almost entirely one of avoiding development in shade, and not a direct flowering (or sink strength) stimulus: leaves on reproductive tillers pinned down (and so developing in shade) show the same low photosynthetic capacity as those on vegetative tillers developing in shade in a dense sward in spring. Unshaded leaves of vegetative tillers in spring show similar photosynthetic characteristics to leaves elevated on reproductive tillers (Woledge 1978, 1979).

There are also changes in the partition of assimilates that increase the harvest index. There is a decline in the *proportion* of carbon invested in root growth, although it is very important to note that during spring is the time of the maximum *amount* of root growth (see Garwood 1967 in Jones & Lazenby 1990: p. 146). This is made possible by the high overall carbon fixation of the grass canopy in spring. Moreover, much of the carbon retained in the shoot is invested into the elongating stem internodes (if the management system allows these to continue to develop). This not only increases the harvest index directly, but stems and flag leaves do not show the same rapid turnover as do leaves on vegetative tillers, and so their formation leads to a net accumulation of tissues that would otherwise have been lost by death. In swards of Italian ryegrass (*Lolium multiflorum*), for example, harvested for the first time late in spring, elongated stems plus ears can make up some 80% of the crop, perhaps 6 t OM ha or 50% of a total annual net accumulation of about 12 t OM ha^{-1}.

3.16 Seasonal pattern(s) of grass production and utilization

The preceding sections show that management has a profound effect on the amount of grass grown as well as on the degree to which the amount grown is harvested. As a result, it is not possible to describe a single, seasonal pattern of production. Every management, under its unique circumstances, will generate its own unique seasonal pattern of gross production, level of utilization and so yield. There is absolutely no logic to regarding the yield achieved under any one management (e.g. cutting, or in cut areas in cages within a grazed sward) as representing the amount potentially available for harvest under another (e.g. grazing), as has often been done. However, it still pays to describe the seasonal pattern of production that has routinely been observed under some standard management regimes, and it is interesting and insightful to compare these.

Classic accounts that convey general seasonal patterns in the growth of grass and of grass/clover swards have been provided, e.g. by Brougham (1959) and Alberda & Sibma (1968) (and see Fig. 3.14a, b). These, like other presentations (e.g. Corrall & Fenlon 1978), use systems of overlapping cutting (on replicate paddocks) to provide a general picture of the response of net accumulation to the seasonal changes in the environment. In well-watered and well-fertilized (with NPK) pastures, the mean rate of net accumulation follows the general improvement in light and temperature, but with a bias toward greater production in spring than in autumn. This is largely explained by the seasonal changes in the physiology of the crop, associated with flowering, described above. This is notably so for relatively infrequent cutting regimes, as these allow a substantial expression of reproductive development.

However, one of the most fundamental explanations for why the sward displays a longer 'linear' phase of DM accumulation, and achieves a higher ceiling yield in single growth periods in spring than at other times of the year (Leafe *et al.* 1974, Parsons & Robson 1982), is concerned with the pattern of light energy receipt during each single regrowth. This is often overlooked.

During regrowth in spring, average daily light-energy receipt increases (as both light intensity and day length increase). This (notably day length) allows the rate of gross canopy photosynthesis to continue to rise beyond the time when the sward intercepts virtually all the available light. Thus, the period during which gross tissue production exceeds ('keeps ahead of') the rate of tissue death is also extended, and a greater ceiling yield can be achieved. In autumn, the converse applies. Mathematical models (Johnson & Thornley 1983) confirm that the seasonal *pattern* of light energy *per se* can account for much of the observed seasonal differences in the shape of regrowth curves of herbage production.

The use of overlapping sequences of cuts, to provide a smoother impression of the seasonal pattern of production, disguises one serious feature of seasonality under infrequent defoliation. Under infrequent cutting, the production of herbage benefits considerably from reproductive development. However, the

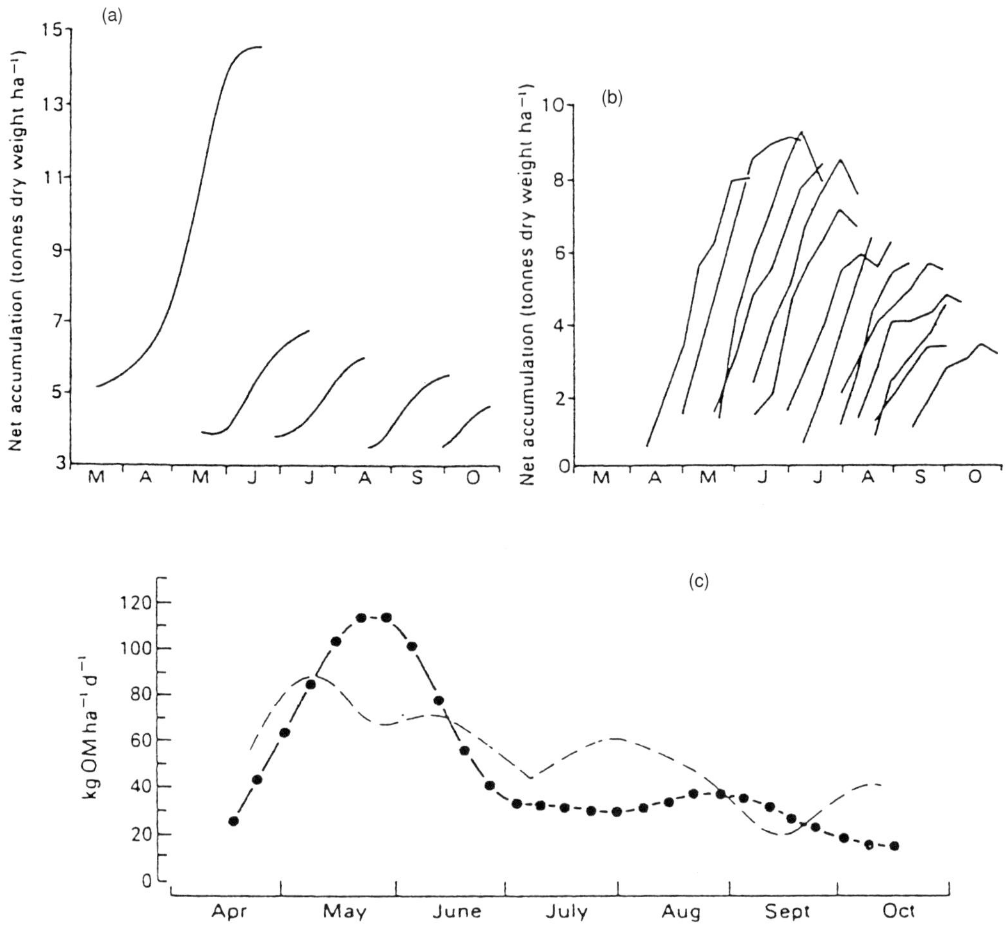

Fig. 3.14 Seasonal patterns of production of perennial ryegrass under infrequent cutting managements. (a) Net accumulation above ground under a four-cut system (Leafe *et al.* 1974). (b) A succession of regrowths each following nitrogen application (Alberda and Sibma 1968). (c) Seasonal patterns of grass production: the amount harvested ha^{-1} under a standard 4-weekly overlapping sequence of cuts (•–•) (from Corrall & Fenlon 1978), and the amount harvested ha^{-1} (intake by ewes) under continuous grazing in swards maintained at close to the optimum sward surface height (---) (after Orr *et al.* 1988).

removal of not only all the leaf, but also the apices (meristems) of these same tillers during harvest in late spring, causes very poor regrowth and a severe depression of net production in mid-summer. In a situation where a large proportion of the grass area is growing at one time in a single extended period of regrowth (e.g. for a forage conservation crop) this will be very apparent. Where multiple paddocks are in use (e.g. in a rotational grazed system) this depression will be apparent on each paddock in sequence, but the effects will be averaged out and less apparent at the farm scale.

Recent developments have made it possible to measure intake from pasture directly and so to determine seasonal patterns of production (herbage harvested) even under continuous grazing. Swards have been maintained at a set (the proposed optimal) height and LAI (see later). The seasonal pattern of production (the amount harvested per hectare) of swards maintained by such a system of continuous grazing is far more uniform than under cutting (Lantinga 1986, Orr *et al.* 1988), largely because reproductive development is suppressed (Fig. 3.14b). Production in spring is lower. The large numbers of vegetative tillers present at this time show fewer of the physiological advantages of reproductive tillers, but because there is a new population of small (and so leaf-retaining) vegetative tillers to maintain the continuity of the crop through the summer, there is less evidence of any 'mid-summer depression' in yield in watered and fertilized crops.

3.17 Theory into practice – a basis for management

Having established the theoretical basis of grass growth and utilization, we can now consider how to apply this understanding on the farm. It is a sweeping generalization, but one we do not have space to justify or enlarge on here, that very similar guidelines apply for grass/clover swards as for all-grass swards to maintain both total output and the desired clover content. The question of what *is* the optimum clover content, judged by a range of requirements and criteria, and how this can be achieved is discussed by Chapman *et al.* (1996) and Clark & Harris (1996). Here we concentrate on grass growth and utilization, but can be confident that this, in principle, applies also to grass/clover swards.

It is clear from the foregoing sections that the state of the sward, notably the amount of leaf area it is possible to sustain over time and the structure of the vegetation, is fundamental to both growth and utilization. Consequently, there has been increasing interest in using attributes of sward state as practical guidelines for monitoring and controlling the supply of grass on the farm, as well as to characterize grazing managements in research (Hodgson 1985, Parsons & Johnson 1985). Grass management calendars based on such research (Parsons 1984, 1987, Hodgson *et al.* 1985), and which encourage monitoring and controlling sward state, are now widely used on farms in northern Europe and New Zealand where they feature on many a farm office wall (e.g. ADAS 1987). The approach has particular value in that sward state is also more closely linked to the behaviour (bite size, bite rate, time spent grazing) and so the intake of grazing animals than are measures of stocking rate or herbage allowance (see Chapter 9).

Under 'continuous' grazing (or where animals have access to one or more paddocks for long periods), practical guidelines have been developed (notably in the UK and NZ) based on monitoring and maintaining a given *mean* sward state – variously measured in terms of a sward surface height or its equivalent in a trained 'eye-estimate' of herbage mass, or of LAI. Maximum production has been achieved from swards maintained at a mean height of 4–6 cm for sheep, and similar guidelines are given for dairy and beef cattle, with the optimum height increased slightly to 5–8 cm. To increase flexibility to meet changing demands for intake, and to allow for some accumulation of feed, e.g. for dry periods or improving ewe or lamb condition, the guidelines specify a desired seasonal profile of sward state. To control sward height on the grazed area requires that some grass area is closed from grazing in spring, e.g. to be mown for conservation or as a 'buffer' feed. In this way the system not only ensures high utilization and nutritive value from the grazed areas, but provides a rational basis for integrating grazing and conservation. The closed areas are returned to grazing as required. For management guidelines see Chapter 10.

Under rotational grazing, the principles described (Sections 3.5–3.11), including those relating to avoiding a deterioration in sward structure (Sections 3.14–3.16), dictate that where possible each paddock should be defoliated close to the ground (*c.* 5 cm). This not only ensures a high degree of utilization of the standing material, but also the best chance of achieving the greatest maximum average growth rate, as well as the sward structure to sustain it in the subsequent regrowth. However, it is important to recognize that a single criterion such as 'residual sward state' is insufficient to define the regrowth and performance of the pasture, as a huge range of average rates of yield can be achieved depending on how long the sward is allowed to regrow (see Fig. 3.11). This explains the poor relationship between residual sward state and, for example, milk yield seen in practical trials. It has proved difficult to identify a signal/measure for the timing of the maximum average growth rate (but note that, provided defoliation is severe, the maximum average growth rate changes little beyond an initial period (Fig. 3.9)). Given this flexibility, some long-standing guidelines – notably harvesting the sward soon after close to full light interception and before any serious decline in sward tiller density – is still valid (Tainton 1974).

If the grazing fails to remove a substantial proportion of the material accumulated at a given harvest, and defoliation leaves a larger mean residual sward state, then the principles described earlier (Sections 3.9 and 3.10) dictate how the duration of the subsequent regrowth on that paddock should be reduced.

Clearly, under rotational grazing, monitoring residual sward state and attempting to harvest each paddock at the optimum timing will dictate a 'speed of rotation' (note rotation speed is conventionally the number of days of regrowth *plus* the time taken to graze the area down). For a given grass area this will determine the optimal number of paddocks. Beyond this, despite the familiarity with rotational grazing, there is no recognized fundamental principle for determining the optimum number of paddocks or the speed of rotation for a given stock density. Hence, farmers and researchers alike experience uncertainty about how to manage intensive rotation systems outside the limits set by simple guidelines (e.g. Doyle *et al.* 1996, Jelbart 1996). Just as under continuous grazing, and as explained in Fig. 3.12 (see solid circles), animals may be unable to harvest sufficient material to sustain the optimum sward conditions. Varying the total area grazed again provides the best tool for controlling utilization (Curtis & O'Brien 1994). This is arguably better than allowing all paddocks in succession to become poorly utilized. Hence stock density, the number of paddocks and the 'speed' of rotation are really *output* variables – i.e. values that should be determined by, and so emerge from, the main driving factors of weather and plant growth. The more these are predetermined, the more the system is constrained.

Despite long-standing controversy over the relative merits of continuous *vs.* rotational grazing, it is now widely recognized that *in principle*, growth and utilization under these seemingly contrasting managements are similar. The similarity in production is all the more expected as we increasingly recognize that even continuous grazing is 'rotational' grazing at the bite or plant scale. Disputes over this are more to do with the basis on which the comparison is made. Biologically, the yield of both systems is similar when compared at the same long-term average leaf area (Grant *et al.* 1988, Parsons *et al.* 1988b). This comparison avoids the long-standing dilemma of whether the systems should be compared at the same stock density (when it can be argued that any benefits to one system might not be apparent), or at different stock densities (when it can be argued that the outcome might be predetermined) (McMeekan 1960, Raymond 1981). At the practical level, continuous and rotational grazing systems offer alternative, rather than fundamentally different, ways of monitoring sward state and so of controlling feed supply. They also offer quite different opportunities for managing the seasonality of feed supply from pasture to meet animal feeding objectives, such as the practice of building high average pasture cover (LAI) on the farm at calving in dairy systems in NZ and Australia by rationing the intake of non-lactating cows in slow (long) rotations during winter, but the major difference between them is in the way heterogeneity arises and is distributed spatially and temporally, and its implications for utilization.

3.18 Heterogeneity and its impact on utilization

Spatial heterogeneity (patchiness in sward state) is inevitable under grazing because animals are constrained to harvesting the sward with their mouths, bite by bite, and the proportion removed at each bite is considerably greater than the proportion that could be removed sustainably, averaged across the area of the sward as a whole (see Parsons & Chapman 1998, Schwinning & Parsons 1999). To maintain the optimal LAI, the amount removed averaged homogeneously across the entire grazed area is equivalent to only some 2.5% of the standing mass at any point in the sward (e.g. 50 kg ha^{-1} day^{-1}, relative to 2000 kg ha^{-1} standing biomass). Given that both sheep and dairy cattle remove closer to 35–50% at each point they graze, it is readily apparent that animals must confine their grazing to only a small proportion of the total grassland area in any one day (i.e. some 2.5/35 = 7%). This is in keeping with the few measurements made of the spatial distribution of grazing across fields under a wide range of managements (e.g. Wade *et al.* 1989, Wade 1991).

The fact that, at a given time, some portions of the total grazed area are tall, while others are short, does not in itself imply poor utilization. In a rotationally grazed system, all paddocks might be grazed optimally in succession at precise intervals, e.g. 14 days; this would mean, therefore, that there is no spatial or temporal *variance* in the mean defoliation interval. However, at a farm scale there would be extreme heterogeneity with paddocks at a whole range of heights at any instant. The fact that some areas are tall while others are short simply reflects that the defoliations are out of phase. Likewise, the presence of tall and short areas in a continuously grazed pasture need not imply poor utilization (Schwinning & Parsons 1999) – it may even be beneficial in reducing the problems of locating preferred food when grazing selectively (Edwards *et al.* 1994, 1996a, b, 1997).

What matters in either grazing system is if some areas (be they paddocks or patches) are grazed too soon, or too late, relative to the optimum timing of harvest, and so there is a temporal, and spatial, component to *variance* about the mean defoliation interval. Major reductions in utilization arise if some areas reach a ceiling yield and contribute nothing to net growth rates. One phenomenon that distinguishes detrimental from benign heterogeneity, then, is simply that tall areas do not 'move around'.

The most deleterious form of spatial heterogeneity arises, ironically, from preference. If patches/paddocks are, for whatever reason, not grazed down at the end of one period of regrowth, the deterioration in quality (notably if they come to contain reproductive stems) may mean that animals show a preference for avoiding these same areas in subsequent grazings. It has been noted that swards under continuous grazing may develop into a 'bi-modal' state, where large areas remain tall and 'less frequently defoliated', while other areas in the same sward become short and 'frequently defoliated' (Gibb *et al.* 1997). Different stock densities have been shown to alter the proportion of the total area in each category. Just as in a rotational grazing system, this phenomenon requires that animals must be 'encouraged' to eat down the tall and repeatedly poorly utilized areas. Rotational grazing makes this easier *if* the tall areas are confined to particular whole paddocks, although losses are incurred as time passes, with animals grazing reluctantly from poor quality forage (Rattray 1978, and see Chapter 9). Moreover animals recognize, in rotational grazing systems, that there are better paddocks to come and this can exacerbate the tendency not to graze a paddock to a low residual state. In practice, poor and patchy utilization within paddocks is a feature of sub-optimal grazing in both continuous and rotational grazing. Cutting, e.g. for a conservation crop, or 'clean-up' grazing by less critical stock are valuable management tools.

3.19 References

ADAS (Agricultural Development and Advisory Service) (1987) *Grassland Management Calendar Dairy Cows Continuous Grazing*. HMSO, London.

Alberda T. & Sibma L. (1968) Dry matter production and light interception of crop surfaces. 3. Actual herbage production in different years as compared with potential values. *Journal of the British Grassland Society* **23**, 206–15.

Arnott R.A. (1984) An analysis of the uninterrupted growth of white clover swards receiving either biologically fixed nitrogen or nitrate solution. *Grass and Forage Science* **39**, 305–10.

Barker D.J., Chu A.C.P. & Korte C.J. (1985) Some effects of spring defoliation and drought on perennial ryegrass swards. *Proceedings of the New Zealand Grassland Association* **46**, 57–63.

Begon M., Harper J.L. & Townsend C.R. (1990) *Ecology: Individuals, Populations and Communities*. Blackwell Scientific Publications, Oxford.

Beinhart G. (1963) Effects of environment on meristematic development leaf area and growth of white clover. *Crop Science* **3**, 209–13.

Briske D.D. (1996) Strategies of plant survival in grazed systems: A functional interpretation. In Hodgson J. & Illius A.W. (eds) *The Ecology and Management of Grazing Systems*, 37–67. CAB International, Wallingford.

Brock J.L. (1988) Evaluation of New Zealand bred white clover cultivars under rotational grazing and set stocking with sheep. *Proceedings of the New Zealand Grassland Association* **49**, 203–6.

Brock J.L. & Caradus J.R. (1995) Influence of grazing management and drought on white clover population performance and genotypic frequency. In Woodfield D.R. (ed) *White Clover: New Zealand's Competitive Edge. Symposium of the New Zealand Grassland Association, Lincoln, NZ*, 79–82.

Brock J.L. & Hay M.J.M. (1995) A review of the role of grazing management on the growth and performance of white clover cultivars in lowland New Zealand pastures. In Woodfield D.R. (ed) *White Clover: New Zealand's Competitive Edge. Symposium of the New Zealand Grassland Association, Lincoln, NZ*, 65–70.

Brock J.L. & Moon Chul Kim (1994) Influence of the stolon/soil surface interface and plant morphology on the survival of white clover during severe drought stress. *Proceedings of the New Zealand Grassland Association* **56**, 187–91.

Brock J.L. & Thomas V.J. (1991) The pasture ryegrass plant. What is it? *Proceedings of the New Zealand Grassland Association* **53**, 111–16.

Brock J.L., Hume D.E. & Fletcher R.H. (1996) Seasonal variation in the morphology of perennial ryegrass (*Lolium perenne*) and cocksfoot (*Dactylis glomerata*) plants and populations in pastures under intensive sheep grazing. *Journal of Agricultural Science, Cambridge* **126**, 37–51.

Brock J.L., Hay M.J.M., Thomas V.J. & Sedcole J.R. (1988) Morphology of white clover (*Trifolium repens* L.) plants in pastures under intensive sheep grazing. *Journal of Agricultural Science, Cambridge* **111**, 273–83.

Brougham R.W. (1958) Interception of light by the foliage of pure and mixed stands of pasture plants. *Australian Journal of Agricultural Research* **9**, 39–52.

Brougham R.W. (1959) The effect of season and weather on the growth rate of a ryegrass and clover pasture. *New Zealand Journal of Agricultural Research* **2**, 283–96.

Burke M.J.W. & Grime J.P. (1996) An experimental study of plant community invasibility. *Ecology* **77**, 776–90.

Cacho O.J. (1993) A practical equation for pasture growth under grazing. *Grass and Forage Science* **48**, 387–94.

Cain M.L., Pacala S.W., Silander J.A. & Fortin M.J. (1995) Neighborhood models of clonal growth in the white clover *Trifolium repens*. *The American Naturalist* **6**, 888–917.

Caraco T. & Kelly C.K. (1991) On the adaptive value of physiological integration in clonal plants. *Ecology* **72**, 81–93.

Caradus J.R. (1990) The structure and function of white clover root systems. *Advances in Agronomy* **43**, 1–46.

Caradus J.R., Woodfield D.R. & Stewart A.V. (1996) Overview and vision for white clover. In *White Clover: New Zealand's Competitive Edge. Symposium of the New Zealand Grassland Association, Lincoln, NZ*, 1–6.

Caradus J.R., Clifford P.T.P., Chapman D.F., Cousins G.R., Williams W.M. & Miller (1997) Breeding and description of 'Grasslands Sustain', a medium-large-leaved white clover (*Trifolium repens L.*) cultivar. *New Zealand Journal of Agricultural Research* **40**, 1–7.

Carlson G.E. (1966) Growth of clover leaves – developmental morphology and parameters at ten stages. *Crop Science* **6**, 293–4.

Chapman D.F. (1983) Growth and demography of *Trifolium repens* stolons in grazed hill pastures. *Journal of Applied Ecology* **20**, 597–608.

Chapman D.F. (1986) Development removal and death of white clover leaves under three grazing managements in hill country. *New Zealand Journal of Agricultural Research* **29**, 39–47.

Chapman D.F. & Clark D.A. (1984) Pasture responses to grazing management in hill country. *Proceedings of the New Zealand Grassland Association* **45**, 168–76.

Chapman D.F. & Hay M.J.M. (1993) Translocation of phosphorus acquired by nodal roots of contrasting genotypes of white clover (*Trifolium repens* L.). *Physiologia Plantarum* **89**, 323–30.

Chapman D.F. & Lemaire G. (1993) Morphogenetic and structural determinants of plant regrowth following defoliation. *Proceedings of the 17th International Grassland Congress, Palmerston North, NZ*, 95–104.

Chapman D.F., Clark D.A., Land C.A. & Dymock N. (1983) Leaf and tiller growth of *Lolium perenne* and *Agrostis* spp. and leaf appearance rates of *Trifolium repens* in set-stocked and rotationally grazed hill pastures. *New Zealand Journal of Agricultural Research* **26**, 159–68.

Chapman D.F., Robson M.J. & Snaydon R.W. (1991) Quantitative carbon distribution in clonal plants of white clover: Source–sink relationships during undisturbed growth. *Journal of Agricultural Science, Cambridge* **116**, 229–38.

Chapman D.F., Robson M.J. & Snaydon R.W. (1992) Physiological integration in the clonal perennial herb *Trifolium repens* L. *Oecologia* **89**, 338–47.

Chapman D.F., Clark D.A., Land C.A. & Dymock N. (1994) Leaf and tiller or stolon death of *Lolium perenne, Agrostis* spp. and *Trifolium repens* in set-stocked and rotationally grazed hill pastures. *New Zealand Journal of Agricultural Research* **27**, 303–12.

Chapman D.F., Parsons A.J. & Schwinning S. (1996) Management of clover in grazed pastures: Expectations limitations and opportunities. In Woodfield D.R. (ed) *White Clover: New Zealand's Competitive Edge. Symposium of the New Zealand Grassland Association, Lincoln, NZ*, 55–64.

Charnov E.L. (1976) Optimal foraging: the marginal value theorem. *Theoretical Population Biology* **9**, 129–36.

Clark D.A. & Harris S. (1996) White clover or nitrogen fertilizer for dairying? In Woodfield D.R. (ed) *White Clover: New Zealand's Competitive Edge. Symposium of the New Zealand Grassland Association, Lincoln, NZ*, 107–14.

Clark D.A., Chapman D.F., Land C.A. & Dymock N. (1984) Defoliation of *Lolium perenne* and *Agrostis* spp. tillers and *Trifolium repens* stolons in set-stocked and rotationally grazed hill pastures. *New Zealand Journal of Agricultural Research* **27**, 289–301.

Clark H., Newton P.C.D. & Barker D.J. (1998) Physiological and morphological responses to elevated CO_2 and soil moisture deficit of temperate pasture species growing in an established plant community. *Journal of Experimental Botany* **50**, 233–42.

Cooper J.P. (1951) Studies on growth and development in *Lolium*. II. Pattern of bud development of the shoot apex and its ecological significance. *Journal of Ecology* **39**, 228–70.

Cooper J.P. (1964) Climatic variation in forage grasses. 1. Leaf development in climatic races of *Lolium* and *Dactylis*. *Journal of Applied Ecology* **1**, 45–61.

Corrall A.J. & Fenlon J.S. (1978) A comparative method for describing the seasonal distribution of production from grasses. *Journal of Agricultural Science, Cambridge* **91**, 61–7.

Crawley M.J. (1990) The population dynamics of plants. *Philosophical Transactions of the Royal Society London, Series B* **330**, 125–40.

Curtis A. & O'Brien G. (1994) *Pasture management for Dairy Farmers, 'Target 10'*, 68 pp. Victorian Department of Agriculture, Melbourne.

Davidson J.L. & Donald C.M. (1958) The growth of swards of subterranean clover with particular reference to leaf area. *Australian Journal of Agricultural Research* **9**, 53–72.

Davidson J.L. & Milthorpe F.L. (1966a) Leaf growth in *Dactylis glomerata* following defoliation. *Annals of Botany* **30**, 173–84.

Davidson J.L. & Milthorpe F.L. (1966b) The effect of defoliation on the carbon balance in *Dactylis glomerata. Annals of Botany* **30**, 185–98.

Davidson I.A. & Robson M.J. (1985) Effects of nitrogen supply on the grass and clover components of simulated mixed swards grown under favourable environmental conditions. 1. Carbon assimilation and utilization. *Annals of Botany* **55**, 685–95.

Davidson I.A., Robson M.J. & Drennan D.S.H. (1986) Effects of temperature and nitrogen supply on the growth of perennial ryegrass and white clover. 1. Carbon and nitrogen economies of mixed swards at low temperature. *Annals of Botany* **57**, 697–708.

Davies A. (1977) Structure of the grass sward. In Gilsenan B. (ed) *Proceedings of an International Meeting on Animal Production from Temperate Grassland*, 36–44. An Foras Taluntais, Dublin.

Davies A. (1988) The regrowth of grass swards. In Jones M.B. & Lazenby A. (eds) *The Grass Crop – The Physiological Basis of Production*, 85–127. Chapman & Hall, London.

De Kroon H. & Kwant R. (1991) Density-dependent growth responses in two clonal herbs: Regulation of shoot density. *Oecologia* **86**, 298–304.

Dennis W.D. & Woledge J. (1985) The effect of nitrogenous fertilizer on the photosynthesis and growth of white-clover/perennial ryegrass swards. *Annals of Botany* **55**, 171–8.

Donald C.M. (1961) Competition for light in crops and pastures. In Milthorpe F.L. (ed) *Mechanisms in Biological Competition*, 283–313. Cambridge University Press, Cambridge.

Doyle P.T., Stockdale C.R. & Lawson A.R. (1996) *Pastures for Dairy Production in Victoria*, 56 pp. Agriculture Victoria, Kyabram, Australia.

Edwards G.R., Newman J.A., Parsons A.J. & Krebs J.R. (1994) Effects of scale and spatial distribution of the food resource and animal state on diet selection: An example with sheep. *Journal of Animal Ecology* **63**, 816–26.

Edwards G.R., Parsons A.J., Newman J.A. & Wright I.A. (1996a) The spatial pattern of vegetation in cut and grazed grass/white clover pastures. *Grass and Forage Science* **51**, 219–31.

Edwards G.R., Newman J.A., Parsons A.J. & Krebs J.R. (1996b) The use of spatial memory by grazing animals to locate food patches in spatially heterogeneous environments: An example with sheep. *Applied Animal Behaviour Science* **50**, 147–60.

Edwards G.R., Newman J.A., Parsons A.J. & Krebs J.R. (1997) Use of cues by grazing animals to locate food patches: An example with sheep. *Applied Animal Behaviour Science* **51**, 59–68.

Enquist B.J., Brown J.H. & West G.B. (1998) Allometric scaling of plant energetics and population density. *Nature* **395**, 163–5.

Evans L.T., Wardlaw I.F. & Williams C.N. (1964) Environmental control of growth. In Barnard C. (ed) *Grasses and Grasslands*, 102–25. Macmillan, London.
Fothergill M. & Davies D.A. (1993) Morphological changes in three clover (*Trifolium repens* L.) cultivars under continuous stocking. *Proceedings of the 17th International Grassland Congress, New Zealand*, 153–4.
Fothergill M., Davies D.A., Morgan C.T. & Jones J.R. (1996) White clover crashes. In Younie D. (ed) *Legumes in Sustainable Farming Systems*. British Grassland Society Occasional Symposium, No. 30, 172–6.
Garwood E.A. (1967) Studies on the roots of grasses. *Annual Report of the Grassland Research Institute, Hurley*, 1966, 72–9.
Gastal F., Belanger G. & Lemaire G. (1992) A model of the leaf extension rate of tall fescue in response to nitrogen and temperature. *Annals of Botany* **70**, 437–42.
Gibb M.J., Huckle C.A., Nuthall R. & Rook A.J. (1997) Effect of sward surface height on intake and grazing behaviour by lactating Holstein Friesian cows. *Grass and Forage Science* **52**, 309–21.
Gordon A.J., Ryle G.J.A. & Powell C.E. (1977) The strategy of carbon utilization in uniculm barley. I. The chemical fate of photosynthetically assimilated ^{14}C. *Journal of Experimental Botany* **28**, 1258–69.
Grant S.A., Barthram G.T., Torvill L., King J. & Smith H.K. (1983) Sward management lamina turnover and tiller population density in continuously stocked *Lolium perenne* dominated swards. *Grass and Forage Science* **38**, 333–44.
Grant S.A., Barthram G.T., Torvell L., King J. & Elston D.A. (1988) Comparison of herbage production under continuous stocking and intermittent grazing. *Grass and Forage Science* **43**, 29–39.
Harper J.L. (1977) *Population Biology of Plants*. Academic Press, London.
Hart A.L. (1987) Physiology. In Baker M.J. & Williams W.M. (eds) *White Clover*, 125–47. CAB International, Wallingford.
Hay M.J.M. (1983) Seasonal variation in the distribution of white clover (*Trifolium repens* L.) stolons among three horizontal strata in two grazed swards. *New Zealand Journal of Agricultural Research* **26**, 29–34.
Hay M.J.M. & Newton P.C.D. (1996) Effect of severity of defoliation on the viability of reproductive and vegetative axillary buds of *Trifolium repens* L. *Annals of Botany* **78**, 117–23.
Hay M.J.M., Chapman D.F., Hay R.J.M., Pennell C.G.L., Woods P.W. & Fletcher R.H. (1987) Seasonal variation in the vertical distribution of white clover stolons in grazed swards. *New Zealand Journal of Agricultural Research* **30**, 1–8.
Hay M.J.M., Brock J.L., Thomas V.J. & Knighton M.V. (1988) Seasonal and sheep grazing management effects on branching structure and dry weight of white clover plants in mixed swards. *Proceedings of the New Zealand Grassland Association* **49**, 197–201.
Hay M.J.M., Brock J.L. & Thomas V.J. (1989a) Characteristics of white clover plants in grazed swards. *Proceedings of the 16th International Grassland Congress, Nice*, 1051–2.
Hay M.J.M., Brock J.L. & Thomas V.J. (1989b) Density of *Trifolium repens* plants in mixed swards under intensive grazing by sheep. *Journal of Agricultural Science, Cambridge* **113**, 81–6.
Hay M.J.M., Thomas V.J. & Brock J.L. (1990) Frequency distribution of shoot weight of plants in populations of *Trifolium repens* persisting by clonal growth in grazed pastures. *Journal of Agricultural Science, Cambridge* **115**, 41–7.
Hay M.J.M., Newton P.C.D. & Thomas V.J. (1991) Nodal structure and branching of *Trifolium repens* in pastures under intensive grazing by sheep. *Journal of Agricultural Science, Cambridge* **116**, 221–8.
Haynes R.J. & Goh K.M. (1978) Ammonium and nitrate nutrition of plants. *Biological Reviews* **53**, 465–510.
Hodgson J. (1985) The significance of sward characteristics in the management of temperate sown pastures. *Proceedings of the 15th International Grassland Congress, Kyoto, 1985*, 31–4.

Hodgson J., Mackie C.K. & Parker J.W.G. (1985) Sward surface heights for efficient grazing. *Grass Farmer* **24**, 5–10.

Hoglund J.H. & Williams W.M. (1984) Genotypic variation in white clover growth and branching in response to temperature and nitrogen. *New Zealand Journal of Agricultural Research* **27**, 19–24.

Hollowell E.A. (1966) White clover (*Trifolium repens* L.): Annual or perennial? *Proceedings of the 10th International Grassland Congress, Helsinki*, 184–7.

Hume D.E. & Brock J.L. (1997) Morphology of tall fescue (*Festuca arundinacea*) and perennial ryegrass (*Lolium perenne*) plants in pastures under sheep and cattle grazing. *Journal of Agricultural Science, Cambridge* **129**, 19–31.

Hunt L.A. (1965) Some implications of death and decay in pasture production. *Journal of the British Grassland Society* **20**, 27–31.

Hunt L.A. & Brougham R.W. (1967) Some changes in the structure of a perennial ryegrass sward frequently but leniently defoliated during the summer. *New Zealand Journal of Agricultural Research* **10**, 397–404.

Hutchings M.J. & de Kroon H. (1994) Foraging in plants: The role of morphological plasticity in resource acquisition. *Advances in Ecological Research* **25**, 159–238.

Jelbart M. (1996) Putting it all together – integrating efficient management, environmental sustainability and maximum productivity. *Proceedings of the 37th Annual Conference of the Grassland Society of Victoria*, 88–93.

Jewiss O.R. (1972) Tillering in grasses, its significance and control. *Journal of the British Grassland Society* **27**, 65–82.

Jewiss O.R. (1993) Shoot development and number. In Davies A., Baker R.D., Grant S.A. & Laidlaw A.S. (eds) *Sward Measurement Handbook*. 2nd edn, 99–120. British Grassland Society, Reading.

Johnson I.R. & Parsons A.J. (1985a) A theoretical analysis of grass growth under grazing. *Journal of Theoretical Biology* **112**, 345–67.

Johnson I.R. & Parsons A.J. (1985b) Use of a model to analyse the effects of continuous grazing managements on seasonal patterns of grass production. *Grass and Forage Science* **40**, 449–58.

Johnson I.R. & Thornley J.H.M. (1983) Vegetative crop growth model incorporating leaf area expansion and senescence and applied to grass. *Plant Cell and Environment* **6**, 721–9.

Jones M.B. & Lazenby A. (1990) *The Grass Crop – The Physiological Basis of Production*, 25–83. Chapman & Hall, London.

Kays S. & Harper J.L. (1974) The regulation of plant and tiller density in a grass sward. *Journal of Applied Ecology* **62**, 97–105.

Kemp D.R. (1981) Comparison of growth rates and sugar and protein concentrations of the extension zone of main shoot and tiller leaves of wheat. *Journal of Experimental Botany* **32**, 151–8.

King J., Sim E.M. & Grant S.A. (1984) Photosynthetic rate and carbon balance of grazed ryegrass pastures. *Grass and Forage Science* **39**, 81–92.

Langer R.H.M. & Ryle G.J.A. (1958) Vegetative proliferation in herbage grasses. *Journal of the British Grassland Society* **13**, 29–33.

Lantinga E.A. (1986) Seasonal patterns of grass assimilation and net herbage production under continuous stocking. In Frame J. (ed) *Grazing*, 32–8. British Grassland Society Occasional Symposium, No. 19. BGS, Hurley

Leafe E.L., Stiles W. & Dickinson S.A. (1974) Physiological processes influencing the pattern of productivity of the intensively managed grass sward. *Proceedings of the 12th International Grassland Congress, Moscow*, **1**, 442–57.

Lemaire G. (1988) Sward dynamics under different management programmes. *Proceedings of the 12th General Meeting of the European Grassland Federation, Dublin*, 7–22.

Lemaire G. & Chapman D.F. (1996) Tissue flows in grazed plant communities. In Hodgson J. & Illius A.W. (eds) *The Ecology and Management of Grazing Systems*, 3–36. CAB International, Wallingford.

Lerdau M. (1972) Future discounts and resource allocation in plants. *Functional Ecology* **6**, 371–5.

Lotka A.J. (1932) The growth of mixed populations: Two species competing for a common food supply. *Journal of the Washington Academy of Sciences* **22**, 461–9.

Lotscher M. & Hay M.J.M. (1996) Distribution of mineral nutrient from nodal roots of *Trifolium repens*: Genetic variation in intra-plant allocation of ^{32}P and ^{45}Ca. *Physiologia Plantarum* **97**, 269–76.

Marriott C.A., Bolton G.R. & Duff E.I. (1997a) Factors affecting the stolon growth of white clover in ryegrass/clover patches. *Grass and Forage Science* **52**, 147–55.

Marriott C.A., Fisher J.M., Hood K.J. & Smith M.A. (1997b) Persistence and colonization of gaps in sown swards of grass and clover under different sward managements. *Grass and Forage Science* **52**, 156–66.

Matthew C., Lemaire G., Sackville-Hamilton N.R. & Hernandez-Garay A. (1995) A modified self-thinning equation to describe size/density relationships for defoliated swards. *Annals of Botany* **76**, 579–87.

Matthew C., Hernandez-Garay A. & Hodgson J. (1996) Making sense of the link between tiller density and pasture production. *Proceedings of the New Zealand Grassland Association* **57**, 83–7.

May R.M. (1974) *Stability and Complexity in Model Ecosystems*. Monographs in Population Biology 6. Princeton University Press, Princeton, NJ.

May R.M. (1981) Models for two interacting populations. In May R.M. (ed) *Theoretical Ecology: Principles and Applications*. 2nd edn. 78–104. Blackwell Scientific Publications, Oxford.

Mazzanti A. & Lemaire G. (1994) Effect of nitrogen fertilization on herbage production of tall fescue swards continuously grazed by sheep. 2. Consumption and efficiency of herbage utilization. *Grass and Forage Science* **49**, 352–9.

Mazzanti A., Lemaire G. & Gastal F. (1994) The effect of nitrogen fertilization on the herbage production of tall fescue swards grazed continuously with sheep. 1. Herbage growth dynamics. *Grass and Forage Science* **49**, 111–20.

McCree K.J. (1970) An equation for the rate of respiration of white clover plants grown under controlled conditions. In Malek I. (ed) *Prediction and Measurement of Photosynthetic Productivity*, 221–9. Pudoc, Wageningen.

McCree K.J. (1974) Equations for the rate of respiration of white clover and grain sorghum as functions of dry weight photosynthetic rate and temperature. *Crop Science* **14**, 509–14.

McCree K.J. & Troughton J.H. (1966a) Prediction of growth rate at different light levels from measured photosynthesis and respiration rates. *Plant Physiology* **41**, 559–66.

McCree K.J. & Troughton J.H. (1966b) Non-existence of an optimum leaf area index for the production rate of white clover grown under constant conditions. *Plant Physiology* **41**, 1615–22.

McMeekan C.P. (1960) Grazing management. *Proceedings of the 8th International Grassland Congress, Reading*, 21–7.

Mitchell K.J. (1953) Influence of light and temperature on the growth of ryegrass (*Lolium* spp.) I. Pattern of vegetative development. *Physiologia Plantarum* **6**, 21–46.

Mitchell K.J. & Lucanus R. (1960) Growth of pasture species under controlled environment. II. Growth at low temperature. *New Zealand Journal of Agricultural Research* **3**, 647–55.

Morley F.H.W. (1968) Pasture growth curves and grazing management. *Australian Journal of Experimental Agriculture and Animal Husbandry* **30**, 40–5.

Murphy J.S. & Briske D.D. (1994) Density-dependent regulation of ramet recruitment by the red:far-red ratio of solar radiation: A field evaluation with the bunchgrass *Schizachyrium scoparium*. *Oecologia* **97**, 462–9.

Nee S. & May R.M. (1992) Dynamics of metapopulations: Habitat destruction and competitive coexistence. *Journal of Animal Ecology* **61**, 37–40.

Nelson C.J. (1982) Physiology of leaf growth in grasses. *Proceedings of the 14th General Meeting of the European Grassland Federation, Lahti, Finland*, 175–9.

Neuteboom J.H. & Lantinga E.A. (1989) Tillering potential and relationship between leaf and tiller production in perennial ryegrass. *Annals of Botany* **63**, 265–70.

Newton P.C.D. & Hay M.J.M. (1994) Patterns of nodal rooting in *Trifolium repens* (L.) and correlations with stages in the development of axillary buds. *Grass and Forage Science* **49**, 270–6.

Newton P.C.D. & Hay M.J.M. (1996) Clonal growth of white clover: Factors influencing the viability of axillary buds and the outgrowth of a viable bud to form a branch. *Annals of Botany* **78**, 111–15.

Newton P.C.D., Hay M.J.M., Thomas V.J. & Dick H.B. (1992) Viability of axillary buds of white clover (*Trifolium repens*) in grazed pasture. *Journal of Agricultural Science, Cambridge* **119**, 345–54.

Oborny B. (1994) Growth rules in clonal plants and environmental predictability – a simulation study. *Journal of Ecology* **82**, 341–51.

Ong C.J. (1978) The physiology of tiller death in grasses. 1. The influence of tiller age size and position. *Journal of the British Grassland Society* **33**, 197–203.

Orr R.J., Parsons A.J., Treacher T.T. & Penning P.D. (1988) Seasonal patterns of grass production under cutting or continuous stocking managements. *Grass and Forage Science* **43**, 199–207.

Parsons A.J. (1984) Guidelines for management of continuously grazed swards. *Grass Farmer* No. 17, 5–9. British Grassland Society, Hurley.

Parsons A.J. (1987) The management of grass/clover swards for sheep. *Grass Farmer* No. 26, 26–31. British Grassland Society, Hurley.

Parsons A.J. (1988) The effects of season and management on the growth of grass swards. In Jones M.B. & Lazenby A. (eds) *The Grass Crop – The Physiological Basis of Production*, 129–77. Chapman & Hall, London.

Parsons A.J. & Chapman D.F. (1998) Principles of grass growth and pasture utilization. In Cherney J.H. & Cherney D.J.R. (eds) *Grass for Dairy Cattle*, 283–310. CAB International, Wallingford.

Parsons A.J. & Johnson I.R. (1985) The physiology of grass growth under grazing. In Frame J. (ed) *Grazing*. British Grassland Society Occasional Symposium, No. 19, 3–13. BGS, Hurley.

Parsons A.J. & Penning P.D. (1988) The effect of the duration of regrowth on photosynthesis, leaf death and the average rate of growth in a rotationally grazed sward. *Grass and Forage Science* **43**, 15–27.

Parsons A.J. & Robson M.J. (1980) Seasonal changes in the physiology of S24 perennial ryegrass (*Lolium perenne* L.). 1. Response of leaf extension to temperature during the transition from vegetative to reproductive growth. *Annals of Botany* **46**, 435–44.

Parsons A.J. & Robson M.J. (1981a) Seasonal changes in the physiology of S24 perennial ryegrass (*Lolium perenne* L.). 2. Potential leaf and canopy photosynthesis during the transition from vegetative to reproductive growth. *Annals of Botany* **47**, 249–58.

Parsons A.J. & Robson M.J. (1981b) Seasonal changes in the physiology of S24 perennial ryegrass (*Lolium perenne* L.). 3. Partition of assimilates between root and shoot during the transition from vegetative to reproductive growth. *Annals of Botany* **48**, 733–44.

Parsons A.J. & Robson M.J. (1982) Seasonal changes in the physiology of S24 perennial ryegrass (*Lolium perenne* L.). 4. Comparison of the carbon balance of the reproductive crop in spring and the vegetative crop in autumn. *Annals of Botany* **50**, 167–77.

Parsons A.J., Leafe E.L., Collett B. & Stiles W. (1983a) The physiology of grass production under grazing. 1. Characteristics of leaf and canopy photosynthesis of continuously grazed swards. *Journal of Applied Ecology* **20**, 117–26.

Parsons A.J., Leafe E.L., Collett B., Penning P.D. & Lewis J. (1983b) The physiology of grass production under grazing. 2. Photosynthesis crop growth and animal intake of continuously grazed swards. *Journal of Applied Ecology* **20**, 127–39.

Parsons A.J., Collett B. & Lewis J. (1984) Changes in the structure and physiology of a perennial ryegrass sward when released from a continuous stocking management – implications for the use of exclusion cages in continuously stocked swards. *Grass and Forage Science* **39**, 1–9.

Parsons A.J., Johnson I.R. & Williams J.H.H. (1988a) Leaf age structure and canopy photosynthesis in rotationally and continuously grazed swards. *Grass and Forage Science* **43**, 1–14.

Parsons A.J., Johnson I.R. & Harvey A. (1988b) Use of a model to optimise the interaction between the frequency and severity of intermittent defoliation and to provide a fundamental comparison of the continuous and intermittent defoliation of grass. *Grass and Forage Science* **43**, 49–59.

Parsons A.J., Harvey A. & Woledge J. (1991a) Plant/animal interactions in continuously grazed mixtures. 1. Differences in the physiology of leaf expansion and the fate of leaves of grass and clover. *Journal of Applied Ecology* **28**, 619–34.

Parsons A.J., Harvey A. & Johnson I.R. (1991b) Plant/animal interactions in continuously grazed mixtures. 2. The role of differences in the physiology of leaf expansion and of selective grazing on the performance and stability of species in a mixture. *Journal of Applied Ecology* **28**, 635–58.

Parsons A.J., Thornley J.H.M., Newman J.A. & Penning P.D. (1994a) A mechanistic model of some physical determinants of intake rate and diet selection in a two-species temperate grassland sward. *Functional Ecology* **8**, 187–204.

Parsons A.J., Newman J.A., Penning P.D., Harvey A. & Orr R.J. (1994b) Diet preference of sheep: Effects of recent diet, physiological state and species abundance. *Journal of Animal Ecology* **63**, 465–78.

Peacock J.M. (1975a) Temperature and leaf growth in *Lolium perenne*. I. The thermal microclimate, its measurement and relation to crop growth. *Journal of Applied Ecology* **12**, 99–114.

Peacock J.M. (1975b) Temperature and leaf growth in *Lolium perenne*. II. The site of temperature perception. *Journal of Applied Ecology* **12**, 115–23.

Peacock J.M. (1975c) Temperature and leaf growth in *Lolium perenne*. III. Factors affecting seasonal differences. *Journal of Applied Ecology* **12**, 685–97.

Peel S. & Matkin E.A. (1982) The productivity of grassland farms in seven climatic zones in England and Wales. *Grass and Forage Science* **37**, 299–310.

Penning P.D. (1986) Some effects of sward conditions on grazing behaviour and intake by sheep. In Gudmundsson O. (ed) *Grazing Research at Northern Latitudes*, 219–26. Plenum, New York.

Penning P.D., Parsons A.J., Orr R.J., Harvey A. & Champion R.A. (1995) Intake and behaviour responses by sheep in different physiological states when grazing monocultures of grass or white clover. *Applied Animal Behaviour Science* **45**, 63–78.

Penning de Vries F.W.T. (1972) Respiration and growth. In Rees A.R., Cockshull K.E., Hand D.W. & Hurd R.G. (eds) *Crop Processes in Controlled Environments*, 327–47. Academic Press, London.

Pollock C.J. & Jones T. (1979) Seasonal patterns of fructosan metabolism in forage grasses. *New Phytologist* **83**, 9–15.

Prioul J.L. (1971) Réaction des feuilles de *Lolium multiflorum* à l'éclairement pendant la croissance et variation des résistances aux éxchange gazeux photosynthétiques. *Photosynthetica* **5**, 364–75.

Prioul J.L., Reyss A. & Chartier P. (1975) Relationship between carbon dioxide transfer resistances and some physiological and anatomical features. In R. Marcelle (ed) *Environmental and Biological Control of Photosynthesis*, 17–28. Junk, The Hague.

Prioul J.L., Brangeon J. & Reyss A. (1980a) Interaction between external and internal conditions in the development of photosynthetic features in a grass leaf. I. Regional responses along a leaf during and after low-light or high-light acclimation. *Plant Physiology* **66**, 762–9.

Prioul J.L., Brangeon J. & Reyss A. (1980b) Interaction between external and internal conditions in the development of features in a grass leaf. II. Reversibility of light-induced responses as a function of development. *Plant Physiology* **66**, 770–4.
Rattray P.V. (1978) Pasture constraints to sheep production. *Proceedings of the Agronomy Society of New Zealand* **8**, 103–8.
Raymond W.F. (1981) Grassland research. In Cooke G.W. (ed) *Agricultural Research 1931–1981*, 311–23. Agricultural Research Council, London.
Rhodes I. (1969) The yield canopy structure and light interception of two ryegrass varieties in mixed culture and mono-culture. *Journal of the British Grassland Society* **24**, 123–7.
Richards J.H. (1993) Physiology of plants recovering from defoliation. *Proceedings of the 17th International Grassland Congress, Palmerston North, New Zealand*, **I**, 85–94.
Rickard D. S. & McBride S.D. (1986) Irrigated and non-irrigated pasture production at Winchmore. *Technical Report 21*. Winchmore Irrigation Research Station.
Robin Ch., Hay M.J.M. & Newton P.C.D. (1994) Effect of light quality (red:far-red ratio) and defoliation treatments applied at a single phytomer on axillary bud outgrowth in *Trifolium repens* L. *Oecologia* **100**, 236–42.
Robson M.J. (1972) The effect of temperature on the growth of S170 tall fescue (*Festuca arundinacea*). I. Constant temperature. *Journal of Applied Ecology* **9**, 647–57.
Robson M.J. (1973a) The growth and development of simulated swards of perennial ryegrass. I. Leaf growth and dry weight changes as related to the ceiling yield of a seedling sward. *Annals of Botany* **37**, 487–500.
Robson M.J. (1973b) The growth and development of simulated swards of perennial ryegrass. II. Carbon assimilation and respiration in a seedling sward. *Annals of Botany* **37**, 501–18.
Robson M.J. (1973c) The effect of temperature on the growth of S170 tall fescue (*Festuca arundinacea*). II. Independent variation of day and night temperature. *Journal of Applied Ecology* **10**, 93–105.
Robson M.J. (1982) The growth and carbon economy of selection lines of *Lolium perenne* cv S23 with differing rates of dark respiration. 2. Grown as young plants from seed. *Annals of Botany* **49**, 331–9.
Robson M.J., Ryle G.J.A. & Woledge J. (1988) The grass plant – its form and function. In Jones M.B. & Lazenby A. (eds) *The Grass Crop – The Physiology Basis of Production*, 25–83. Chapman & Hall, London.
Ryle G.J.A. & Langer R.H.M. (1963) Studies on the physiology of flowering in timothy. I. Influence of daylength and temperature on initiation and differentiation of the inflorescence. *Annals of Botany* **27**, 213–31.
Ryle G.J.A., Powell C.E. & Gordon A.J. (1979) The respiratory costs of nitrogen fixation in soyabean, cowpea and white clover. *Journal of Experimental Botany* **30**, 145–53.
Sackville-Hamilton N.R. & Harper J.L. (1989) The dynamics of *Trifolium repens* in a permanent pasture. 1. The population dynamics of leaves and nodes per shoot axis. *Proceedings of the Royal Society of London, B* **237**, 133–7.
Schnyder H. (1986) Carbohydrate metabolism in the growth zone of tall fescue leaf blades. *Current Topics in Plant Biochemistry and Physiology* **5**, 47–60.
Schwinning S. & Parsons A.J. (1996a) Analysis of the coexistence mechanisms for grasses and legumes in grazing systems. *Journal of Ecology* **84**, 799–814.
Schwinning S. & Parsons A.J. (1996b) A spatially explicit population model of stoloniferous N-fixing legumes in mixed pasture with grass. *Journal of Ecology* **84**, 815–26.
Schwinning S. & Parsons A.J. (1996c) Interaction between grasses and legumes: Understanding variability in species composition. In Younie D. (ed) *Legumes in Sustainable Farming Systems*. British Grassland Society Occasional Symposium, No. 30, 153–63. BGS, Reading.
Schwinning S. & Parsons A.J. (1999) The stability of grazing systems revisited: Spatial models and the role of heterogeneity. *Functional Ecology* **13**, 737–47.

Simon J.C. & Lemaire G. (1987) Tillering and leaf area index in grasses in the vegetative phase. *Grass and Forage Science* **42**, 373–80.

Smetham M.L. (1975) The influence of herbage utilization on pasture production and animal performance. *Proceedings of the New Zealand Grassland Association* **37**, 91–103.

Soussana J.F., Vertès F. & Arregui M.C. (1995) The regulation of clover shoot growing point density and morphology during short-term clover decline in mixed swards. *European Journal of Agronomy* **4**, 205–15.

Spalinger D.E. & Hobbs N.T. (1992) Mechanisms of foraging in mammalian herbivores: New models of functional response. *American Naturalist* **140**, 325–48.

Steele K.W. & Shannon P. (1982) Concepts relating to the nitrogen economy of a Northland intensive beef farm. In Gandar P. (ed) *Nitrogen Balances in New Zealand Ecosystems*. DSIR New Zealand, 85–9.

Stephens D.W. & Krebs J.R. (1986) *Foraging Theory.* Princeton University Press, Princeton, NJ.

Tainton N.M. (1974) Effect of different grazing rotations on pasture production. *Journal of the British Grassland Society* **29**, 191–202.

Thomas R.G. (1987a) Vegetative growth and development. In Baker M.J. & Williams W.M. (eds) *White Clover*, 31–62. CAB International, Wallingford.

Thomas R.G. (1987b) Reproductive development. In Baker M.J. & Williams W.M. (eds) *White Clover*, 63–123. CAB International, Wallingford.

Thompson L. & Harper J.L. (1988) The effect of grasses on the quality of transmitted radiation and its influence on the growth of white clover *Trifolium repens*. *Oecologia* **75**, 343–7.

Thornley J.H.M. & Johnson I.R. (1990) *Plant and Crop Modelling*. Clarendon Press, Oxford.

Thornley J.H.M., Bergelson J. & Parsons A.J. (1995) Complex dynamics in a carbon–nitrogen model of a grass–legume pasture. *Annals of Botany* **75**, 79–94.

Tilman D. (1982) *Resource Competition and Community Structure*. Princeton University Press, Princeton, NJ.

Treharne K.J. & Eagles C.F. (1969) Effect of growth at different light intensities on photosynthetic activity of two contrasting populations of *Dactylis glomerata* L. In Metzner H. (ed) *Progress in Photosynthesis Research*, 377–82. International Union of Biological Science, Tubingen.

Ungar E.D. (1996) Ingestive behaviour. In Hodgson J. & Illius A.W. (eds) *The Ecology and Management of Grazing Systems*, 185–218. CAB International, Wallingford.

Volenec J.J. (1986) Non-structural carbohydrates in stem base components of tall fescue during regrowth. *Crop Science* **26**, 122–7.

Volenec J.J. & Nelson C.J. (1981) Cell dynamics in leaf meristems of contrasting tall fescue genotypes. *Crop Science* **22**, 531–5.

Volenec J.J. & Nelson C.J. (1982) Diurnal leaf elongation of contrasting tall fescue genotypes. *Crop Science* **21**, 381–5.

Volenec J.J. & Nelson C.J. (1983) Response of tall fescue leaf meristems to N fertilization and harvest frequency. *Crop Science* **23**, 720–4.

Volenec J.J. & Nelson C.J. (1984a) Carbohydrate metabolism in leaf meristems of tall fescue. 1. Relationship to genetically altered leaf elongation rates. *Plant Physiology* **74**, 590–4.

Volenec J.J. & Nelson C.J. (1984b) Carbohydrate metabolism in leaf meristems of tall fescue. 2. Relationship to leaf elongation rates modified by nitrogen fertilization. *Plant Physiology* **74**, 595–600.

Volterra V. (1926) Variations and fluctuations of the numbers of individuals in animal species living together. In Chapman R.N. (ed) *Animal Ecology*. McGraw Hill, New York.

Wade M.H. (1991) *Factors Affecting the Availability of Vegetative* Lolium Perenne *to Grazing Dairy Cows with Special Reference to Sward Characteristics, Stocking Rate and Grazing Method*. Doctoral Thesis. University of Rennes, France.

Wade M.H., Peyraud J.L., Lemaire G. & Comeron E.A. (1989) The dynamics of daily area and depth of grazing and herbage intake of cows in a five-day paddock system. *Proceedings of the 16th International Grassland Congress, Nice France*, 1111–12.

West G.B., Brown J.H. & Enquist B.J.A. (1997) A general model for the origin of allometric scaling laws in biology. *Science* **276**, 122–6.

Westoby M. (1984) The self-thinning rule. *Advances in Ecological Research* **14**, 167–225.

Wilson D. & Cooper J.P. (1969) Effect of light intensity during growth on leaf anatomy and subsequent light-saturated photosynthesis among contrasting *Lolium* genotypes. *New Phytologist* **68**, 1125–35.

Woledge J. (1971) The effect of light intensity during growth on the subsequent rate of photosynthesis in leaves of tall fescue (*Festuca arundinacea* Schreb.). *Annals of Botany* **35**, 311–22.

Woledge J. (1972) The effect of shading on the photosynthetic rate and longevity of grass leaves. *Annals of Botany* **36**, 551–61.

Woledge J. (1973) The photosynthesis of ryegrass leaves grown in a simulated sward. *Annals of Applied Biology* **73**, 229–37.

Woledge J. (1977) Effects of shading and cutting treatments on the photosynthetic rate of ryegrass leaves. *Annals of Botany* **41**, 1279–86.

Woledge J. (1978) The effect of shading during vegetative and reproductive growth on the photosynthetic capacity of leaves in a grass sward. *Annals of Botany* **42**, 1085–9.

Woledge J. (1979) Effect of flowering on the photosynthetic capacity of ryegrass leaves grown with or without natural shading. *Annals of Botany* **44**, 197–207.

Woledge J. (1988) Competition between grass and clover in spring as affected by nitrogen fertilizer. *Annals of Applied Biology* **112**, 175–86.

Woledge J. & Dennis W.D. (1982) The effect of temperature on photosynthesis of ryegrass and white clover leaves. *Annals of Botany* **50**, 25–35.

Woledge J. & Jewiss O.R. (1969) The effect of temperature during growth on the subsequent rate of photosynthesis in leaves of tall fescue (*Festuca arundinacea* Schreb.). *Annals of Botany* **33**, 897–913.

Woledge J. & Leafe E.L. (1976) Single leaf and canopy photosynthesis in a ryegrass sward. *Annals of Botany* **40**, 773–83.

Woledge J. & Parsons A.J. (1986) Temperate grasslands. In Baker N.R. & Long S.P. (eds) *Photosynthesis in Contrasting Environments*, 173–97. Elsevier, Amsterdam.

Woledge J. & Pearse J.P. (1985) The effect of nitrogen fertilizer on the photosynthesis of leaves of a ryegrass sward. *Grass and Forage Science* **40**, 305–9.

Woodward S.J.R. (1998) Quantifying different causes of leaf and tiller death in grazed perennial ryegrass swards. *New Zealand Journal of Agricultural Research* **41**, 149–59.

Chapter 4
Herbage Production

A. Hopkins

4.1 Introduction

The previous chapter described the physiological processes controlling the production of leaves and branches in grasses and herbage legumes. These ultimately determine the levels of herbage accumulation, and thus the potential herbage production available for grazing or mowing. Site conditions, including soil nutrient status, temperature and water availability, and the various management inputs employed by the farmer, collectively determine the levels of production. This chapter examines the role of these factors.

Throughout much of the latter half of the twentieth century there was encouragement for farmers to increase production, largely by overcoming the factors limiting herbage production. In practice, this was achieved by sward improvement, by increasing the nitrogen (N) supply, and by correcting other soil nutrient and drainage conditions that affect grass growth or utilization. There is now generally less emphasis on managing grassland for maximum production and more on greater efficiency in the use of inputs within limits set by the need to meet environmental targets and the overall demand for grassland products.

4.2 Measurement of herbage production

The term herbage production here refers to the mass of herbage that can be removed either by animals under grazing (herbage consumed) or by mechanical harvesting (herbage harvested) (Hodgson 1979). Herbage production is normally expressed on an area basis for a given period, e.g. weight of dry matter (DM) or organic matter (OM) in $\mathrm{kg\,ha^{-1}}$ or in $\mathrm{kg\,OM\,ha^{-1}\,year^{-1}}$.

There is a long history of recording grassland production from cut and weighed field samples, although there is still relatively little standardization of sward measurement techniques. The small-plot cutting method is, however, the most widely used technique for assessing herbage production. The basic operation involves cutting and weighing a sample of fresh herbage from a precisely measured area at a specified cutting height. After weighing, the sample (or a representative sub-sample) is dried in a forced-draught oven to determine the

DM percentage, so enabling the DM yield harvested per unit area to be calculated. Field sampling typically involves the use of a mower with a reciprocating blade, either an autoscythe mower or a Haldrup mechanical plot harvester, at a cutting width of 1.0–1.5 m. An area of approximately 5–10 m^2 is usually sampled from each plot. Rotary or hover mowers are not suitable for herbage production assessments. Further details, including the logistics and degree of precision required, are given in Davies *et al.* (1993).

The procedures outlined above have been widely used at research centres in trials to investigate, for example, the response to cutting frequencies, fertilizer rates and pesticide treatments, and for comparing swards of different age or botanical composition (Morrison *et al.* 1980, Clements *et al.* 1990, Hopkins *et al.* 1990, 1995). They are also used in varietal evaluation trials for National List or Recommended List purposes (Weddell *et al.* 1997). Herbage production assessments from small plots attempt to simulate grazing or forage conservation management depending on the cutting frequency and severity used. This may include mowing at intervals of 3–4 weeks to simulate rotational grazing, or of 3 or 4 'silage-stage' cuts per year, although the methodology may be based on other parameters, such as cutting at a particular growth stage. Where assessments of seasonal production are required, it is necessary to include a series of plots harvested in rotation. For example, the methodology of Corrall & Fenlon (1978) has been widely used, and involves four series of plots harvested in rotation spaced a week apart, thereby enabling seasonal production to be determined on a weekly basis.

Although this approach is of great value for evaluating mown swards, it can never adequately simulate grazing conditions; for example, under grazing, a high proportion of N is returned through excreta, and the leaf area of the sward is constantly affected by fouling, treading and grazing. Under sheep grazing there is little herbage yield response above 200 kg fertilizer N ha^{-1} (Orr *et al.* 1990), although intensive dairy systems may use rates of 300–350 kg N ha^{-1}, depending on the soil N status (MAFF 1994). Mown swards can respond to N (although not necessarily an economic response) at rates of up to about 500 kg ha^{-1}, although 400 kg would normally be the advised maximum amount (MAFF 1994). In mixed-species swards, e.g. grass/white clover, there is the additional complication of selective grazing. Large-scale grazing trials in which production is evaluated in terms of animal output are extremely costly to conduct, and treatments need to be adequately replicated. Alternative methods have been developed which involve measurements on small cut areas within larger plots. These may be either grazed (mob-stocked) immediately after mowing and the sward resampled (so called 'difference' techniques), or involve the use of moveable exclosure cages. Results from assessments based on caged areas need to be interpreted with caution, as the presence of the cage affects net herbage accumulation and the sward beneath the cage does not fully represent the grazed sward outside the cage (Parsons *et al.* 1984, Frame 1993).

4.3 Climatic factors

The main environmental factors affecting growth and herbage production are temperature, light and soil moisture. Temperature affects enzyme-controlled processes such as photosynthesis and respiration, and rates of growth and senescence depend on the temperature pattern, including the diurnal range. Temperature determines the start of spring growth, heading date, rate of senescence etc. Temperate (C_3) grasses make little growth below 6°C, and optimum growth occurs at about 20°C (Cooper & Tainton 1968). Increases in grass growth rates of 13 kg DM ha^{-1} day^{-1} for each degree rise in temperature between 8°C and 15°C, and of 21 kg DM ha^{-1} day^{-1} for each degree rise between 14°C and 20°C, have been reported (Colman *et al.* 1974).

In temperate grasslands, the most important effect of temperature is on the length of the growing season. Taking 6°C as the threshold soil temperature for grass growth, and correcting for altitude, maps have been constructed showing the mean length of the growing season. In Britain, the lowland and coastal areas of southwest England and west Wales have above 300 growing days, while large areas of lowland England have around 250 growing days, decreasing to around 200 days in the more elevated northern areas (Lazenby & Down 1982, see also Chapter 1, Fig. 1.2). The latitudinal range in mainland Britain extends from 50°N to 59°N, and the mean temperature during the growing season decreases by about 4°C between south and north. However, climatic differences between west and east (decreasing oceanicity) and those associated with altitude are of greater agricultural significance. Within Europe, temperature (influenced by the greater latitudinal and altitudinal range) is an important factor affecting grass growth. In a 32-site trial in 16 countries (Corrall 1984, Peeters & Kopec 1996), up to three-fold differences in herbage production remained between similarly managed sites, even with the limiting effect of water deficit removed by irrigation. The effects of light (solar radiation) may be considered in terms of its wavelength, density (photons) and duration (daylength). Absorbed photosynthetically active solar radiation enables the transformation of CO_2 into biomass, and can thus affect the level of herbage production. Light interception and photosynthesis have been the subject of much research (e.g. Sheehy & Johnson 1988; the subject is also considered in Chapter 3).

On grassland that receives fertilizer N, temperature is an important factor affecting the production response to fertilizer, and in avoiding N losses through leaching. Spring temperature has been used as a basis for recommending application dates, and research in The Netherlands suggested that the best date for applying fertilizer N was that on which the accumulated mean air temperature reached 200°C, or T-sum 200 (Van Burg *et al.* 1981). In the UK the applicability of this technique is not entirely reliable, but in practice fertilizer N application is appropriate from mid-February onwards if ground conditions are suitable and a T-sum of 180–200°C, accumulated from 1 January, has been reached (Baker 1986, Harkess & Frame 1986).

Soil moisture, which is affected by the amount and distribution of precipitation as well as by temperature and soil conditions, is a most important factor affecting levels of herbage production, particularly in summer and in areas that experience seasonal drought. Water stress reduces the transformation of light energy into chemical energy (the '*t*' coefficient), and the recovery of sward growth after cutting (Peeters & Kopec 1996). All green plants require water, since nutrients are taken up, and physiological processes take place, in solution. Stomata remain open during daylight if there is adequate water within the leaf. They close, bringing photosynthesis to a halt, if a shortage of water occurs. As long as they remain open, transpiration of water from the plant continues. The rate of transpiration is a function of the evaporative capacity of the environment, and can be calculated from the duration of bright sunshine, mean air temperature, humidity and wind speed (MAFF 1954), assuming a continuous cover of green vegetation with adequate soil moisture in the root zone. Transpiration so calculated is known as the potential evapotranspiration (PT). Actual transpiration can be reduced by low soil moisture conditions, or by an incomplete cover of green vegetation. Both are measured in terms of millimetres of water. PT peaks in mid-summer, but it varies less than rainfall, and in the UK annual mean values range from 600 mm in southeast England to less than 300 mm on the uplands in northern Britain (Jones & Thomasson 1985).

For grass growth, the available water capacity (AWC) in the rooting zone has a significant effect on herbage growth. The field capacity (FC) is a concept that describes the maximum amount of soil water retained in a freely draining situation, against the influence of gravity. The duration of FC ranges from about 250 days in high rainfall areas of northern England to about 100 days in East Anglia. Between the end and return dates of FC, soils are in a state of moisture deficit: the soil moisture deficit (SMD) is the amount of water required to restore FC. Temperature and SMD are the main factors that determine the number of grass growing days. For further information on these concepts see MAFF (1976) and Jones & Thomasson (1985).

4.4 Irrigation

In areas of low rainfall, particularly on light textured soils, irrigation may be used to reduce the SMD when it has reached a stage that is critical for grass growth. In the UK there has been relatively little adoption of irrigation on commercial farms to overcome the limitations of water deficits on herbage production, although the situation is different in some other countries, including The Netherlands and in southern Europe. Research has shown little differential response to irrigation between *Lolium* spp. and the other main sown grasses *Dactylis glomerata* and *Phleum pratense* (Stiles 1966), but *Festuca arundinacea* has been shown to be less susceptible to drought (Garwood & Sheldrick 1978). White clover is more drought-susceptible than most grass species, and responds better to irrigation. Lucerne (*Medicago sativa*) shows little response to irrigation in most summers, and it is unlikely that droughts of a severity to affect lucerne occur in the UK with sufficient regularity to warrant irrigation. Growth is greatly restricted when the SMD exceeds 40–50 mm, a quantity of water available in the top 300 mm of many soils. Grass roots can extract water from a further 300–600 mm depth (Garwood & Williams 1967).

The response to irrigation of perennial ryegrass–white clover swards varies widely within the range 15–25 kg DM mm^{-1} of water ha^{-1} in years when the potential SMD exceeds 100 mm. At the former Grassland Research Institute, Hurley, a site considered typical for southeast England, seasonal production was determined over a 20-year period for both irrigated and non-irrigated first-harvest-year swards of perennial ryegrass (Corrall 1984). Without irrigation the mean annual production was 8.0 t DM ha^{-1}, and the between-years coefficient of variation was 19.4%. Irrigation increased DM production by about 25% to 9.9 t ha^{-1}, and the coefficient of variation was reduced to 9.5%. In addition to increased herbage production, irrigation enables levels of herbage production to be obtained with greater reliability, thus allowing better planning of stock-carrying capacity and the allocation of areas available for forage conservation. However, the cost of irrigation is high, and it is seldom economical to install it for grassland-based enterprises alone. Furthermore, there may be restrictions on water abstraction in dry years when the response to irrigation is likely to be greatest.

4.5 Soil drainage

Satisfactory grass production can be obtained under a wide range of soil drainage conditions, and grassland can tolerate a degree of seasonal waterlogging or occasional flooding that would be unacceptable for arable crops. There has been a long history of land drainage in Britain and in many other areas of Europe, with a particularly active period in the 1940s–1980s (Thomasson 1975, Baldock 1984). This was aimed at improving the agricultural value of grassland or enabling it to be brought into arable cropping. In recent years this has also been one of the sharpest points of conflict between agricultural and conservation interests, because of the importance of wetland habitats for wildlife (RSPB *et al.* 1997).

Poorly drained soils are a characteristic of much of the permanent grassland of western Britain, where high rainfall often occurs in association with impermeable soils (Thomasson 1982). Grassland with poor drainage may pose problems of utilization, including grazing and machinery access, and be susceptible to compaction and poaching by livestock (Scholefield & Hall 1985). Swards on poorly drained soils tend to contain lower proportions of ryegrasses, white clover and the other commonly sown species, and surveys have shown that drainage status is a factor contributing to the loss of sown species as swards age

(Hopkins & Green 1979). These are all factors that may be expected to affect herbage production and utilized output, although relatively few experimental investigations have been carried out to determine the agricultural benefits or the environmental impact associated with drainage installation. In one major long-term experiment, with replicated drained and undrained grazed plots on a clay soil in southwest England (Fig. 4.1), investigations of the effects of drainage on both herbage production and cattle liveweight gain have been shown to be relatively small (Tyson *et al.* 1992). The main benefit was in spring, when herbage DM production was 11% greater on the drained plots. This was reduced to 3% on an annual basis due to the effect of the larger soil water deficits sustained by the drained areas in summer, and the overall benefits were considered unlikely to be economically justifiable in the short term.

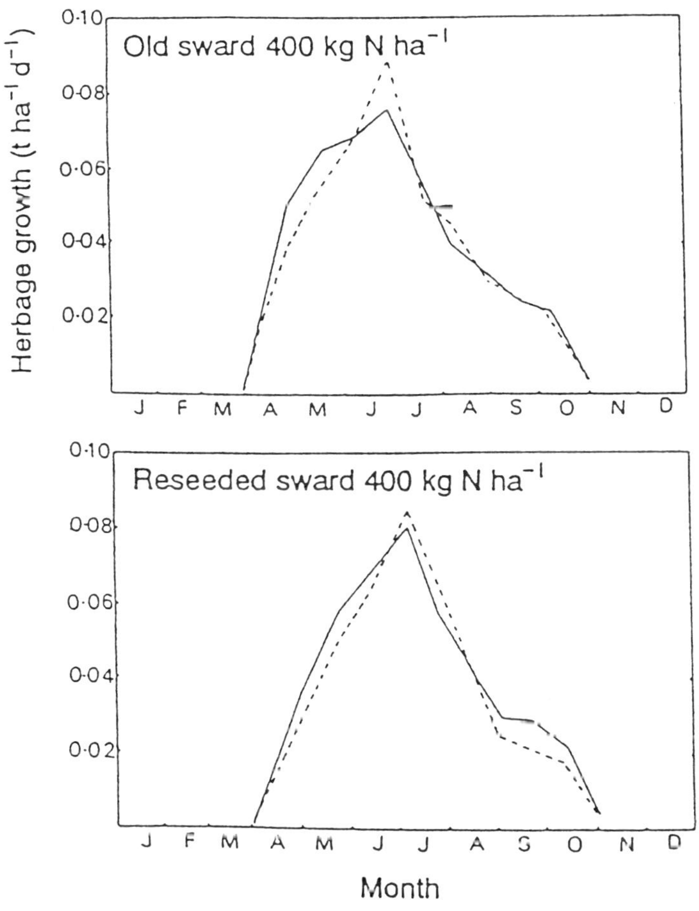

Fig. 4.1 Seasonal patterns of grass growth averaged over 5 years from drained (—) and undrained (----) swards receiving 400 kg N ha^{-1} year (source: Tyson *et al.* 1992, reproduced from *Grass and Forage Science* **47** p 295).

Furthermore, rarely is all the grassland on a farm poorly drained, and much can be done through the use of better drained fields to carry the livestock during wet periods.

4.6 Soil pH and liming

The pH of soils under grassland in Britain ranges from markedly acidic (pH <4.0) to alkaline (pH >7.0). Plant production is reduced at the extremes of this range through an excess of toxic ions (e.g. aluminium (Al), and manganese (Mn) under acid conditions) and reduced nutrient availability of some elements under alkaline conditions (Goulding & Annis 1998). Calcium is the dominant ion that controls soil pH (Rowell 1988). The most productive grass species require a minimum pH of 5.5, and current recommendations are that grassland soils are limed to maintain pH 6.0, especially for swards in which white clover is an important component. A higher pH (6.5) is usually recommended for lucerne, partly because pH affects its susceptibility to disease (Frame *et al.* 1998).

Calcium is removed from the soil by leaching and in harvested herbage. Acidification of the soil profile occurs in situations where high rates of fertilizers, particularly N, are applied, and also as a result of acid deposition. Ground limestone (which is predominantly calcium carbonate) is the main liming material used in the UK. A range of other calcareous materials are, or have been, used: these include ground magnesian limestone, ground chalk, shell sand, calcified seaweed, marls, bone meal and industrial by-products such as sugar beet factory sludge and basic slag. Farmyard manure also supplies calcium as well as other nutrients.

A number of trials have shown the agricultural benefits of applying lime to grassland in terms of increased animal production and changes to the botanical composition of the sward (Jones 1967, Cromack *et al.* 1970). In a review of five experiments, Skinner (1997) showed that mean herbage DM increased from about $6\,t\,ha^{-1}$ at pH 4.5 to about $9\,t\,ha^{-1}$ at pH 6.0–6.5. These increases showed clearly the financial benefits of applying lime to grassland soils of low pH.

4.7 Soil nutrient status and the role of fertilizers

The major nutrients that affect herbage production are nitrogen (N), phosphorus (P) and potassium (K). Gaseous N is the most abundant element in the atmosphere, but the supply of N to grass (and other plants) must be in the form of ammonium or nitrate. The concentration of N in herbage is typically in the range $10–50\,g\,kg^{-1}$ DM, and therefore large quantities of N can be removed when herbage is harvested. Grassland soils typically contain large resources of non-available N in the soil organic matter, but only a small proportion is mineralized to soil-available N in any year, and herbage production can be increased considerably by increasing the N supply, particularly when other environmental factors are not restricting herbage growth. The main sources are mineral N fertilizers, N-fixation (mainly by rhizobial bacteria associated with legumes in the sward), and animal manures and excreta.

Herbage production and the response of swards to N will also be limited if other essential elements, particularly P and K, are in short supply. Farmers have traditionally been recommended to base fertilizer applications on the results of soil analyses. A simplified system in which the quantities of available nutrients are expressed as indices, ranging from 0 (low), 1 (moderately low), 2 (adequate) and upwards, is often used in the UK. At a soil P and K index of 2 or above, it is unlikely that economic responses of grassland to applied P and K would occur. Under these conditions, the recommended applications are designed to maintain soil nutrient status. However, herbage analysis may be a more efficient indicator of nutrient requirements for the healthy growth of both plants and animals. Requirements for nutrients will also depend upon whether swards are established or are being resown, whether they are mown (with the consequent removal of nutrients) or grazed, and, if the swards are cut, whether nutrients are being returned in livestock manures.

Phosphorus, which is involved in many chemical reactions within plants, is present in grassland herbage in concentrations of about 2–4 g P kg^{-1} DM. In many areas of the world, including New Zealand, soils that support grassland have low concentrations of P, but UK grassland soils, except in the uplands, are generally not P-deficient. Animal manures and, indirectly, purchased feeds are important sources of P. These sources need to be taken into account (see values given in MAFF 1994) when calculating required inputs of fertilizer P (usually as triple superphosphate).

Potassium also has an important metabolic role and is vital for many plant functions, and in grassland herbage it is typically present at about 20 g K kg^{-1} DM. Potassium exists in the soil in several forms: soluble, exchangeable and non-exchangeable (Syers 1998). The amounts available from the soil may be adequate for low-yielding crops, but on productive N-fertilized swards the K removal rate can be high, especially under mowing. Thus, its replacement, either by efficient recycling of manures within the farm (see MAFF 1994) or by fertilizer applications (commonly as potassium chloride), is needed to prevent potential soil K deficiency.

A number of other plant nutrients can be deficient in grassland, e.g. magnesium and copper, although problems from these deficiencies are more likely to be ones affecting animal health rather than herbage production. Their importance is considered elsewhere in this book (see Sections 7.9.2, 7.9.3 and 10.11). One plant nutrient whose deficiency can also adversely affect herbage production is sulphur.

Sulphur is essential for both plants and animals. Reduced industrial emissions of sulphur dioxide (and therefore decreased inputs to soils), coupled with the increased use at high rates of high-analysis N fertilizers containing little or no S, may result in widespread imbalances, especially in wet years. A number of proprietary compound fertilizers are available that contain S in addition to N–P–K. The greatest responses to S-containing fertilizers have been reported in trials done in areas where sulphur deposition is low (e.g. southwest England) and for grass at the second and third cuts of multi-cut silage (McGrath *et al.* 1996).

4.7.1 Production response to nitrogen

When water and other minerals are non-limiting, grass production responds markedly to N application. Fertilizer N has been used in increasing amounts to increase forage production, and its use has enabled the benefits associated with the adoption of improved ryegrass cultivars and silage technology to be realized. It has therefore had a pivotal role in the development of intensively managed livestock systems in the UK (Frame *et al.* 1995) and elsewhere (Van Burg *et al.* 1981, Desroches *et al.* 1984). The use of nitrogen in grassland has also contributed to a number of environmental problems. These issues have been extensively reviewed (Whitehead 1995, Jarvis & Pain 1997).

The response of grass swards to fertilizer N has been investigated in trials throughout Europe. Most trials have been made under cutting, many have been based on perennial ryegrass, and the fertilizer usually applied at intervals throughout the growing season has been used in amounts ranging from nil to rates in excess of 500 kg ha^{-1} year^{-1}. Basal yields, with no added fertilizer N, vary greatly depending on site conditions: under a monthly cutting regime in the UK, annual herbage production from sown perennial ryegrass may exceed 6 t DM ha^{-1}, but be less than 1 t DM ha^{-1} on sites with poor grass-growing conditions (Morrison *et al.* 1980, Hopkins *et al.* 1990). The slope of the response curve to fertilizer N is also influenced by factors including available water, season and the sward itself (legume content, grass tiller density and root development). However, the general nature of the response curve is found when either herbage or animal production variables are recorded (Van Burg *et al.* 1981, Baker 1986). Herbage response follows an initial linear phase of 15–30 kg DM kg N^{-1}, usually up to an application rate within the range 250–400 kg N ha^{-1}. As the rate of fertilizer N is increased, the response diminishes until a maximum yield is reached. The point at which the herbage response falls to a specified value, e.g. 10 kg DM kg^{-1} N, has been used to define the concept of optimum fertilizer N rate (Fig. 4.2), in this case the N_{10} value (Morrison *et al.* 1980). For an individual site, the calculation of this input is necessarily imprecise as other factors affect herbage production in different seasons. The value of the herbage is also dependent on its nutritional quality, the method of utilization, fertilizer costs, the enterprise, the price of alternative feeds and product prices. The rate of application that coincides with this response optimum may also be excessive in terms of the potential for N losses to the environment.

Results from several multi-site N-response trials support the general form of response noted in Fig. 4.2, but they also show great variation from site to site (Van Steenbergen 1977, Morrison *et al.* 1980, Hopkins *et al.* 1990, 1995). In the UK 'GM20 trials', which were carried out on uniformly managed perennial ryegrass at 21 sites and cut at monthly intervals (Morrison *et al.* 1980), the mean maximum yield of 11.9 $t\,ha^{-1}$ was achieved with an input of 624 $kg\,N\,ha^{-1}$ (Table 4.1). In practice, maximum yield is of little interest because of the low

Table 4.1 Mean annual yields and responses to fertilizer N (4-year averages) (source: Morrison *et al.* 1980).

	Mean	Range
Maximum DM Yield ($t\,ha^{-1}$)	11.9	6.5–15.0
N required for maximum yield ($kg\,ha^{-1}$)	624	446–750
DM yield at 10:1, kg DM:kg N ($t\,ha^{-1}$)	10.9	5.4–14.4
N required for 10:1 limit ($kg\,ha^{-1}$)	386	260–530
Response kg^{-1} N at 300 $kg\,ha^{-1}$ (kg DM)	23	14–29
DM yield with no N ($t\,ha^{-1}$)	2.6	0.6–5.7
% recovery of fertilizer N at 300 $kg\,ha^{-1}$	70	51–87

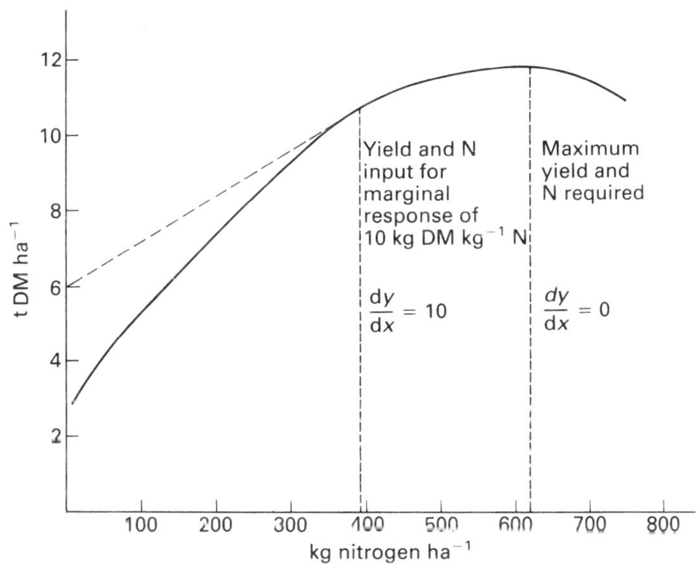

Fig. 4.2 Mean DM yields of grass and grass–clover swards in response to N fertilizer. Curve for grass sward (——) and curve for grass–white clover sward (---) (after Morrison *et al.* 1980).

DM returns for N applied as the maximum is approached. As N input was increased from low values, the DM:N ratio fell, reaching 10:1 at an input of 386 kg N ha^{-1} and a yield of 10.9 t ha^{-1}. From this point to the maximum yield resulted in only 1 t additional DM, but it required an extra 238 kg N. Data from national trials such as this have been used as a basis for the prediction of herbage production, and potential stocking rates, for grassland classified by site suitability (Thomas *et al.* 1991).

The response values given above are based on evidence from sown perennial ryegrass swards. A high proportion of long-term grassland is of mixed composition, with swards that typically contain common pasture species such as *Holcus lanatus*, *Agrostis* spp., *Festuca* spp., *Poa* spp., legumes and other forbs, in addition to variable amounts of *Lolium perenne* or other sown grasses. The response of old grassland swards (which contained <30% perennial ryegrass) to fertilizer N was compared with sown perennial ryegrass at 16 sites in England and Wales (Fig. 4.3) (see Hopkins *et al.* 1990). The results showed that increased DM production was obtained from the reseeded swards in the first year after sowing. In subsequent years the production advantage of the ryegrass was sustained only at the higher fertilizer N rates, in effect a steeper response curve on the reseeded swards. The relationship between herbage production and sward type is discussed further in Section 4.8.

The pattern and timing of fertilizer N application can also affect both the overall response per kg of N applied, and the seasonal pattern of production. The timing of early season N applications in relation to temperature has been mentioned (Section 4.3). Poor responses (and loss of N through leaching) can also occur when N is applied late in the growing season. Fertilizer applications

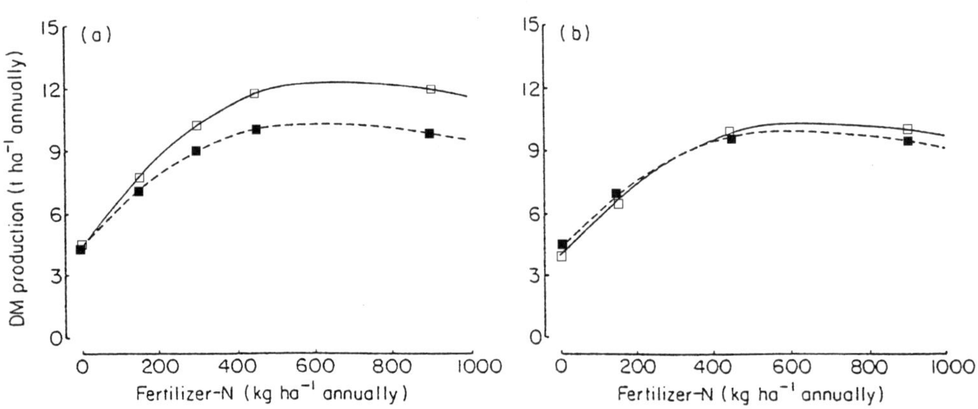

Fig. 4.3 Fitted response curves showing dry matter production from permanent swards (--■--) and reseeded perennial ryegrass (—□—) in relation to fertilizer-N. Mean of 16 sites under 4-weekly cutting: (a) 3-year mean (1985–1986), and (b) 4-year mean (1983–1986) including the establishment-phase of the reseeded sward. (Reproduced from *Grass and Forage Science* **45** p. 47.)

can be adjusted according to the amount of mineral N in the soil, as determined by a rapid field test (Scholefield & Titchen 1995) and a mechanism by which the measured levels of mineral N can be interpreted in relation to a fertilizer requirement (Brown *et al.* 1997). Thus, modelling combined with soil testing has the potential to establish the links between fertilizer, grassland production and environmental impact in a way that farmers can understand and manipulate. Indications are that this approach can enable inputs of fertilizer N to be reduced by about 25% without affecting annual herbage production.

Swards based on grass/legume mixtures show a lower response to fertilizer N than grass monocultures. In a series of trials on 22 sites in the UK (the 'GM23' trials), the fertilizer N response of perennial ryegrass and ryegrass/white clover was measured under 4-weekly cutting (Baker 1986). At inputs of up to about 300 kg N ha^{-1} the grass/clover swards had higher DM yields (especially at nil-fertilizer N when average yields were *c.* 7 t DM ha^{-1}, compared with *c.* 3 t from the ryegrass). However, the response of the grass/white clover to fertilizer N was only about 10 kg DM per kg N, compared with 20–25 kg DM per kg N for the pure ryegrass. One of the benefits of including white clover and other legumes is, of course, the potential to increase overall sward production through N fixation, thereby reducing, or completely eliminating, the need for fertilizer N. Seasonal deficiencies in production, for example in spring, can be overcome by tactical doses of fertilizer N (Morrison *et al.* 1983). Much research on N responses has been based on cut systems. Under grazing, responses may be much influenced by livestock treading and by the returns of N in dung and urine.

4.8 Herbage production and botanical composition

The need to have the most appropriate sward composition in order to maximize the productivity and herbage quality of grassland, for a given site and level of inputs, is a topic that has long held interest for grassland agronomists (Charles & Haggar 1979, Lazenby 1981). The recognition that some of the most productive pastures in the English Midlands contained a high proportion of perennial ryegrass and white clover was one of the factors which led to the development of the ley farming doctrine (see also Chapter 1). The ploughing of old pastures and their replacement with ryegrass-based leys was one of the key features of grassland improvement in the twentieth century. The major plant breeding developments have been with these species, and most of the herbage seed sold for agriculture consists of perennial ryegrass (MAFF 1998). However, botanical composition *per se* is not a major determinant of sward productivity, but it reflects fertility and site conditions. Permanent swards dominated by perennial ryegrass are invariably in fertile situations, and on most farms sown swards tend to receive relatively high inputs of fertilizers or other manures to sustain their productivity. For example, fertilizer N rates on leys in England and

Wales are *c.* 200 kg N ha^{-1}, whereas 40% of over 20-year-old swards receive less than 50 kg N ha^{-1}.

4.8.1 Permanent swards or leys?

A large proportion of productive grassland supports permanent swards, which despite having been modified by management inputs have, in many instances, no recent history of cultivation. In England and Wales, for example, approximately 50% of the enclosed grassland area has swards over 20 years old (Hopkins & Hopkins 1994). A number of studies have been conducted, using multi-site trials on farms as well as on-farm recording methods, to determine the production of permanent grassland and swards of differing botanical composition. In a comprehensive survey of 500 permanent-grassland farms (Forbes *et al.* 1980, Peel & Matkin 1982) the factors limiting output (as utilized metabolizable energy, though not actual herbage production) were investigated. The main factors affecting output were stocking rate, fertilizer N rate and land manageability, whilst sward age and botanical composition had relatively little effect. These findings are consistent with several trials carried out previously (e.g. Mudd 1971) or subsequently (see also Section 4.7.1). A multi-site trial to compare the herbage production of permanent swards with *L. perenne*-reseeded swards (see Fig. 4.3) showed that responses to reseeding were considerable in the first harvest year, with a 30–40% increase in DM yield at the same fertilizer treatment (Hopkins *et al.* 1990). Relative differences were much reduced in subsequent years. On nine sites that were recorded for 7 years, and received 300 kg N ha^{-1} $year^{-1}$, there was a mean increase in DM production of 6% from the reseeded swards, compared with permanent swards, during the 6-year period following the first harvest year (Hopkins *et al.* 1992). The value of reseeding to the farmer needs to be considered in relation to the costs involved, the loss of production during the establishment phase, and the number of years over which benefits from reseeding can be obtained.

Permanent swards also have a number of potential benefits associated with sward density, which can enable production to be utilized in conditions when leys would be vulnerable to poaching. There may also be some nutritional benefits associated with species diversity in permanent swards, e.g. higher concentrations of certain minerals. However, it must also be recognized that improvements in plant breeding have provided opportunities for sward improvement beyond simple measures of herbage production for a given level of fertilizer N. These include improvements in herbage quality, including higher D value and water-soluble carbohydrate content to optimize the protein/energy balance in the rumen (Humphreys & Thomas 1998), and improvements in the ensilability of grasses (see Chapter 8, Table 8.10). Reseeding also offers opportunities for introducing white clover or other legumes where these were previously absent.

4.8.2 Production from legume-based swards

White clover is the principal forage legume in the British Isles and in many other temperate regions, and New Zealand, in particular, has developed an internationally competitive pastoral agriculture in which white clover has a pivotal role. It has four major benefits: it fixes nitrogen, it improves sward quality, it complements the seasonal growth of most grass species, and it improves forage intake and utilization (Caradus *et al.* 1996). White clover is normally sown with perennial ryegrass, or oversown (or otherwise encouraged by appropriate management) into existing permanent swards (see Chapter 2). With the widespread use of fertilizer N and concentrate feeds, the potential role of white clover and other legumes was disregarded for many years in both research and farming practice in the UK and western Europe, except for extensive grazing systems and hill sheep pastures (Frame *et al.* 1995). The environmental and economic consequences of high N-fertilizer usage, combined with the nutritional advantages of white clover, have led to a growing realization that clover-based swards have a potential role in dairy production (Ryan 1989, Bax & Schills 1993) as well as for beef and sheep (Davies & Hopkins 1996). White clover-based swards have the potential to replace, either wholly or in part, the fertilizer N needed to meet production targets and improve the sustainability of livestock production systems. Much of the earlier research was based on varieties with limitations on persistence. Varieties of white clover have been released with significantly improved agronomic characteristics, and these will increase the potential for its wider use (Abberton *et al.* 1998).

The theoretical potential annual production of grass/white clover swards (in a 50:50 mixture) has been estimated at 18.5–22.5 t DM ha^{-1}, compared with 27–30 t ha^{-1} for pure grass swards (Frame & Newbould 1984). Actual herbage productivity is usually much lower, e.g. 12–15 t DM ha^{-1} from perennial ryegrass swards receiving maximum rates of fertilizer N on good grass-growing sites (Hopkins *et al.* 1995). For grass–white clover swards, typical production values are 5–10 t DM ha^{-1}, the lower figure associated with adverse growing conditions (e.g. moisture stress) or with a low mean clover content in the sward. These values are derived from cutting experiments (e.g. National List Trials, Table 4.2)

Table 4.2 Herbage production and white clover content from National List trials, nil-fertilizer N (from Frame & Newbould 1984).

	DM (t/ha)		White clover (%)
	Total	White clover	
Highest	15.53	12.12	78
Lowest	2.03	0.49	24
Mean	8.31	4.16	50

which attempt to simulate the defoliation frequency of rotational grazing. Annual herbage production from mown swards containing *c.* 30% white clover has been found to be comparable to that from all-grass swards receiving *c.* 200 kg fertilizer N ha^{-1} (Morrison 1981). In the hills and uplands, productivity of grass/white clover swards is lower than in the lowlands owing to the shorter growing season and lower temperatures. Under a cutting management at a site at 305 m elevation in mid-Wales, mean production from grass/white clover swards was 4.5 and 3.1 t DM ha^{-1} over 4 harvest years on fertile brown earth and impoverished stagnogley soils, respectively (Munro & Davies 1974). These values were equivalent to herbage production from perennial ryegrass swards receiving 129 and 181 kg N ha^{-1}, respectively. Averaged over the 4 years, the clover contribution to herbage production was 28% and 46% on the brown earth and stagnogley soils, respectively.

The benefits of clover-based pastures cannot be assessed in terms of DM productivity alone. Higher nutritive value, particularly digestibility and crude protein content, must also be taken into account. In the upland experiment reported above, the replacement values of grass/white clover swards in terms of N harvested in the herbage were 156 and 268 kg N ha^{-1}, respectively. This highlights the fact that on soils having a poor N supply, the benefits of grass/legume swards are relatively greater than on soils with a good supply of mineralized N.

The ultimate production benefit of incorporating legumes into grassland systems is their contribution to animal production. Numerous reviews of grazing studies have been published showing the benefits of white clover in grazed swards. The greatest benefits are with sheep (Table 4.3). For further examples, see Davies & Hopkins (1996), Hopkins (1998) and also Chapter 10, Section 10.4.6.

The other main temperate forage legumes, lucerne (*Medicago sativa*), red clover (*Trifolium pratense*) and the birdsfoot trefoils (*Lotus corniculatus* and *L. pedunculatus*) occupy relatively minor roles in terms of their contribution to the forage resources of the British Isles, but they are of considerable importance internationally. In Britain, lucerne is usually grown only on calcareous soils as a specialized cutting crop without a companion grass (see Chapter 2). When harvested at the mid-bud stage (*D*-value of 55–60), a succession of three to four harvests will provide an annual production of about 7–15 t DM ha^{-1}. Red clover

Table 4.3 Daily lamb growth rate (g day^{-1}) on grass/white clover and N-fertilized grass (200 kg N ha^{-1}) pastures at upland and lowland sites (from Davies & Munro 1988).

	Pre-weaning		Post-weaning	
	Grass/clover	Grass	Grass/clover	Grass
Upland	201	186	112	86
Lowland	232	212	140	81

is normally sown with a companion grass and the crop managed as a ley for forage conservation. Annual production is similar to that of lucerne crops. The *Lotus* species have a potential role in grazed swards, particularly in situations where the soil nutrient status is sub-optimal for white clover. Annual production of about 8–9 t DM ha^{-1} has been obtained in trials in southwest England from sown grass/*Lotus* swards under low-input management (Hopkins *et al.* 1998). For further information on forage legumes, see Sheldrick *et al.* (1995) and Frame *et al.* (1998), and for specific information on *Lotus* species see the *Lotus Newsletter* website (http://www.psu.missouri.edu/lnl).

4.8.3 Production from rough and hill grazings

Rough and hill grazings occupy land above the limits of cultivation, and infertile heath soils at lower elevations. The UK has about 6 million hectares of rough grazing, and equivalent vegetation communities are widespread in Ireland and northern Europe. In the UK, rough grazings include *Festuca–Agrostis* grassland and communities based on *Molinia caerulae*, *Nardus stricta*, bracken, and heather/bilberry moor (see Chapter 1, Table 1.4). The combination of soils and climate has a more decisive influence on the vegetation of rough grazing land than on permanent grassland at lower elevations. In some areas, modification of the vegetation has occurred as a result of management, including lime applications on the more accessible areas, controlled burning of heather, and grazing, which is mainly with sheep. The utilization under grazing is discussed in Chapter 10 (Section 10.1).

Agrostis–Festuca pastures provide good quality rough grazing, and are found on better-drained soils with pH of 4.5 or higher, where they are most responsive to management improvements, and they also occur on some wetter gley soils. Levels of herbage production are in the range 2–4 t DM ha^{-1} and forage quality is good. However, on areas where management has been poor (including an excessive number of sheep without cattle), invasion by bracken (*Pteridium aquilinum*) has led to loss of pasture (Smith & Taylor 1986) (see also Section 6.2.5.).

Swards dominated by *N. stricta* mainly occur on drier hill slopes and podsols and peaty podsols. Net annual production is generally lower than that from *Festuca–Agrostis* (around 2–3 t DM ha^{-1}). Extensive areas dominated by *M. caerulea* occur on wet peaty gley soils. *M. caerulea* has a short period of rapid growth (June to August) and it has a high feeding value in summer, with production about 2–3 t DM ha^{-1}. Heather moor (composed of *Calluna vulgaris* in association with other heath, forb and grass species) is found on freely drained peaty podsols. Net production is around 1.5–3 t DM ha^{-1}, with growth concentrated in June to September, and utilization mainly during autumn and winter (see Section 10.10.1). On deep peat soils in areas of high rainfall, blanket bog communities (*Sphagnum* mosses, with *Trichophorum* and *Eriophorum* spp.) develop. Net production is low (around 2 t DM ha^{-1}), feed value is low and sensitivity to overgrazing is high. For further information on the characteristics and management of these communities, see Hodgson & Grant (1981) and Newbould (1981).

4.8.4 Production from semi-natural grasslands

Lowland semi-natural grasslands have become vastly depleted and fragmented as a result of agricultural improvements (Hopkins & Hopkins 1994), and the importance for nature conservation of the remaining areas is now recognized through national and European policies and international Biodiversity Action Plans (see Chapter 11). A key requirement for the continued survival of the conservation interest of these grasslands is that they remain in agricultural production but are utilized under appropriate management. This typically involves nil use or very low inputs of fertilizer or manure, and timing of defoliation that does not prejudice the survival of key species. High inputs of fertilizers can result in greatly increased production, but the conservation value is lost through changes in the species balance and sward structure. There is a paucity of agronomic data on the productivity of most semi-natural grassland communities but, in a review of available evidence, Tallowin (1997) concluded that DM production was typically 40–80% of that of agriculturally improved grassland. On wet heath and fen meadow the utilized output was estimated at only *c.* 25% of that of intensively managed permanent pasture, but some mesotrophic grasslands were shown to support individual animal performance comparable to agriculturally improved pastures.

4.9 References

Abberton M.T., Michaelson-Yeates T.P.T., Macduff J.H., Marshall A.H. & Rhodes I. (1998) New approaches to legume improvement for sustainable agriculture. In Boller B. & Stadelmann F.J. (eds) *Breeding for a Multi-Functional Agriculture*, 12–15. Swiss Federal Research Station for Agroecology and Agriculture, Zurich–Reckenholz.

Baker R.D. (1986) Efficient use of nitrogen fertilizers. In Cooper J.P. & Raymond W.F. (eds) *Grassland Manuring*. British Grassland Society Occasional Symposium, No. 20, 15–27. BGS, Hurley.

Baldock D. (1984) *Wetland Drainage in Europe – The Effects of Agricultural Policy in Four EEC Countries*, 119–57. Institute for European Environmental Policy/International Institute for Environment and Development, London.

Bax J.A. & Schills R.L.M. (1993) Animal responses to white clover. *White Clover in Europe – State of the Art*, 7–23. REUR Technical Series, **29**.

Brown L., Scholefield D., Jewkes E.C. & Preedy N. (1997) Integrated modelling and soil testing for improved fertilizer recommendations for grassland. In Van Cleemptut O. (ed) *Fertilization for Sustainable Plant Production and Soil Fertility. Proceedings 11th CIEC World Congress, September 1997, Gent*, 248–55.

Caradus J.R., Woodfield D.R. & Stewart A.V. (1996) Overview and vision for white clover. In *White Clover: New Zealand's Competitive Edge. Symposium of the New Zealand Grassland Association, Lincoln, NZ*, 1–6.

Charles A.H. & Haggar R.J. (eds) (1979) *Changes in Sward Composition and Productivity*. British Grassland Society Occasional Symposium, No. 10, 253 pp. BGS, Hurley.
Clements R.O., Murray P.J., Bentley B.R., Lewis G.C. & French N. (1990) The impact of pests and diseases on the herbage yield of permanent grassland at eight sites in England and Wales. *Annals of Applied Biology* **117**, 349–57.
Colman R.L., Lazenby A. & Grierson J. (1974) Nitrogen fertilizer responses and seasonal production of temperate and warm climate grasses on the Northern Tablelands of New South Wales. *Australian Journal of Experimental Agriculture and Animal Husbandry* **14**, 362–72.
Cooper J.P. & Tainton (1968) Light and temperature requirements for the growth of temperate and tropical grasses. *Herbage Abstracts* **38**, 167–76.
Corrall A.J. (1984) Grass growth and seasonal pattern of production under varying climatic conditions. In *The Impact of Climate on Grass Production and Quality. Proceedings of the 10th General Meeting of the European Grassland Federation*, 36–45.
Corrall A.J. & Fenlon J.S. (1978) A comparative method for describing the season distribution of production from grasses. *Journal of Agricultural Science, Cambridge* **91**, 61–7.
Cromack H.T.T., Mudd C.H. & Strickland M.J. (1970) A comparison of types of lime and their frequency of application to grassland. *Experimental Husbandry* **19**, 40–8.
Davies A., Baker R.D., Grant S.A. & Laidlaw A.S. (eds) (1993) *Sward Measurement Handbook*. 2nd edn. British Grassland Society, Reading.
Davies D.A. & Hopkins A. (1996) Production benefits of legumes in grassland. In Younie D (ed) *Legumes in Sustainable Farming Systems*, 234–46. British Grassland Society Occasional Symposium, No. 30. BGS, Reading.
Davies D.A. & Munro J.M.M. (1988) Assessment of grass–clover pastures for lowland and upland lamb production. *Proceedings of 12th General Meeting European Grassland Federation, Dublin, 1988*, 164–7.
Desroches R., Gayraud P. & de Pontèves B. (1984) Donnees nouvelles concernant les praries cultivees. *Fourrages* **100**, 129–59.
Forbes T.J., Dibb C., Green J.O., Hopkins A. & Peel, S. (1980) *Factors Affecting the Productivity of Permanent Grassland – A National Farm Study*, 140 pp. GRI/ADAS Permanent Pasture Group, Hurley.
Frame J. (1993) Herbage mass. In Davies A., Baker R.D., Grant S.A. & Laidlaw A.S. (eds) *Sward Measurement Handbook*. 2nd edn. 39–68. British Grassland Society, Reading.
Frame J. & Newbould P. (1984) Herbage production from grass/white clover swards. In Thomson D.J. (ed) *Forage Legumes*. British Grassland Society Occasional Symposium, No. 16, 15–35. BGS, Hurley.
Frame J., Baker R.D. & Henderson A.R. (1995) Advances in grassland technology over the past fifty years. In Pollot G.E. (ed) *Grassland into the 21st Century: Challenges and Opportunities*. British Grassland Society Occasional Symposium, No. 29, 31–63. BGS, Reading.
Frame J., Charlton J.F.L. & Laidlaw A.S. (1998) *Temperate Forage Legumes*. CAB International, Wallingford.
Garwood E.A. & Sheldrick R.D. (1978) Dry matter production by tall fescue under dry conditions. *Journal of the British Grassland Society* **33**, 67–8.
Garwood E.A. & Williams T.E. (1967) Soil water use and growth of a grass sward. *Journal of Agricultural Science* **68**, 281–92.
Goulding K.W.T. & Annis B. (1998) Lime, liming and the management of soil acidity. *Proceedings of the Fertilizer Society* No. 410.
Harkess R.D. & Frame J. (1986) Efficient use of fertilizer nitrogen on grass swards: Effects of timing, cutting management and secondary grasses. In Van der Meer H.G., Ryden J.D. & Ennik E.C. (eds) *Nitrogen Fluxes in Intensive Systems*, 29–37. Martinus Nijhoff, Dordrecht.

Hodgson J. (1979) Nomenclature and definitions in grazing studies. *Grass and Forage Science* **34**, 11–18.

Hodgson J. & Grant S.A. (1981) Grazing animals and forage resources in the hills and uplands. In Frame J. (ed) *The Effective Use of Forage and Animal Resources in the Hills and Uplands.* British Grassland Society Occasional Symposium, No. 12, 41–57. BGS, Hurley.

Hopkins A. (1998) The potential of white clover. In Lane G.P.F. & Wilkinson J.M. (eds) *Alternative Forages for Ruminants*, 35–46. Chalcombe, Lincoln.

Hopkins A. & Green J.O. (1979) The effect of soil fertility and drainage on sward changes. In Charles A.H. & Haggar R.J. (eds) *Changes in Sward Composition and Productivity*. British Grassland Society Occasional Symposium, No. 10, 115–29. BGS, Hurley.

Hopkins A. &. Hopkins J.J. (1994) UK grasslands now: Agricultural production and nature conservation. In Peel S. & Haggar R.J. (eds) *Grassland Management and Nature Conservation.* British Grassland Society Occasional Symposium, No. 28, 10–19. BGS, Reading.

Hopkins A., Gilbey J., Dibb C., Bowling P.J. & Murray P.J. (1990) Response of permanent and reseeded grassland to fertilizer nitrogen. 1. Herbage production and herbage quality. *Grass and Forage Science* **45**, 43–55.

Hopkins A., Bowling P.J. & Johnson J. (1992) Site-specific variability in the productivity and nutrient uptake of permanent and sown swards. *Proceedings of 14th General Meeting European Grassland Federation, Lahti, 1992*, 199–203.

Hopkins A., Murray P.J., Bowling P.J., Rook A.J. & Johnson J. (1995) Productivity and nitrogen uptake of ageing and newly sown swards of perennial ryegrass (*Lolium perenne* L.) at different sites and with different nitrogen fertilizer treatments. *European Journal of Agronomy* **4**, 65–75.

Hopkins A., Davies D.A., Johnson R.H. & Sheldrick R.D. (1998) Recent and on-going research with *Lotus* (birdsfoot trefoil) at IGER. In Lane G.P.F. & Wilkinson J.M. (eds) *Alternative Forages for Ruminants*, 134–7. Chalcombe Publications, Lincoln.

Humphreys M.O. & Thomas H. (1998) Breeding to improve the protein value of grass. In Boller B. & Stadelmann F.J. (eds) *Breeding for a Multi-Functional Agriculture*, 12–15. Swiss Federal Research Station for Agroecology and Agriculture, Zurich–Reckenholz.

Jarvis S. & Pain B.F. (eds) (1997) *Gaseous Nitrogen Emissions from Grasslands*. CAB International, Wallingford.

Jones LL. I. (1967) Studies on hill land in Wales. *Welsh Plant Breeding Station Technical Bulletin* No. 2.

Jones R.J.A. & Thomasson A.J. (1985) An agroclimatic databank for England and Wales. *Soil Survey Technical Monograph No. 16*. Soil Survey, Harpenden.

Lazenby A. (1981) British grasslands: Past, present and future. *Grass and Forage Science* **36**, 243–66.

Lazenby & Docon (1982) Realizing the potential of British grasslands: Some problems and possibilities. *Applied Geography* **2**, 171–88.

MAFF (Ministry of Agriculture, Fisheries & Food) (1954) *The Calculation of Irrigation Needs. Technical Bulletin No. 4.* HMSO, London.

MAFF (Ministry of Agriculture, Fisheries & Food) (1976) *The Agricultural Climate of England and Wales. Technical Bulletin No. 33.* HMSO, London.

MAFF (Ministry of Agriculture, Fisheries & Food) (1994) *Fertilizer Recommendations. Reference Book 209.* HMSO, London.

MAFF (Ministry of Agriculture, Fisheries & Food) (1998) *Seed Traders Annual Return Summary, Year Ended June 1998.* MAFF, Cambridge.

McGrath S.P., Zhao F.J. & Withers P.J.A. (1996) Development of sulphur deficiency in crops and its treatment. *Proceedings of the Fertilizer Society*, No. 379.

Morrison J. (1981) The potential of legumes for forage production. In *Legumes and Fertilizers in Grassland Systems*. British Grassland Society Winter Meeting 1981, 1.1–1.10.

Morrison J., Jackson M.V. & Sparrow P.E. (1980) The response of perennial ryegrass to fertilizer nitrogen in relation to climate and soil. *GRI Technical Report No. 27*. Grassland Research Institute, Hurley.
Morrison J., Denehy H. & Chapman P.F. (1983) Possibilities for the strategic use of fertilizer N on white clover/grass swards. In Corrall A.J. (ed) *Efficient Grassland Farming*. British Grassland Society Occasional Symposium, No. 14, 227–31. BGS, Hurley.
Mudd C.H. (1971) Yields of natural and artificial grassland under five levels of fertilizer treatment. *Proceedings of the 4th General Meeting of the European Grassland Federation, Lausanne, 1971*, 69–72.
Munro J.M.M. & Davies D.A. (1974) Potential pasture production in the uplands of Wales. 5. The nitrogen contribution of white clover. *Journal of the British Grassland Society* **29**, 213–23.
Newbould P. (1981) The potential of indigenous plant resources. In Frame J. (ed) *The Effective Use of Forage and Animal Resources in the Hills and Uplands*. British Grassland Society Occasional Symposium, No. 12, 1–16. BGS, Hurley.
Orr R.J., Parsons A.J., Penning P.D. & Treacher T.T. (1990) Sward composition, animal performance and the potential production of grass/white clover swards continuously stocked with sheep. *Grass and Forage Science* **45**, 325–36.
Parsons A.J., Collett B. & Lewis J. (1984) Changes in the structure and physiology of a perennial ryegrass sward when released from a continuous stocking management – implications for the use of exclusion cages in continuously stocked swards. *Grass and Forage Science* **39**, 1–9.
Peel S. & Matkin E.A. (1982) The productivity of grassland farms in seven climatic zones of England and Wales. *Grass and Forage Science* **37**, 299–310.
Peeters A. & Kopec S. (1996) Production and productivity of cutting grasslands in temperate climates of Europe. In *Grassland and Land Use Systems. Proceedings of the 16th General Meeting of the European Grassland Federation, Grado, 1996*, 59–73.
Rowell D.L. (1988) Soil acidity and alkalinity. In Wild A. (ed) *Russell's Soil Conditions and Plant Growth*. 11th edn, 844–98. Longman, Harlow.
RSPB, EN & ITE (1997) *The Wet Grassland Guide: Managing Floodplain and Coastal Wet Grasslands for Wildlife*. Royal Society for the Protection of Birds, Sandy.
Ryan M. (1989) Development of a legume-based dairy system. In *Developments in Plant and Soil Science* **37**, 159–67.
Scholefield D. & Hall D.M. (1985) A method to measure the susceptibility of pasture soils to poaching by cattle. *Soil Use and Management* **1**, 134–8.
Scholefield D. & Titchen N.M. (1995) Development of a rapid field test for soil mineral nitrogen and its application to grazed grassland. *Soil Use and Management* **11**, 33–43.
Sheehy J.E. & Johnson I.R. (1988) Physiological models of grass growth. In Jones M.B. & Lazenby A. (eds) *The Grass Crop*, 243–75. Chapman & Hall, London.
Sheldrick R., Newman G. & Roberts D. (1995) *Legumes for Milk and Meat*. 2nd edn, 15–74. Chalcombe Publications, Canterbury.
Skinner R.J. (1997) Lime – who needs it? *Grass Farmer* **57**, 13.
Smith R.T. & Taylor J.A. (eds) (1986) *Bracken: Ecology, Land Use and Control Technology*. 464 pp. Parthenon, Carnforth.
Stiles W. (1966) Ten years of irrigation experiments. *GRI Annual Report 1965*, 57–66. Grassland Research Institute, Hurley.
Syers J.K. (1998) Soil and plant potassium in agriculture. *Proceedings of the Fertilizer Society*, No. 411.
Tallowin J.R.B. (1997) The agricultural productivity of lowland semi-natural grassland: A review. *English Nature Research Reports* No. 233, Peterborough.
Thomas C., Reeve A. & Fisher G.E.J. (1991) *Milk from Grass*. 2nd edn, 112 pp. BGS, Reading.
Thomasson A.J. (ed) (1975) Soils and field drainage. *Soil Survey Technical Monograph No. 7*. Soil Survey, Harpenden.
Thomasson A.J. (1982) The distribution and properties of British soils in relation to land drainage. In Gardner M.J. (ed) *Land Drainage*, 3–10. Balkema, Rotterdam.

Tyson K.C., Garwood E.A., Armstrong A.C. & Scholefield D. (1992) Effect of field drainage on the growth of herbage and the liveweight gain of grazing beef cattle. *Grass and Forage Science* **47**, 290–301.

Van Burg P.F.J., Prins W.H., Den Boer D.J. & Sluiman W.J. (1981) Nitrogen and intensification of livestock farming in EEC countries. *Proceedings of the Fertilizer Society* No. 199.

Van Steenbergen T. (1977) Influence of type of soil and year on the effect of nitrogen fertilization on the yield of grassland. *Stikstof, Dutch Nitrogen Fertilizer Review* **20**, 29–35.

Weddell J.R., Gilliland T.J. & McVittie J. (1997) Evaluation procedures: Past, present and future. In Weddell J.R. (ed) *Seeds of Progress*, 202–25. British Grassland Society Occasional Symposium, No. 31. BGS, Hurley.

Whitehead D.C. (1995) *Grassland Nitrogen*. CAB International, Wallingford.

Chapter 5
Herbage Seed Production

A.H. Marshall and D.H. Hides

5.1 Introduction

The production of herbage seed is a relatively small and specialized component of the world's grassland production systems, but is of considerable value in some countries and localities. Although seed production can arguably be regarded as an arable enterprise, it is of critical importance in providing the source material for grassland reseeding in both agricultural and amenity situations. This chapter considers the historical development of the demand and provision for herbage seed, primarily with reference to the UK and EU situations, and outlines the management requirements of herbage seed crops and the certification arrangements that apply.

Before 1940, herbage seed production in England and Wales was confined to catch-cropping of red and white clover fields, and apart from some Italian ryegrass, little ryegrass seed was produced in the UK outside of Northern Ireland and Scotland. In 1940, when supplies from European countries were stopped and importation from the US and New Zealand became impossible, the impetus was created for a new herbage seed industry in the UK. Variety evaluation procedures were established (Weddell *et al.* 1997) and herbage seed production developed rapidly, reaching a peak in 1964 when some 20 000 ha of herbage seed crops, predominantly Aberystwyth (S-strain) varieties covering a wide range of species, were inspected. During the next 10 years, seed consumption remained steady, although there was a gradual decline in the use of Aberystwyth varieties from nearly 100% to 54% of the seed produced, with a rapid increase in the production of seed of foreign-bred varieties.

Since the mid-1970s there has been a near 50% reduction in the consumption of grass seed in the UK, and a commensurate decrease in the area used for grass seed production (from 21 201 ha in 1975 to 10 646 ha in 1995). There has also been a decline in the production of seed of UK-bred varieties, peaking in the late 1980s when over 80% of the seed growing area was used for non-UK bred varieties. Since then, a resurgence in forage plant breeding has taken place within the UK, particularly of perennial and Italian ryegrasses, and by the late 1990s UK-bred varieties accounted for 30–40% of the total grass seed produced. Further information on the forage seed trade in the UK and elsewhere is reviewed by Burgon *et al.* (1997).

5.2 Use and production of herbage seed

The main sown grass species in the UK is perennial ryegrass (*Lolium perenne*), and it accounts for about 70% of the forage grass seed sown. Approximately 10 250 t of seed of perennial ryegrass are used each year for agricultural use, and 2440 t of Italian ryegrass (*L. multiflorum*) (MAFF 1996). This compares with 15 446 t of perennial ryegrass seed in 1971 (MAFF 1976). Hybrid ryegrasses (*Lolium* × *boucheanum* Kunth.) were first developed in the UK in the 1960s, but by the late 1990s their use was still less than 1000 t per year. Sowings of cocksfoot (*Dactylis glomerata*) and fescues (*Festuca pratensis*, *F. arundinacea*) have decreased markedly, in response to the greater emphasis on herbage quality which ryegrasses can provide, and as plant breeders have developed ryegrass varieties with improved yield and persistence (Camlin 1997).

The predominant forage legume sown in the UK is white clover (*Trifolium repens*); usage amounts to approximately 800 t per year. Sowings of red clover (*T. pratense*) seed decreased from some 3000 t in 1960 to 53 t in 1995 (MAFF 1996). Its decline is due to a combination of factors, including the replacement of mixed rotational farming with continuous cereal production, problems of silage making, the adoption of simpler grass–legume mixtures and the greatly increased use of nitrogenous fertilizers during this period.

Seed production and consumption in the European Union was reviewed by Kley (1995). Perennial ryegrass is the most important species for agricultural use. Some 131 000 t of grass seed are used per year, of which 73 600 t are for forage, the remainder being for amenity use. Seeds of meadow fescue (*F. pratensis*), Italian ryegrass, timothy (*Phleum pratense*) and cocksfoot are almost entirely used for forage, and some 50% of the seed of perennial ryegrass and tall fescue (*F. arundinacea*), and 20% of smooth-stalked meadow grass (*Poa pratensis*) are also used for forage. Around 23 000 t of legume seed are produced each year, lucerne accounting for the majority by weight (54%), followed by white clover (16%) (Table 5.1).

The main grass-seed-producing country in Europe is Denmark (Kley 1995, Burgon *et al.* 1997), with a 40% share of the total seed growing area (51 000 ha), followed by Germany (20%) and The Netherlands (Table 5.2). The total EU production is generally equivalent to the total amount required to satisfy demand. The UK production of grass seed, typically 10 000 t per annum, is approximately 8% of the total EU production and is insufficient to satisfy an annual demand of some 27 000 t.

The main production of legume seed in Europe is concentrated in the southern European countries: Italy (38%), France (28%) and Spain (18%). Lucerne seed accounts for a high proportion of the total legume seed produced. Seed production of white and red clover is mainly carried out in Denmark and Germany, with Denmark being the main producer of white clover seed within the EU. Relatively little seed of white clover (*c.* 40 t per annum) is produced within the UK apart from some locally produced types, e.g. Kent Wild White.

Table 5.1 Average annual grass and legume seed consumption and production (1994) in the European Union (source: Kley 1995).

Species	Consumption (tonnes)	Production (tonnes)
GRASSES		
Festuca pratensis	4500	4000
Lolium perenne	60000	54000
Lolium multiflorum	31000	25000
Poa pratense	2700	1000
Dactylis glomerata	3200	4300
Festuca arundinacea	2800	2500
Poa pratensis	4900	14000
Festuca rubra	14400	27000
Others	7500	5100
Total	131000	137000
LEGUMES		
Trifolium pratense	2900	2200
Trifolium repens	3600	2400
Trifolium incarnatum	1200	1500
Medicago sativa	12400	13000
Others	2900	1300
Total	23000	20400

Table 5.2 Herbage seed growing areas in the main seed-producing countries of the European Union (1994) (source: Kley 1995).

Grasses		Legumes	
Country	Hectares	Country	Hectares
Denmark	51000	Italy	21000
Germany	27000	France	16000
Netherlands	20000	Spain	10000
France	14000	Denmark	3500
UK	10000	Germany	2600
Others	5000	Greece	2100
Total	127000	Others	200
		Total	55400

This is due primarily to climatic conditions within the UK, which make it a relatively unreliable crop for the farmer, and seed production of UK-bred white clover varieties is carried out primarily in New Zealand and in Oregon, USA. However, the availability of varieties more amenable to seed production within the UK environment, allied to management systems specifically suited to these varieties, does offer the potential to increase UK seed production in the future.

New Zealand has long been a major producer and supplier of white clover seed to the UK, and the New Zealand-bred variety Grasslands Huia had a dominant role in the UK and many other world markets because of the favourable climatic conditions and ready seed supply (Lancashire 1990). Thus, although the EU probably produces enough legume seed to satisfy total demand, there are some important differences within some species and countries.

5.3 Herbage seed production in the UK

Herbage seed production is carried out primarily on arable farms where harvesting equipment for combinable crops is available. Climate is particularly important in determining the area best suited for seed production, since rainfall and its seasonal distribution affect plant growth, seed setting and seed harvesting. However, it is also important to meet the isolation requirements to comply with seed certification regulations (Kelly & Bowring 1990). This means that seed production of out-crossing forage species tends to be located away from grassland areas. The agronomic requirements for good quality seed production and high yields can be summarized as follows: moderate and well distributed rainfall during the vegetative growth period in spring and early summer, a fine spell of weather at the time of pollination followed by moderate rainfall during seed filling, and a dry and sunny period in July and August for ripening and harvesting of the seed. These conditions are met in the southern and eastern counties of England, and to a lesser extent in the Midlands and Welsh Border counties. The average annual rainfall in these areas ranges from 500 mm in the east to 750 mm in the western counties. Soil type is also critical, and conditions which promote excessive water retention, leading to excessive leaf growth at harvest, need to be avoided. Excessive soil moisture is a particular problem in seed crops of legumes such as white clover, leading to considerable variability in harvested seed yields, and therefore soils and climatic areas with high levels of moisture need to be avoided.

5.4 Seed crop management

Management of seed crops of forage species is not the same as that for forage production, as there is a requirement to maximize inflorescence production and the other components of seed yield. Harvested yields from grass and legume seed crops are often low or unpredictable, as a result of varieties having been bred for forage production and quality rather than seed production (Hides & Desroches 1990). Nevertheless, the ability of a variety to produce adequate seed yields is essential for its commercial success. Specific management systems have been developed to maximize seed production of different forage species. Management includes attention to detail at establishment, crop management in spring, weed control, fertilizer regime, harvesting, and the place of the seed crop in the farm rotation. Whilst it is outside the scope of this chapter to discuss the biology and management of seed crops in detail (the reader is referred to a specialist text, e.g. Fairey & Hampton 1997), some general comments follow.

5.4.1 Weed control

This can be a problem in herbage seed crops. In addition to the competition effect of weeds, some weed seeds cannot be removed during the cleaning process, and therefore the weeds themselves must be removed using suitable herbicides. This is becoming more difficult as the legislation on herbicide usage tightens, and there are fewer chemicals approved for use in herbage seed crops. Information on the herbicides approved for use in the UK is given in the UK Pesticide Guide, revised annually (e.g. PSD 1998). Alternative methods of weed control are also being investigated (Rolston *et al.* 1997).

5.4.2 Nutrient requirements

Research has identified the optimal levels of nutrients for seed production. Grass seed crops require nitrogen (N) to encourage tillering, and also lodging to reduce seed shed (though not before pollination has occurred). Applications of excessive levels of N can cause premature lodging and secondary tillering, which makes harvesting difficult. In forage legumes, phosphorus and potassium are the most important major nutrients, and micro-nutrients such as boron can also be important, particularly in ensuring satisfactory pollination.

5.4.3 Spring management

Seed crops may be cut or grazed in spring, depending upon the species. Seed crops of perennial ryegrass are generally neither cut nor grazed, as they do not have the inherent capacity for regrowth to produce good seed yields. Italian ryegrass seed crops can be cut for silage, although they should not be cut too late as that can reduce seed yields. Some hybrid ryegrasses ('Italian' or 'intermediate' types) can be cut for silage, but cutting of 'perennial' types has a detrimental effect on seed yield. Hence, Italian and hybrid ryegrasses are often grown for seed on farms where livestock can utilize high-quality silage and grazing from these species. Forage legumes such as white clover must be cut or grazed in spring to remove excessive leaf growth and to stimulate flowering. However, this should not be carried out later than the bud-emergence stage, otherwise seed yield is reduced.

5.4.4 Pollination

Grass seed crops are wind-pollinated, whereas legume species such as white and red clover are insect-pollinated. Seed producers may, therefore, place beehives on legume seed crops to ensure satisfactory pollination. Hence, there is a need to contain crops to avoid pollen contamination from adjoining crops.

5.4.5 Seed-crop harvesting

Herbage seed crops are generally grown as an arable crop, in a cereal rotation, by specialist growers who develop an understanding of the specialized techniques required for herbage seed production, and who have the equipment for handling the crop. This is particularly true of harvesting, which can be problematic as flowering and seed ripening in forage crops take place over a long period. At any one time the crop will contain seeds at several different stages of ripeness, and therefore include a proportion of seeds which are immature. Harvesting secures a mix of ripe and immature seeds, and many ripe seeds may already have fallen to the ground. Some loss is therefore inevitable, but the aim is to harvest when the number of mature seeds is greatest. The skill and experience of the seed producer, as well as the weather conditions, are important factors in determining the success of the harvesting operation. In grasses, this can be judged by the colour and consistency of the seeds, and the appearance of the inflorescences and stem, but is probably best decided on the basis of seed moisture content. In legumes, the colour of the inflorescence and stem is a good indicator of harvestability. It should be noted that if herbage seed does not meet the statutory requirements for germination it is worthless, and unlike arable seed, it has no alternative use as stock feed.

5.5 Seed certification

Seed of forage varieties can only be sold in the UK if the varieties are on the National List or the EU Common Catalogue. For this they must be distinct, uniform and stable (DUS), and thus can be distinguished from other varieties, and must also have acceptable agronomic value for cultivation and use (VCU). In order to ensure that seed sold to farmers is of high quality, systems have been adopted to provide some control over the seed industry. This involves a complex system of monitoring seed quality throughout the various stages of production, processing and marketing, through the OSTS (Official Seed Testing Station), and a certification authority to ensure and maintain genetic integrity. In England and Wales, seed certification is administered by the National Institute of Agricultural Botany, on behalf of the Ministry of Agriculture, Fisheries and Food (MAFF) who ensure varietal purity and authenticity. Details of the certification process are available (MAFF 1983), but some general comments on the principles used in the certification system are appropriate here. The system ensures that checks are made on the authenticity of the variety during seed multiplication. For example, when breeders' seed stock of a variety of perennial ryegrass is multiplied, a sample of seed is submitted to the NIAB, and samples of subsequent seed generations (pre-basic, basic and certified) are then requested by the NIAB and grown outside with the original sample to check that varietal purity and authenticity is maintained in subsequent generations of seed multiplication.

The certification system also regulates the seed multiplication process. Regulations limit the number of generations of seed multiplication, the number of seed harvests of perennial crops, and the years that must elapse before crops of the same species can be grown in the same field. Seed crops are also closely monitored during seed multiplication, and passed for harvest only if certain criteria are met. These relate specifically to the satisfactory isolation of the seed crop, which is particularly important for out-breeding species, and crop uniformity. Seed crops may be 'rogued' and any off-types removed to comply with the maximum number permissible per unit area. Crop inspection is supplemented by reference to a control plot. These regulations are different for the different stages of seed multiplication, but all seek to ensure varietal purity and authenticity. Bags of seed, certified as having acquired a satisfactory level of purity, germination and genetic integrity, are strictly labelled to ensure that the user knows exactly what is being bought. All these statutory regulations ensure that the labelled bag of seed passed to the end user contains high genetic quality seed of acceptable purity and germination.

5.6 References

Burgon A., Bondesen O.B., Verburgt W.H., Hall A.G., Bark N.S., Robinson M. & Timm G. (1997) The forage seed trade. In Fairey D.T. & Hampton J.G. (eds) *Forage Seed Production. Vol. 1. Temperate Species*. CAB International, Wallingford.

Camlin M.S. (1997) Grasses. In Weddell J.R. (ed) *Seeds of Progress*. British Grassland Society Occasional Symposium, No. 31, 2–14. BGS, Reading.

Fairey D.T. & Hampton J.G. (eds) (1997) *Forage Seed Production. Vol. 1. Temperate Species.* CAB International, Wallingford.

Hides D.H. & Desroches R. (1990) Role of seeds in forage production – factors limiting optimal utilization. *Proceedings of the XVI International Grassland Congress, Vol. III*, Nice, France, 1989, 1777–84.

Kelly A.F. & Bowring J.D.C. (1990) The development of seed certification in England and Wales. *Plant Varieties and Seeds* **3**, 139–50.

Kley G. (1995) Seed production in grass and clover species in Europe. *Proceedings of the Third International Herbage Seed Conference*, Germany, 12–22.

Lancashire J.A. (1990) 150 years of grassland development in New Zealand. *Proceedings of the New Zealand Grassland Association* **52**, 9–15.

MAFF (Ministry of Agriculture, Fisheries and Food) (1976) *Annual Returns of Stocks of Seeds in the Hands of Seedsmen and Disposals*. MAFF, Cambridge.

MAFF (Ministry of Agriculture, Fisheries and Food) (1983) *Certification of Seed of Grasses and Herbage Legumes*. MAFF, Cambridge.

MAFF (Ministry of Agriculture, Fisheries and Food) (1996) *Seed Traders Annual Return Summary. Year Ended 30 June 1996*. MAFF, Cambridge.

PSD (Pesticides Safety Directorate) (1998) *Pesticides 1998*. Stationery Office, London.

Rolston M.P., Rowarth J.S., Young W.C. & Mueller-Warrant G.W. (1997) Grass seed crop management. In Fairey D.T. & Hampton J.G. (eds) *Forage Seed Production. Vol. 1. Temperate Species.* CAB International, Wallingford.

Weddell J.R., Gilliland T.J. & McVittie J. (1997) Evaluation procedures: Past, present and future. In Weddell J.R. (ed) *Seeds of Progress*. British Grassland Society Occasional Symposium, No. 31, 202–20. BGS, Reading.

Chapter 6
Weeds, Pests and Diseases of Grassland

G.C. Lewis and A. Hopkins

6.1 Introduction

The aim of this chapter is to highlight the main weeds, pests and diseases of grassland, the extent of damage caused, and the control measures available, with particular reference to the British Isles. Emphasis is given to perennial ryegrass and white clover because they are the most widely sown forage grass and legume species in the British Isles. Also, most of the information about the effects of pests and diseases on productivity of grassland is in relation to these two species. Many of the weeds, pests and diseases described in this chapter occur elsewhere in Europe, and in other regions of temperate grassland throughout the world, although they may not have the same status as in the British Isles.

Research activity on weeds, pests and diseases in grassland in the British Isles was much reduced in the 1990s, partly reflecting concerns about the effects of agrochemicals on the environment, including non-target flora and fauna. Nevertheless, reports by the Pesticide Usage Survey Group (Thomas & Garthwaite 1994, Garthwaite *et al.* 1998) showed marked increases in the area of grassland treated with agrochemicals between 1989 and 1997. Therefore, the increasing area of grassland receiving agrochemicals has been matched by a decrease in research on the efficacy of treatments and the necessity for their use.

Loss of production or persistence due to weeds, pests and diseases is likely to be of greater significance, and control measures more economic, under intensive rather than extensive management. Losses often go unnoticed, because in farming practice it is difficult to compare the yields of damaged and undamaged grassland. Damage by some species is known to reduce yield, persistency or forage quality, but the effects of many others are uncertain and difficult to assess. Lenné (1989) and Price (1993) discussed some of the difficulties associated with the evaluation of the economic importance of diseases in grassland, and these apply equally to pests and weeds. One difficulty is the need to assess output in terms of animal production rather than the grass product itself. Another difficulty is the often diverse botanical composition of grassland, with a mixture of different species and cultivars. Monocultures of a single cultivar are likely to be more at risk from pests and diseases than mixtures of cultivars or species. Many studies on the effects of weeds, pests and diseases in grassland have been done using monocultures in order to eliminate the confounding effects of mixtures. In mixtures, any reduction in vigour of one component is likely to be exaggerated by competition from other components. In addition to the complexity of mixtures of plant species, there is that of interactions between species of weeds, pests and diseases in forage crops. Usually, there are more than one species present at any one time, but research has concentrated on individual species in isolation. Examples of interactions that have been reported in ryegrass are a synergistic effect between fungicide seed treatment and insecticide sprays to control frit fly (Lewis 1988), and a reduction in crown rust infection when plants were also infected with ryegrass mosaic virus (Latch & Potter 1977). A greater understanding is required of interactions between species of plants, invertebrates and micro-organisms in grassland.

Newly sown grassland is more vulnerable to weeds, pests and diseases than established grassland. Sowings following previous grassland are particularly prone to attack because of the carryover of weeds, pests and diseases. A main objective when establishing new grassland is to obtain an adequate population of sown species over the whole field. Many more seeds of grass and clover are sown per unit area than the number of plants that could be sustained on that area, and losses of seed and seedlings are greater than would be tolerated with sowings of arable crops (see Chapter 2).

Descriptions and illustrations of the wide range of weeds, pests and diseases of grassland in the British Isles can be found in the following publications. For seedling weeds, consult Hanf (1983). For pests, consult Edwards & Heath (1964) and Gratwick (1992). For fungal diseases, consult O'Rourke (1976) and Priestley *et al.* (1988).

6.2 Weeds of grassland

The definition of a 'weed' in grassland is not straightforward. Unlike the situation with arable or horticultural crops, unsown species in grassland can make some contribution to the yield and quality of the product (Barber 1985, Dibb 1985). The presence of broad-leaved (dicotyledonous) species in grass swards can extend the range of concentration of some major elements (Wilman & Derrick 1994). Also, some weeds are a food source or host for many species of insects and birds (see Chapter 11). Therefore, a case can be made for restricting weed control to the prevention or eradication of serious infestations in more productive swards. Control is required for some weeds that cause physical injury to grazing animals or are toxic to them when ingested. If herbicide treatment is required, spot treatment should be used where possible, or if the weeds are standing proud of the crop they can be treated separately by applying glyphosate with a wick 'wiper'. In the UK, herbicide use can be enforced when an infestation of a perennial weed listed as an 'injurious weed' under the Weeds Act 1959 is deemed to present a risk to neighbouring farms. Most of the perennial weeds described below are in this category.

When an area of grassland is sown, a choice is made of species or cultivars appropriate for the site and its intended use. To maintain the presence of the sown species, cultural and management practices are required to prevent the ingress of other species. Such practices should be continued through the life of the sward in order to avoid reducing the competitiveness of forage species and creating gaps that can be colonized by weeds. Cultural control of perennial weeds is based largely on maintaining a continuous cover of vigorously growing forage species to reduce sites of invasion by weeds. One management factor that can be utilized for weed control is grazing, and this factor, together with the concept of integrated weed management, has been reviewed by Popay & Field (1996). Deep ploughing will eliminate many weeds when established grassland is ploughed and resown, but some weeds, e.g. docks and thistles, can reappear from buried propagules. The potential for biological control of weeds (e.g. Frantzen 1994) has yet to be exploited in the British Isles.

6.2.1 Annual dicotyledonous weeds

Of the numerous species of annual dicotyledonous weeds found in grassland in the British Isles, chickweed (*Stellaria media*) has been considered of particular importance in new sowings (Haggar *et al.* 1985). Chickweed can form large patches that smother grass and clover seedlings (Haggar *et al.* 1985, Standell & Marshall 1991). Various herbicide treatments have been approved for annual weed control in grassland (Whitehead 1999), and often these are combinations of two or more chemicals in order to broaden the spectrum of activity. The choice of treatment depends on the particular weed species present, and whether legumes have been sown with the grass. Defoliation of the crop can reduce competition from chickweed and other annual weeds, but may be less effective on prostrate or rosette types such as knotgrass (*Polygonum aviculare*) and shepherd's purse (*Capsella bursa-pastoris*).

6.2.2 Docks

Docks, predominantly broad-leaved dock (*Rumex obtusifolius*) and curled dock (*R. crispus*), are widespread in sown grass leys and permanent grassland in the British Isles. They are arguably the most important weed problem affecting lowland grassland, particularly on dairy farms (Hopkins & Peel 1985). Frequently, the level of infestation is sufficient to reduce both forage production and herbage digestibility (Courtney 1985). Surveys have shown that grassland management affects the incidence of docks (Haggar 1980, Hopkins & Peel

1985). Infestations are encouraged by inputs of nitrogen and slurry that increase soil fertility, and by silage cutting and other management that reduces sward density. The main means of dock colonization in grassland is by seed. Mature plants can produce up to 60000 seeds per year, and seeds are capable of remaining viable for several years when buried in the soil (Cavers & Harper 1964). Seeds germinate rapidly after soil disturbance (Roberts & Totterdell 1981). Once established in a grass sward, individual dock plants persist and expand through the development of deep tap roots and a phalanx growth system, resulting in a dense population of clones (Pino *et al.* 1995).

Various herbicide treatments have been approved for dock control in the UK (Whitehead 1999), of which asulam and mecoprop have been most widely used. Non-chemical methods have been developed in line with policies to reduce pesticide use in agriculture and to meet the needs of organic farming systems. These methods depend on understanding and exploiting the relative competitiveness of docks within the sward, and include frequent defoliation and mechanical aeration of the soil (Courtney 1985, Foster 1989, Hopkins *et al.* 1997).

6.2.3 Thistles

Creeping thistle (*Cirsium arvense*) and, to a lesser extent, the spear thistle (*C. vulgare*) are widespread in grassland in the British Isles. The prickly leaves and stems reduce both grazing and utilization of herbage in their immediate vicinity (Oswald 1985). Thistles are associated particularly with fields that are never mown, and are grazed by sheep alone, sheep with cattle, or by horses, with a grazing management involving only lax defoliation in summer (Hopkins 1986). Also, high levels of phosphorus and potassium in the soil increase the risk of infestations. The herbicide clopyralid, alone or in a mixture, has been approved for thistle control in the UK (Whitehead 1999) and MCPA has been widely used for thistles and other perennial weeds. Mechanical control is by mowing, which weakens dock plants and prevents further seeding; this may be the only appropriate control in some situations, e.g. organic farms (Chapter 14) or conservation management (Chapter 11).

6.2.4 Ragwort

Common ragwort (*Senecio jacobaea*), which is widespread in the British Isles, and marsh ragwort (*S. aquaticus*), which is locally abundant, are amongst the most poisonous of grassland plant species. Their significance and control were reviewed by Forbes (1985). Few cases of livestock poisoning by ragwort are reported, but the risk is greatly increased when the plant is cut and dried, as in hay making. A common situation in which ragwort thrives is in the overgrazed open swards and areas of rejected herbage typical of many horse paddocks. Cattle-grazed fields, particularly when grazed at low densities, can also become infested, and there can be a considerable variability between years in the scale of the problem. The herbicides MCPA and 2, 4–D have been approved for ragwort control in the UK (Whitehead 1999). Cutting does not kill the plant, and there is a risk of cut foliage being consumed by grazing animals. On small areas, or where herbicide use is inappropriate, hand-pulling (with all ragwort herbage being removed and destroyed) is an alternative method, albeit labour-intensive.

6.2.5 Bracken

The bracken fern (*Pteridium aquilinum*) is the most important weed species affecting grazing land in upland and marginal areas of the British Isles. Indeed, it is one of the commonest and most widespread plants in the world. There are two main reasons for the status of bracken as a weed. Firstly, its aggressive growth and dense shading reduces the quality and availability of land for grazing, as well as for conservation and amenity interests. Secondly, it is toxic when ingested by livestock, with both acute and chronic effects being reported (Taylor 1989, Hopkins 1990). There are also possible human health implications arising from bracken toxins in milk in some parts of the world, and from inhalation of bracken spores (Taylor 1989). In the British Isles, the rapid spread of bracken by rhizomatous growth, particularly on brown-earth soils in upland areas, poses a serious land-use problem. Annual rates of spread of up to 3% per year have been reported for some localities (Taylor 1985). This spread is attributed to the decline in cattle, relative to sheep, in the uplands, and to the reduction of labour-intensive management practices, which historically would have contained bracken to an acceptable level. The bracken problem has been the subject of a wide-ranging international literature (e.g. Smith & Taylor 1986, 1995, Thomson & Smith 1990).

The herbicide asulam has been approved for bracken control in the UK (Whitehead 1999), including aerial application, which is the only suitable means for spraying large areas in the uplands. Asulam is most effective when applied in late July/early August in the UK, when the fronds are fully open but relatively unlignified (Lowday 1987). This treatment results in severe local damage to the fronds, but an integrated programme of follow-up management is required to prevent regeneration from the rhizomes.

6.2.6 Weed problems and their control in horse paddocks

The growth in popularity of keeping horses for recreation has lead to increased use of grassland for horse grazing. This is a significant and economically important land-use activity: in Britain it is estimated that over 550 000 horses and ponies are kept (Peat Marwick McLintock 1988). Horse pastures present special problems for weed management, and weedy swards on frequently over-stocked horse paddocks are a common site on the fringes of many towns and cities. Similar weed problems can also occur on farms where there are opportunities to integrate horse grazing with other types of grassland utilization, unless pasture management appropriate for horses is implemented. The development of infestations of weeds, including ragwort, thistles and docks, is associated with the highly selective grazing behaviour of the horse and its preference for separate grazing and latrine areas within the same paddock.

Control of the major problem weeds can be effected by the appropriate herbicidal and mechanical treatments described above, with localized infestations of weeds such as thistles and docks controlled using a knapsack sprayer or 'wick' wiper. However, there are sound reasons for avoiding more general use of herbicides. The ideal horse pasture is likely to contain a range of pasture species, including legumes and other dicotyledonous species, and botanically diverse swards have potential nutritional advantages and medicinal properties (Allison 1995). Old permanent pastures used as horse paddocks can also be important wildlife refugia (see Section 11.3.5). Pasture management is needed that prevents the development of rank herbage around latrine areas, and avoids overgrazing which opens the sward to ruderal and rosette-forming species (Cooper *et al.* 1981). This should include either a mixed grazing or a rotational management system, in conjunction with other grassland livestock or mowing, or if neither of these is possible, a rotational paddock system within the horse-grazed area. Particularly in those situations where horses are the only grazing animals, any tall herbage should be mown to a sward height of about 5–7 cm after each grazing to prevent the heading of weeds and tall grass. Harrowing at the end of each grazing season will help distribute dung patches over a wider area. Heavily grazed areas may also develop low pH and nutrient-deficient soils, resulting in thin swards and bare patches which provide opportunities for weed establishment. These should be corrected by liming or by applying fertilizers where most needed. However, recognizing that the horse keeper wants to avoid lush herbage because of the risk of laminitis, where fertilizer N is applied this should be in small frequent doses that allows herbage growth to be matched to grazing demand.

6.3 Pests of grassland

Pest damage to established grass has been reviewed by Curry (1994) and Clements & Cook (1996), with summaries of the extensive work in the UK on the yield response to pesticide treatments. The main pest species associated with responses to pesticide treatment are described below.

The extent of damage to white clover in England and Wales by pests (and diseases) has been assessed by Lewis & Thomas (1991) and Clements & Murray (1993). Damage was attributed mainly to slugs and Sitona weevils (see below for descriptions of these pests). Drawings of the characteristic damage to white clover leaves by different pests are given by Wiech & Clements (1992). Little is known of the effect of pest damage on white clover production. Experimental pesticide treatments applied to grass/clover swards have reduced pest populations and increased clover content (e.g. Mowat & Shakeel 1989a, Lewis 1991, Pederson *et al.* 1991). Control strategies are based largely on breeding for resistance. One approach is to exploit the cyanogenic compounds produced by white clover, which reduce feeding by some pests (Pederson & Brink 1998). Cultivars vary in their cyanogenic potential (Mowat and Shakeel 1989b). Another possible means of control is the incorporation of genes from the insect pathogen *Bacillus thuringiensis* into white clover (Voisey *et al.* 1994).

6.3.1 Frit fly larvae

Frit fly is the common name for a complex of Dipterous flies, including *Oscinella frit* (Mowat 1975). The larvae of these flies hatch from eggs laid on or near seedlings or mature grass plants and bore into the base of seedlings and tillers, causing death or greatly reduced vigour. Populations of larvae in new sowings can reach several thousand per m^2 (e.g. Clements *et al.* 1990) and exceed the number of seedlings present. Several factors that influence the risk of serious damage have been identified. Italian ryegrass seedlings are more susceptible to damage than those of perennial ryegrass, and although established crops are less affected, the relatively low tiller population of Italian ryegrass can result in significant damage (Clements *et al.* 1991). The nature of the previous crop is important because when grassland is ploughed and resown, larvae can migrate from buried plants to seedlings. The shorter the interval between ploughing and sowing, the greater the survival rate of larvae, and direct drilling poses the greatest risk. The population of adult frit flies is likely to be greater if the sowing is in a grassland rather than an arable area. Finally, frit flies lay eggs at certain times of the year, and seedlings emerging at these times are at particular risk of attack (Clements *et al.* 1990).

Yield responses to pesticides applied to control frit fly have been reviewed by Clements & Cook (1996). Chlorpyrifos has been approved for use in the UK (Whitehead 1999), and an integrated control system has been developed using low doses and parasitoids (Clear Hill *et al.* 1990). However, chlorpyrifos can reduce populations of carabid beetles, predators of many pest species (Asteraki *et al.* 1992). Resistance to frit fly has been reported in Italian ryegrass, which may be associated with the arrangement of silica bodies in the leaf sheath (Moore 1984). Cultural control methods include sowing outside the main periods of egg-laying and allowing a gap of 4 weeks or more between ploughing or sward destruction and sowing to avoid larval migration to seedlings.

6.3.2 Slugs

Slugs (particularly the grey field slug *Deroceras reticulatum*) feed on seeds, seedlings and plants of many species. Leaf tissue is rasped away in strips between the veins, and a key for assessing the area of white clover leaf removed was devised by Lewis & Thomas (1991). Slugs prefer white clover to grasses (Barker *et al.* 1983), and consequently white clover suffers greater seedling and yield losses than grass (Ferguson 1984, Cottam 1986). Cultivars of white clover vary in their susceptibility to slug damage (Glen *et al.* 1991). Seedling morphology and age, and the presence of seedlings of other plant species, may be factors in determining the extent of slug feeding (Glen *et al.* 1991, Hanley *et al.* 1995). White clover slot-seeded into grass swards is highly vulnerable to damage from slugs moving along the slots (Ferguson & Barratt 1983). Large populations of slugs are likely on heavy soils and where the soil is loosely tilled, providing crevices in which slugs can hide. On established white clover plants, slug feeding on leaves probably has little effect, but feeding on leaf buds, as observed by Barker (1989), may cause significant damage in spring, when growth commences.

The molluscides metaldehyde and methiocarb have been approved for slug control in grassland in the UK (Whitehead 1999). Use of methiocarb when slot-seeding white clover has increased the number of seedlings per drill row from 3 to 24 (Clements & Bentley 1983). Trampling of slugs by using very high stocking rates of sheep has reduced populations by 90% in New Zealand (Ferguson *et al.* 1988). Indications of cultivar resistance to slugs in white clover are available (NIAB 1998).

6.3.3 Leatherjackets

Leatherjackets (larvae of craneflies, *Tipula* spp.) are soil-dwelling and feed on many grass, legume and other plant species. There can be several hundred leatherjackets per m^2, particularly in long-established grassland. The species found in grassland are mainly *T. paludosa* and *T. oleracea*, and some indication of their relative occurrence has been reported for Scotland and Northern Ireland (Humphreys *et al.* 1993). Leatherjackets feed mainly on root tissue, but will consume leaf tissue where it is accessible. Damage is greatest in spring, when the larvae are feeding actively. Plants are severed at or just below soil level, causing patches of yellowing plants that later die. Factors influencing the size of leatherjacket populations have been studied (Blackshaw 1990, McCracken *et al.* 1995).

Chlorpyrifos and gamma HCH have been approved for use in the UK (Whitehead 1999). It is possible to estimate whether populations are sufficiently high to warrant treatment by using a technique described by Clements & Moore (1989). A linear relationship between response to chlorpyrifos and leatherjacket numbers, with a maximum response in excess of 1 t DM ha^{-1}, has been reported

(French *et al.* 1990). No effect of chlorpyrifos treatment to control leatherjackets was detected on grazing geese (Clements *et al.* 1992). Biological control of leatherjackets using *Bacillus thuringiensis* may have some potential (Evans & Keatinge 1996).

6.3.4 *Sitona* weevil

Sitona weevils (*Sitona* spp.) cause characteristic notching of leaf margins of clovers and lucerne. A key for assessing the area of leaf removed was devised by Lewis & Thomas (1991). Three species have been recorded in grassland in England, *S. lineatus*, *S. hispidulus* and *S. flavescens*. These species vary in their preferences for feeding on different leguminous plants (Murray & Clements 1994) and different white clover cultivars (Murray 1996). Populations of up to 373 adult weevils per m^2 have been recorded (Murray & Clements 1995), and weevils are more abundant during warm, dry periods, in contrast to slugs. Sowings in April and August are particularly susceptible. *Sitona* weevils can cause severe losses of seedlings through consumption of cotyledons and first leaves (Murray & Clements 1992). Also, weevil larvae feeding on root nodules reduce the growth of white clover seedlings, and this damage is a factor in the transfer of nitrogen to the grass component of a sward (Murray & Hatch 1994). Wounds caused by larval feeding can predispose legumes to various crown- and root-rotting fungi (Kalb *et al.* 1994). Removal of leaf tissue by adult weevils probably has little impact on plants with a substantial amount of leaf, but may impede the growth of plants in spring, when little leaf is present.

No insecticide treatments have been approved specifically for control of *Sitona* weevils in the UK, but when white clover is grown with grass, treatments used for the control of grass pests such as frit fly and leatherjackets are likely to reduce populations of adult weevils.

6.3.5 Nematodes

Various species of pathogenic nematode are found in grassland, sometimes in great numbers (Cook & Yeates 1993), but little is known of their effect on grass yield and persistence in the British Isles. More research has been done on certain nematode species infesting white clover, in particular stem nematode (*Ditylenchus dipsaci*). This pest causes stunting and swelling of stem bases, nodes and petioles in most forage legumes, and infestation typically forms patches of stunted plants. However, patches are difficult to detect when clover is grown in a mixture with grass. Different races of the nematode have different host ranges. The nematode can survive for many years in the absence of the host, especially on heavy soils, and can be a contaminant of seed. Stem nematode was detected in 41% of 56 grass/white clover swards examined in England and Wales (Cook *et al.* 1992a). Measures to control stem nematode in new sowings resulted in a three-fold increase in forage yield at the first cut (Cook *et al.* 1992b).

Cultivars of white clover resistant to stem nematode are being developed (Cook & Clifford 1997). Indications of resistance to stem nematodes in cultivars of red clover and lucerne are available (NIAB 1998). Cultural control measures include the avoidance of animal or machinery movement from old to new swards, and crop rotation, although breaks of eight or more years between clover crops may be necessary. Biological control by nematophagous fungi may have some potential (Hay & Bateson 1997).

6.4 Diseases of grassland

Studies of diseases in grassland in the British Isles have concentrated mainly on seedling or foliar pathogens. Knowledge of the impact of root pathogens is very limited.

6.4.1 Pre-emergence seedling death

Fusarium culmorum is particularly pathogenic to grass seedlings before emergence (Holmes 1983). The critical stage for infection is between seed germination and seedling emergence; factors that increase the length of time at this stage, such as insufficient soil moisture content or sowing too deeply, will increase the risk of seedling death (Lewis 1989). Little is known of the impact of seedling diseases in white clover and other forage legumes.

Seed treatment containing thiram has been approved for control of grass seedling diseases in the UK (Whitehead 1999). The use of experimental fungicide seed treatments has increased seedling emergence and early forage yield in small-scale experiments (Lewis 1991), but there is a lack of information about the responses to treatment in farm sowings in the British Isles.

6.4.2 Foliar fungal diseases of ryegrass

Fungal diseases are common on leaves of all forage grasses. In ryegrass, the consensus from results of studies on the incidence and severity of these diseases is that *Drechslera* leaf spot (*Drechslera* spp.), *Rhynchosporium* leaf blotch (*Rhynchosporium* spp.), crown rust (*Puccinia coronata*) and powdery mildew (*Blumeria graminis*) are the most important (Lam 1983a, Thomas 1991, Lewis 1992). Keys for the scoring of severity of disease have been developed (Lam 1983b, Thomas 1985, Birckenstaedt *et al.* 1994). Crown rust occurs mainly in early autumn, but the other three diseases are probably most important in spring, before the first cut for silage or hay. The longer the period before this first cut, the greater is the opportunity for disease spread. Thus, hay crops are more at risk than silage crops, but seed crops are at the greatest risk. Rust and mildew are essentially diseases of drier regions, whereas *Drechslera* and *Rhynchosporium* are more common in wetter ones. Higher levels of N fertilizer tend to increase foliar diseases, although reports have not been consistent.

Crown rust is the most visually striking of the foliar fungal diseases of grasses. The distinctive orange pustules on the leaves normally appear from late summer to early autumn. Epidemics occur during periods of warm, dry days, which favour spore dispersal, and dewy nights, which favour spore germination and infection of leaves. Infection of ryegrass reduces forage yield and quality and the competitive ability of plants (Potter 1987), and can render forage unpalatable to livestock, although this is not well documented. Powdery mildew attacks most grasses and is apparent on the leaves as a greyish white superficial fungal growth. Grass yield and quality can be reduced (Thomas 1991), and infection is most severe in dense crops and following dry periods. *Drechslera* leaf spot is common throughout the year on ryegrass and fescues. Five species of *Drechslera* have been reported on ryegrass in England and Wales, and it is difficult to distinguish between them (Lam 1985a), although isozyme analysis has proved successful (Burhenne *et al.* 1994). *Drechslera* is typically a disease of high incidence but low severity, although the brown spots vary in size and frequency depending on the species of host and fungus, and on environmental conditions. Infection can reduce forage quality (Lam 1985b) and has been associated with yield responses to fungicide treatment in the spring (Thomas 1991). Infection can be seed-borne (Labruyère 1977). *Rhynchosporium* leaf blotch causes large brown lesions with lighter coloured centres in ryegrasses and cocksfoot. Infection is most severe in Italian ryegrass and during cool, moist weather in spring and autumn, and it can reduce forage quality (Lam 1985b).

Data for the reduction in forage yield and quality caused by foliar fungal diseases have been obtained largely from comparisons between swards with and without fungicide treatment. Often it has proved difficult to relate the response to fungicides to the level of diseases present. The fungicide propiconazole has been approved in the UK for control of all four aforementioned diseases (Whitehead 1999), and reductions in infection have been reported (Lewis *et al.* 1996). Also, triadimefon has been approved for control of rust, mildew and *Rhynchosporium* (Whitehead 1999). Breeding for resistance to diseases in ryegrass in the UK has concentrated on the foliar fungal diseases, and indications of resistance in cultivars are available (NIAB 1998). The relationship between ratings for resistance to foliar diseases and the degree of infection is described by Thomas (1997). In the case of crown rust, a continued effort in breeding is required to counter the appearance of new strains of the fungus able to break down plant resistance (Roderick 1997). Management of the crop can have a large influence on disease levels and can be manipulated to reduce the effects of disease. Defoliation removes much of the disease inoculum and reduces the moist microclimate within a tall standing crop that favours disease spread.

6.4.3 Leaf-spot diseases of white clover

Several species of fungi cause leaf spots of white clover, including black (or sooty) blotch (*Cymadothea trifolii*), *Pseudopeziza* leaf-spot (*Pseudopeziza trifolii*) and pepper spot (*Leptosphaerulina trifolii*) (Lewis & Thomas 1991). Black blotch appears as large, circular, shiny black lesions on the underside of leaves. *Pseudopeziza* appears as large brown lesions with a star-shaped margin on the upper surface of leaves. Mature lesions produce yellowish, glistening, dish-shaped fruiting bodies in the centre. Pepper spot appears as small, abundant brown lesions on both leaf surfaces, and on the petiole. A technique for assessing the area of white clover leaf covered by lesions was described by Skipp & Lambert (1984). *Pseudopeziza* and pepper spot increase the oestrogenic activity of white clover (Saba *et al.* 1972), which may affect the reproductive performance of grazing animals. No control measures have been established for leaf spot diseases in the UK.

6.4.4 Clover rot

Clover rot *(Sclerotinia trifoliorum)* is a serious disease of red clover and lucerne, but although forage yield of white clover can be greatly reduced by clover rot in small plots (Scott & Evans 1980), the extent of damage in the field situation is not known. Infection begins in the autumn as necrotic spots on the leaves, and during mild weather spreads within the stems into the crown, causing plant death. Sclerotia develop in diseased tissue, which fall onto the soil and germinate in the following year to produce spores and continue the infection cycle. If conditions are unsuitable for germination, sclerotia can survive in the soil for many years before germinating. Infection can be introduced to new sowings by infected seed or seed contaminated by sclerotia.

There are no fungicide treatments approved for the control of clover rot in the UK, and breeding for resistance is the best option (Cook & Clifford 1997). Indications of resistance in cultivars to clover rot in red and white clover are available (NIAB 1998). Cultural control by strategic defoliation of clover has been suggested (Pratt 1991). Crop rotation will reduce the risk from infection by germinating sclerotia, but the longevity of sclerotia in soil necessitates an interval of eight or more years.

6.4.5 Virus diseases

In contrast to the foliar fungal diseases, virus diseases are systemic, and infected plants remain infected even when the plant is defoliated. In fact, cutting and grazing can spread some virus diseases. In Europe, 26 viruses infecting grass species have been identified (Huth 1994), but only the two described below are economically important in the British Isles. Although the detection of viruses in plants has become more rapid and sensitive with the introduction of serological and molecular methods (e.g. Webster *et al.* 1996), there is still a lack of information on the extent of virus infection in grassland.

Ryegrass mosaic virus (RgMV)

Surveys have shown RgMV to be common in ryegrass crops in UK (Heard *et al.* 1974, Cooper & McDowell 1978, Holmes, 1980), but no recent information is available. RgMV causes greater damage in Italian than perennial ryegrass (Catherall 1987). Symptoms of RgMV infection are light-green streaks in the leaves, that may become necrotic; in severe cases the whole plant turns brown. Infection can spread rapidly in Italian ryegrass (Heard *et al.* 1974) and yield can be reduced severely (Holmes 1980). Infection is introduced to new sowings by the wind-dispersed mite vector *Abacarus hystrix*, and mite populations increase rapidly in autumn. Colonies of several hundred mites have been recorded on a single leaf (Lewis & Heard 1981). The importance of sap transmission in the spread of infection within crops is uncertain, but mowing machinery does not appear to be an effective vector (Heard & Chapman 1986).

Cultivars of Italian ryegrass with improved resistance to RgMV have been developed, and indications of resistance in cultivars are available (NIAB 1998). The introduction and subsequent spread of RgMV may be reduced by sowing in autumn rather than spring, to delay ingress of the mite, and by defoliating in autumn to curtail the increase in mite populations.

Barley yellow dwarf virus (BYDV)

BYDV infects many grass species in southwest England (Kendall *et al.* 1996) and probably elsewhere in the British Isles. BYDV causes greater damage in perennial than in Italian ryegrass, in contrast to RMV (Catherall 1987). Three strains of BYDV occur in England and Scotland (Dempster & Holmes 1995, Kendall *et al.* 1996). Almost all perennial ryegrass crops examined in England, Wales and Scotland were infected (Doodson 1967, Holmes 1985, Dempster & Holmes 1995). Infection, which is spread by several species of aphid, is mostly symptomless and its impact on grass production in the British Isles has yet to be ascertained. However, the potential for loss of yield and persistence has been demonstrated (Eagling *et al.* 1991). BYDV is an important disease of cereal crops, and grass areas in the vicinity can constitute a reservoir of viruliferous aphids, although the risk of cross infection may be small (Dempster & Holmes 1995).

Control of BYDV is likely to depend on the development of resistant cultivars (Eagling *et al.* 1991), but none are available in the UK at present. Insecticides used to control aphids in grassland could reduce the spread of BYDV infection, but the likely impact is unknown.

White clover viruses

Of the various viruses that infect white clover, white clover mosaic virus (WClMV) and clover yellow vein virus (ClYVV) were the most frequently detected at 49 upland sites in the UK and France (Scott & Hughes 1980). Over 50% of plants examined contained these two viruses. However, the results may have overestimated the level of infection (Potter 1993), and very little infection was detected in a later survey (Lewis & Thomas 1991). Both viruses reduce forage yield (Campbell & Moyer 1983, Potter 1993, Dudas *et al.* 1998), and WCMV also reduces nodulation (Guy *et al.* 1980). Both viruses are sap-transmitted and ClYVV is also aphid-transmitted. Little is known of the effects of virus diseases in white clover in the British Isles, but viruses are associated with loss of production and lack of persistence in the USA (McLaughlin *et al.* 1992), Australia (McKirdy & Jones 1997) and New Zealand (Dudas *et al.* 1998).

Control of viruses in white clover is likely to depend on the development of resistant cultivars, but none are available in the UK at present. In the USA, virus-resistant white clover has yielded substantially more than virus-susceptible cultivars (Taylor *et al.* 1995).

6.5 Fungal endophytes

Many species of temperate grasses are infected by non-pathogenic fungi of the genus *Neotyphodium* (formerly *Acremonium*). These fungi are known as 'endophytes' because their entire life cycle is spent within the plant. *Neotyphodium* endophytes have a symbiotic association with their grass host, although these endophytes are descended from the genus *Epichloë* (Wilkinson & Schardl 1997), which contains species that cause 'choke' disease in grasses. The significance of the *Neotyphodium*/grass association to agriculture is largely as a result of the alkaloid compounds that are produced in infected plants (Powell & Petroski 1992). The main groups of alkaloids produced are ergopeptines (e.g. ergovaline), indole diterpenes (e.g. lolitrem B), pyrrolopyrazines (e.g. peramine) and pyrrolozidines (e.g. N-formyl and N-acetyl loline). Ergovaline and lolitrem B are associated with toxicity to grazing animals, and peramine and the lolines are associated with deterrence of feeding by insects (Latch 1997). However, there is some overlap in effects, and it is difficult to define individual alkaloids as beneficial or detrimental. The type and amount of alkaloid produced varies seasonally and with the grass and fungus genotype (e.g. Roylance *et al.* 1994). Endophyte infection (primarily *N. lolii*) in perennial ryegrass is widespread in the British Isles and elsewhere in Europe (Lewis *et al.* 1997), and cases of toxicity to animals have been reported (Lewis 1997). Some beneficial effects of infection have been reported (Lewis 1994), but the main pests of ryegrass in the British Isles, frit fly and leatherjackets, appear to be unaffected by the presence of endophyte (Lewis & Clements 1986, Lewis & Vaughan 1997).

6.6 Nutritional disorders

An imbalance of nutrients in grasses and legumes can cause symptoms that resemble disease. Descriptions and illustrations of some of these symptoms are given by Bergmann (1992).

6.7 References

Allison K. (1995) *A Guide to Herbs for Horses*, 48 pp. J.A. Allen, London.

Asteraki E.J., Hanks C.A. & Clements R.O. (1992) The impact of two insecticides on predatory ground beetles (Carabidae) in newly sown grass. *Annals of Applied Biology* **120**, 25–39.

Barber W.P. (1985) The nutritional value of common weeds. In Brockman J.S. (ed) *Weeds, Pests and Diseases of Grasses and Herbage Legumes*. British Grassland Society Occasional Symposium, No. 18, 104–11. BGS, Hurley.

Barker G.M. (1989) Slug problems in New Zealand pastoral agriculture. In Henderson I. (ed) *Slugs and Snails in World Agriculture.* British Crop Protection Council Monograph No. 41, 59–68.

Barker G.M., Willoughby B.E. & Pottinger R.P. (1983) Feeding by slugs on some pasture plant species: Laboratory experiments. *Proceedings of 36th New Zealand Weed and Pest Control Conference*, 207–11.

Bergmann W. (ed) (1992) *Nutritional Disorders of Plants: Development, Visual and Analytical Diagnosis*, 741 pp. Gustav Fischer, Jena.

Birckenstaedt E., Eickel P. & Paul V.H. (1994) Scoring of grass diseases for the evaluation of varieties. *International Conference on Harmful and Beneficial Microorganisms in Grassland, Pastures and Turf, Paderborn, Germany*, 193–200.

Blackshaw R.P. (1990) Observations on the distribution of leatherjackets in Northern Ireland. *Annals of Applied Biology* **116**, 21–6.

Burhenne S., Hein D., Paul V.H. & Kettrup A. (1994) Rapid identification of gramineous *Drechslera* species by isozyme analysis. *International Conference on Harmful and Beneficial Microorganisms in Grassland, Pastures and Turf, Paderborn, Germany*, 83–91.

Campbell C.L. & Moyer J.W. (1983) Effects of clover yellow vein virus and *Codinaea fertilis* on growth of white clover. *Plant Disease* **67**, 70–3.

Catherall P.L. (1987) Effects of barley yellow dwarf and ryegrass mosaic viruses alone and in combination on the productivity of perennial and Italian ryegrasses. *Plant Pathology* **36**, 73–8.

Cavers P.B. & Harper J.L. (1964) Biological flora of the British Isles: *Rumex obtusifolius* L. and *R. crispus* L. *Journal of Ecology* **52**, 737–66.

Clear Hill B.H., van Emden H.F. & Clements R.O. (1990) Control of frit fly (*Oscinella* spp.) in newly sown grass using a combination of low doses of pesticide, resistant grass cultivars and indigenous parasitoids. *Crop Protection* **9**, 97–100.

Clements R.O. & Bentley B.R. (1983) The effect of three pesticide treatments on the establishment of white clover (*Trifolium repens*) sown with a slot-seeder. *Crop Protection* **2**, 375–8.

Clements R.O. & Cook R. (1996) Pest damage to established grass in the UK. *Agricultural Zoology Reviews* **7**, 157–79.

Clements R.O. & Moore D. (1989) A rapid field sampling kit for leatherjackets. *Proceedings of the 5th Australasian Conference on Grassland Invertebrate Ecology, Melbourne*, 341–5.

Clements R.O. & Murray P.J. (1993) *Sitona* damage to clover in the UK. *Proceedings of 6th Australasian Conference on Grassland Invertebrate Ecology, Hamilton, New Zealand*, 260–4.

Clements R.O., Bentley B.R. & Jackson C.A. (1990) Influence of date of sowing on frit-fly damage to newly sown Italian ryegrass. *Crop Protection* **9**, 101–4.

Clements R.O., Bentley B.R., Murray P.J. & Henderson I.F. (1991) Improvement of persistence and yield of Italian ryegrass (*Lolium multiflorum*) through pesticide use. *Annals of Applied Biology* **119**, 513–19.

Clements R.O., Murray P.J. & Tyas C. (1992) The short-term effect on wild geese behaviour of chlorpyrifos application to permanent pasture. *Annals of Applied Biology* **120**, 17–23.

Cook R. & Clifford B.C. (1997) Clover resistance to stem nematode and *Sclerotinia*. *Seeds of Progress*: British Grassland Society Occasional Symposium, No. 31, 77–8. BGS, Reading.

Cook R. & Yeates G.W. (1993) Nematode pests of grassland and forage crops. In Evans K., Trudgill D.L. & Webster J.M. (eds) *Plant Parasitic Nematodes in Temperate Agriculture*, 305–50. CAB International, Wallingford.

Cook R., Evans D.R., Williams T.A. & Mizen K.A. (1992a) The effect of stem nematode on establishment and early yields of white clover. *Annals of Applied Biology* **120**, 83–94.

Cook R., Mizen K.A., Plowright R.A. & York P.A. (1992b) Observations on the incidence of plant parasitic nematodes in grassland in England and Wales. *Grass and Forage Science* **47**, 274–9.

Cooper J.P, Green J.O. & Haggar R.J. (1981) *The Management of Horse Paddocks – A Booklet of Instructions*, 9 pp. Horserace Betting Levy Board, London.

Cooper P. & McDowell J.B. (1978) Ryegrass mosaic virus in Northern Ireland. *Record of Agricultural Research, Northern Ireland* **26**, 1–6.

Cottam D.A. (1986) The effects of slug-grazing on *Trifolium repens* and *Dactylis glomerata* in monoculture and mixed sward. *Oikos* **47**, 275–9.

Courtney A.D. (1985) Impact and control of docks in grassland. In Brockman J.S. (ed) *Weeds, Pests and Diseases of Grasses and Herbage Legumes*. British Grassland Society Occasional Symposium, No. 18, 120–7. BGS, Hurley.

Curry J.P. (1994) *Grassland Invertebrates*, 437 pp. Chapman & Hall, London.

Dempster L.C. & Holmes S.J.I. (1995) The incidence of strains of barley yellow dwarf virus in perennial ryegrass crops in south-west and central Scotland. *Plant Pathology* **44**, 710–17.

Dibb C. (1985) Problems and benefits of grass weeds. In Brockman, J.S. (ed) *Weeds, Pests and Diseases of Grasses and Herbage Legumes*. British Grassland Society Occasional Symposium, No. 18, 112–19. BGS, Hurley.

Doodson J.K. (1967) A survey of barley yellow dwarf virus in S.24 perennial ryegrass in England and Wales, 1966. *Plant Pathology* **16**, 42–5.

Dudas B., Woodfield D.R., Tong P.M., Nicholls M.F., Cousins G.R., Burgess R., White D.W.R., Beck D.L., Lough T.J. & Forster R.L.S. (1998) Estimating the agronomic impact of white clover mosaic virus on white clover performance in the North Island of New Zealand. *New Zealand Journal of Agricultural Research* **41**, 171–8.

Eagling D.R., Halloran G.M. & Sward R.J. (1991) Barley yellow dwarf luteovirus in pasture grasses. *Review of Plant Pathology* **70**, 133–43.

Edwards C.A. & Heath G.W. (1964) *The Principles of Agricultural Entomology*. 418 pp. Chapman & Hall, London.

Evans K.A. & Keatinge R. (1996) The control of leatherjackets in grassland using *Bacillus thuringiensis*. *Proceedings of Crop Protection in Northern Britain 1996*, 231–6.

Ferguson C.M. (1984) Slug feeding on seeds and seedlings of ryegrass and white clover. *Proceedings of 37th New Zealand Weed and Pest Control Conference*, 64–7.

Ferguson C.M. & Barratt B.I.P. (1983) Slug damage to pasture renovated by direct drilling. *Proceedings of 36th New Zealand Weed and Pest Control Conference*, 212–15.

Ferguson C.M., Barratt B.I.P. & Jones P.A. (1988) Control of the grey field slug (*Deroceras reticulatum* (Muller)) by stock management prior to direct-drilled pasture establishment. *Journal of Agricultural Science, Cambridge* **111**, 443–9.

Forbes J.C. (1985) The impact and control of ragwort (*Senecio jacobaea* and *S. aquaticus*) in grassland. In Brockman J.S. (ed) *Weeds, Pests and Diseases of Grasses and Herbage Legumes*. British Grassland Society Occasional Symposium, No. 18, 147–54. BGS, Hurley.

Foster L. (1989) The biology and non-chemical control of dock species *Rumex obtusifolius* and *R. crispus. Biological Agriculture and Horticulture* **6,** 11–25.

Frantzen J. (1994) An epidemiological study of *Puccinia punctiformis* (Str) Rohl as a stepping-stone to the biological control of *Cirsium arvense* (L.) Scop. *New Phytologist* **127**, 147–54.

French N., Nichols D.B.R. & Wright A.J. (1990) Yield response of improved upland pasture to the control of leatherjackets under increasing rates of nitrogen. *Grass and Forage Science* **45**, 99–102.

Garthwaite D.G., Thomas M.R., Banham A.R. & De'ath A. (1998) Grassland and fodder crops in Great Britain 1997. *Pesticide Usage Survey Report 151.* MAFF, London.

Glen D.M., Cuerden R. & Butler R.C. (1991) Impact of the field slug *Deroceras reticulatum* on establishment of ryegrass and white clover in mixed swards. *Annals of Applied Biology* **119**, 155–62.

Gratwick M. (1992) *Crop Pests in the UK: Collected Edition of MAFF Leaflets*, 490 pp. Chapman & Hall, London.

Guy P., Gibbs A. & Harrower K. (1980) The effect of white clover mosaic virus on nodulation of white clover (*Trifolium repens* L. cv. Ladino). *Australian Journal of Agricultural Research* **31**, 307–11.

Haggar R.J. (1980) Survey on the incidence of docks (*Rumex* spp.) in grassland in 10 districts in the United Kingdom in 1972. *ADAS Quarterly Review* **39**, 256–70.

Haggar R.J., Standell C.J. & Birnie J.E. (1985) Occurrence, impact and control of weeds in newly sown leys. In Brockman J.S. (ed) *Weeds, Pests and Diseases of Grasses and Herbage Legumes*. British Grassland Society Occasional Symposium, No. 18, 11–19. BGS, Hurley.

Hanf M. (1983) *The Arable Weeds of Europe, with their Seedlings and Seeds*, BASF, 494 pp. Germany.

Hanley M.E., Fenner M. & Edwards P.J. (1995) The effect of seedling age on the likelihood of herbivory by the slug *Deroceras reticulatum. Functional Ecology* **9**, 754–9.

Hay F.S. & Bateson L. (1997) Effect of the nematophagous fungi *Hirsutella rhossiliensis* and *Verticillium balanoides* on stem nematode (*Ditylenchus dipsaci*) in white clover. *Australasian Plant Pathology* **26**, 142–7.

Heard A.J. & Chapman P.F. (1986) A field study of the pattern of local spread of ryegrass mosaic virus in mown grassland. *Annals of Applied Biology* **108**, 341–5.

Heard A.J., A'Brook J., Roberts E.T. & Cook R.J. (1974) The incidence of ryegrass mosaic virus in crops of ryegrass grown for seed in some southern counties of England. *Plant Pathology* **23**, 119–27.

Holmes S.J.I. (1980) Field studies on the effects of ryegrass mosaic infection on the yield of Italian ryegrass cv. S22. *Annals of Applied Biology* **96**, 209–17.

Holmes S.J.I. (1983) The susceptibility of agricultural grasses to pre-emergence damage caused by *Fusarium culmorum* and its control by fungicide seed treatment. *Grass and Forage Science* **38**, 209–14.

Holmes S.J.I. (1985) Barley yellow dwarf virus in ryegrass and its detection by ELISA. *Plant Pathology* **34**, 214–20.

Hopkins A. (1986) Botanical composition of permanent grassland in England and Wales in relation to soil, environment and management factors. *Grass and Forage Science* **41**, 237–46.
Hopkins A. (1990) Bracken (*Pteridium aquilinum*): Its distribution and animal health implications. *British Veterinary Journal* **146**, 316–26.
Hopkins A. & Peel S. (1985) Incidence of weeds in permanent grassland. In Brockman J.S. (ed) *Weeds, Pests and Diseases of Grasses and Herbage Legumes*. British Grassland Society Occasional Symposium, No. 18, 93–103. BGS, Hurley.
Hopkins A., Jones E.L., Bowling P.J. & Johnson R.H. (1997) Cultural methods of dock control in permanent pasture. *Proceedings British Grassland Society Fifth Research Conference, Newton Abbot*. 39–40, BGS, Reading.
Humphreys I.C., Blackshaw R.P., Stewart R.M. & Coll C. (1993) Differentiation between larvae of *Tipula paludosa* and *Tipula oleracea* (Diptera, Tipulidae) using isoelectric focusing, and their occurrence in grassland in northern Britain. *Annals of Applied Biology* **122**, 1–8.
Huth W. (1994) Significance of virus infections on pasture and turf grasses in Central Europe. *International Conference on Harmful and Beneficial Microorganisms in Grassland, Pastures and Turf, Paderborn, Germany*, 21–6.
Kalb D.W., Bergstrom G.C. & Shields E.J. (1994) Prevalence, severity, and association of fungal crown and root rots with injury by the clover root curculio in New York alfalfa. *Plant Disease* **78**, 491–5.
Kendall D.A., George S. & Smith B. (1996) Occurrence of barley yellow dwarf viruses in some common grasses (Graminae) in south west England. *Plant Pathology* **45**, 29–37.
Labruyère R.E. (1977) Contamination of ryegrass seed with *Drechslera* species and its effect on disease incidence in the ensuing crop. *Netherlands Journal of Plant Pathology* **83**, 205–15.
Lam A. (1983a) A survey of ryegrass swards for the presence of fungal pathogens in relation to management. *Grass and Forage Science* **38**, 55–65.
Lam A. (1983b) Leaf area guides for assessing fungal leaf diseases of ryegrasses (*Lolium* spp.). *Plant Pathology* **32**, 213–15.
Lam A. (1985a) *Drechslera andersenii* sp. nov. and other *Drechslera* spp. on ryegrass in England and Wales. *Transactions of the British Mycological Society* **85**, 595–602.
Lam A. (1985b) Effect of fungal pathogens on digestibility and chemical composition of Italian ryegrass (*Lolium multiflorum*) and tall fescue (*Festuca arundinacea*). *Plant Pathology* **34**, 190–9.
Latch G.C.M. (1997) An overview of *Neotyphodium*–grass interactions. In Bacon C.W & Hill N.S. (eds) *Neotyphodium/Grass Interactions*, 1–11. Plenum, New York.
Latch G.C.M. & Potter L.R. (1977) Interaction between crown rust (*Puccinia coronata*) and two viruses of ryegrass. *Annals of Applied Biology* **87**, 139–45.
Lenné J.M. (1989) Problems associated with evaluation of diseases of perennial pasture plants – some recommendations. *Proceedings of XVI International Grassland Congress, Nice, France*, 695–6.
Lewis G.C. (1988) Improvements to newly sown ryegrass by use of combined fungicide and insecticide treatment. *Crop Protection* **7**, 34–8.
Lewis G.C. (1989) Factors affecting the effects of fungicide treatment on seedling emergence of perennial ryegrass (*Lolium perenne*). *Grass and Forage Science* **44**, 417–22.
Lewis G.C. (1991) Interactions between pesticide treatment, cutting frequency and rate of fertiliser on white clover grown in mixture with perennial ryegrass. *Grass and Forage Science* **46**, 399–403.
Lewis G.C. (1992) Foliar fungal diseases of perennial ryegrass at 16 sites in England and Wales. *Crop Protection* **11**, 35–8.
Lewis G.C. (1994) Incidence of infection of grasses by endophytic fungi in the UK, and effects of infection on animal health, pest and disease damage and plant growth.

International Conference on Harmful and Beneficial Microorganisms in Grassland, Pastures and Turf, Paderborn, Germany, 161–7.
Lewis G.C. (1997) Significance of endophyte toxicosis and current practices for dealing with the problem in Europe. In Bacon C.W. & Hill N.S. (eds) *Neotyphodium/Grass Interactions*, 377–82. Plenum, New York.
Lewis G.C. & Clements R.O. (1986) A survey of ryegrass endophyte (*Acremonium lolii*) in the UK and its apparent ineffectuality on a seedling pest. *Journal of Agricultural Science* **107**, 633–8.
Lewis G.C. & Heard A.J. (1981) The incidence in ryegrass (*Lolium* spp.) crops of eriophyid mites, vectors of ryegrass mosaic virus. *Proceedings of 3rd Conference on Virus Diseases of Graminae in Europe, Rothamsted Experimental Station, Hertfordshire, 1980*, 89–98.
Lewis G.C. & Thomas B.J. (1991) Incidence and severity of pest and disease damage to white clover foliage at 16 sites in England and Wales. *Annals of Applied Biology* **118**, 1–8.
Lewis G.C. & Vaughan B. (1997) Evaluation of a fungal endophyte (*Neotyphodium lolii*) for control of leatherjackets (*Tipula* spp.) in perennial ryegrass. *Tests of Agrochemicals and Cultivars No. 18 (Annals of Applied Biology,* ***130****, Supplement)*, 34–5.
Lewis G.C., Lavender R.H. & Martyn T.M. (1996) The effect of propiconazole on foliar fungal diseases, herbage yield and quality of perennial ryegrass. *Crop Protection* **15**, 91–5.
Lewis G.C., Ravel C., Naffa W., Astier C. & Charmet G. (1997) Occurrence of *Acremonium* endophytes of wild populations of *Lolium* spp. in European countries and a relationship between level of infection and climate in France. *Annals of Applied Biology* **130**, 227–38.
Lowday J.E. (1987) The effects of cutting and asulam on numbers of frond buds and biomass of fronds and rhizomes of bracken *Pteridium aquilinum*. *Annals of Applied Biology* **110,** 175–84.
McCracken D.I., Foster G.N. & Kelly A. (1995) Factors affecting the size of leatherjacket (Diptera: Tulipidae) populations in pastures in the west of Scotland. *Applied Soil Ecology* **2**, 203–13.
McKirdy S.J. & Jones R.A.C. (1997) Further studies on the incidence of virus infection in white clover pastures. *Australian Journal of Agricultural Research* **48**, 31–7.
McLaughlin M.R., Pederson G.A., Evans R.R. & Ivy R.L. (1992) Virus diseases and stand decline in a white clover pasture. *Plant Disease* **76**, 158–62.
Moore D. (1984) The role of silica in protecting Italian ryegrass (*Lolium multiflorum*) from attack by stem-boring larvae (e.g. *Oscinella frit*). *Annals of Applied Biology* **104**, 161–6.
Mowat D.J. (1975) The dipterous shoot-flies inhabiting grassland in Northern Ireland. *Record of Agricultural Research, Northern Ireland* **23**, 11–17.
Mowat D.J. & Shakeel M.A. (1989a) The effect of some invertebrates species on persistence of white clover in ryegrass swards. *Grass and Forage Science* **44**, 117–24.
Mowat D.J. & Shakeel M.A. (1989b) The effect of different cultivars of clover on numbers of, and leaf damage by, some invertebrate species. *Grass and Forage Science* **44**, 11–18.
Murray P.J. (1996) *The biology of Sitona weevils in forage legumes*. PhD Thesis, University of Bristol, 171 pp.
Murray P.J. & Clements R.O. (1992) Studies on the feeding of *Sitona lineatus* L. (Coleoptera: Curculionidae) on white clover (*Trifolium repens* L.) seedlings. *Annals of Applied Biology* **121**, 233–8.
Murray P.J. & Clements R.O. (1994) Investigations of the host feeding preferences of *Sitona* weevils found commonly on white clover (*Trifolium repens*) seedlings. *Entomologia Experimentalis et Applicata* **71**, 73–9.

Murray P.J. & Clements R.O. (1995) Distribution and abundance of three species of *Sitona* (Coleoptera, Curculionidae) in grassland in England. *Annals of Applied Biology* **127**, 229–37.

Murray P.J. & Hatch D.J. (1994) *Sitona* weevils (Coleoptera: Curculionidae) as agents for rapid transfer of nitrogen from white clover (*Trifolium repens* L.) to perennial ryegrass (*Lolium perenne* L.). *Annals of Applied Biology* **125**, 29–33.

NIAB (National Institute of Agricultural Botany) (1998) *NIAB Recommended Lists of Grasses and Herbage Legumes 1998/99: Grasses and Herbage Legumes Variety Leaflet*, 48 pp. NIAB, Cambridge.

O'Rourke C.J. (1976) *Diseases of Grasses and Forage Legumes in Ireland.* An Foras Taluntais, Oak Park Research Centre, Carlow, Ireland, 115 pp.

Oswald A. (1985) Impact and control of thistles in grassland. In Brockman J.S. (ed) *Weeds, Pests and Diseases of Grasses and Herbage Legumes*. British Grassland Society Occasional Symposium, No. 18, 128–36. BGS, Hurley.

Peat Marwick McLintock (1988) *The Economic Contribution of the British Equine Industry*, 54 pp. British Horse Society, Stoneleigh.

Pederson G.A. & Brink G.E. (1998) Cyanogenesis effect on insect damage to seedling white clover in a Bermudagrass sod. *Agronomy Journal* **90**, 208–10.

Pederson G.A., Windham G.L., Ellsbury M.M., McLaughlin M.R., Pratt R.G. & Brink G.E. (1991) White clover yield and persistence as influenced by cypermethrin, benomyl, and root-knot nematode. *Crop Science* **31**, 1297–302.

Pino J., Haggar R.J., Sans F.X., Masalles R.M. & Sackville-Hamilton R.N. (1995) Clonal growth and fragment regeneration of *Rumex obtusifolius* L. *Weed Research* **35**, 141–8.

Popay I. & Field R. (1996) Grazing animals as weed control agents. *Weed Technology* **10**, 217–31.

Potter L.R. (1987) Effect of crown rust on regrowth, competitive ability and nutritional quality of perennial and Italian ryegrasses. *Plant Pathology* **36**, 455–61.

Potter L.R. (1993) The effects of white clover mosaic virus on vegetative growth and yield of clones of S.100 white clover. *Plant Pathology* **42**, 797–805.

Powell R.G. & Petroski R.J. (1992) Alkaloid toxins in endophyte-infected grasses. *Natural Toxins* **1**, 163–70.

Pratt R.G. (1991) Evaluation of foliar clipping treatments for cultural control of *Sclerotinia* crown and stem rot in crimson clover. *Plant Disease* **75**, 59–62.

Price T.V. (1993) Problems and progress in quantifying the losses due to pasture grass diseases. In Delfosse E.S. (ed) *Pests of Pastures: Weed, Invertebrate and Disease Pests of Australian Sheep Pastures*, 80–84. CSIRO Information Services, Melbourne.

Priestley R.H., Thomas J.E. & Sweet J.B. (1988) *Diseases of Grasses and Herbage Legumes*, 37 pp. NIAB, Cambridge.

Roberts E.H. & Totterdell S. (1981) Seed dormancy in *Rumex* species in response to environmental factors. *Plant, Cell and Environment* **4**, 97–106.

Roderick H.W. (1997) Ryegrass resistance to crown rust. In Weddell J.R. (ed) *Seeds of Progress.* British Grassland Society Occasional Symposium, No. 31, 75–6. BGS, Reading.

Roylance J.T., Hill N.S. & Agee C.S. (1994) Ergovaline and peramine production in endophyte-infected tall fescue: Independent regulation and effects of plant and endophyte genotype. *Journal of Chemical Ecology* **20**, 2171–83.

Saba N., Drane H.M., Herbert C.N., Newton J.E. & Betts J.E. (1972) Effect of disease on the oestrogenic activity and coumestrol content of white clover and lucerne. *Journal of Agricultural Science* **78**, 471–5.

Scott S.W. & Evans D.R. (1980) *Sclerotinia trifoliorum* Erikss. on white clover (*Trifolium repens* L.). *Grass and Forage Science* **35**, 159–63.

Scott S.W. & Hughes S. (1980) Survey of viruses in upland white clovers. *Annual Report of the Welsh Plant Breeding Station, 1979.* 173–4. WPBS, Aberystwyth.

Skipp R.A. & Lambert M.G. (1984) Damage to white clover foliage in grazed pastures caused by fungi and other organisms. *New Zealand Journal of Agricultural Research* **27**, 313–20.

Smith R.T. & Taylor J.A. (eds) (1986) *Bracken: Ecology, Land Use and Control Technology*, 464 pp. Parthenon, Carnforth.

Smith R.T. & Taylor J.A. (eds) (1995) *Bracken: An Environmental Issue.* International Bracken Group, Special Publication No. 2, 227 pp.

Standell C.J. & Marshall E.J.P. (1991) Weeds in newly sown grassland. *Strategies for Weed, Disease and Pest Control in Grassland: Proceedings of British Grassland Society Conference, Gloucester*, 4.1–4.11.

Taylor J.A. (1985) Bracken encroachment rates in Britain. *Soil Use and Management* **1**, 53–60.

Taylor J.A. (1989) *Bracken Toxicity and Carcinogenicity as Related to Animal and Human Health.* International Bracken Group, Special Publication, 79 pp.

Taylor N.L., Ghabriel S.A., Pederson G.A. & McLaughlin M.R. (1995) Quantification of yield benefits from incorporation of virus-resistant white clover germ plasm into grass–legume systems. *Plant Disease* **79**, 1057–61.

Thomas J.E. (1991) Diseases of established grassland. *Strategies for Weed, Disease and Pest Control in Grassland: Proceedings of British Grassland Society Conference, Gloucester*, 3.1–3.12. BGS, Hurley.

Thomas J.E. (1997) The significance of disease as a limiting factor to variety performance potential. In Weddell J.R. (ed) *Seeds of Progress.* British Grassland Society Occasional Symposium, No. 31, 114–23. BGS, Reading.

Thomas M.R. (1985) Assessment of disease in grassland. In Brockman J.S. (ed) *Weeds, Pests and Diseases of Grasses and Herbage Legumes*. British Grassland Society Occasional Symposium, No. 18, 188–94. BGS, Hurley.

Thomas M.R. and Garthwaite D.G. (1994) Grassland and fodder crops in Great Britain 1993. *Pesticide Usage Survey Report 119*, MAFF, London.

Thomson J.A. & Smith R.T. (eds) (1990) *Bracken Biology and Management.* Australian Institute of Agricultural Science, Sydney, Occasional Publication No. 40, 341 pp.

Voisey C.R., White D.W.R., Wigley P.J., Chilcott C.N., McGregor P.G. & Woodfield D.R. (1994) Release of transgenic white clover plants expressing *Bacillus thuringiensis* genes – an ecological pespective. *Biocontrol Science and Technology* **4**, 475–81.

Webster D.E., Forster R.L.S., Sinclair L. & Guy P.L. (1996) Distribution of ryegrass mosaic virus in New Zealand perennial ryegrass pastures as determined by ELISA and RT-PCR. *New Zealand Journal of Agricultural Research* **39**, 405–12.

Weich K. & Clements R.O. (1992) Studies on the *Sitona* spp. and *Apion* spp. weevils feeding on white clover foliage at a site in SE England. *Journal of Applied Entomology* **113**, 437–40.

Whitehead R. (ed) (1999) *The UK Pesticide Guide*. 736 pp. CAB International/British Crop Protection Council, Wallingford.

Wilkinson H.H. & Schardl C.L. (1997) The evolution of mutualism in grass–endophyte associations. In Bacon C.W. & Hill N.S. (eds) *Neotyphodium/Grass Interactions*, 13–25. Plenum, New York.

Wilman D. & Derrick R.W. (1994) Concentration and availability to sheep of N, P, K, Ca, Mg and Na in chickweed, dandelion, dock, ribwort and spurrey, compared with perennial ryegrass. *Journal of Agricultural Science, Cambridge* **122**, 217–23.

Chapter 7
The Feeding Value of Grass and Grass Products

D.E. Beever, N. Offer and M. Gill

7.1 Introduction

The feeding value of a forage, defined as its capacity to promote animal production, depends upon its ability to supply nutrients to the animal. It has three main components: the amount of forage the animal will eat (voluntary intake), the content of nutrients in the forage (nutrient content) and the ability of the animal to absorb and utilize the nutrients (nutrient availability). The ruminant's ability to extract nutrients from forages is mainly dependent on digestive processes carried out by microbes resident in the reticulo-rumen. This highly active continuous fermentation vessel contains around 10^{10} bacteria and 10^6 protozoa per ml of contents and is equipped with a range of sophisticated control systems to regulate factors including temperature, osmotic pressure, pH, substrate and waste product concentration, and to achieve mixing and physical processing of the food. Microbial digestion in the rumen allows the ruminant to utilize β-linked polysaccharides, such as cellulose and hemicellulose, which provide most of the energy in forages (Hungate 1966). Thus, feeding value is not solely a feed characteristic, but depends on a complex three-way interaction between the ruminant animal, its feed and the microbial population of its rumen. This chapter covers the impact of microbial digestion on the nutrients absorbed by the ruminant and their subsequent metabolism by the tissues, and also considers the chemical composition and nutritive value of forages, the prediction of animal performance on forage-based diets and the use of supplements to overcome nutrient limitations.

7.2 Chemical and physical characteristics of forage

The growth characteristics of grasses and legumes, with particular reference to their distinct phases of vegetative and reproductive development, have been highlighted in Chapter 3. The importance of factors such as forage species and variety, water, light and nutrient availability, and other environmental effects have been emphasized. The factors that influence forage growth are dynamic, and will change significantly with time. Consequently the chemical and physical characteristics of forages are influenced by the stage and rate of growth achieved by the crop, as well as previous management of the sward. Because of this extensive variation it is not possible to provide a complete assessment of forage composition, but some of the major features, changes and differences will be highlighted.

7.2.1 Cell wall constituents

The classical partition of feedstuffs into crude protein, crude fibre, ether extract, ash and nitrogen-free extractives did not facilitate an adequate assessment of feeding value. Attempts to provide a better chemical characterization of forages led to the concept of a division of forage DM into *cell contents* and *cell walls*, and on this basis, Van Soest & Wine (1967) established the neutral detergent fibre (NDF) technique. In this fractionation, the cell walls are considered to consist of pectic substances, the structural polysaccharides (hemicellulose and cellulose) and lignin. With advancing maturity, the concentrations of cellulose, hemicellulose and lignin in grass increase, whilst the proportion of forage organic matter digested in the whole digestive tract (OMD) decreases. For a particular forage at varying stages of maturity, there is usually a highly significant inverse relationship between NDF content and OMD, and between lignin content and OMD. However, the equations which describe these relationships rarely apply across different forages, because of between-forage differences in the detailed structure of the cell wall material. Of particular importance is the nature and development of linkages between lignin and the polysaccharides, which have a profound effect on the rate and extent of digestion of the forage. For example, compared with grasses at comparable growth stages, legumes are characterized by reduced amounts of cell walls with lower hemicellulose levels, but higher concentrations of pectic substances and lignin (see Fig. 7.2), and this in part may be related to their increased rate of breakdown in the rumen.

7.2.2 Cell contents

The cell contents comprise the cell nucleus and cytoplasm, and account for a major part of the proteins, peptides, nucleic acids, lipids, sugars and starches contained in the whole plant. In the early stages of growth, cell contents may represent at least two-thirds of forage DM, with protein being a major component. As forages mature, cell contents, as a proportion of plant DM, decrease as cell wall constituents increase. This is generally accompanied by a pronounced decline in protein concentration, due in part to an increased stem:leaf ratio, and in some cases, increased concentrations of sugars (glucose, fructose, sucrose, fructans). However, the sugar fraction of grasses and other forages is highly labile, and the amounts present in the plant at any stage of growth depend on prevailing environmental conditions, especially light and temperature. Relative to grasses, legumes contain higher proportions of crude protein, organic acids and minerals, but lower proportions of water-soluble carbohydrates. In both grasses and legumes, the content of starch or α-linked glucose polymer rarely exceeds 3–4% of the DM. A schematic representation of the effects of growth stage on grass composition is given in Fig. 7.1.

Plant breeding programmes have led to improvements in the crude protein content of forages, but protein quality is equally important. Generally, true proteins (high molecular weight polypeptides) account for over 80% of the forage crude protein, with fraction 1 protein, the principal photosynthetic protein (enzyme), being most common. During periods of impaired growth, however, non-protein nitrogen components (nitrates, amines and amides) may increase markedly. These are of lower nutritional value since they are rapidly and extensively degraded in the rumen (see Section 7.3.5), and ultimately contribute little in terms of absorbed amino acids. In particular, autumn growth may contain significant amounts of nitrate N, with true protein comprising only about 50% of total crude protein. Increased nitrogen fertilization will promote the synthesis of crude protein, but if this is associated with an increased yield of DM, then

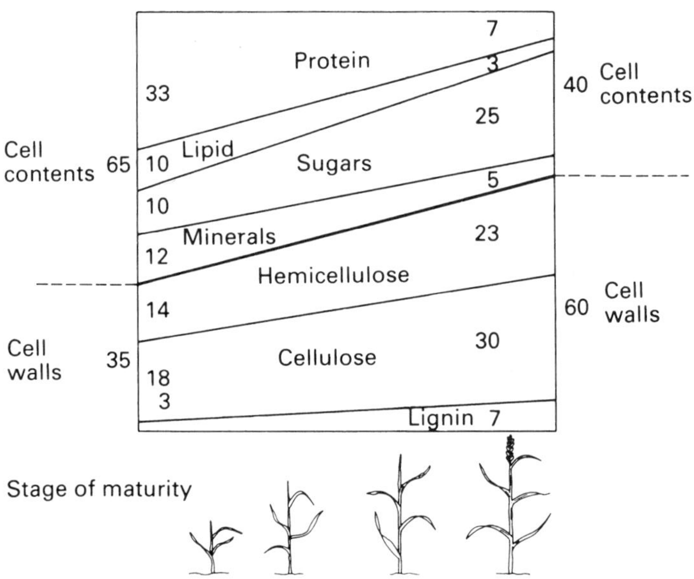

Fig. 7.1 Schematic representation of the effect of maturity on the chemical composition of grasses. As the grass plant matures, the proportion of cell walls and their constituent fractions increases and the cell content fraction decreases. The proportion of protein, lipid and mineral matter in the DM decreases, but the non-structural carbohydrates (mainly fructans, in the stem, stem base and inflorescence) increase. True protein and amino acids vary between about 90% of total crude protein in the young plant to 70% in the mature plant, but maturity has little effect on the amino acid composition of either grasses or legumes.

the change in crude protein content may be minimal. Equally, irrigation has been shown to have only marginal effects on chemical composition, whilst periods of high light intensity and temperature will increase the synthesis of both water-soluble carbohydrates and cell walls, but unfortunately often lead to increased cell wall lignification and reduced digestibility.

7.2.3 Plant structure

The developmental stage of the plant (i.e. vegetative or reproductive phase) will affect the proportions of leaf to stem in the plant, and will have a major effect on chemical composition. It is known to be affected by forage species, variety, environment and, in particular, previous management of the sward. For example, at all stages of growth, Italian ryegrass (*Lolium multiflorum*) has a higher proportion of leaf to stem than other grass species, whilst timothy (*Phleum pratense*), meadow fescue (*Festuca pratensis*) and cocksfoot (*Dactylis glomerata*) are characterized by higher cell wall contents. The prevailing environment will influence the rate of development of leaves and stems. Regrowths usually have a higher proportion of leaf, although this is influenced by the stage at which the primary growth was harvested, and whether or not the inflorescence primordia were removed.

Thus, whilst forage species and variety can affect the chemical and physical characteristics of the crop, many of the changes which occur are due to alterations in the stem:leaf ratio of the crop as influenced by environment. Hence, management decisions regarding harvesting of the crop, either by grazing or cutting, have a major bearing on the quality of forage harvested and also on the annual yield of digestible nutrients obtained from forage.

7.3 Nutritive value of forage

All animals have a minimum (maintenance) requirement for nutrients to maintain essential processes and prevent loss of body weight. Additional nutrients are required by the growing animal to synthesize muscle, adipose tissue and bone, and by the lactating animal to synthesize fat, protein and lactose in milk. While the major requirement is for energy-yielding nutrients, there are specific requirements for amino acids, glucose, fatty acids, minerals and vitamins which depend upon the stage of development, size, type and level of production of the animal.

The description of chemical composition of forages in Section 7.2 characterized the constituents of a forage, but gave little information about their availability to the animal. Methods of characterizing the availability of nutrients for fermentation in the rumen, overall digestibility and efficiency of utilization of nutrients are considered in the following sub-sections.

7.3.1 Digestibility in the whole digestive tract

The first assessments of the ability of forages to supply nutrients were made by measuring the losses of feed components across the whole digestive tract, which were then assumed to represent the nutrients available to the animal. For energy-yielding nutrients and minerals, this approach remains the fundamental method for assessing nutritive value, although modern nutritional systems require additional detailed information on digestion in various sections of the gut.

The digestibility coefficient of a nutrient is the proportion of that nutrient eaten that disappears across the whole gut, e.g. for organic matter (OM), the digestibility coefficient (OMD) is defined as

$$\text{OMD} = (\text{OM consumed} - \text{OM in faeces})/\text{OM consumed}$$

This value can either be expressed as a proportion, e.g. 0.75, or a percentage, e.g. 75%. However, digestibility coefficients calculated in this way are considered to be 'apparent' values only, since no allowance is made for the contribution of undigested microbes and endogenous secretions in faeces. These inevitable metabolic faecal losses do not arise directly from the feed, although they are to some extent affected by diet. Equations to estimate metabolic faecal losses have been proposed and can be used to correct apparent to true digestibilities (e.g. Hogan & Weston 1970). However, since these calculations are themselves approximations, apparent digestibility values are most frequently quoted.

Organic matter digestibility is a better estimate of nutritive value than DM digestibility, as it is not affected by the ash content of the forage, which can be quite variable, usually due to soil contamination. Also, OMD is considered to be more closely correlated with energy availability. Since intakes are normally given in terms of DM per day, an often-quoted parameter for calculation of total digestible organic matter intake is digestible organic matter in the DM (DOMD or D-value), expressed as

$$\text{DOMD (\%)} = ((\text{OM consumed} - \text{OM in faeces})/\text{DM consumed}) \times 100$$

However, the use of OM fails to allow for the different gross energy (GE) values of feed components. Whilst sugars, starch, cellulose and hemicellulose have GE values (MJ kg^{-1} DM) of approximately 16–18, values for protein (24–26) and oils (36–39) are higher. The use of digestible energy (DE) overcomes this problem:

$$\text{DE}(\text{MJ kg}^{-1}\,\text{DM}) = (\text{GE eaten} - \text{GE in faeces})/\text{DM consumed}$$

7.3.2 Measurement and prediction of digestibility

Reference measurements of digestibility comprise *in vivo* balance experiments involving total collection of faeces from animals eating measured amounts of feed (usually close to maintenance requirements) over a 7–10-day period (Cammell 1977), with subsequent analysis of the feeds and collected faeces. This method is time-consuming and not appropriate for routine evaluations, but its use is essential for the creation of reference populations of forages of known digestibility which can then be used to calibrate and validate indirect methods for predicting digestibility. Some prediction methods attempt to simulate digestion in the rumen, including the *in vitro* procedure of Tilley & Terry (1963) which uses live rumen microbes, or the cellulase method of Jones & Hayward (1975) which uses fungal enzymes. Both methods also involve incubation with pepsin, although this is used as a preliminary treatment in the latter case. Other predictions of digestibility are based on relationships with levels of cell wall constituents such as acid detergent fibre. These methods have the advantage that laboratory measurements are easier and cheaper to undertake, but the predictions are often less accurate than the *in vitro* or cellulase methods and less robust in terms of their ability to evaluate different forage species and varieties. Examples of prediction equations are shown below.

Fresh grass (primary growth) (Givens *et al.* 1990)

DOMD = 29.9 + 0.571 NCD	$R^2 = 0.77$
DOMD = 12.0 + 0.857 IVD	$R^2 = 0.67$
DOMD = 967 – 0.950 MADF	$R^2 = 0.67$

High-temperature dried grass (Givens *et al.* 1992)

DOMD = 59.9 + 0.828 NCD	$R^2 = 0.90$
DOMD = 27.3 + 0.998 IVD	$R^2 = 0.77$
DOMD = 1012 – 1.243 MADF	$R^2 = 0.73$

Grass hays (field-cured) (Moss & Givens 1990)

DOMD = 27.6 + 0.559 NCD	$R^2 = 0.49$
DOMD = 17.2 + 0.710 IVD	$R^2 = 0.83$
DOMD = 1023 – 1.222 MADF	$R^2 = 0.67$

Grass silage (Givens *et al.* 1989)

DOMD = 32.1 + 0.560 NCD	$R^2 = 0.59$
DOMD = 10.0 + 0.870 IVD	$R^2 = 0.74$
DOMD = 996 – 1.04 MADF	$R^2 = 0.56$

where NCD is the digestibility (%) of neutral detergent fibre measured using the cellulase method as described by MAFF (1993), IVD is DOMD (%) measured using the *in vitro* method of Tilley & Terry (1963), and MADF is the modified acid detergent fibre (mainly cellulose and lignin, $g\,kg^{-1}$ DM) measured using the method of Clancy & Wilson (1966). R^2 describes the proportion of the variation in the calibration population explained by the prediction.

Near infrared spectroscopy (NIRS) has replaced many of the prediction methods described above, especially in laboratories where large numbers of

Table 7.1 Regression statistics for the prediction of *in vivo* OMD of grass silages (Barber *et al.* 1990).

	Method				
	MADF	LIGA	PCOMD	IVOMD	NIRS
Calibration n = *122*					
R^2	0.34	0.52	0.55	0.74	0.85
SEC	5.1	4.4	4.2	3.2	2.5
Validation n = *48*					
R^2	0.20	0.14	0.40	0.64	0.76
SEP	5.1	5.3	4.7	3.6	2.6
Bias	–0.59	1.18	2.33	–1.85	–0.79

MADF, modified acid detergent fibre; LIGA, acetyl bromide lignin; PCOMD, pepsin–cellulase OMD (Jones & Hayward 1975); IVOMD, *in vitro* OMD (Tilley & Terry 1963); NIRS, near infrared diffuse reflectance spectroscopy; SEC, standard error of calibration; SEP, standard error of prediction.

forage samples are received (Offer *et al.* 1996). It is a physical method which depends on measurements of the absorption of light by the surface of the sample at individual wavelengths in the infrared region of the spectrum (1100–2500 nm). The absorption (or reflectance) spectrum depends on the chemical bonding within the forage, and is affected by factors such as cell wall content and composition. In principle, derivation of NIRS-based predictions is no different from the methods used for the original 'wet chemistry' techniques involving calibration against reference (*in vivo*) values, followed by validation or 'blind-testing' on a separate population of reference samples. NIRS has proved to be as accurate or better than other indirect prediction techniques (Table 7.1), whilst being substantially quicker and more cost-effective.

7.3.3 Factors affecting digestibility of forages

Digestibility decreases with advancing maturity as cell wall content increases, and thus, for a particular forage, D-value can be predicted from a knowledge of stage of growth. However, a constant relationship between cell wall content and digestibility does not hold across species, since legumes tend to have a lower proportion of more lignified (and less digestible) cell wall than grasses of the same digestibility (Fig. 7.2). For the grass it can be estimated that about 58% of the total digestible energy (DE) is derived from cell walls (cellulose and hemicellulose), but only 35% for white clover, for which protein and oil (cell contents) become relatively important.

Typical patterns of decline in D-value of primary growth as the season progresses are 3–5 units per week, as shown in Fig. 7.3. The rate of decline of subsequent vegetative growth is lower, being approximately 1.5–2.0 units D-value per week.

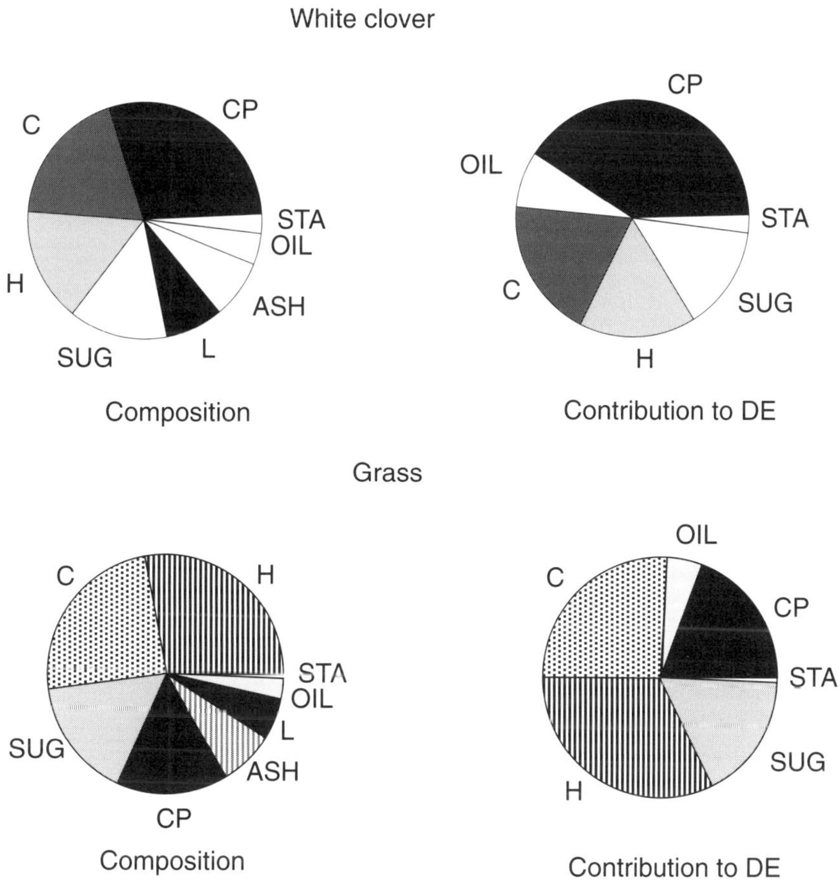

Fig. 7.2 Composition and contribution to digestible energy (DE) for highly digestible (70D) grass and white clover. C, cellulose; H, hemicellulose; L, lignin; SUG, sugar; STA, starch; CP, crude protein.

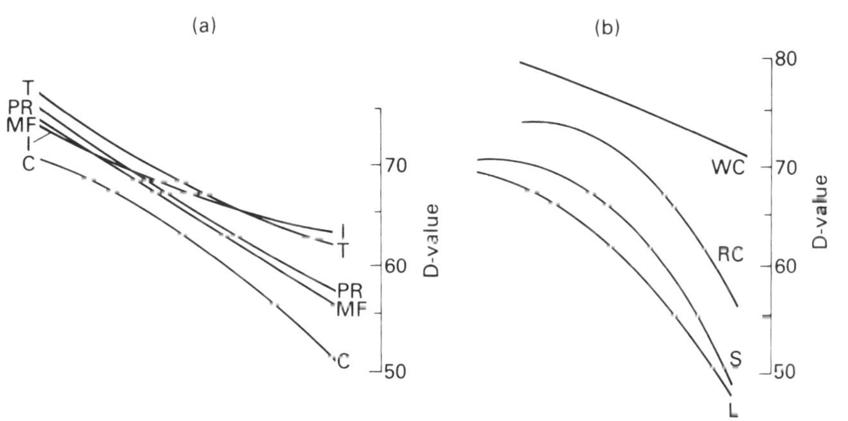

Fig. 7.3 Changes in the D-value of (a) grass species (Italian ryegrass (I); timothy (T); perennial ryegrass (PR); meadow fescue (MF); cocksfoot (C)) and (b) legume species (white clover (WC); red clover (RC); lucerne (L); sainfoin (S)) during primary growth.

Cocksfoot is characterized by a lower initial D-value than the other grass species, and this difference is maintained as the season progresses. With the other grasses, initial values for Italian ryegrass are lower than for meadow fescue, timothy and perennial ryegrass, but along with timothy, this forage has a slower rate of decline and eventually assumes a value closer to that for timothy. As for legumes, white clover has the highest initial D-value and, with a slow rate of decline (0.8 units per week), it is superior to all other legumes at all times. In contrast, D-values for red clover, sainfoin and lucerne decline at 2.5, 2.5 and 2.8 units per week, respectively, so that differences between the legumes are accentuated as the season progresses. In the tropics, changes in D-value during the growing season are less marked, but average digestibility of tropical forages is generally lower than for those grown in temperate regions, largely because tropical cell walls tend to contain more lignin.

7.3.4 Metabolizable energy value

The digestibility measurements described in Section 7.3.2 provide a useful estimate of the total energy available for absorption, since faecal losses represent the largest and most variable component of energy loss in the conversion of ingested energy to absorbed energy. Measurements on 244 fresh grass samples showed a mean faecal energy/forage GE ratio of 0.26 (s.d. 0.08), with a range of 0.09–0.48 (MAFF 1990). However, not all the digestible energy is available to the tissues, since in ruminants, energy is lost from the rumen as methane produced during microbial fermentation. Feed energy values for ruminants are therefore described as metabolizable energy (ME), defined as digestible energy content minus the losses of energy in methane and urine, as depicted in Fig. 7.4. Whilst urinary energy losses are included in the definition and thus estimation of ME, it must be recognized that this was largely based on the relative ease with which they can be quantified, for in reality they represent

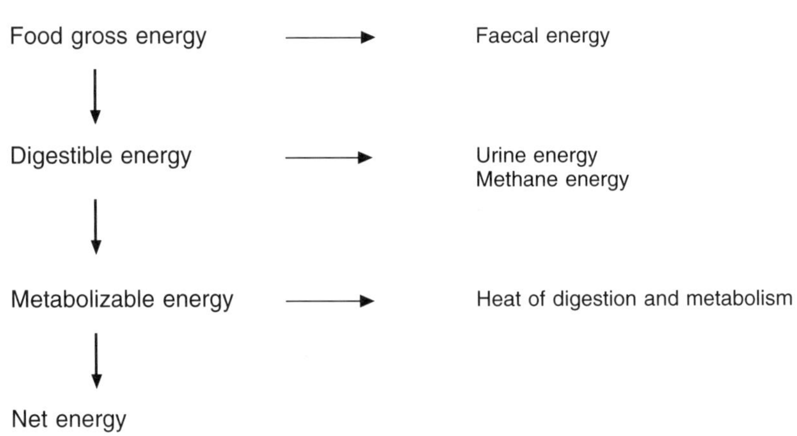

Fig. 7.4 The partition of energy during digestion and metabolism according to the ME system.

losses as a consequence of post-absorptive nutrient metabolism. A full description of the UK ME system as currently used is given in Alderman & Cottrill (1993).

In many situations, the combined losses of energy as methane and urine vary little between 16 and 20% of digestible energy content, and thus ME content is frequently assumed to equal to 0.81 × DE content. Values for 243 measurements on all species of grass showed a range of ME/DE of 0.74–0.88 (s.d. 0.03) (MAFF 1990), and similar values have been found for grass silages. However, the conversion factor for high-protein clovers may be as low as 0.77 because of higher energy losses in urine arising from the higher protein intake. Furthermore, there is mounting evidence that ME/DE may be influenced by the production level of the animal and the diet being fed, where many values measured for dairy cows are in the range 0.84–0.88 (Beever *et al.* 1991, Sutton *et al.* 1991, Yan *et al.* 1997).

Measurements of ME *in vivo* are more difficult than measuring digestibility because total collections of urine and methane are also required, and this is particularly the case with dairy cows at high levels of intake. Consequently, most ME values used to formulate rations in advisory work are based on indirect predictions of OMD or DOMD because reference populations of forages of measured ME are insufficient for calibration purposes. Conversion of DOMD (%) to ME ($MJ\,kg^{-1}\,DM$) involves assumption of the GE value of digestible organic matter (DOM) as well as the extent of losses in urine and methane. For fresh and dried grass and hay, a factor of 0.15 (i.e. ME = 0.15 DOMD) is appropriate, but for fermented forages or high protein-containing legumes the factor is considerably higher because of the higher GE contents of these forages. The mean measured conversion factor for grass silages is 0.168, although in practice a lower value of 0.16 is used to allow for losses of volatiles between sampling and consumption. However, the use of a single constant value could be a significant source of error in ME prediction. It has been shown that with individual silages the conversion value may vary between 0.13 and 0.21, mainly as a result of the variable GE content of silages arising principally from differences in the contents of volatile fermentation products (Givens *et al.* 1989). This issue was demonstrated by Cammell *et al.* (1998), who evaluated maize silage harvested at two stages of maturity. Analysis of the silages by eight different laboratories indicated in each instance that the later harvested crop had the highest ME content. However, *in vivo* measurement of ME content with lactating dairy cows revealed a higher value with the earlier cut crop, a difference which was wholly attributable to the higher GE content, due to the more extensive fermentation which this crop had undergone during the ensiling process. This issue may be worthy of further consideration if reliable predictions of ME content are to be achieved.

As indicated, ME is a measure of the feed energy available for metabolism but, because heat is produced as a result of digestion and metabolism of feed, not all the ME can be used for maintenance or productive purposes. The efficiency with which ME is utilized by the tissues depends on the process for which the energy is used. Separate equations to predict efficiency values (NE/ME) from the metabolizability (q_m = ME/GE) of diets have been provided by Alderman & Cottrill (1993) for maintenance (k_m), growth and fattening (k_f) and lactation (k_l). Values for k_f are affected most by diet metabolizability, and are substantially reduced on forages of low digestibility and for forages compared with concentrate feeds. This effect, which has a major impact on the feeding value of forages, is discussed in more detail in Section 7.6.1. In contrast, whilst relationships between k_l and q_m have been proposed, it is generally assumed that k_l varies little between 0.60 and 0.62, which probably reflects the reduced variation which exists in the overall composition of dairy cow diets.

7.3.5 Degradability in the rumen

Measurements of metabolizable energy provide valuable information about the total energy-yielding nutrients supplied by the feed, and evaluation of the ME system has shown that it accurately predicts overall energy balance. Two major limitations of the ME system are that it fails to predict partition of nutrients to competing processes within the animal, and the composition of animal products, namely milk and meat. Thus, with a known ME intake for a particular animal, it is not possible to predict how much energy will be used for different functions, such as liveweight gain or lactation. Related to this problem is the failure to predict dietary effects on the relative amounts of protein and fat contained in animal products. It is now accepted that resolution of these issues can be best achieved through the development of more detailed nutritional models which qualitatively and quantitatively describe the digestion, absorption and utilization of individual nutrients. As most of the OM apparently digested in the whole tract is digested initially in the rumen, prediction of the overall outcome of the processes of degradation, fermentation, absorption and passage within the rumen is essential. Of particular importance is prediction of the ruminal production of individual volatile fatty acids (VFA; mainly acetic, propionic and butyric acids) as these have different metabolic fates in the animal, as well as the yield of microbial protein, as this makes a significant contribution to the overall supply of amino acids to the animal (see Sections 7.5, 7.6 and 7.8 for further discussion). Studies of the rates and extents of digestion of food components within the rumen are thus considered to be essential in the ultimate prediction of individual nutrient supply.

The kinetics of feed digestion in the rumen have been estimated by placing samples (about 5 g DM) of test feed in polyester bags and incubating them in the rumen under defined conditions (Ørskov & McDonald 1979) for different time periods. Subsequently, the feed residues contained in the bags are analysed and the percentage disappearance of any feed fraction from the bag is plotted against the duration of incubation. Examples of degradation curves obtained by the *in situ* method are shown in Fig. 7.5. Various mathematical models have been proposed to describe the degradation curves and to derive a series of

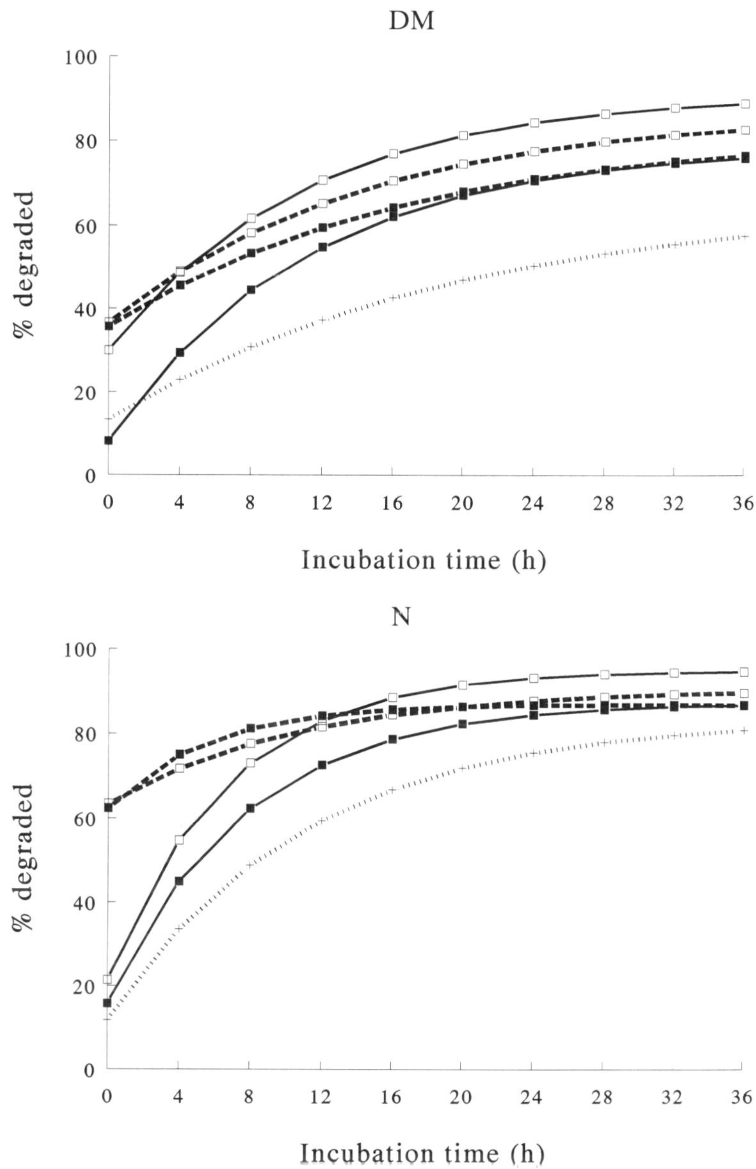

Fig. 7.5 Typical degradability curves from *in situ* measurements of forages.

	Forage	DOMD (%)	ED (DM)*	ED (N)*
—■—	Fresh grass	56	54	68
—□—	Fresh grass	76	69	68
----■----	Grass silage	66	62	79
----□----	Grass silage	72	65	81
····+····	Grass hay	54	39	58

* ED, effective degradability (%) calculated for a rumen outflow rate of 5% h^{-1}.

kinetic values which could be used in the metabolizable protein scheme proposed by Alderman & Cottrill (1993) and other similar schemes. However, incubation of feeds within bags does not represent the real situation within the rumen, since particles of partially digested feed continuously leave the rumen at rates which depend upon their size and density, as well as the level and frequency of feeding. The true extent of digestion, known as the effective degradability (ED%), therefore depends on the two competing processes of digestion within, and passage from, the rumen. Outflow rate of feeds from the rumen is ideally estimated by measuring the outflow of feed particles which have been labelled with markers that have made them indigestible (e.g. using a chromium mordant), or sometimes simply by use of an indigestible marker *per se*. The marked feed or the marker is added to the rumen as a single dose, and the decrease in marker concentration in the rumen is measured relative to time after dosing. Equally, it is possible to use the appearance of the marker in the faeces and appropriate mathematical modelling (Blaxter *et al.* 1956); either method can be used to provide estimates of rumen outflow rate. This is normally expressed as the proportion of the marker present at the start of each hour, which leaves in the following hour. Rumen outflow rates are used to calculate effective degradability values from the *in situ* kinetic constants and increase with increasing level of feeding, but decrease when the diet contains high levels of finely ground concentrate.

The ED values for DM of the forages in Fig. 7.5 reflect their DOMD values, which is to be expected as most DM and OM digestion takes place in the rumen. However, substantial differences in the slopes of the curves and the losses at zero-time (i.e. not incubated) are evident. The grass silages and the high D-value fresh grass have high initial DM losses because of their high content of soluble components (fermentation products and soluble N compounds in silages; sugars in the grass). The lower D-value fresh grass reaches a similar level of DM degradation as the lower D-value grass silage value after 20h incubation, but the higher initial degradation of the silage is the reason for the higher ED value and DOMD of the latter forage. However, the degradability curves for N are very different for the silages compared with the other forages. The high initial N loss (60%) for the silages is characteristic of extensively fermented forages, caused by the breakdown of the true protein of grass to soluble non-protein nitrogen constituents, including free amino acids and ammonia. This results from the action of plant enzymes after cutting, as well as microbial degradation of protein during the ensiling process. The reduced plateau value for the low digestibility hay sample reflects a higher content of unavailable N bound to cell walls.

Measurement of gas production *in vitro* has been proposed as an alternative technique to describe the dynamics of rumen fermentation. In this procedure, the feed sample is incubated with buffered rumen fluid, and cumulative gas production is recorded (Menke *et al.* 1979, Theodorou *et al.* 1994). Potential advantages of this approach are that it requires less use of animals whilst a greater number of feed samples can be processed at any one time. In addition, as this is a batch culture, the soluble and small particles that are frequently 'lost' from polyester bags used in the *in situ* process are retained, and can be subjected to normal digestive processes. However, it does have some major limitations. The measured production of gas is derived from fermentation of the feed (as carbon dioxide and methane) and from the bicarbonate-containing medium (as carbon dioxide). Furthermore, rate of gas production is affected by the pattern of fermentation end-products produced (i.e. the particular metabolic pathways used by the rumen microbes) and by substrate concentration. Also, the method assumes complete fermentation of hexose derived from the breakdown of dietary carbohydrates, when it is recognized that a significant proportion may be used directly by the microbes in the synthesis of microbial biomass (Beever 1993). Finally, both the *in situ* and gas production methods are susceptible to various animal-dependent factors, including the diet of the test animal and the time of obtaining the rumen fluid or incubating bags in relation to consumption of feed by the animal.

Effective degradability values from the *in situ* technique are used to estimate the amount of feed N available to the rumen micro-organisms, along with the amounts that escape degradation in the rumen to reach the intestines (see Section 7.8.2 for a discussion of the metabolizable protein system). Both the *in situ* and gas production techniques can be used to provide estimates of the energy supply to the microbes, which is usually the limiting factor for microbial protein synthesis in the rumen. Within the current MP system, energy supply to the microbes is described as fermentable metabolizable energy (FME). This is estimated from dietary ME content, with appropriate deductions of the energy content of dietary fat and fermentation products, both of which are assumed to make no contribution to overall energy availability within the rumen. Estimation of FME in this manner is an accepted weakness of the MP system, but data from the *in situ* and gas production techniques will provide improved evaluations. However, this approach is unlikely to be successful until appropriate *in vivo* data are acquired, against which the *in vitro* assessments can be validated. Other possible uses of the techniques include prediction of the pattern and level of VFA production, and as an aid to predicting forage intake.

7.4 Regulation of forage intake

Large differences exist between feeds in the amount that an animal will voluntarily consume, and also between animals in the amount of the same feed that they will consume. It is generally accepted that animals eat principally to supply their tissues with the nutrients required to fuel the physiological processes of maintenance, growth (including fat deposition in mature animals), milk production and work. Thus, for example, consumption of a particular feed will increase with milk yield providing that intake is not first constrained by some characteristic of the feed.

One of the problems with fibrous feeds, such as grasses, is the bulky nature of the feed and the time taken to reduce this volume within the rumen. Conrad *et al.* (1964) suggested that physical distension of the rumen was one of the main factors which limited the intake of forages of medium to low digestibility. Research since Conrad's work has identified other factors which are correlated with low intakes of forages, such as high acidity levels in silages, but it is still accepted that the time required to process forage is a key constraint to forage intake. General aspects of control related to the fibrous nature of forages are discussed in Section 7.4.1, while the subsequent sections consider other factors involved in the control of intake of grazed forage and conserved forage, and the effect of supplementation.

7.4.1 Fibre and the physical control of intake

The degree of physical distension of the rumen is monitored by tension and epithelial receptors which respond to both stretch and mechanical stimulation of the rumen wall. It is not simply the volume or weight of contents in the rumen which limit intake, but also the texture of the contents that come into contact with the rumen wall as the rumen contracts. The texture changes with time as the feed is digested by the rumen microbes, aided by the process of rumination, the latter being a key component in reducing the size of feed particles. It was originally believed that it was the size of particles which limited outflow from the rumen, and thus provided a physical restriction to intake, but fractionation of rumen contents has demonstrated that a major proportion of material in the rumen at any one time was in the form of small particles (Ulyatt *et al.* 1986). Attention then turned to the specific gravity of particles and the function of the raft-like structures which are observed in the rumens of forage-fed animals (Sutherland 1988), but understanding how these structures are involved in controlling digesta flow out of the rumen has been difficult to obtain.

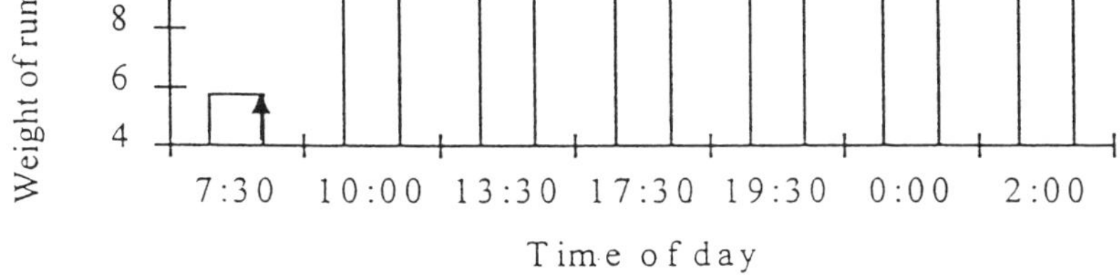

Fig. 7.6 Change in the weight of rumen contents in cattle fed once daily at 8:00 h (Gill *et al.* 1999).

Measurement of the weight of rumen contents in lactating dairy cows (Fig. 7.6) demonstrates that the cows did not eat to maximal rumen fill during their first meal, when a large proportion of the digesta was newly ingested. It was not until later that the highest weight of rumen contents was achieved, by which time the texture of the original digesta would have changed considerably through the processes of digestion and rumination. Findings such as these seriously challenge the simplistic assumption that it is the weight/volume of digesta in the rumen that provides the physical limitation to intake. However, it is clear that the time taken to process the fibre in forages limits their intake, particularly when the inherent digestibility of the feed is low and/or the nutritional requirements of the animal are high.

7.4.2 Intake at grazing

The intake of grazing ruminants is determined by the integration of grazing time, biting rate and dry matter intake per bite. Intake per bite is highly dependent on sward structure, while grazing time and biting rate are influenced by the hunger/satiety state of the animal. In reality, the principles of physical control described above also apply to grazing animals, providing ingestion is not limited by the availability of forage or the animal's ability to harvest it.

It has long been established that sward characteristics are a major determinant of intake in tropical pastures (Chacon & Stobbs 1976), but similar evidence for temperate pastures is more recent. Research has shown that sward surface height is a key determinant of intake per bite (Penning *et al.* 1991), and while this is irrespective of bulk density for swards of 200 mm height or more, bulk density can become an important factor as sward height decreases (Cushnahan *et al.* 2000).

When herbage availability is not limiting, the animals' demands for nutrients can have a major impact on forage intake at grazing. Increases in intake of up to 460 g organic matter per kg increase in milk yield have been obtained (Stakelum & Dillon 1991), through increases in intake rate and/or grazing time. Further work remains to be done to ascertain how this relationship is affected by forage quality in order that sward management can be aimed at maximizing forage intake by high-producing animals.

7.4.3 Forage conservation and intake

The conservation of forage generally results in reduced intake levels relative to the fresh forage, but where silage is well preserved, the lower intakes of silage compared with hay observed in earlier studies now appear to be much less of a problem. Ensiled feeds contain varying amounts of fermentation acids, but no individual organic acid has been identified as the major inhibitor of silage intake. Research by Steen *et al.* (1998) suggests that changes in the characteristics of the protein and fibre fractions of grass during ensiling are more likely to be responsible for the effects of ensiling on forage intake.

Pre-wilting of grass usually results in an increased intake of silage compared with unwilted silage from the same sward (Teller *et al.* 1993), but this effect was more apparent for animals with increased energy requirements (R. Sanderson, personal communication, 1998). Studies on rumen digestion suggested that the wetter material (unwilted) had a longer rumen retention time, which could have led to physical limits being reached earlier in animals striving to eat large amounts of forage in order to meet their requirements. It is concluded that the use of methods to control fermentation during ensiling, such as pre-wilting and application of additives or inoculants, is likely to increase forage intake, especially in high-producing animals.

7.4.4 The effects of supplements on forage intake

With respect to forage diets, supplements are considered as feedstuffs that are rich in particular nutrients. Supplements are particularly important in diets for high genetic merit cows, for example, for which high forage diets are not sufficient to meet all nutrient requirements (Veerkamp *et al.* 1994). The most common supplements to forage diets contain high levels of carbohydrate in a form that is more readily digested than the fibre present in forages. In contrast, many protein supplements used with forage diets contain protein which is less rumen-degradable than forage protein, since the rate of nitrogen release from less soluble protein provides a better source of nitrogen for the fibre-digesting microbes in the rumen. These two types of supplement have the greatest impact on forage intake, and discussion in this section is limited to them. However, other supplements may be offered with forage diets to compensate for specific deficiencies, and their effects on productivity are discussed in Sections 7.6 and 7.10.6.

It is accepted that forage intake decreases as the amount of high-energy supplement is increased; this rate of decrease in forage intake per kg increase in concentrate DM intake is referred to as the substitution rate. It is affected by the nature of the forage, and with hay, the addition of a starch-based supplement has a greater depressive effect on forage intake as the digestibility of the forage increases (Fig. 7.7). Silage intake is not affected to the same extent by changes in digestibility, and previous research (Wilkins 1974) has established that substitution rate is more related to the intake of the silage when offered as the sole feed than to the digestibility of the silage *per se*. This is illustrated schematically in Fig. 7.8.

In the grazing situation, supplements can have a marked effect on herbage intake, but substitution rate is dependent on herbage availability. Where availability is high, supplementation causes a reduction in grazing time and hence in intake, but with restricted availability, supplementation has little effect on grazing time, resulting in lower substitution rates (Mayne 1991).

Evidence of the effect of concentrate type on forage intake is not conclusive. There are research data (e.g. Thomas *et al.* (1986) for silage, and Meijs (1986)

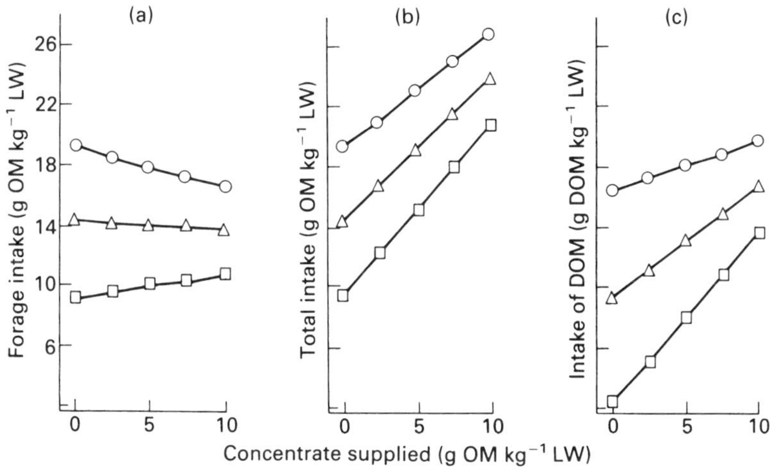

Fig. 7.7 Interaction between the level of concentrate consumption by cattle and the quality of hay. Effects of level of concentrate consumption by finishing cattle on (a) forage OM intake, (b) total OM intake, and (c) total digestible OM intake. High (○), medium (△), and low (□) digestibility of hay (after Vadiveloo & Holmes 1979).

for grazed grass) which suggest that forage intakes are higher with fibre-based, as compared with starch-based, concentrates. However, there is also evidence that the type of energy source has no effect on forage intake (e.g. Castle *et al.* 1981).

With respect to the effect of protein-containing supplements on forage intake, the situation is much clearer. Increasing protein content of the supplement generally has a beneficial effect on silage, and hence on total DM intake, especially in those instances where the protein content of the control diet is considered to be limiting animal productivity (e.g. Aston *et al.* 1998; see also Table 7.2).

Table 7.2 Daily intake of DM (kg per day) and digestible energy (DE; MJ per day) by lactating cows given grass silage *ad libitum*, supplemented with 5 kg cereal-based concentrate supplying 156 (L), 247 (M) or 338 (H) g crude protein per kg DM (extra protein supplied as 3:1 soya:fishmeal) (source: Aston *et al.* 1998).

	L	M	H
Weeks 4–12			
DM intake	12.5	13.9	13.7
DE intake	246	267	268
Weeks 13–21			
DM intake	11.9	13.5	13.5
DE intake	237	260	266

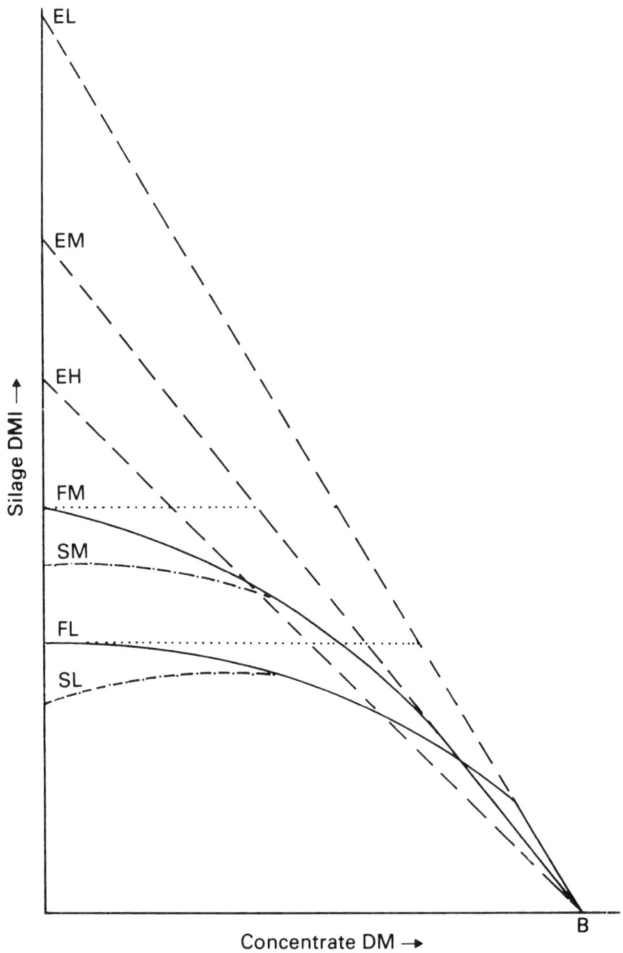

Fig. 7.8 Schematic representation of the relationship between silage intake and concentrate supplementation (Thomas 1987). Low (L), medium (M) and high (H) digestibility. Broken lines EL, EM and EH to B represent the intakes of silage of low, medium and high digestibility, respectively, assuming that the animal could eat to maintain constant energy intake. Solid curves FM and FL to B represent the effect of concentrate in depressing cellulolytic activity in the rumen and hence increasing rumen 'fill', while the dotted lines (FM and FL to B) predict the intake of M and L if the concentrate had no effect on fill. Finally, the broken lines SM and SL represent the control of intake by factors related to fermentation characteristics of the silage.

7.5 Digestion in the ruminant

7.5.1 Physical and microbiological interactions in the gastrointestinal tract of ruminants

All ruminant animals, with the exception of the milk-fed neonate, possess a highly developed forestomach, the main function of which is the digestion of fibrous materials by the resident microbial population. The ruminant forestomach is an anaerobic fermentation system in which the end products of digestion are either absorbed through the rumen wall to enter the blood stream, or propelled to the small intestines for further digestion. There are four distinct parts to the ruminant foregut, namely the reticulum, the dorsal and ventral sacs of the rumen, and the omasum, which are all anterior to the abomasum, the latter being analogous to the true stomach of non-ruminant species. At feeding, ingested food rapidly enters the reticulum via the oesophagus although, depending upon the nature of the feed, some physical disruption may occur as a consequence of chewing within the mouth. Salivation rate increases during ingestion, and when the ingested feed enters the oesophagus, it will have been partially wetted in preparation for subsequent digestion. The reticulum is not the largest organ in the ruminant forestomachs, but it plays an important role in the initiation of muscular contractions which occur approximately every 2 minutes. These transfer in a wave-like motion through the dorsal and ventral sacs of the rumen and are responsible for the movement and mixing of digesta. It is in the rumen *per se* that the bulk of the ingested feed resides and, depending upon the nature of the feed consumed, rafts of feed will form between the two parts of the rumen. With each muscular contraction, freshly ingested feed is brought into the fermentation system, whilst smaller particles of feed which have undergone significant digestion will be washed out and moved towards the omasal orifice for possible onward transmission to the intestines. Thus, digestion in the rumen is a coordination of microbial and physical activity, the latter being further enhanced by the process of rumination in which distinct boli of partially digested feed are transported by reverse peristalsis up the oesophagus for further comminution in the mouth. Rumination is stimulated by the presence of coarse fibrous materials in the rumen, and adult ruminants can spend at least one-third of the day ruminating, although this depends upon the nature of the feed being consumed. Subsequently, the feed bolus re-enters the rumen, where it will be subjected to further microbial attack.

As a consequence of these processes the ruminant forestomach can be broadly considered as a batch fermentation, with ingested feed residing there for variable periods of time dependent upon the nature of the feed consumed, and in particular the level and composition of dietary fibre. In adult ruminants total rumen contents, including partially digested feed, consumed water and saliva, can be as much as 100 kg, or over 20% of the total liveweight of the animal. Given that volatile fatty acids are one of the major products of rumen fermentation, and that microbial enzyme activity is affected by the pH of rumen contents, it is important that the pH of the rumen is controlled within reasonable boundaries. Removal of VFA by absorption across the rumen wall will reduce the overall acid load in the rumen. In those environments where there has been no ingestion of feed for several hours (e.g. overnight on fixed feeding regimes) it is possible to record pH values in excess of 7.0, probably as a function of the resting secretion of saliva which is known to contain buffer salts. However, following the ingestion of feed, pH will fall, despite an increase in salivation rate, and on high-grain diets it is quite common to observe post-feeding pH values below 6.0. Unless this is controlled, overall microbial activity will decline, which usually results in feed inappetance, with the long-term effect of acidosis being severe loss of rumen activity, permanent damage to the rumen epithelium and ultimately death of the animal.

Following extensive digestion/fermentation in the rumen, feed particles at different stages of digestion, but of a sufficiently small size to pass through the omasal orifice (<1.0 mm), will flow out of the rumen into the abomasum for ultimate transmission to the small intestine. Accompanying this passage of digesta will be a significant quantity of microbial biomass, which may be attached to the feed particles or free floating in the liquid which accompanies the expulsion of digesta contents. Thus, the digesta finally presented to the intestines are comprised of a mixture of undigested feed particles and microbial biomass of variable composition. The small intestine is the site of host enzymatic digestion, in which amino acids, glucose and long-chain fatty acids are the major end-products. There is virtually no microbial activity in this part of the alimentary tract, due to the reduced pH (<2.75) conditions as a consequence of acidic secretions during passage of digesta through the abomasum. However, as digesta passes along the small intestine, pH increases as a consequence of pancreatic secretions, and the occurrence of small amounts of VFA in the intestines immediately prior to the terminal ileum is indicative of limited microbial activity. Finally, undigested material of feed and microbial origin passes into the large intestines, principally the caecum and colon. In the caecum there is a small microbial population capable of digesting a small, but on some diets not insignificant, amount of fibre which has escaped rumen degradation, although this fermentation process does not compare with the rate and extent to which ingested feeds are fermented in the forestomachs.

7.5.2 Microbiological/biochemical interactions in nutrient digestion within the gastrointestinal tract

Bacterial species constitute the largest proportion of the microbial population in the rumen, at least in terms of enzymatic activity, and are responsible for the digestion and fermentation of dietary carbohydrates and proteins. Their principal role is the digestion of carbohydrate, and irrespective of the structural complexity of dietary carbohydrates, this process can be considered as two parts, namely the dissimulation of carbohydrate polymers and the subsequent metabolism of the constitutive sugars. The different carbohydrate fractions of forages have been discussed in Section 7.2. Water-soluble carbohydrates are comprised of simple carbohydrates of relatively low molecular weight (limited polymerization). Generally the content of free sugars is minimal, whilst fructose-containing polymers (fructans) and sucrose can occur in significant quantities, dependent upon forage species and prevailing environmental conditions. All water-soluble carbohydrates are rapidly digested by the rumen microbes (Beever *et al.* 1972). Starch is the next carbohydrate with respect to structural complexity, and under most circumstances will be quantitatively reduced to glucose. However, the form of the starch (e.g. maize *vs.* wheat) and the degree to which it has been processed can have a significant bearing upon the rate and extent of dissimulation to the constitutive sugars. The most complex carbohydrate fraction is fibre, comprising cellulose, hemicellulose and lignin. Inherently, both cellulose and hemicellulose are highly digestible, but their spatial distribution within the feed in relation to lignin can significantly reduce the rate and extent to which they are degraded in the rumen. If cellulose is degraded, the sole end product is glucose, whilst hemicellulose, a carbohydrate of mixed composition, will yield xylose and arabinose, plus other minor pentose and hexose sugars, along with uronic acids. Lignin is considered to be indigestible, and thus the concentration present in gut contents will increase as digesta passes along the alimentary tract.

The second component of carbohydrate dissimulation in the rumen involves the production of ATP to support the energy demands of the microbes for cell maintenance and growth. Given that the rumen environment is anaerobic, the yield of ATP from carbohydrate is considerably reduced compared with that achieved when glucose, for example, is completely oxidized. The principal route of carbohydrate fermentation involves the production of 2 moles of pyruvate, along with some ATP, for each mole of monosaccharide fermented. The resultant pyruvate is metabolized further, but the end products of this process will depend upon the type of microflora present in the rumen, which is affected by the nature of the diet. On high fibre diets, those bacteria which promote acetate production will dominate, whilst on high starch diets, levels of propionic acid are usually increased. The other principal VFA is butyrate, and whilst never found in those amounts recorded for acetate and propionate, butyrate levels approaching 20% (molar basis) can be observed on diets containing significant quantities of rapidly fermentable carbohydrate, especially sucrose. The principal pathways of VFA are presented in Table 7.3 (from Beever, 1993), indicating an average VFA yield of 1.8–1.9 moles per mole fermented carbohydrate. At the same time significant quantities of hydrogen are released during the production of acetate and butyrate, whilst propionate synthesis is a hydrogen-utilizing reaction. The ultimate metabolic fate of surplus ruminally derived hydrogen is the production of methane by the action of the methanogenic bacteria. This reaction will also provide a small amount of extra ATP. However, in estimating methane production on the basis of hydrogen release, it is necessary to take account of any hydrogen which may be used to support the ruminal hydrogenation of dietary unsaturated fatty acids, which on many feeds may be the principal source of long-chain fatty acids consumed. The only other notable product in these reactions is carbon dioxide, which accounts for 0.33 of the pyruvate carbon used in the synthesis of acetate or butyrate. Production of propi-

Table 7.3 The estimated stoichiometric yields (moles per mole) of volatile fatty acids, ATP, carbon dioxide and methane from the fermentation of different dietary carbohydrates.

Diet type	Acetate	Propionate	Butyrate	ATP	Methane	CO_2
High fibre	1.34	0.45	0.11	4.62	0.61	0.92
High cereal	0.90	0.70	0.20	4.38	0.38	0.92
Molasses-containing	0.94	0.40	0.33	4.54	0.54	1.06

onate from pyruvate is not a carbon dioxide-producing reaction, as there is no excess carbon to dispose of. Thus, energetically, propionate production from monosaccharide is more efficient in terms of nutrient conservation than either acetate or butyrate production. On the basis of this description, it is easy to assume that all of the carbohydrate molecules released from the degradation of structural carbohydrates will be metabolized to VFA, methane and carbon dioxide. However, a significant amount of hexose derived from carbohydrate dissimilation is used directly to support the synthesis of microbial biomass (Black *et al.* 1981). It is an important substrate in the production of microbial protein when ammonia is the preferred nitrogen source, but is not required when amino acids are incorporated directly. Hexose is also used for the synthesis of nucleic acids, which account for a significant proportion of microbial biomass, and pathways involving hexose have been identified in the *de novo* synthesis of microbial lipid. An alternative fate of hexose is the synthesis of microbial polysaccharide, but this is considered more as a process designed to dispose of excess hexose, rather than having a significant obligatory role in microbial metabolism. Furthermore, it is likely that hexose utilization in this way is more common in protozoa than bacteria, whilst there is considerable evidence that protozoa engulf whole starch grains, to utilize them later or provide a sustained nutrient source for bacteria, following death and rupture of the protozoa.

The other major function of some bacteria is the dissimilation of dietary protein and associated nitrogenous fractions. Whilst true protein is likely to be the major component of most feeds for ruminants, in some feeds significant quantities of non-protein nitrogen can be present. This is particularly the case with forages, both grazed and ensiled, as well as other fermented feeds and those diets containing supplementary urea. In fresh forages the major protein is ribulose dicarboxylase (fraction 1 protein), the principal photosynthetic enzyme, along with variable amounts of chloroplastic and membrane proteins. However, all forages contain significant amounts of non-protein nitrogen, including amino acids, amines and amides, whilst nitrate levels can be quite substantial, especially in recently fertilized crops where impaired growing conditions have been experienced. In ensiled forages, especially grass silage, levels of true protein are generally reduced compared with those present in the original forage, due to events occurring within the silo. During this period of conservation, proteolysis can be quite substantial, especially when the silage fermentation is not controlled by use of an appropriate additive. Thus in such feeds, levels of amino acids, amines and amides can increase substantially, and significant levels of ammonia can be detected, in the worst cases often accounting for 20% or more of the total forage nitrogen.

Based on these likely inputs of nitrogen, most of the non-protein nitrogen will become available to support microbial biomass synthesis, with the most common substrate being ammonia. On the other hand, the rate and extent of true protein degradation will be a function of the chemical characteristics of that protein. Thus, in the case of forages, fraction 1 protein is highly susceptible to microbial attack, whilst other proteins, especially those which have been heat-treated or sprayed with formaldehyde, are broken down more slowly. During the process of dietary protein and non-protein nitrogen degradation, there will be a continuous, although not necessarily constant, supply of nitrogen-containing intermediates, some of which can be used to support microbial protein synthesis. In all biological systems, protein synthesis is an energy-consuming process; in this case with ATP generated from the fermentation of dietary carbohydrate being used to support the synthesis of amino acids from ammonia and proteins by the formation of peptide bonds between adjoining amino acids. Thus on an energy basis, microbial protein synthesized from pre-formed amino acids will be less energy-demanding than when the amino acids are synthesized *de novo* from ammonia and a suitable carbohydrate skeleton. On the basis of ^{15}N tracer studies, it has been shown that both processes occur in the rumen, in part regulated by the relative availabilities of different nitrogen-containing substrates, whilst recognizing that some bacteria are obligate utilizers of either pre-formed amino acids or ammonia.

The largest issue associated with optimizing the conversion of degraded dietary nitrogen is related to the need to synchronize the relative availabilities of energy and nitrogen. To illustrate this point, reference to studies undertaken with ensiled forages provides an excellent example. During ensiling, a significant part, if not all, of the readily available carbohydrate, namely the free and water-soluble sugars, will be fermented to organic acids, whilst a significant part of the forage protein will be converted to non-protein nitrogen. Thus, following ingestion, when the readily available nitrogen is released and made available for microbial protein synthesis, there is likely to be a substantial lack of available carbohydrate. Consequently, microbial biomass synthesis is affected. The net result is a pronounced increase in rumen ammonia levels, which leads to significant amounts of ammonia being absorbed across the rumen wall to enter the circulatory system, where it is transported to the liver and converted into urea. Whilst part may be recycled to the rumen through saliva, a significant part is excreted in urine, and this represents a significant loss of a potential source of nitrogen for the ruminal synthesis of microbial protein. Rooke *et al.* (1987) demonstrated the impact of this, where, from a relatively low estimate of the efficiency of microbial biomass synthesis for a grass silage diet fed to cows, it was only possible to increase overall synthetic rate and efficiency once supplemental energy was provided, and this in turn promoted a further response when true protein supplements were added to the energy supplement. However, addition of urea or casein alone had only a marginal effect on microbial protein synthesis.

A similar situation also exists with fresh forages, where the imbalance between readily available nitrogen and carbohydrate can lead to substantial increases in rumen ammonia levels some 2–3h after feed ingestion. Thus, duodenal protein supply on fresh forages (as with many silages, excluding formaldehyde-treated silages) is often less than crude protein intake, suggesting a major loss of N from the rumen as ammonia and an associated inefficiency of N utilization. On high-N white clover diets, Beever *et al.* (1986) and Ulyatt *et al.* (1988) found duodenal non-ammonia nitrogen (NAN) flows equivalent to only 60% of N intake, and for a range of fresh forage diets of contrasting N contents, Ulyatt *et al.* established the following relationship:

$$Y = 1.430 - 0.0169\,x \qquad (r = 0.72, \text{RSD } 0.140)$$

where Y = NAN flow/N intake (g/g), and x = N content in crop (g kg^{-1} OM).

In contrast, the above equation shows that on low-N diets duodenal NAN flow may exceed N intake due to the inflow of endogenous N and an enhanced efficiency of capture of degraded N by the microbes. Research with a range of forage diets established that the efficiency of microbial protein synthesis was highly variable (McAllan *et al.* 1987), and on many silage diets may be considerably less than the mean value proposed by ARC (1984) of 32g microbial N kg^{-1} organic matter apparently digested in the rumen (OMADR). In Table 7.4, values ranging from 13 to 28 g kg^{-1} OMADR for ensiled diets are presented. Those below 20 g kg^{-1} are particularly interesting, and on the basis of other data point to opportunities for enhancing microbial metabolism, and hence rumen function, through the possible use of ruminally degradable carbohydrate or protein sources. In contrast, values for fresh forages are much higher (range 33–58 g kg^{-1}), which may be related to an enhanced ruminal energy supply in the form of soluble carbohydrates.

Lipid components are unlikely to exceed 3–4% of forage dry matter, and undergo less extensive transformation within the rumen than either carbohydrates or protein. Bacterial hydrolysis occurs within the rumen, with the associated production of long-chain fatty acids and glycerol. The former then undergo extensive, although not necessarily complete, saturation due to the reducing conditions in the rumen, whilst the small amount of glycerol released is fermented with the production of VFA. Consequently, the profile of long-chain fatty acids entering the small intestine can be substantially different from

Table 7.4 Estimates of the efficiency of rumen microbial N synthesis for ensiled and fresh forages fed to sheep or cattle.

	Animal species	g microbial N per kg OMADR
Ensiled forages		
(All grasses)		
Unwilted	S	28.1
Wilted	S	21.6
Formic, unwilted	S	19.1
Formic, wilted	S	21.6
Formic, unwilted	C	19.0
F + F, unwilted	S	13.1
F + F, unwilted	C	23.7
F + F, wilted	S	27.2
Fresh forages		
Perennial ryegrass	C	46.9
White clover	C	58.1
Lucerne	S	40.9
Subterranean clover	S	33.2
Perennial ryegrass	S	33.1

S, sheep; C, cattle; F + F, formic and formaldehyde-containing additives.

that in the diet, whilst total fatty acid supply is usually increased, albeit marginally, as a consequence of microbial lipid synthesis.

7.5.3 Intestinal absorption of amino acids and lipids

Duodenal protein comprises protein of microbial and undegraded feed origin, plus an endogenous contribution arising from gut secretions and gut wall sloughings. Apparent digestibility of duodenal protein is generally high, and often exceeds 700 g kg^{-1}. Thus, the contribution of absorbed protein to total ME supply will be significantly influenced by duodenal supply. On low-N diets or those containing high quantities of readily degradable N, the amount of absorbed protein may be as low as 5.5 g MJ^{-1} ME intake, whilst on high N diets, including legumes or those containing quantities of undegradable protein, values may exceed 11.0 g MJ^{-1} ME (Beever 1980).

There is still little information on the contribution of lipids to host-animal nutrition. As indicated, duodenal long-chain fatty acid supply generally exceeds dietary intake, reflecting microbial lipid synthesis within the rumen plus the non-destructive transformation (i.e. hydrolysis and saturation) of ingested lipid. On a series of cocksfoot and red clover diets offered to sheep, Outen *et al.* (1974, 1975) concluded that long-chain fatty acids absorbed from the small intestine accounted for some 7–10% of apparently digested energy. On lipid-supplemented diets, this figure would rise, but levels above 15% of ME intake should be avoided (Reynolds & Beever 1995).

7.5.4 Digestion in the hindgut

In most situations digestion in the hindgut is insignificant, amounting to less than 75 kJ MJ^{-1} DEI. One notable exception is when dried forages are fed in ground and pelleted form. On such diets, passage of partially digested material from the rumen is accelerated, and Beever *et al.* (1972) were able to account for 70% of the increased flow of duodenal OM as cellulose and hemicellulose. At the same time, extensive compensatory digestion in the hindgut was observed, with as much as 25% of the digestible OMI being digested therein. However, when the intake levels of such feeds are increased, the extent of digestive compensation in the hind gut is compromised, and whole-tract digestibility of the principal nutrients will decline.

7.6 Tissue utilization of forage energy and protein

The level of production achieved by ruminants depends not only on the total amount of energy and protein reaching the tissues, but also on the efficiency with which these nutrients are used for maintenance, growth and milk production purposes. The concept of ME as the energy available to the tissues was introduced in Section 7.3.4. Conversion of ME to net energy, defined as the energy retained in the tissues or secreted as milk, is achieved by using efficiency terms to correct for the heat losses associated with the oxidation of nutrients to provide energy to maintain the tissues (k_m), as well as during the synthesis of fat and protein for growth (k_f) or fat, protein and lactose for milk (k_l). The efficiency of energy use for maintenance is considered to vary with diet composition, and is recognized to be lower for forage only compared with mixed diets (Table 7.5). However, the efficiency of utilization for growth and

Table 7.5 Variation in the efficiency of use of metabolizable energy (ME) with physiological function and diet type (when metabolizability ME/GE = 0.50 or 0.60).

	Efficiency of use of ME					
	Maintenance k_m		Growth k_f		Lactation k_l	
Metabolizability	0.50	0.60	0.50	0.60	0.50	0.60
Forages	0.66	0.68	0.34	0.47	0.60	0.63
Pelleted feeds	0.67	0.70	0.48	0.48	0.60	0.63
Mixed diets	0.73	0.75	0.47	0.51	0.60	0.63

Values from ARC (1980).

fattening is more sensitive to diet, as discussed below. During lactation, the dairy cow attempts to overcome dietary deficiencies in energy supply by utilizing her body reserves, and whilst some variations in k_l have been reported, the range is quite limited, suggesting that the efficiency of utilization of ME is relatively independent of diet. However, both growth and milk production are also sensitive to protein supply. The utilization of energy and protein for these processes will be discussed, followed by consideration of the effect of forage diets on the composition of liveweight gain and milk.

7.6.1 The form of energy supplied in forage diets

The efficiency values for growth (k_f) proposed by ARC (1980) are lower for forages than for pelleted feeds or mixed diets (Table 7.5). This phenomenon is well known but poorly understood. Blaxter & Clapperton (1965) suggested it could be due to the higher proportion of acetate produced in the rumen on forage diets. Their evidence, based on the effect of ruminal infusions of individual VFAs on heat production, indicated a higher heat production (lower efficiency) for acetate compared with propionate or butyrate; subsequent work, however, has not always confirmed this observation (Hovell *et al.* 1976, Ørskov *et al.* 1979). These conflicting results may be explained by differences in the basal nutrients supplied by the diet, since Tyrrell *et al.* (1979) found that the efficiency of acetate utilization was greater when acetate was added to a diet of forage and concentrates than to a diet of forage alone, whilst Armstrong (1965) suggested that the rate of conversion of acetate to fatty acids on high fibre diets may be limited by low availability of the reduced co-factor NADPH, which is essential for fatty acid synthesis. The main precursor of NADPH is glucose, which is absorbed in negligible amounts on most forage diets, so ruminants are obliged to rely almost exclusively on gluconeogenesis from absorbed propionate and amino acids. This led MacRae & Lobley (1986) to suggest that those diets where efficiency of acetate utilization was high were most likely to be those that supplied sufficient glucose precursors to ensure an adequate NADPH supply. Subsequently, Black *et al.* (1987) examined this hypothesis using a mathematical model of nutrient utilization, and suggested that with high protein diets additional acetate would be used to supply the energy required for increased protein synthesis. Hence, it is probable that the ratio of individual nutrients supplied on forage diets may contribute to the low efficiency observed, but this is unlikely to be the sole explanation.

It has been suggested that the energy required to eat, ruminate and subsequently move fibrous digesta through the gut may be elevated on fibrous diets, but experimental data suggest the direct cost is small (Webster 1980). The possibility of an indirect cost on forage diets through the stimulation of increased metabolism within the gut was also examined by Webster, but the evidence is insufficient to draw any definite conclusions.

7.6.2 Protein utilization by the tissues

Undoubtedly, increased levels of animal production can result from protein supplements provided to both grazing animals and those offered silage, suggesting that protein supply may limit production on many forages. Part of this limitation is due to the effect of low protein diets in limiting intake (see Section 7.4.4), but in those cases where protein supplementation has no effect on intake, responses in animal production may still be observed (Gill *et al.* 1987). The effect of protein supplements on production will be considered in Sections 7.7.2 and 7.7.4, the aim here being to consider differences in how absorbed protein is utilized.

The efficiency of utilization of absorbed amino acids is partly dependent on the ratio of amino acids absorbed relative to the requirements for specific amino acids, and the relative proportions of these amino acids in the proteins synthesized by the animal. Initially, ARC (1980) assumed the efficiency of utilization to be 0.75 of amino acids apparently absorbed, but subsequent revision of the system (ARC 1984) considered truly absorbed amino acids and the efficiency value was increased to 0.80. However, there is increasing evidence to suggest that the overall utilization is much lower on forage diets. Reasons for this are as yet unknown, but it may be due to the different ratio of non-protein-containing nutrients absorbed from forage diets having a negative effect on the utilization of amino acids for protein synthesis.

7.6.3 Effects of forage on composition of meat and milk

Since the profile of nutrients absorbed from forage diets differs from those absorbed on concentrate or mixed diets, it is perhaps not surprising that the compositions of both liveweight gain and milk tend to differ also. The well-known effects of high fibre diets in increasing the fat content of milk have been attributed to the higher contribution of acetate to total absorbed energy (Fig. 7.9, after Broster *et al.* 1979). The protein and lactose contents in milk are much less sensitive to dietary manipulation, although there is evidence that on forage-based diets where ME intake may be compromised, significant reductions in milk protein content can be observed. In this respect, studies by Phipps *et al.* (1995) showed that replacement of part of the grass silage component of dairy cows' diets with alternative forage sources, including whole-crop wheat and maize silage, led to increased energy intake associated with improved contents of milk protein, whilst milk fat levels were more influenced by specific dietary situations. In relation to carcass composition in growing animals, there appears to be less opportunity for change, given that the percentages of water and protein in the fat-free empty body are relatively constant, while the amount of fat associated with the fat-free empty body tends to be higher on, for example, non-supplemented silage diets (Lonsdale 1976).

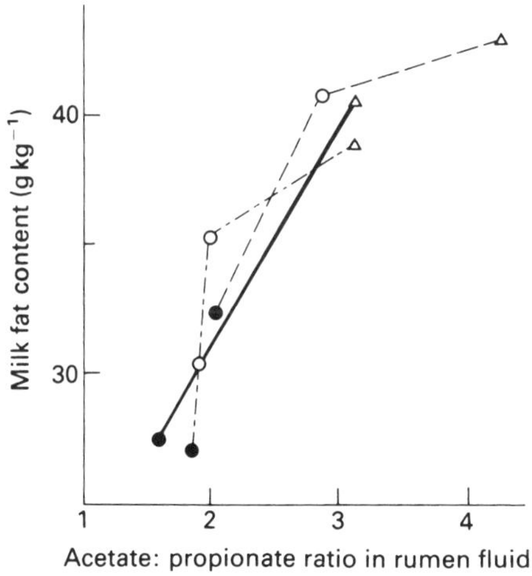

Fig. 7.9 Relationships between milk fat content and the ratio of acetic acid to propionic acid in the rumen. Ratio of concentrates: hay, 90:10 (●); 75:25 (○); 60:40 (△). Intake *ad libitum* (—); moderate intake (– – –); high intake (–·–·–·) (after Broster *et al.* 1979).

However, it is interesting to note that Thomson *et al.* (1983) found an increase in lean/fat ratio in lambs fed white clover at the same intake rate as fresh ryegrass. However, when a further group of lambs was allowed to eat the white clover diet *ad libitum*, most of the extra nutrients consumed were deposited as carcass fat.

7.7 The use of supplements and nutritional manipulants

In systems of animal production that depend on forages, the provision of feed supplements may be based on different criteria. If overall ME supply from the forage is inadequate to meet desired production targets, energy-rich feeds to increase dietary ME concentration can be used. Additionally, specific feed supplements may be used to provide desirable changes in the composition of the final product (i.e. milk or meat), whilst some feed supplements may be required to meet specific nutritional deficiencies (e.g. minerals, vitamins) which have been identified. As the use of supplements that fit into the first two categories is easier with indoor feeding, energy and protein supplementation of conserved feeds is considered separately from grazing.

7.7.1 Energy supplementation of conserved feeds

Given the levels of milk production which are achieved or expected of today's dairy cow, it is common practice to feed a significant proportion (60%+) of the total diet as concentrates. These may be purchased compound feeds which are usually defined in terms of their nutrient (fibre, oil and protein) contents rather than as their constituent feeds. However, since the late 1980s, the use of individual components (straights) has increased, and considerable quantities of wheat, sugar beet pulp, brewers grains and maize gluten, as well as barley and some oats, are fed.

The overall effect of supplementation on the performance of the cow will be influenced by the amount and nature of the supplement used, as well as by the composition of the basal forage and the lactational stage of the cow. The effect of forage composition on the substitution rate of concentrate for forage intake has already been discussed (Section 7.4.4). In addition to increasing total ME intake through the use of concentrate supplements, where an increased proportion of ME is derived from concentrates, improvements in the efficiency of utilization of ME for growth and fattening (k_f), may occur (see Table 7.5). The origin of the effect probably resides at the tissue level, possibly associated with nutrition/hormonal interactions which as yet are not fully understood.

7.7.2 Protein supplementation of conserved feeds

Protein supplements have been used for several years in forage-based diets, especially silages, for both dairy and beef cattle. The high-producing dairy cow and the young growing animal both have high protein requirements, and the discrepancy between supply (i.e. small intestinal absorption) and requirement is exacerbated on silage diets where, due to extensive proteolysis during harvesting and conservation, the supply of amino acids is often less than expected on the basis of dietary crude protein content. Use of formaldehyde-containing silage additives and minimization of the wilt period will reduce this shortfall, but there is a need to improve the preservation of protein within the conservation process, possibly through the use of more specific silage additives. In the meantime, use of protein supplements of inherently low rumen degradability, e.g. fishmeal, at levels equivalent to, or in excess of, the recommendations established by ARC (1980, 1984) has led to sizeable responses in milk output and milk protein secretion in dairy cows (Castle 1982), and in liveweight gain and carcass protein/fat ratio in growing cattle (Gill *et al.* 1987). However, with beef cattle receiving mixed diets of highly digestible silage plus concentrates, the response to fishmeal has been poor. Moreover in some instances, some or all of the apparent benefits in liveweight gain over the winter period have been lost over the subsequent grazing season. Furthermore, the BSE crisis in Britain has resulted in greater public concern over the use of animal-based proteins, and the use of soya, treated to reduce the susceptibility of the protein to degradation in the rumen, has increased. Only time will tell if this trend will continue, given the concern which exists with respect to genetically modified crops.

7.7.3 Energy supplementation at grazing

As mentioned earlier (Section 7.4.4), supplementation of highly digestible pasture with energy-rich supplements is associated with a high substitution rate. However, substitution rate varies with herbage availability, and increases as the amount of offered herbage increases (Meijs & Hoekstra 1984). Thus, supplementation can have a beneficial effect when herbage availability is limited (Le Du & Newberry 1982), whilst other forages can be used as supplements when herbage availability is low, with hay or silage being offered in yards at night.

Forages may also be used to supplement tropical pastures during the dry season, although responses are difficult to predict. Other possibilities include the use of crop by-products such as molasses or vegetable waste, but the nutritive value of many can be quite low or variable, whilst cost must remain an important consideration.

7.7.4 Protein supplementation at grazing

Protein supplementation of lambs at pasture has been shown to increase rates of liveweight gain and protein: fat ratio in the resultant carcass (Black *et al.* 1982), but the problem remains to ensure complete consumption of the supplement. Urea–molasses blocks will improve the intake of non-protein N, but uneven consumption of the blocks leads to poor utilization. The use of legumes (e.g. *Leucaena leucocephala*) has led to increased production on tropical pastures (Pound & Martinez-Cairo 1983), but whilst grazing of predominantly legume pastures may enhance animal productivity, the risk of bloat is increased (see Section 7.9.5).

7.7.5 Mineral and vitamin supplementation

When cattle are housed and fed conserved feeds, it is relatively simple to supplement the diet with minerals as proprietary mixtures or blocks. Supplementation at pasture is, however, more difficult. Vitamin and mineral deficiencies that may occur in either housed or grazing animals are discussed in Sections 7.9.1–7.9.3.

7.8 Prediction of animal performance

Prediction of the level of animal performance with any particular forage requires an estimation of both nutritive value and voluntary intake. The main parameters used to define energy and protein value are DOMD, ME, crude protein and protein degradability (Section 7.3). The aim in this section is to discuss ways of predicting both nutritive value and voluntary intake. It concludes with a discussion on alternative methods of prediction which may become important in the future.

7.8.1 Energy value

As indicated earlier, the energy value of a forage is related to the maturity of the crop and will change throughout the growth period. Each grass or legume variety has its own characteristic pattern of change in D-value during primary growth in the spring, and this is related to stage of growth and calendar date within broad bands of latitude (e.g. Fig. 7.10). Charts to predict D-value, N content and moisture content for a range of herbage species and varieties are available in the UK as NIAB Farmers publications (e.g. NIAB 1998) and GRI Technical Reports (Green *et al.* 1971, Corrall *et al.* 1979). Predictions may need to be modified to take account of unusually early or late, wet or dry seasons, and, as they are based on a particular management system during the previous autumn, further modification needs to take account of deviations from this (Walters 1976).

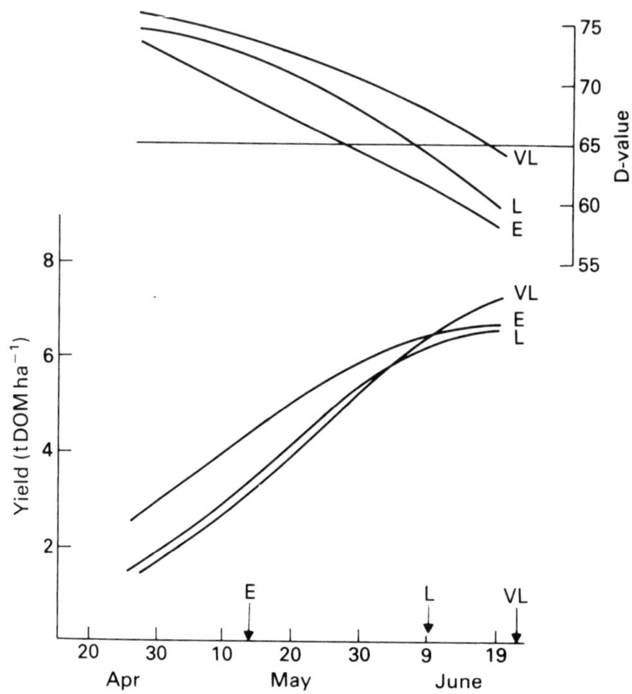

Fig. 7.10 Changes in the D-value and yield of DOM (t ha^{-1}) of early (E), late (L) and very late (VL) flowering varieties of perennial ryegrass during primary growth.

Prediction of D-value is particularly important for planning the winter feeding programme and estimating the level of performance that conserved feed will support. Thus, while approximate predictions can be made as to when to cut the crop for conservation to achieve a desired D-value, more accurate assessments for rationing purposes should be based on laboratory analysis of the conserved crop rather than relying on date of harvest. In acquiring samples for analysis, care should be taken that these are representative of the whole, with each batch of forage sampled from at least 12 random points, with the samples thoroughly mixed, before a representative sub-sample is taken.

The method most commonly used for advisory purposes to assess fibre content determines the content of modified acid detergent fibre (MADF), and equations to estimate DOMD content from MADF content have been given previously (Section, 7.3.2), whilst those to predict ME content are given below.

For hay (MAFF 1987)

ME (MJ kg^{-1} DM) = 16.5 – 0.21 MADF

For grass silage (MAFF 1987)

ME (MJ kg^{-1} TDM) = 15.33 – 0.152 MADF (corrected)
(where TDM is ODM + 1.9)

Corrected MADF = (MADF on ODM basis) × (ODM/TDM)

For fresh grass (Morgan 1973)

ME (MJ kg^{-1} DM) = 15.9 – 0.19 MADF

(TDM = toluene DM, ODM = oven DM)

To establish such equations, the ME content is generally estimated with mature wether sheep offered feed at a level sufficient to maintain body weight (maintenance). Daily intake and faecal and urine excretion of energy are measured for individual animals over a 10-day period, and the loss of energy as methane is either determined in calorimeters, or estimated from published equations (Blaxter & Clapperton 1965). As mentioned earlier, an alternative laboratory method for estimating digestibility was developed by Jones & Hayward (1975) using a pepsin/cellulase digestion. However, comparison of this method with the *in vitro* method (Tilley & Terry 1963) and with the summative equation for fibre analysis (Van Soest & Wine 1967), carried out on tropical forages by Thomas *et al.* (1980), showed the *in vitro* method to be the most accurate.

7.8.2 Prediction of protein supply

Appropriate protein supplementation of forage-based diets relies on a protein rationing system to predict accurately both the supply of nitrogen to the rumen microbes and amino acids to the animal's tissues. Figure 7.11 is a schematic representation of the metabolizable protein (MP) system, as currently used in the UK (Alderman & Cottrill 1993).

Generally a significant part (usually 60–80%) of the feed CP entering the rumen is potentially degradable (see Section 7.3.5) (RDP, rumen degraded protein), with the principal end product being ammonia, and to a lesser extent amino acids and peptides. However, effective rumen degradable protein (ERDP) is a measure of the degraded protein available for microbial protein synthesis, recognizing that some ammonia diffuses into blood, whilst part of the RDP fraction may escape degradation in the rumen, especially at high rumen fractional outflow rates. ERDP availability in relation to FME supply will influence the net synthesis of microbial protein (MCP), which passes to the intestine and is digested by small intestinal enzymes. The digestible component (DMTP) is absorbed and usually constitutes the main source (60–70% of total) of amino acids (MP, metabolizable protein) available to the animal. Dietary CP which escapes ruminal degradation (UDP) is also digested in the intestines, and the digestible component (DUP) contributes to MP supply. The MP requirements of animals at low production levels can be met entirely by MCP, whilst

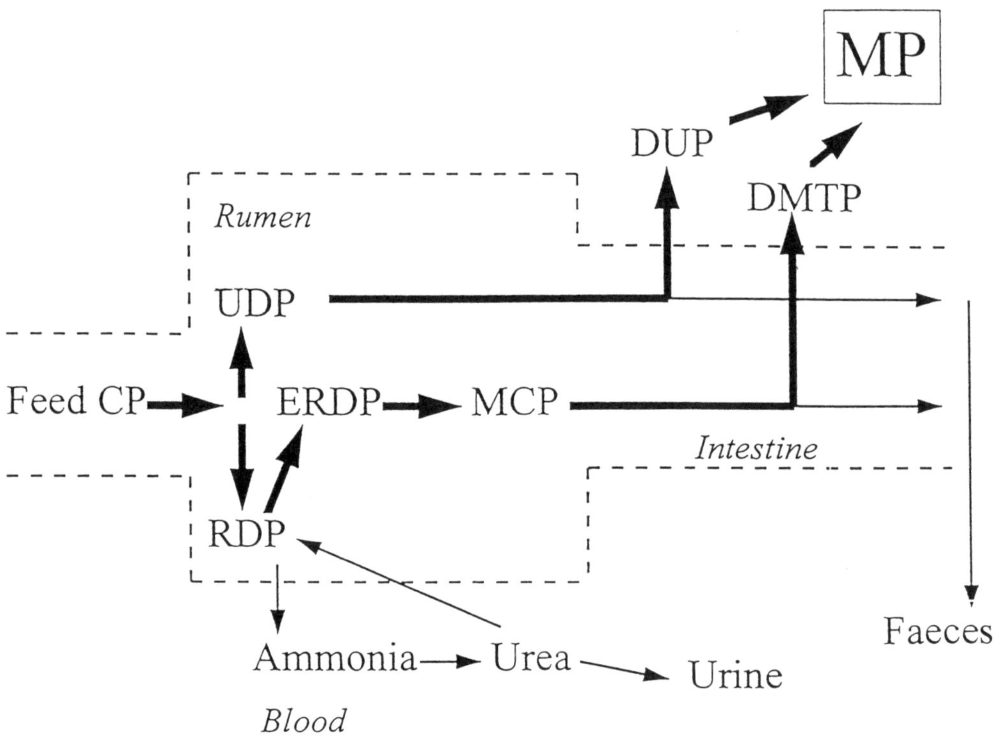

Fig. 7.11 Schematic representation of the UK Metabolizable Protein system.

additional DUP will be required for animals with higher levels of performance, such as high-yielding dairy cows.

Thus, the MP system provides a set of relationships to permit prediction of protein (amino acid) supply from a detailed characterization of dietary crude protein. Effective protein degradability, calculated with an appropriate rumen outflow rate for the feeding regime in question, allows calculation of the RDP and UDP proportions. The supply of ERDP is then calculated by assuming 100% and 80% efficiency of capture of slowly (SDP) and quickly (QDP) degraded CP, respectively, QDP being the fraction of the feed CP that is instantaneously lost *in situ*. Synthesis of MCP is estimated from the amount of ERDP available in situations when ERDP is limiting, but this should rarely be the case for normal forage-based diets, as microbial synthesis is usually limited by energy supply. Thus, MCP is calculated from FME supply (FME, see Section 7.3.5) using the following relationships:

MCP (g/d) = 9 FME (MJ/d) for low outflow rates (maintenance)
MCP (g/d) = 10 FME (MJ/d) for medium outflow rates (twice maintenance)
MCP (g/d) = 11 FME (MJ/d) for high outflow rates (three times maintenance)

Higher levels of feeding increase rumen outflow rates, which are known to increase the efficiency of microbial protein synthesis as the shorter residence times of microbes in the rumen reduces their maintenance energy costs. Consequently, higher outflow rates indicate that more ERDP (per unit of FME) will be required in the diet (equal to the MCP synthesized) to meet microbial needs. Failure to provide sufficient ERDP will impair microbial growth rate and net MCP yield, whilst possibly reducing fermentation rates in the rumen, leading to lowered diet digestibility and voluntary intake.

The supply of digestible microbial true protein (DMTP) is calculated as

$$\text{DMTP (g/d)} = 0.6375\ \text{MCP (g/d)}$$

The coefficient (0.6375) is based on the assumption that microbial true protein is 0.75 of MCP with a true digestibility in the intestines of 0.85.

DUP supply from feed protein escaping degradation in the rumen (UDP) is estimated as

$$\text{DUP (g/kg DM)} = 0.9\ (\text{UDP} - 6.25\text{ADIN})$$

assuming a digestibility of 0.9 for feed protein not present as acid detergent insoluble N (ADIN). This latter fraction is tightly bound to cell walls and assumed to be completely unavailable to the animal. Total DUP supply (g/d) is then calculated by multiplying DUP concentration (g/kg DM) for each feed by its DM intake (kg). The total supply of metabolizable protein (MP amino acids available to the tissues) is then the summation of the digestible microbial and undegraded feed components:

$$\text{MP (g/d)} = \text{DMTP} + \text{DUP}$$

The MP system is an improvement over previous systems because it recognizes the central part played by rumen microbes in protein digestion, and attempts to provide some representation of energy and protein interactions, albeit only with respect to rumen metabolism. Areas of weakness, including the use of calculated FME values to describe energy supply to the microbes, have been identified and are receiving attention. However, it is anticipated that future improvements can be made without major alteration to the structure of the system.

Table 7.6 provides an assessment of a range of typical forages as sources of N for rumen microbes. The low FME contents of the silages relative to their ME values is noteworthy, and results from their content of fermentation end-products which, by definition, is unavailable to the microbes. Forages with ERDP values below 9 g MJ^{-1} (low-quality grass and hay) are ERDP-deficient for all rumen outflow rates, whilst those with values less than 11 g MJ^{-1} are deficient only for animals at high feeding levels. The intake and efficiency of utilization of ERDP-deficient forages can be improved by supplementation with an ERDP source, either as true protein (e.g. soya) or possibly non-protein (e.g. urea), although the limitations of urea as an alternative 'protein' source must be recognized.

7.8.3 Prediction of intake

Interest in the ability to predict voluntary feed intake has led to the derivation of a range of equations based on different feed and animal characteristics. However, it is important to differentiate between equations which *predict* intake

Table 7.6 Use of the MP system to assess the ability of typical forages to meet microbial requirements for N.

Forage	ME* (MJ)	FME* (MJ)	CP* (g)	ERDP*,† (g)	ERDP : FME (g/MJ)
Fresh grass	7.5	6.9	97	52	7.5
Fresh grass	9.5	8.8	120	72	8.2
Fresh grass	10.7	10.0	135	87	8.7
Fresh grass	11.6	10.7	150	103	9.6
Fresh grass	12.3	11.4	190	141	12.4
Fresh grass	12.6	11.7	190	134	11.5
Grass silage	10.3	7.6	140	97	12.8
Grass silage	11.7	8.6	174	120	14.0
Grass hay	9.2	8.6	81	44	5.1
Lucerne hay	8.5	8.1	183	131	16.2
Dried grass	10.7	9.4	199	117	12.4
Dried lucerne	8.8	7.8	199	123	15.8

* Units per kg DM.
† ERDP for a rumen outflow rate of 5% per hour.

and those which simply represent correlations between intake and defined characteristics of the feed. Prediction implies the application of an equation or a set of equations to estimate the intake of feeds by animals not included in the data set from which the equations were derived, while correlations are generally derived from analysis within limited data sets. Correlations are usually derived for the purpose of understanding the factors which control intake, while prediction equations can be used by farmers and advisers in rationing programmes. Recent interest in the *in vitro* gas production technique and analysis of the data to predict intake can be considered as an example of the correlation approach rather than a genuine attempt to develop practical prediction equations.

The accuracy of prediction equations depends in part on the degree of fit between the data set from which the equation(s) was derived and the animals and feeds to which it is to be applied. Early prediction equations for dairy cows tended to be general, i.e. to relate to more than one type of conserved feed, and thus to be based on an extensive database. In the 1980s, separate equations were derived for the prediction of silage and hay, leading to improved accuracy.

A plethora of equations to predict forage intake has been derived from a variety of data sets, usually based on intakes measured under experimental conditions. Digestibility and animal liveweight are parameters common to many prediction equations, and for dairy cows, milk yield has frequently been included as a proxy for the genetic potential of the cow. However, this can have a confounding effect, as milk yield can also be considered as a response to feed intake.

Alderman & Cottrill (1993) reviewed a number of prediction equations of which only one is quoted here, that presented by AFRC (1991) to predict silage intake (SDMI).

$$\text{SDMI} = -3.74 - 0.387C + 0.1055Y + 0.0066W_n + 0.0136\,[\text{DOMD}]$$

where C is concentrate DM intake (kg/day), Y is actual milk yield (kg/day), W_n is actual cow liveweight in lactation week n and DOMD is the digestible organic matter content of the dry matter (g/kg).

Where there is a practical need for intake prediction, DOMD is often measured by near infrared reflectance spectroscopy, and a number of studies have been carried out to assess the accuracy of using NIRS spectra for intake prediction. Offer *et al.* (1998) quoted prediction errors for NIRS to be only one-third of errors for equations based on animal characteristics and traditional chemical analysis, but it is not clear how this would change if NIR spectra measured in different laboratories had been used. It should also be remembered that the accuracy of prediction equations is usually assessed with reference to data obtained under control conditions, while in practice, management factors are also likely to affect the actual intake achieved.

7.8.4 Future predictions

Most approaches currently available to predict nutrient supply (i.e. ME or MP) or animal performance in terms of feed intake, milk yield or liveweight rely on empirical equation(s) which include appropriate assessments of feed and/or animal characteristics. They were derived from data sets that were developed through exhaustive research effort. However, the derived equations are largely best mathematical fits, and do not necessarily contain explicit representations of the biology. This is certainly the case when ME content is predicted from DOMD content, and for silage intake predicted from silage DOMD content and level of concentrate feeding, along with some cow characteristics. In reality the same situation prevails with the MP system. Whilst it has some biological representation, the scheme is a factorial approach based on a series of empirical equations to describe specific functions such as level of feeding effect on rumen outflow rate and the yield of microbial protein in relation to FME supply. Clearly, as ruminant livestock feeding becomes more demanding with respect to maximizing the efficiency of utilization of dietary nutrients, and producing milk and meat of predetermined composition, so systems which provide a more comprehensive representation of biological mechanisms will be required. In this respect, considerable progress is already evident with respect to mechanistic models which represent the biological control of major processes, including feed intake (Forbes 1995), rumen digestion (Dijkstra *et al.* 1992) and nutrient metabolism in lactating (Baldwin *et al.* 1987) or growing (Black *et al.* 1987) animals. Further refinement of these models should ultimately lead to the development of more comprehensive models capable of being used for on-farm decision support in relation to feeding and management.

7.9 Nutritional disorders

Nutritional disorders may result from either a deficiency of specific nutrients such as vitamins or minerals, the presence of a toxin in the diet, or the production of an abnormal state (e.g. bloat) by a particular diet. These three states are discussed separately in the following sub-sections.

7.9.1 Vitamins

The vitamin content of herbage has little significance in the feeding of ruminants. With the possible exception of vitamin E, fresh herbage contains adequate levels of the fat-soluble vitamins (A, E and K) to meet the requirements of ruminant animals. Furthermore, bacteria resident in the rumen of the adult ruminant animal normally synthesize adequate quantities of the B vitamins, whilst the animal will synthesize its vitamin C and vitamin D requirements provided there is adequate exposure to sunlight in the latter case.

According to the suggested requirements (ARC 1980) for β-carotene (vitamin A precursor), a lactating cow with suckled calf requires 390 μg kg^{-1} liveweight which, assuming 450 kg body weight, estimates a daily requirement of 175 mg. This compares with contents of 9–127 mg kg^{-1} for grass, 3.9–18 mg kg^{-1} for hay and 8–13 mg kg^{-1} for silage (INRA 1980).

For vitamin D, the requirement of a lactating cow is given (ARC 1980) as 0.25 μg kg^{-1}, or 112 μg for a 450 kg cow. INRA (1980) quotes values of 400–800 mg kg^{-1} for hay, which suggests there should be no problem in meeting their requirements from forage. The suggested requirements for vitamin A, as β-carotene and retinol, and vitamin D as proposed by ARC (1980) for other classes of livestock, are given in Tables 7.7 and 7.8.

Because of the complexity of the inter-relationship between vitamin E, selenium deficiency and the quantities of unsaturated fats in diets for cattle, which interact to cause muscular dystrophy, no requirements for vitamin E are quoted. The effective prophylactic dose for calves would seem to be at least 20 mg

Table 7.7 Requirements for vitamin A for cattle and sheep (ARC 1980).

	Daily requirements (μg kg^{-1} LW)	
	As retinol	As β-carotene
Cattle		
Maintenance and growth	20	120
Conception and pregnancy	30	180
Lactation (minimum)	25	150
Lactation with suckled calf	65	390
Sheep		
Maintenance and growth	10	60
Pregnancy	20	120
Lactation	15	90

6 g carotene assumed to be equal to 1 g retinol or 3.33 iu in vitamin A.

Table 7.8 Requirements for vitamin D for cattle and sheep (ARC 1980).

	Daily requirement (μg kg^{-1} LW)
Cattle	
Calves	0.10
Growing cattle	0.10
Pregnant and lactating cows	0.25
Sheep	0.13

α-tocopherol daily. Because of the toxicity of excess selenium, the use of vitamin E to treat muscular dystrophy is recommended. The effects of conservation on vitamin contents are given in Table 7.9 (from INRA 1980), indicating that β-carotene, the main precursor of vitamin A, is well preserved by dehydration, whilst losses during haymaking and ensiling can be considerable. Vitamin D is not present in herbage, and is only formed when either the ergosterol in plants or the dehydrocholesterol in the animal's skin is irradiated to give ergocalciferol (vitamin D_2) or cholecalciferol, (vitamin D_3).

7.9.2 Minerals: major elements

The mineral elements essential for normal functioning of both plants and animals are phosphorus, potassium, sodium, calcium, magnesium, sulphur, chlorine, iron, manganese, zinc, copper and cobalt. In addition, animals require iodine and selenium, and possibly chromium, vanadium, nickel and tin, and may be harmed by excesses of molybdenum, fluorine, silica, selenium, lead and other heavy metals.

The mineral composition of grassland herbage summarized by McDonald *et al.* (1995) is given in Table 7.10, but it should be noted that the supply of

Table 7.9 The content ($mg\,kg^{-1}$) of fat-soluble vitamins in fresh and conserved feeds (after INRA 1980).

	Vitamin A β-carotene	Vitamin D_2	Vitamin E as total tocopherols
Lucerne			
Fresh	30–60	—	114–280
Hay	15–8	400–2000	38
Dehydrated	200–320+	160–400	148–301
Grass			
Fresh	9–127	—	105–166
Hay	3.9–18	400–800	200
Silage	8–13	—	38–470

Table 7.10 The mineral content ($g\,kg^{-1}$ DM) of pasture herbage (after McDonald *et al.* 1995).

Element	Low	Normal	High
Calcium	<2.0	2.5–5.0	>6.0
Phosphorus	<2.0	2.0–3.5	>4.0
Sodium	<1.5	1.5–3.0	>3.0
Magnesium	<1.0	1.2–2.0	>2.5
Potassium	<12	15–30	>35
Sulphur	<2.0	2.0–3.5	>4.0

minerals may be further augmented by ingestion of soil, either indirectly through contamination of harvested crops, or directly through the grazing/overgrazing of pastures. The extent of soil ingestion by grazing animals is influenced by the severity of defoliation and prevailing weather conditions, and under extreme conditions up to 1.5 kg (in cattle) and 0.3 kg (sheep) of soil may be consumed daily. However, knowledge of mineral intake is not sufficient to assess the ability of forages to supply an animal's mineral requirements, since the mineral elements are not all available to the animal. Apparent availabilities of 0.3–0.4 are commonly reported for calcium, phosphorus and magnesium in herbage, but true availabilities (accounting for endogenous losses) may be higher.

The effect of availability on requirement is considered in detail in several texts (ARC 1980, Ammerman *et al.* 1995, Underwood & Suttle 1999). When animals are fed indoors, it is relatively simple to supplement their diet with the necessary minerals, but grazing animals may suffer from a number of disorders arising from deficiencies, imbalances or excesses of mineral elements in the diet. Clinical signs of deficiency in the major elements are usually restricted to calcium, phosphorus and magnesium.

Hypocalcaemia results in 'milk fever' in cows and 'lambing sickness' in ewes. It generally occurs in cows in winter and early spring and is not directly attributable to a reduction in calcium availability in the diet. Rather it is the animal's inability to mobilize calcium rapidly enough from the skeleton in response to increased lactational demands which precipitates the symptoms, although the animal's previous nutritional history with respect to calcium and phosphorus may have some predisposing effect. Prevention or minimization of the occurrence of the disease can be achieved by feeding low-calcium diets prior to parturition, which stimulate calcium intake and mobilization at the onset of lactation. This approach, combined with an oral calcium prophylactic treatment for 2 days post-calving, reduced a 35% incidence of the condition to zero on one badly affected farm.

Phosphorus content in forages is influenced by the level of available phosphorus in the soil, and phosphorus deficiencies will reduce both forage yield and phosphorus content. This may lead to reduced animal performance and poor fertility especially if the forage Ca : P ratio increases above 2, when phosphorus availability declines. Hypomagnesaemia or 'grass staggers' is attributable to low blood magnesium levels, and is generally found in dairy cows following turnout to spring pasture, and in suckler cows in late autumn when under-nutrition may coincide with climatic stress. Excessive mobilization of body fat in postpartum cows can induce hypomagnesaemia. During active grass growth in spring, magnesium levels and the amount of available magnesium in grass decreases. At such times, magnesium supplementation is needed, either through magnesium-enriched supplements (e.g. concentrates or blocks), by magnesium oxide dusting of pasture, or by the addition of magnesium acetate to the drinking water. The use of ammonium and potassium fertilizers in early spring is another contributory cause to low forage magnesium levels ('t Hart 1956,

Paterson & Crichton 1960). Failure to detect and treat animals suffering from hypomagnesaemia with a subcutaneous injection of magnesium sulphate leads to rapid death.

7.9.3 Minerals: minor elements

The content of minor elements in forage is extremely variable depending on underlying geology, forage variety and species, stage of growth, fertilizer application, soil acidity and contamination. The threshold levels for ruminants are summarized in Table 7.11. Iodine deficiency is generally confined to particular geographical areas and to specific crops (e.g. brassicas), which contain goitrogens that limit the uptake of iodine by the thyroid gland, and may give rise to thyroid deficiency and goitre. In the UK the incidence of such diseases is low, due to the use of iodized salt licks which minimize the effects in susceptible areas. In countries such as Australia, where iodine deficiency is quite common, and with extensive grazing systems which minimize the effectiveness of mineral licks, slow-release intra-ruminal devices may be used to provide long-term

Table 7.11 Threshold concentrations of trace elements in forage for adult ruminants (mg kg^{-1} DM).

	Threshold	Exceptions	Cattle	Sheep
Co	Adequate		>0.11	>0.11
	Marginal		0.05–0.07	0.07–0.10
	Deficient		<0.05	<0.07
Cu	Adequate	Mo <1.0	>10.0	>5.0
	Marginal	Mo >3.0	>10.0	>5.0
	Deficient	Mo >3.0	<10.0	<5.0
Fe	Adequate		>30	>30
	Adequate	Calves <150 kg	100	
	Adequate	Cows in late pregnancy	40	
I	Adequate	Normal diet	0.50	0.50
	Adequate	Goitrogenic diet	2.0	2.0
Mn	Adequate		25	25
Se	Adequate		>0.10	>0.10
	Marginal		0.03–0.05	0.03–0.05
	Deficient		<0.03	<0.03
Zn	Adequate		50	50
	Marginal		20–40	30–50
	Deficient		<20	<30

NB: Because of the extreme susceptibility of some sheep breeds to Cu toxicity, it is possible even with forage Cu levels within the normal range of 8–15 mg kg^{-1} DM if forage Mo is also low (0.2–0.8 mg kg^{-1} DM). The critical Cu : Mo ratio for toxicity is 20 or greater.

protection (>5 years). Intramuscular injection with iodised poppy-seed oil has also been used quite successfully.

Of the trace elements, copper-related disorders are the most common. At concentrations in forage dry matter of less than 5 mg Cu kg^{-1}, frank copper deficiency will occur with resulting ill thrift of all affected animals. In sheep, the most common disease is swayback (neonatal ataxia) in lambs, whilst in adult sheep symptoms of impaired wool growth can often be seen. Cattle manifest copper deficiency as scouring, anaemia, impairment of bone development and coat pigmentation. Deficiency symptoms also occur on forages with apparently adequate copper content, where high levels of molybdenum (alone or in combination with high levels of inorganic sulphate) markedly reduce Cu availability to the animal. Infertility in cattle is a possible consequence of molybdenum-induced copper deficiency. It has been proposed that the availability of copper (ACu) can be predicted by the following equation:

$$ACu(\%) = 5.72 - 1.297S - 2.785\log_e Mo + 0.227\,(Mo \times S)$$

where S and Mo represent sulphur and molybdenum concentrations in grass (g kg^{-1} DM), respectively. High iron concentrations, due to soil contamination of pasture or silage, can also reduce copper availability.

Pasture treatment with $CuSO_4$ will not provide long-term elevation of forage copper concentrations, so direct animal treatment is required to counteract deficiency. Injectable copper preparations, slow-release boluses or capsules containing copper oxide needles are all viable alternatives. However, the homeostatic mechanisms controlling plasma copper levels appear to be less well developed in sheep than in cattle. Consequently, indiscriminate use of copper-containing salt licks or free access minerals for sheep should be avoided, otherwise copper poisoning may result. Furthermore, housed sheep tend to accumulate copper in the liver and other vital organs, and in such situations dietary copper levels must be controlled. As an added complication, the susceptibility of sheep to high levels of copper, and hence the manifestation of copper poisoning, varies between breeds, with Texels being particularly susceptible when dietary molybdenum levels are low.

Pine is a nutritional disorder arising from cobalt insufficiency in the diet. It is related to the animal's failure to synthesize sufficient vitamin B_{12} in the rumen, and is manifested as inappetance and progressive emaciation. Lambs after weaning are most susceptible, whilst adult cattle are less susceptible. The condition has been recognized for many years and its occurrence controlled by the application of cobalt fertilizer to pasture or by direct treatment of the animal with slow-release cobalt boluses. Although the incidence of the disease in Britain is declining, continuing vigilance is required as the disease is endemic in many areas and relaxation of the control measures will inevitably lead to its recurrence.

Several trace mineral deficiencies (including cobalt, manganese, zinc and copper) are aggravated on calcareous soils, or after lime has been applied. One final aspect of trace-element deficiency that is worth noting is its effect on immune competence. This renders the newborn less viable and the adult more susceptible to infectious diseases. Further consideration of mineral nutrition in relation to grazing management is given in Section 10.11.

7.9.4 Toxins

The presence of oestrogens (e.g. formononetin) has been observed in some pasture legumes, especially red clover, at variable concentrations related to the growing conditions of the plant. The effects may be observed as increased udder development in heifers, and teat length extension in both female and male sheep. The presence of oestrogens has little effect on cattle fertility, but in sheep it is advisable to remove all breeding stock from pastures containing significant amounts of red clover at least 3 weeks prior to commencement of the breeding season. In growing animals, plant oestrogens may have beneficial effects on carcass protein deposition, which may be worthy of exploitation whilst the ban on hormonal implants continues. However, the quantitative effect of naturally occurring oestrogens on growth has not been fully assessed. White clover, unlike red clover, does not have high oestrogen contents unless the plants are diseased. Much of the oestrogenic activity of legumes is lost during the process of artificial dehydration using high temperatures.

The presence of thiaminases in the diet can reduce thiamin absorption, producing a simple deficiency in horses but rather more complex disorders in sheep and cattle. Bracken (*Pteridium aquilinum*) and horsetails (*Equisetum* spp) contain thiaminases (see also Chapter 6), whilst it has been suggested that cerebrocortical necrosis, which occurs sporadically in sheep and young cattle, is caused by fungal infections of ryegrass which produce thiaminases.

The occurrence of toxins in tropical forages has been reviewed by Crowder & Chheda (1982), who concluded that legumes generally contain a wider range of deleterious substances than grasses. For example, *Leucaena leucocephala*, a leguminous shrub, will induce toxicity due to high concentrations of mimosine (amino acid). However, in some areas of the world, animals have adapted to mimosine due to the occurrence of mimosine-metabolizing bacteria in their rumens, whilst new low-mimosine species of *Leucaena* have also been bred.

7.9.5 Bloat

Ruminant bloat is the abnormal distension of the reticulo-rumen, caused by excessive retention of fermentation gases within the ruminal cavity. Bloat can occur in stock grazing grass pastures and brassica forage crops, but is most common on pastures containing high proportions of legumes such as red clover, lucerne and particularly white clover. In frothy bloat, which is typical on such diets, the gas bubbles remain dispersed within the rumen contents, and it is the failure of the bubbles to coalesce which inhibits removal of gases from the rumen by eructation.

At moderate stages of bloat, as measured by distension of the upper portion of the left flank immediately behind the ribs, practical advice is to monitor the affected animals but take no further action unless the position deteriorates. If the severity of bloat increases, the animal will suffer increased discomfort, followed by inappetance. In such cases, treatment is essential, otherwise in extreme situations death will occur, often quite rapidly. In New Zealand, where heavy reliance is placed on the use of legume-containing pastures, the annual death rate due to legume bloat has been estimated at between 0.3 and 1.2% of the dairy population.

Many possible explanations for bloat have been advanced, but the sporadity of its occurrence has hindered precise identification of the factors involved and prediction of its occurrence. It is clearly related to the extensive ruminal fermentation which occurs following the ingestion of a significantly high quantity of legumes, exacerbated in those instances when the animals were hungry at the time of access to the pasture. In this regard, the animal's ability to positively select for legumes in a grass: legume sward will worsen the situation, whilst dairy cows, which have to tolerate an interrupted grazing pattern due to the milking routine, are more susceptible than beef cattle which are more continuous grazers. Bloat in sheep occurs infrequently, and these animals can be considered as bloat-safe on most legume-containing diets.

There are several options available to reduce the risk of bloat. Minimizing the interruptions to grazing is recommended wherever feasible, and limiting forage availability when animals are introduced on to legume-containing pastures is advisable. Provision of limited quantities of hay or straw may also reduce the incidence of bloat, whilst cutting and partially wilting the pasture prior to 'grazing' has been effective. Alternatively, use of proprietary antifoaming agents and vegetable oils is effective, either through spraying on to the pasture or oral drenching. The practice of flank painting with vegetable oil is not considered to be a guaranteed way of minimizing the occurrence of bloat.

7.10 Potential for forages

Forages are perceived by the general public to be the 'natural' feed of ruminants, and following the BSE crisis, consumers in the UK and elsewhere are demanding a diet that is healthy and naturally produced, yet at reasonably low cost. Young fresh grass has a high potential feeding value, but herbage yield is low at this time, whilst the efficiency of utilization of such feeds by the animal is often compromised when compared with cereal-grain feeding. If the potential feeding value of forages could be maintained for longer (while yield increases) and nutrient utilization improved, then it may be possible for forages to make a more significant contribution to ruminant meat and milk production. However, one major limitation to animal productivity from grass-dominant diets is the quantity of forage which animals can eat and process by digestion. This is particularly relevant with high genetic-merit cows where, even if forage quality is optimal, it will never be consumed in sufficient quantities to meet their nutrient requirements, especially energy.

In the following paragraphs some suggestions which could lead to increased use of forages are discussed, recognizing that better progress in these areas will be achieved through closer relationships between animal scientists, plant breeders and farmers, whilst also recognizing the importance of food purveyors (e.g. supermarkets) and consumers.

7.10.1 Forage mixtures at pasture

The introduction of legumes into pastures for grazing ruminants confers significant benefits in terms of animal performance, provided the occurrence of bloat can be minimized or controlled. With grazed grass, optimal DM intakes rarely exceed 2.5% of body weight, with many instances when intakes will be substantially lower. On the other hand, with swards containing high proportions of white clover, intakes exceeding 3.0% of body weight have been observed, and nutrient intake is further enhanced by the slower rate of decline in clover digestibility compared with grass. However, maintaining mixed swards of grass and legumes is a management challenge, although research and plant breeding are increasingly providing technical advice to support appropriate grazing systems. In support of the increased feed intakes on legume-containing swards, Beever *et al.* (1986) found, when comparing spring, summer and autumn growths of both perennial ryegrass and white clover, that in all instances the rate and extent of OM digestion in the rumen was higher on the clover diets. This was despite all forages having similar overall digestibilities. This and other research has helped to identify and quantify the mechanisms which contribute to the increased levels of feed intake seen on legumes, and may provide the necessary impetus for grass breeders to reconsider the selection criteria currently used.

In contrast, one of the major objectives of legume breeding has been to reduce the propensity for bloat. The trefoils (*Lotus* and *Medicago* spp.) are known to be non-bloating, and the same is true of sainfoin (*Onobrychis viciifolia*) due to the presence of tannins in the crops which control the rate of rumen proteolysis. Suggestions of introducing tannins into bloating legumes, such as red and white clover and lucerne, have been made for many years, with relatively little success. However, with the advent of more sophisticated techniques in molecular genetics, the possibility of bloat-free legumes becoming commercial realities must have increased considerably.

7.10.2 Conservation techniques

Silage DM intake invariably fails to reach that which could be achieved with the same forage fed fresh, whilst silage N utilization is generally poor (Beever 1980). To date, emphasis on the development of silage additives has been to optimize the proportion of harvested DM which is ultimately fed, with an associated interest in silage stability during storage and feed out. However, such approaches are unlikely to improve nutritive value. Use of formaldehyde-containing additives, which had significant beneficial effects on silage N utilization, has declined in popularity since the 1980s, and whilst acid-based additives are still used, there has been a remarkable increase in the use of biological additives (inoculants). If applied at the correct rate, such products will undoubtedly promote a homolactic fermentation which is energetically more efficient, but evidence of increased efficiency of nutrient utilization by the animal is difficult to establish unequivocally. With interest in all aspects of the ensiling process, coupled with a greatly expanded range of silage additives, it is likely that new ensiling techniques, which pay greater attention to the nutritive value of the resulting silage, will be developed in conjunction with the need to improve silo management. In this respect, the progress achieved in the 1990s has improved the predictability of silage making, but considerable scope remains to improve the intake and nutritive value characteristics of silage when fed.

7.10.3 Mixtures of conserved forages

As with mixed swards, it is possible to predict that nutritional benefits are likely to occur when appropriate mixtures of conserved forages are included in the total mixed ration. As indicated earlier, Phipps *et al.* (1995) showed positive responses in intake and animal performance to mixtures of grass and maize silage fed to lactating cows, although earlier work using legume silages as supplements to maize silage for growing calves showed inconsistent effects (Bowden *et al.* 1980). Responses in forage intake can also be observed when lucerne, especially in the dried (artificially or sun-cured) form, is included in the ration. Also, many farmers with high genetic merit cows aim to feed at least three different forages in the total ration in order to achieve maximum intake of nutrients from the forage, and hence the whole diet. There are other ways in which mixtures of conserved forages can be used, the main advantage being that such crops can be grown separately and mixed at the time of feeding, thus avoiding the complications of managing mixed forage swards. Finally, with respect to meeting the protein requirements of forage-fed ruminants, there is considerable interest in alternative crops which have significant levels of protein in the seed or grain, notably peas, beans and soya, which could all become important replacements for proteins of animal origin.

7.10.4 Chemical processing of low quality diets

Whilst much of this chapter has concentrated on the subject of maximizing forage nutritive value, and hence the level of animal productivity achieved from forages, there are undoubtedly some situations where poorer quality forages can be used to advantage. This would be particularly the case with heifer replacements and beef store cattle, where it may be possible to include a certain quantity of cereal straw in the diet. In such situations it is tempting to consider upgrading the nutritive value of such feeds by chemical treatment (e.g. with ammonia or caustic soda), which can be readily undertaken on the farm using specialist equipment. Substantial improvements in forage digestibility and total nutrient intake have been reported, but the economic implications of such processes should be fully considered before placing too much reliance upon them. Furthermore, ammoniation of poorer quality hays has given rise to animal health problems with the sporadic occurrence of unpredictable hyperactivity, possibly related to the presence of 4-methyl imidiazole, thought to be produced during ammoniation when the treated forage contains significant quantities of residual sugars.

7.10.5 The use of feed additives

The use of feed additives, including ionophores (monensin, lasalocid etc.), will manipulate the production of VFA towards propionate at the expense of acetate, with consequent beneficial effects on efficiencies of production. Such additives will also reduce the extent of methane production in the rumen, which will result in more energy being available to the animal. However, use of such products is not permitted for dairy cows in the EU, whilst its long-term use in beef and sheep is questionable. However, given the benefits that they are known to confer, there is now research to find suitable alternatives which are publicly acceptable.

7.10.6 Protein supplementation

With respect to the digestion of forages, two important aspects have emerged from research studies. First, due to the high solubility of dietary N in most silages and fresh forages, ruminal digestion of protein on such diets is extensive. Second, high rates of ammonia absorption from the rumen give rise to reduced supplies of duodenal protein, particularly with wilted grass silages (Siddons *et al.* 1982), whilst high forage diets tend to promote high acetate production rates in the rumen. This aspect is clearly desirable for the successful production of milk fat by the lactating dairy cow, but it may lead to high levels of fat deposition in the carcass of growing animals, which conflicts with current consumer preference. In both situations, protein supplements have been used to advantage (Castle 1982, Gill *et al.* 1987). Such supplements, provided in small amounts with low- to medium-quality forages, will have a reduced depressive effect on forage intake relative to that observed with energy-rich supplements. Thus protein supplements can be used effectively to optimize both forage intake and utilization, and their use is likely to continue. However, further work is required to fully understand the long-term effects of periods of protein supplementation on whole-lactation yields of cows and the lifetime growth of beef cattle.

7.10.7 'The future'

Undoubtedly grass will continue as an important feedstock for ruminants, provided the associated industries of meat and milk production continue to be practised. However, in the context of the British Isles – although similar arguments apply elsewhere – it is the more fundamental debate over extensification vs. intensification that will provide the main framework by which grassland will be exploited during the early decades of the twenty-first century. Undoubtedly there are strong pressures towards extensification of all systems of animal production, but in this respect it is difficult to visualize major changes with respect to sheep production. Until suitable markets are re-established, bringing greater financial stability, it is likely that there will be little further investment or change within current systems of lamb production which rely primarily on pasture, both improved and unimproved. In contrast, if financial margins remain, at best, modest, then lamb production is likely to intensify, with greater use of supplementary feeding to ensure the rapid production of heavy lambs with desirable carcass traits. In this respect there will be more opportunity for supplementary 'forages', either grazed *in situ* (e.g. stubble turnips, lucerne aftermaths) or fed indoors (e.g. maize silage), than for any major initiative involving the production of new grass or legume varieties.

In relation to beef cattle, where a significant part of their life is spent indoors, there may be more opportunities for the use of improved forages, but once again, before any need is established for 'superior' grasses and legumes, considerable scope still exists in relation to the production of better quality silage, especially improving the predictability of big-bale ensiling, as well as the extended use of forage alternatives, with maize silage being an obvious candidate.

Finally, with dairy cattle it is reasonable to conclude that the extensive/intensive debate is already occurring. Many dairy farmers in the UK have listened with interest to the New Zealand approach to optimizing grass use by dairy cows, and some have adopted such approaches, with variable success. Some of these converts are set to continue in this manner, and these will be the farmers who are able to benefit most from improved forage varieties, especially in the grazing situation. It is indisputable that the major reason why dairy farmers move to other forage sources and supplementary feeds is because at best, grazed grass cannot support the production of more than 27–30 kg milk per day. Improved genetics have provided cows capable of peak yields of over 50 kg per day. For those who wish to follow the extensive route, then issues such as improved grass varieties for both grazing and ensiling will need to be addressed, and without doubt the focus must be towards improving forage intake by whatever means available. This will be more important than raising soluble sugar levels, unless this leads to some improvement in intake, or the introduction of stay-green genes. At the same time, if the nature of forage proteins, with respect to their rate of degradation in the rumen, could be controlled, the use of protein supplements could be reduced. However, in this respect it would be more appropriate to consider legumes as a more efficient means of supplying protein, in both the grazing and ensilage situation.

In contrast, there will be many farmers who will retain milk output per cow as their major focus, and undoubtedly this will lead to more intensification, with increased use of forages other than those based on grass. However, in such situations, it will be difficult to see any major increase in the forage content of the ration, especially when cereals can be purchased for less than the cost of ensiled forage, and greater demands are placed upon cows through genetic improvement, and with extended lactations replacing the 365-day calving interval. Thus, it is likely that more cows will spend less time at pasture, and be housed with total mixed rations, based in part on a variety of feeds that can be grown on the farm and fed with full traceability. Grass will have a role to play in this scenario, but other forages along with cereal grains and pulses will achieve greater importance in the ration of lactating dairy cows.

7.11 References

AFRC (1991) Voluntary intake of cattle. *Nutrition Abstracts and Reviews* **61**, 815–23.

Alderman G. & Cottrill B. (1993) *Energy and Protein Requirements of Ruminants*, 176 pp. CAB International, Wallingford.

Ammerman C.B., Butler D.H. & Lewis A.J. (eds) (1995) *The Bioavailability of Nutrients for Animals.* Academic Press, San Diego, CA.

ARC (Agricultural Research Council) (1980) *The Nutrient Requirements of Ruminant Livestock.* Commonwealth Agricultural Bureaux, Farnham.

ARC (Agricultural Research Council) (1984) *Report on the Protein Group of the Agricultural Research Council Working Party on the Nutrient Requirements of Ruminants.* (Supplementary report to Chapter 4, ARC 1980). Commonwealth Agricultural Bureaux, Farnham.

Armstrong (1965) Carbohydrate metabolism in ruminants and energy supply. In Dougherty R.W., Allen R.S., Burroughs W., Jacobson N.L. & McGilliard A.D. (eds) *Physiology of Digestion in the Ruminant*, 272–88. Butterworth, London.

Aston K., Fisher W.J., McAllan A.B., Dhanoa M.S. & Dewhurst R.J. (1998) Supplementation of grass silage-based diets with small quantities of concentrates: Strategies for allocating concentrate crude protein. *Animal Science* **67**, 17–26.

Baldwin R.L., France J., Beever D.E., Gill M. & Thornley J.H.M. (1987) Metabolism of the lactating cow. 3. Properties of mechanistic models suitable for evaluation of energetic relationships and factors involved in the partition of nutrients. *Journal of Dairy Research* **54**, 133–45.

Barber G.D., Givens D.I., Kridis M.S., Offer N.W. & Murray I. (1990) Predicting the organic matter digestibility of grass silage. *Animal Feed Science and Technology* **28**, 115–28.

Beever D.E. (1980) The utilization of protein in conserved forage. In Thomas C. (ed) *Forage Conservation in the 80s.* British Grassland Society Occasional Symposium, No. 11, 131–43. BGS, Hurley.

Beever D.E. (1993) Rumen function. In Forbes J.M. & France J. (eds) *Quantitative Aspects of Ruminant Digestion and Metabolism*, 187–221. CAB International, Wallingford.

Beever D.E., Coehlo da Silva J.F., Prescott J.H.D. & Armstrong D.G. (1972) The effect in sheep of physical form and stage of growth on the sites of digestion of a dried grass. 1. Sites of digestion of organic matter, energy and carbohydrate. *British Journal of Nutrition* **28**, 347–56.

Beever D.E., Dhanoa M.S., Losada H.R., Evans R.T., Cammell S.B. & France J. (1986) The effect of forage species and stage of harvest on the processes of digestion occurring in the rumen of cattle. *British Journal of Nutrition* **56**, 439–54.

Beever D.E., Rook A.J., France J., Dhanoa M.S. & Gill M. (1991) A review of empirical and mechanistic models of lactational performance by the dairy cow. *Livestock Production Science* **29**, 115–30.

Black J.L., Beever D.E., Faichney G.J., Howarth B.R. & Graham N.McC. (1981) Simulation of the effects of rumen function on the flow of nutrients from the stomach of sheep. Part 1. Description of computer program. *Agricultural Systems* **6**, 195–219.

Black J.L., Faichney G.J., Beever D.E. & Howarth B.R. (1982) Alternative systems for assessing the nitrogen value of feeds for ruminants. In Thompson D.J., Beever D.E. & Gunn R.G. (eds) *Forage Protein in Ruminant Animal Production*. British Society of Animal Production Occasional Publication No. 6, 107.

Black J.L., Gill M., Beever D.E., Thornley J.H.M. & Oldham J.D. (1987) Simulation of the metabolism of absorbed energy-yielding nutrients in young sheep: Efficiency of utilization of acetate. *Journal of Nutrition* **117**, 105–15.

Blaxter K.L. & Clapperton J.L. (1965) Prediction of the amount of methane produced by ruminants. *British Journal of Nutrition* **19**, 511–22.

Blaxter K.L., Graham N.McC. & Wainman F.W. (1956) Some observations on the digestibility of food by sheep, and on related problems. *British Journal of Nutrition* **10**, 69–91.

Bowden D.M., Osbourn D.F., Gill M. & Gibb B.C. (1980) Legume silages as supplements to a maize silage plus urea diet for young calves. *Animal Production* **30**, 355–64.

Broster W.H., Sutton J.D. & Bines J.A. (1979) Concentrate: Forage ratios for high-yielding dairy cows. In Haresign W. (ed) *Recent Developments in Animal Nutrition – 1978*, 99–126. Butterworths, London.

Cammell S.B. (1977) Equipment and techniques used for research into the intake and digestion of forages by sheep. *GRI Technical Report No. 24*. Grassland Research Institute, Hurley.

Cammell S.B., Dhanoa M.S., Beever D.E., Sutton J.D. & France J. (1998) An examination of the metabolizable energy requirements of lactating dairy cows. *Proceedings of the British Society of Animal Science Winter Meeting*, Scarborough, 197. BSAS, Penicuik.

Castle M.E. (1982) Making high quality silage. In Rook J.A.F. & Thomas P.C. (eds) *Silage for milk production. Technical Bulletin, No. 2*, 127–50. Hannah Research Institute, Ayr.

Castle M.E., Gill M.S. & Watson J.N. (1981) Silage and milk production: A comparison between barley and dried sugar beet pulp as silage supplements to silage of high digestibility. *Grass and Forage Science* **36**, 319–24.

Chacon E. & Stobbs T.H. (1976) Influence of progressive defoliation of a grass sward on the eating behaviour of cattle. *Australian Journal of Agricultural Research* **27**, 709–25.

Clancy M.J. & Wilson R.K. (1966) Development and application of a new method for predicting the digestibility and intake of herbage samples. *Proceedings 10th International Grassland Conference*, Helsinki, 445–53.

Conrad H.R., Pratt A.D. & Hibbs J.W. (1964) Regulation of feed intake in dairy cows. 1. Change in importance of physical and physiological factors with increasing digestibility. *Journal of Dairy Science* **47**, 54–62.

Corrall A.J., Lavender R.H. & Terry C.P. (1979) Grass species and varieties: Seasonal patterns of production and relationships between yield, quality and date of first harvest. *GRI Technical Report No. 26*. Grassland Research Institute, Hurley.

Crowder L.V. & Chheda H.R. (1982) *Tropical Grassland Husbandry*. Longman, London.

Cushnahan A., McGilloway D., Laidlaw A.S., Mayne C.S. & Kilpatrick D.J. (2000) The influence of sward height and other sward parameters on short term herbage intake rates and grazing behaviour of lactating dairy cows. *Animal Science* in press.

Dijkstra J., Neal H.D. St.C., Beever D.E. & France J. (1992) Simulation of nutrient digestion, absorption and outflow in the rumen: Model description. *Journal of Nutrition* **122**, 2239–56.

Forbes J.M. (1995) Voluntary feed intake. In Forbes J.M. & France J. (eds) *Quantitative Aspects of Ruminant Digestion and Metabolism*, 479–94. CAB International, Wallingford.

Gill M., Beever D.E., Buttery P.J., England P.M., Gibb M.J. & Baker R.D. (1987) The effect of oestradiol-17 beta implantation on the response in voluntary intake, live-weight gain and body composition, to fishmeal supplementation of silage offered to growing calves. *Journal of Agricultural Science, Cambridge* **108**, 9–16.

Gill M., Robinson P.H. & Kennelly J.J. (1999) Diurnal patterns in rumen volume and composition of digesta flowing into the duodenum. *Animal Science* in press.

Givens D.I., Everington J.M. & Adamson A.H. (1989) The digestibility and metabolizable energy content of grass silage and their prediction from laboratory measurements. *Animal Feed Science and Technology* **24**, 27–43.

Givens D.I., Everington J.M. & Adamson A.H. (1990) The nutritive value of spring-sown herbage produced on farms throughout England and Wales over 4 years. II. The prediction of apparent digestibility in vivo from various laboratory measurements. *Animal Feed Science and Technology* **27**, 173–84.

Givens D.I., Moss A.R. & Adamson A.H. (1992) The chemical composition and energy value of high temperature dried grass produced in England. *Animal Feed Science and Technology* **36**, 215–28.

Green J.O., Corrall A.J. & Terry R.A. (1971) Grass species and varieties: Relationships between stage of growth, yield and forage quality. *GRI Technical Report No. 8*. Grassland Research Institute, Hurley.

Hart M.L. (1956) Some problems of intensive grassland farming in The Netherlands. In Neale G.J. (ed) *Proceedings of the 7th International Grassland Congress*, Palmerston North, New Zealand, 70–9.

Hogan J.P. & Weslon R.F. (1970) Quantitative aspects of microbial protein synthesis in the rumen. In Phillipson A.T. (ed) *Physiology of Digestion and Metabolism in the Ruminant*, 474–85. Oriel, Newcastle upon Tyne.

Hovell F.D. De B., Greenhalgh J.F.D. & Wainman F.W. (1976) The utilization of diets containing acetate salts by growing lambs as measured by comparative slaughter and respiration calorimetry together with rumen fermentation. *British Journal of Nutrition* **35**, 343–63.

Hungate R.E. (1966) *The Rumen and its Microbes*. Academic Press, London.

INRA (Institut National de Reserche Agronomique) (1980) *Alimentation des Ruminants*. 2nd edn. INRA, Versailles.

Jones D.I.H. & Hayward M.V. (1975) The effect of pepsin pretreatment of herbage on the prediction of dry matter digestibility from solubility in fungal cellulase solution. *Journal of the Science of Food and Agriculture* **26**, 711–18.

Le Du Y.L.P. & Newberry R.D. (1982) Supplementing the grazing dairy cow during periods of pasture restriction. *Grass and Forage Science* **37**, 173–4.

Lonsdale C.R. (1976) *The Effect of Season of Harvest on the Utilization by Young Cattle of Dried Grass given Alone or as a Supplement to Grass Silage.* Ph.D. Thesis, University of Reading.

MacRae J.C. & Lobley G.E. (1986) Interactions between energy and protein. In Milligan L.P., Grovumo W.L. & Dobson A. (eds) *Control of Digestion and Metabolism in Ruminants*, 367–85. Prentice-Hall, Englefield Cliffs, NJ.

MAFF (Ministry of Agriculture, Fisheries and Food) (1987) *Feed Evaluation Unit.* Technical Bulletin.

MAFF (Ministry of Agriculture, Fisheries and Food) (1990) *UK Tables of Nutritive Value and Chemical Composition of Feedstuffs*. Rowett Research Services, Aberdeen.

MAFF (Ministry of Agriculture, Fisheries and Food) (1993) *Prediction of the Energy Values of Compound Feeding Stuffs for Farm Animals.* MAFF, London.

Mayne C.S. (1991) Effects of supplementation on the performance of both growing and lactating cattle at pasture. In Mayne C.S. (ed) *Management Issues for the Grassland Farmer in the 1990s.* British Grassland Society Occasional Symposium, No. 25, 55–71. BGS, Hurley.

McAllan A.B., Siddons R.C. & Beever D.E. (1987) The efficiency of conversion of degraded nitrogen to microbial nitrogen in the rumen of sheep and cattle. In Jarrige R. & Alderman G. (eds) *Feed Evaluation and Protein Requirement Systems for Ruminants*, 111–28. CEC, Luxembourg.

McDonald P., Edwards R.A., Greenhalgh J.F.D. & Morgan C.A. (1995) *Animal Nutrition*. 5th edn. 607 pp. Longman, Harlow.

Meijs J.A.C. (1986) Concentrate supplementation of grazing dairy cows. 1. Effect of concentrate intake and herbage allowance on herbage intake. *Grass and Forage Science* **41**, 229–35.

Meijs J.A.C. & Hoekstra J.A. (1984) Concentrate supplementation of grazing dairy cows. 1. Effect of concentrate intake and herbage allowance on herbage intake. *Grass and Forage Science* **39**, 59–66.

Menke K.H., Raab L., Salewski A., Steingass H., Fritz D. & Schneider W. (1979) The estimation of digestibility and metabolizable energy content of ruminant feedstuffs from the gas production when they are incubated with rumen liquor *in vitro*. *Journal of Agricultural Science, Cambridge* **193**, 217–22.

Morgan D.E. (1973) *Agricultural Development and Advisory Service Annual Report 1972.* ADAS, Leeds.

Moss A.R. & Givens D.I. (1990) Chemical composition and in vitro digestion to predict digestibility of field-cured and barn-cured grass hays. *Animal Feed Science and Technology* **31**, 125–38.

NIAB (National Institute of Agricultural Botany) (1998) Grasses and Herbage Legumes Variety Leaflet 1998/99. NIAB, Cambridge.

Offer N.W., Cottrill B.R. & Thomas C. (1996) The relationship between silage evaluation and animal response. In Jones D.I.H., Jones R., Dewhurst R., Merry R. & Haigh P.M. (eds) *Proceedings 11th International Silage Conference*, Aberystwyth, 1996, 26–38. IGER, Aberystwyth.

Offer N.W., Percival D.S., Dewhurst R.J. & Thomas C. (1998) Prediction of the voluntary intake potential of grass silage by sheep and dairy cows from laboratory silage measurements. *Animal Science* **66**, 357–67.

Ørskov E.R. & McDonald I. (1979) The estimation of protein degradability in the rumen from incubation measurements weighted according to rate of passage. *Journal of Agricultural Science* **92**, 499–504.

Ørskov E.R., Grubb D.A., Smith J.S., Webster A.J.F. & Corrigall W. (1979) Efficiency of interaction of volatile fatty acids for maintenance and energy retention by sheep. *British Journal of Nutrition* **41**, 541–51.

Outen G.E., Beever D.E., Osbourn D.F. & Thomson D.J. (1974) The digestion and absorption of lipids by sheep fed chopped and ground dried grass. *Journal of the Science of Food and Agriculture* **25**, 981–7.

Outen G.E., Beever D.E., Osbourn D.F. & Thomson D.J. (1975) The digestion of the lipids of processed red clover herbage by sheep. *Journal of the Science of Food and Agriculture* **26**, 1381–9.

Paterson R. & Crichton C. (1960) Grass staggers in large scale dairying on grass. *Journal of the British Grassland Society* **15**, 100–105.

Penning P.D., Parsons A.J., Orr R.J. & Treacher T.T. (1991) Intake and behaviour responses by sheep to changes in sward characteristics under continuous stocking. *Grass and Forage Science* **46**, 15–28.

Phipps R.H., Sutton J.D. & Jones A.K. (1995) Forage mixtures for dairy cows: The effect on dry matter intake and milk production of incorporating either fermented or urea-treated whole crop wheat, brewers' grains, fodder beet or maize silage into diets based on grass silage. *Animal Science* **61**, 491–6.

Pound B. & Martinez-Cairo L. (1983) *Leucaena: Its Cultivation and Uses*. Overseas Development Administration, London.

Reynolds C.K. & Beever D.E. (1995) Energy requirements and responses – a UK perspective. In Lawrence T.L.J., Gordon F.J. & Carson A. (eds) *Breeding and Feeding the High Genetic Merit Dairy Cow.* British Society of Animal Science Occasional Publication No. 19, 31–41. BSAP Edinburgh.

Rooke J.A., Lee N.H. & Armstrong D.G. (1987) The effects of intraruminal infusions of urea, casein, glucose syrup and a mixture of casein and glucose syrup on nitrogen digestion in the rumen of cattle receiving grass silage. *British Journal of Nutrition* **57**, 89–98.

Siddons R.C., Beever D.E. & Nolan J.V. (1982) A comparison of methods for the estimation of microbial nitrogen in duodenal digesta of sheep. *British Journal of Nutrition* **48**, 377–89.

Stakelum G. & Dillon P. (1991) Influence of sward structure and digestibility on the intake and performance of lactating and growing cattle. In Mayne C.S. (ed) *Management Issues for the Grassland Farmer in the 1990s.* British Grassland Society Occasional Symposium, No. 25, 30–42. BGS, Hurley.

Steen R.W.J., Gordon F.J., Dawson L.E.R., Park R.S., Mayne C.S., Agnew R.E., Kilpatrick D.J. & Porter M.G. (1998) Factors affecting the intake of grass silage by cattle and prediction of silage intake. *Animal Science* **66**, 115–27.

Sutherland T.M. (1988) Particle separation in the forestomachs of sheep. In Dobson A. & Dobson M.J. (eds) *Aspects of Digestive Physiology in Ruminants*, 43–73. Comstock, Ithaca, NY.

Sutton J.D., Cammell S.B., Beever D.E., Haines M.J., Spooner M.C. & Harland J.I. (1991) The effect of energy and protein sources on energy and nitrogen balances in Friesian cows in early lactation. In Wenk C. & Bossinger M. (eds) *Energy Metabolism of Farm Animals. Proceedings 12th Symposium of European Association of Animal Production*, 288–91. ETH, Zurich.

Teller E., Vanbelle M. & Kamatali P. (1993) Chewing behaviour and voluntary grass silage intake by cattle. *Livestock Production Science* **33**, 215–27.

Theodorou M.K., Williams B.A., Dhanoa M.S., McCallan A.B. & France J. (1994) A simple gas production method using a pressure transducer to determine fermentation kinetics of ruminant feeds. *Animal Feed Science and Technology* **48**, 185–97.

Thomas C. (1987) Factors affecting substitution rates in dairy cows on silage-based rations. In Haresign W. & Cole D.J.A. (eds) *Recent Advances in Animal Nutrition*, 205–18. Butterworth, London.

Thomas C., Gill M. & Austin A.R. (1980) The effects of supplements of fishmeal and lactic acid on the voluntary intake of silage by calves. *Grass and Forage Science* **35**, 275–9.

Thomas C., Aston K.A., Daley S.R. & Bass J. (1986) Milk production from silage. IV. The effect of the composition of the supplement. *Animal Production* **42**, 315–25.

Thomson D.J., Haines M.J., Austin A.R., Cammell S.B., Beever D.E., Dhanoa M.S. & Barnes R.L. (1983) The voluntary intake, gain, tissue retention and efficiency of energy and protein utililization by Friesian steers of fresh perennial ryegrass and white clover. *Animal Production* **36**, 501.

Tilley J.M.A. & Terry R.A. (1963) A two-stage technique for the in-vitro digestion of forage crops. *Journal of the British Grassland Society* **18**, 104–11.

Tyrrell H.F., Reynolds P.J. & Moe P.W. (1979) Effect of diet on partial efficiency of acetate use for body tissue synthesis by mature cattle. *Journal of Animal Science* **48**, 598–606.

Ulyatt M.J., Dellow D.W., John A., Reid C.S.W. & Waghorn G.C. (1986) Contribution of chewing during eating and rumination to the clearance of digesta from the rumino-reticulum. In Milligan L.P., Grovum W.L. & Dobson A. (eds) *Control of Digestion and Metabolism in Ruminants*, 498–515. Prentice-Hall, Hemel Hempstead.

Ulyatt M.J., Thomson D.J., Beever D.E., Evans R.T. & Haines M.J. (1988) The digestion of perennial ryegrass and white clover by grazing cattle. *British Journal of Nutrition* **60**, 137–49.

Underwood E.J. & Suttle N. (ed) (1999) *The Mineral Nutrition of Livestock*. CAB International, Wallingford.

Vadiveloo J. & Holmes W. (1979) The effects of forage digestibility and concentrate supplementation on the nutritive value of the diet and performance of finishing cattle. *Animal Production* **29**, 121–30.

Van Soest P.J. & Wine R.H. (1967) Use of detergents in the analysis of fibrous feeds. IV. Determination of plant cell wall constituents. *Journal of the Association of Official Analytical Chemists* **50**, 50–56.

Veerkamp R.F., Simm G. & Oldham J.D. (1994) Effects of interaction between genotype and feeding systems on milk production, feed intake, efficiency and body tissue mobilization in dairy cows. *Livestock Production Science* **39**, 229–41.

Walters R.J.K. (1976) The field assessment of digestibility of grass for conservation. *Agricultural Development and Advisory Service Quarterly Review* **23**, 323–8.

Webster A.J.F. (1980) Energy costs of digestion and metabolism in the gut. In Ruckebusch Y. & Thivend P. (eds) *Digestive Physiology and Metabolism in Ruminants*, 468–84. MTP Press, Lancaster.

Wilkins R.J. (1974) The nutritive value of silages. In Swan H. & Lewis D. (eds) *Proceedings 8th University of Nottingham Nutrition Conference for Feed Manufacturers*, 167–89.

Yan T., Gordon F.J., Agnew R.E., Porter M.G. & Patterson D.C. (1997) The metabolizable energy requirement and the efficiency of utilization of metabolizable energy for lactation by dairy cows offered grass silage based diets. *Livestock Production Science* **51**, 141–50

Chapter 8
The Conservation of Grass

R.J. Merry, R. Jones and M.K. Theodorou

8.1 Introduction

This chapter describes grass conservation processes and outlines the advances made in recent years, with emphasis on hay and silage quality. The chapter focuses on preservation of protein during conservation, which is highly relevant in the context of the role of home-grown, high protein forage in livestock production systems. The production of silage and hay in Western Europe accounts for about 152 million tonnes of dry matter (DM) of forage, of which 70% is in the form of silage (Wilkinson & Stark 1992). During the period 1970–1994, the amount of conserved forage produced in the UK doubled from 8.2 to 16.3 million tonnes DM per annum with a dramatic (*c.* 6-fold) increase in the amount of grass silage produced and a concomitant (*c.* 2-fold) decline in hay making (Fig. 8.1).

The most popular tool for harvesting forages for silage making in the UK is the precision chop harvester, with ensilage being carried out in large clamp silos, although significant amounts of grass silage are harvested, wrapped and stored in plastic film using big-baler technology. There has been a trend, particularly in western areas of the UK, towards direct-cut or minimum-wilt herbage of high digestibility, resulting in problems of greater effluent production and consequently nutrient losses and environmental pollution, particularly during the first few weeks of ensilage (Jones & Jones 1995).

8.2 Principles of conservation

Haymaking is dependent upon the rapid removal of water from herbage to a level sufficiently low to allow the grass to be stored in a safe condition with minimum losses from leaching, leaf shatter and microbial spoilage. In most of Western Europe, where the climate is predominantly cool, haymaking has always required a lengthy pre-drying period even when the grass has been mown at a mature stage. During this drying period considerable changes can occur in both the nutritive quality and the digestibility of the hay. As hay quality is critically dependent upon the duration and intensity of the drying period, and because of the unpredictability of our temperate summers, it is not difficult to

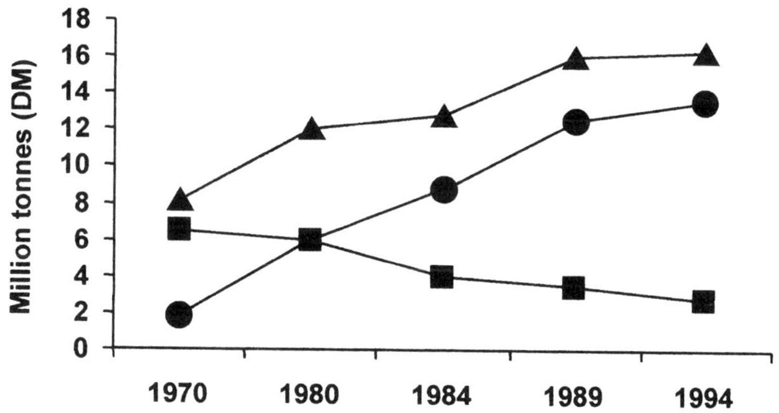

Fig. 8.1 Changes in the amounts of hay (■), silage (●) and total forage (▲) conserved during the period 1970–1994 (source: Wilkinson *et al.* 1996).

account for the decline in hay making in the British Isles and the increase in preservation of grass as silage.

Ensilage is a naturally occurring process which enables the conservation of herbage through an acidification process caused by the production of organic acids, particularly lactic acid, produced by epiphytic bacteria fermenting released plant sugars. As with hay making, the success and efficiency of the ensilage process and the nutritive quality of the preserved crop are dependent upon a number of factors. These include the chemical composition of the harvested crop, type of post-mowing mechanical treatment, and the rate and extent of dry matter losses as a result of respiration, storage and feed-out.

8.3 Crop management

The importance of management of the herbage, in terms of selection of grass varieties, harvesting dates and the rapid wilting of crops prior to ensiling for optimal conservation of nutrients, cannot be over emphasized. Harvesting grass herbage at optimal quality, with crude protein and organic matter digestibility in excess of $160\,g\,kg^{-1}$ DM and $660\,g\,kg^{-1}$ DM, respectively, is essential in order to maximize the nutritive value of hay or silage for economic animal production. The concept of optimizing yields of nutrients rather than DM *per se* has received increasing attention since the 1960s, and research on herbage digestibility has contributed to the improvement in the quality of herbage cut for silage. Ear emergence has been identified as a critical stage of growth, and remains a key reference point beyond which the decline in digestibility and protein content of the grass accelerates. An example of the quality profile of perennial ryegrass with changes in maturity is shown in Fig. 8.2. Although the

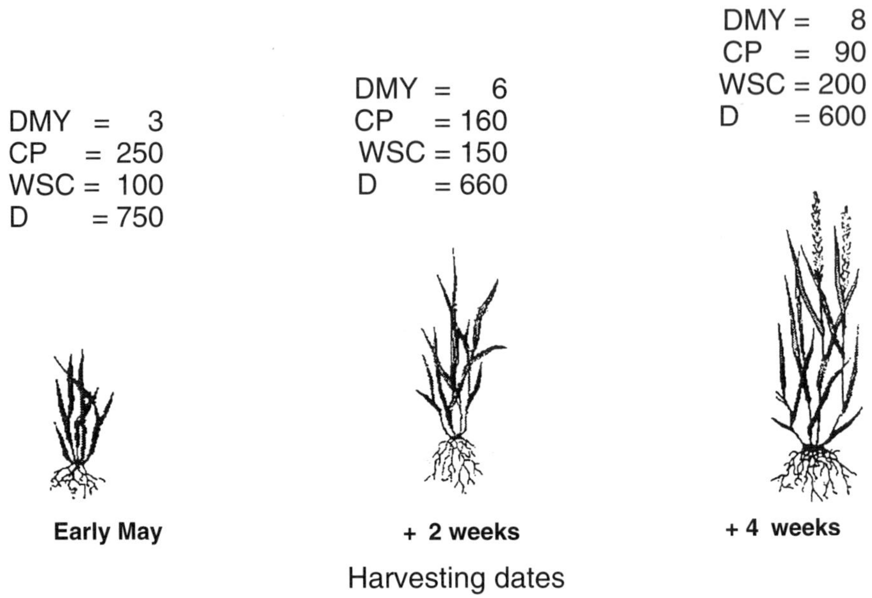

Fig. 8.2 Changes in the production and quality profile of perennial ryegrass during its maturity. DMY, DM yield t ha^{-1}; CP, crude protein g kg^{-1} DM; WSC, water-soluble carbohydrates g kg^{-1} DM; D, digestibility of organic matter g kg^{-1} DM (after Green *et al.* 1971).

rate of decline of protein content varies considerably between grass varieties and species at any given date or growth stage (Harkess & Alexander 1969), the timing of 50% ear emergence for each individual grass species or variety remains relatively constant.

8.4 Haymaking

The aim of haymaking is to preserve grass or other forage by drying to a moisture content of <18%, at which point microbial and plant enzymic activity is minimal. The moisture content at which the crop can safely be stored depends on its composition and its stage of growth at the time of harvesting. Mature grass contains relatively low amounts of sugars that can be oxidized (respired) causing heating. Thus, mature grass may be safely stored at DM contents of 82–85%. By contrast, leafy immature grass harvested for hay must be dried to greater than 88% DM for safe storage without the risk of heating and moulding. The rate of field drying is also dependent on the maturity of the grass crop at harvesting. In immature, leafy grass almost 3.5 tonnes of water per hectare needs be evaporated from a crop at 20% DM in order to produce one tonne of dried hay at 88% DM. By contrast, a mature grass crop at 25% DM will need

Table 8.1 Losses of dry matter and digestible crude protein during haymaking (from Watson & Nash 1960).

Treatment	Loss of DM (%)	Loss of digestible crude protein (%)
No rain – mechanical losses	14.7	22.3
Rain	23.7	34.7
1–2 showers of rain (1–20 mm)	18.9	22.9
5–6 showers of rain (12–63 mm)	27.1	38.3

Mechanical loss: field losses as a result of tedding. Rain: hay subjected to continuous rain for 6 h.

2.5 tonnes of water to be removed to produce dried hay at 85% DM for safe storage. The transpiration of water in uncut grass is a dynamic and continuous process involving the plant's vascular system and stomata located within the tough waxy epidermis of the leaf. Spreading the crop after cutting allows maximum exposure of the foliar surfaces to wind and solar radiation and consequent rapid evaporation of water. However, in freshly mown grass the stomata remain open for 1–2 h, after which time they close, resulting in the residual tissue water being trapped inside the cellular structures (Jones & Palmer 1932). Wilting beyond the first hour after mowing is therefore a slow and ineffective process. Bosma (1987), for example, demonstrated that the rate of evaporation of water from low dry-matter grass crops was five times more rapid during the stomatal opening period compared with periods when they were closed.

When herbage is cut for hay, a swath of varying depth and density will be produced, and drying will be more rapid on the surface and less rapid in the swath interior. Poor circulation of air through the swath contributes to minimizing the rate of drying, particularly in the lower layers of the swath where the humidity is more pronounced, especially during the early stages of the drying process. In practice, therefore, the objective is to create swath conditions conducive to rapid drying of the herbage, thus reducing the nutrient losses caused by plant respiration and the risk of leaching of nutrients by rain. Some types of swath conditioning, such as mechanical treatment in the later stages of drying and increased swath movement following rain, accentuate these losses, as illustrated by the data in Table 8.1. Moreover, in order to minimize nutrient losses, it is important that mechanical treatment should be applied before the leaf becomes brittle, and that the period when hay is exposed to the elements should be reduced as much as possible.

8.4.1 Conditioning, turning and tedding

The sixteenth-century English writer Fitzherbert, in his *Boke of Husbandry* (1523), observed that 'good tedding is the chief poynte to make good hay' – an observation which has been proven repeatedly over the years. Most modern mowers have an integral conditioner system with a high-speed horizontal rotor, fitted with spikes, tines, brushes or ribs, which enables the cut grass to be bruised immediately after cutting, resulting in an increased rate of water loss. Additional treatments of the swath can be applied in order to achieve rapid drying of the crop for quality haymaking. Many investigations have measured the rate of water loss from the swath under differing conditions of weather and management. Mean rates of initial water loss of 0.5–1.0% h^{-1} have been found in undisturbed swaths in a number of countries including UK, Belgium, Denmark, The Netherlands and Sweden. However, if the swath is frequently shaken by turning and tedding to expose the lower layers of the swath to the atmosphere, the rate of water loss can increase to 2.0% h^{-1} (Nash 1985). Whilst rapid rates of drying are of fundamental importance to successful haymaking, uniform drying is also an important factor since hay is highly heterogeneous, and any moist patches will inevitably result in moulding.

8.4.2 Hay for horses

Horses have become increasingly popular for leisure purposes and a new market for high quality hay is developing. Good quality roughage in the form of hay is the foundation of all horse rations, with the possible exception of the young foal. The daily allocation of roughage DM should be no less than 1% of body weight, and for mature horses receiving light exercise this should be increased to 1.5–2.5% of body weight. Horses can digest approximately 30% of the fibre content of hay, whereas cattle and sheep can digest between 60 and 70% of the fibre content. Hay prepared for horses should be harvested at the leafy stage and contain minimal amounts of stem, even though this material will be difficult to wilt.

8.5 Silage making

There have been major developments in the technology of silage making since the 1970s, and a wide range of chemical and biological additives are available to assist the ensilage process and improve silage quality. Some of the biological additives, particularly the inoculants, are used as an effective treatment for conserving the protein fraction in silage (see Section 8.8.2). The trend in additive use bridges the gap left by legislation against the use of animal-based supplements for ruminant feed, in order to maximize forage protein in feeding systems in the post-BSE era.

The importance of harvesting grass crops at the correct stages of maturity for both hay and silage making has been discussed in Section 8.3. Before describing the silage fermentation in more detail, some other aspects of crop management and the processing of herbage prior to ensiling will be outlined.

8.5.1 Traditional wilting techniques for silage

There are strong environmental and economic arguments in favour of high dry-matter silage. Wilting to over 30% DM virtually eliminates the risk of pollution from silage effluent, reduces the characteristic odour associated with wetter silage and reduces the volume of slurry produced by livestock. It also has implications for protein content and the resultant nutritional value of silage. Traditional wilting techniques have relied on mowing the grass, followed by single or multiple tedding, before drawing the cut herbage up into swaths for ensiling when the DM content is between 20% and 60%. A comprehensive review of the effects of wilting technology, based on a coordinated European study (Eurowilt), was published in 1984, and the main conclusions were that field losses were higher, and in-silo losses lower, with wilted than with unwilted herbage (Zimmer & Wilkins 1984). Furthermore, only small improvements (<10%) in DM intake were observed in cows, growing cattle or sheep fed wilted, as opposed to unwilted, silage and this slight increase in intake did not translate to improvements in animal performance. In the Eurowilt study, the range of wilting conditions was extremely variable between trials, as would be expected from large between-country assessments. It has also been shown (Patterson *et al.* 1996) that significant benefits can be achieved in terms of cattle performance when cattle are fed rapidly wilted, compared with unwilted, silage. Rapid wilting (less than 30 h) increased the DM content from 160 to 320 g kg^{-1} at the time of ensiling, which resulted in an increased dry matter intake of 17%, and an increased milk yield of 2.4%. Improvements were also found in milk quality, with higher fat and protein in milk from cows fed the wilted silage.

The rate of both plant and microbial protease activity is affected by the decrease in water content brought about by wilting. Increasing the DM content of silage has been shown to result in higher protein content, with a corresponding decrease in ammonia-N concentration (Table 8.2), presumably

Table 8.2 Influence of wilting on the protein-N and ammonia N components of ryegrass silages (source: Henderson *et al.* 1982).

	Dry matter content of herbage (g kg^{-1})					
	150	200	250	300	350	400
Protein-N (g kg^{-1} TN)	302	319	350	384	370	447
Ammonia-N (g kg^{-1} TN)	101	102	85	93	95	85

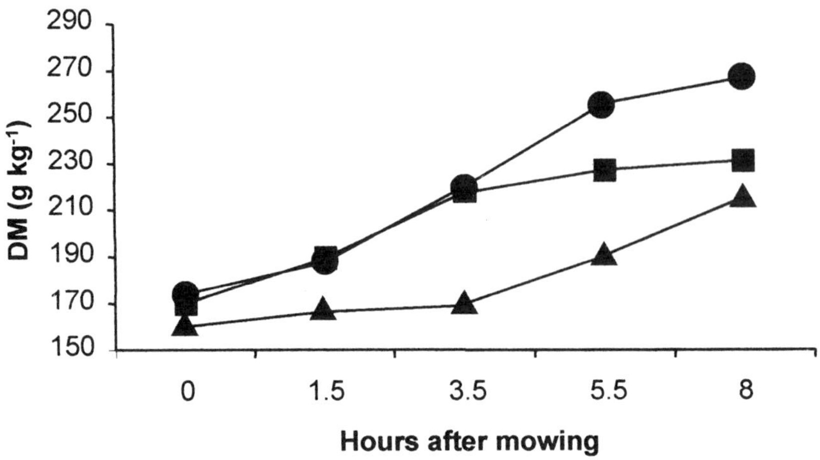

Fig. 8.3 Comparison of wilting efficiency using different mower conditioners (R. Jones, unpublished data, 1999). HPC (●), high performance conditioner, Vicon, Market Drayton, UK; MO + T (■), JF Farer conditioner mower with single-pass tedding; MO – T (▲), JF Farer conditioner mower with no tedding.

because plant protease activity was reduced by the decrease in the water content of the wilted plant cells.

8.5.2 New developments in wilting technology

A new mower/conditioner system was developed in 1994 called the high performance conditioner (or HPC), (Vicon, Market Drayton, UK). The HPC is based on a conventional disc mower with a twin horizontal roller attachment for grass conditioning. The upper roller is constructed from metal and the lower roller is a cylindrical nylon brush. Both rollers extend the full width of the mower and condition the mown grass prior to spreading it evenly along the full width of the swath, using a sophisticated array of deflector plates. Initial trials using this new technology have shown considerable improvements in the rate of wilting, compared with grass mown using a conventional mower conditioner followed by immediate tedding (Fig. 8.3). These trials have shown improvements in wilting efficiency of 55% from the mower conditioner and tedding treatment, and 109% from the HPC treatment, compared with mowing grass using a mower conditioner and no tedding.

8.6 The ensilage process

Immediately after cutting grass for silage, during the wilting period and the early stages of fermentation, changes in the composition of the herbage are brought about by the combined action of plant enzymes and the epiphytic microflora (i.e. naturally occurring on the leaf surface). The primary objective of ensilage is to minimize these changes in plant composition by ensuring rapid development of the anaerobic conditions in the silo that promote the proliferation of lactic acid bacteria naturally occurring on the forage, or that have been added to the herbage as an applied inoculant at the time of harvesting. The anaerobic conditions and availability of water-soluble carbohydrate and other nutrients from cut and compacted herbage encourage growth of the lactic acid bacteria and production of lactic acid, resulting in the characteristic decline in pH of the ensiled herbage. Thus, by lowering the pH of the herbage, plant enzyme activity is reduced and undesirable (spoilage) micro-organisms, which are not tolerant of low pH, are suppressed, ensuring effective crop preservation. The ensilage process can be divided into three phases – the aerobic, fermentative and feed-out phases – each influenced by changing crop and environmental conditions. Although much of the information that follows was determined for silage made in clamp silos where grass is chopped to a length of 2–4 cm, these principles of ensilage are generally applicable to bale silages (see Section 8.10.1). However, it is recognized that the generally higher DM content and longer chop length of bale silage result in a somewhat restricted fermentation in comparison with forage-harvested silage stored in clamps.

8.6.1 The aerobic phase

During this period, air trapped within the freshly ensiled herbage fuels plant respiration. Respiration results in the reduction of plant sugars, and their conversion to water, carbon dioxide and heat (McDonald *et al.* 1991). This process, although undesirable in terms of reducing the availability of nutritionally beneficial sugars in the herbage for animal production purposes, is advantageous in that it 'mops up' residual oxygen and promotes the development of anaerobic conditions suitable for the growth of lactic acid bacteria. If gaseous exchange in clamp or bale silage is not restricted by delayed or poor sealing of the herbage mass, two major effects may ensue. Firstly, respiration will continue and eventually deplete the water-soluble carbohydrates in the herbage to the point where the developing microbial population is limited by the lack of a suitable substrate, thus reducing the production of lactic acid and limiting the pH decline in the silo. Under these conditions, carbohydrases, and particularly proteases, remain active and their degradative activities will contribute to the production of a silage of poor nutritive quality. Secondly, the heat produced from excessive respiration will be retained within the herbage mass causing an increase in temperature, which, if allowed to rise above 40°C, can alter the solubility, and consequently the digestibility, of the grass protein (McDonald *et al.* 1968). Moreover, an extended aerobic phase at enhanced temperature will encourage the growth of some facultative micro-organisms, which may act to spoil the silage. It is essential, therefore, to prevent gaseous exchange in the silo and to remove oxygen from the herbage mass. The aerobic phase in clamp silage is reduced considerably if the herbage is finely chopped and well compacted prior to rapid sealing of the silo.

8.6.2 The fermentative phase

On exclusion of air from the herbage, the fermentation process generally occurs during the first week of ensilage, although sometimes this phase can last one month or more depending on the crop, its composition and the ensiling conditions. The fermentative phase involves the proliferation of a succession of facultative and strictly anaerobic bacteria naturally present on the ensiled herbage or added to the crop as an inoculant at harvesting time. Initially, two populations develop, the Gram-positive, facultatively anaerobic homo-fermentative and hetero-fermentative lactic acid bacteria (*Lactobacillus*, *Enterococcus*, *Pediococcus*, *Lactococcus* and *Leuconostoc* spp.), and the Gram-negative, facultatively anaerobic enterobacteria. A third group, the strictly anaerobic clostridia, may also proliferate later in the fermentation, but this is generally under conditions where carbohydrate supply is limited and moisture content and temperatures are high.

The homo-fermentative lactic acid bacteria are of primary importance in the microbial succession, promoting rapid pH decline through the production of lactic acid as their main fermentation end-product. Homo-fermentative pediococci and enterococci often initiate the fermentation, with lactobacilli consolidating the pH decline owing to their higher acid tolerance. Hetero-fermentative lactic acid bacteria (members of the genera *Lactobacillus* and *Leuconostoc*) are also present in the initial succession, but they have a mixed acid fermentation, producing acetic and lactic acid and other end-products, namely ethanol and carbon dioxide, usually resulting in low pH decline. The enterobacteria are generally regarded as undesirable in silage, primarily because they do not contribute as effectively as lactic acid bacteria to the acidification and preservation process. Although populations of enterobacteria may develop initially, they generally decline in response to lowering of herbage pH (Seale *et al.* 1986, Merry *et al.* 1995a). Clostridia are also undesirable spoilage bacteria in silage, but they appear only under the specific circumstances of low sugar concentration and high moisture and temperature. Clostridia convert both residual water-soluble carbohydrate and lactic acid to butyric acid, resulting in poor quality silage with a characteristically noxious smell. Lowering the pH of silage effectively inhibits the growth of these 'lactate fermenting clostridia'. Proteolytic clostridia are particularly detrimental in silage in that they degrade amino acids to a range of products, including amines and ammonia. Silages where clostridia have been active often have reduced digestibility and intake characteristics, and can increase the incidence of metabolic disturbances in animals to which they are fed. Other groups of micro-organisms, such as yeasts, may also proliferate under certain conditions during the fermentative stage in silage but they are of generally lesser importance. The reader is referred to McDonald *et al.* (1991) for further details of the microbiology of silage.

From the foregoing description of microbial successions and their influence on the biological processes in the silo, it is evident that the size and diversity of the naturally occurring populations that develop on silage can have a major influence on the quality of the finished product. The numbers of lactic acid bacteria and enterobacteria on the freshly harvested herbage vary considerably, with values ranging from $<10^2$ to $>10^7\,g^{-1}$ fresh herbage (Rooke 1990, Merry *et al.* 1993). Numbers of clostridia are generally of lower orders of magnitude (Spoelstra 1990, Rammer *et al.* 1994) and populations are influenced by climatic conditions and particularly by contamination of herbage with soil or slurry.

The increasing quantities of livestock wastes such as slurries and manures applied to grasslands has implications for silage quality and animal and human health. These wastes harbour large populations of undesirable silage bacteria as well as some animal and human pathogens (see Chapter 13). Their presence in livestock waste can act as vectors in silage spoilage and the transmission of disease through the food production chain (Mawdsley *et al.* 1995, Davies *et al.* 1996, Pell 1997). Some of these micro-organisms can survive on herbage for considerable periods after application of waste to land, and often remain viable even after ensilage has occurred (Davies *et al.* 1996, Merry *et al.* 1997).

Silage fermentation and quality

High quality silages made from low dry-matter herbage (<20% DM) tend to have lower pH values and lower concentrations of butyric and acetic acid, but to have higher lactic acid concentrations than lower quality silages. Thus, as a general guide, good and bad silages can be identified by the negative relationships between lactic acid content and butyric acid and ammonia-N concentrations, and the positive relationships between the pH value and the concentration of the latter two products (Table 8.3). Silages made from wilted herbage, where preservation is related to the combined effects of acidity and wilting, are a

Table 8.3 An example showing the negative and positive relationships between silage pH and fermentation end products (after Murdoch 1989)

pH	Lactic acid ($g\,kg^{-1}$ DM)	Butyric acid ($g\,kg^{-1}$ DM)	Ammonia-N ($g\,kg^{-1}$ TN)
3.9	110	0	14
4.1	118	0	22
4.5	63	16	27
4.5	63	16	27
5.2	3	38	51
5.7	1	58	98

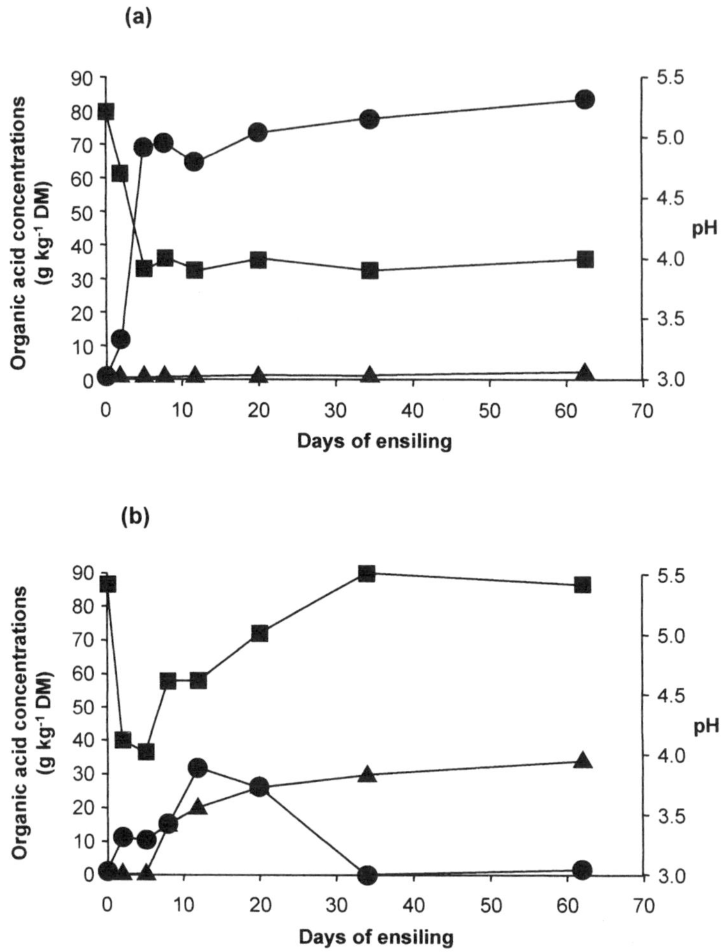

Fig. 8.4 Changes in the pH (■), lactic acid (●) and butyric (▲) concentrations in (a) stable and (b) unstable grass silages (after Langston *et al.* 1958).

notable exception to this rule. They can have relatively high pH values and low lactic acid contents, but contain little or no butyric acid and only a small amount of ammonia-N. Thomas & Fisher (1991) have shown that certain critical levels of acidity must be reached, depending on the dry matter of the crop, to prevent an undesirable fermentation. For example, the pH required for a silage of excellent quality can vary from approximately 3.8 when herbage DM content is 15%, to between 5.0 and 5.5 at 30–40% DM. The pH in the silo is not only related to the ability of the microbial population to produce lactic acid, but also to the buffering capacity of herbage itself. This can vary markedly in herbage; for example, white clover has about twice the buffering capacity of grass. On the other hand, wilting the herbage before ensiling reduces the buffering capacity (Playne & McDonald 1966). The importance of developing sufficient acidity in

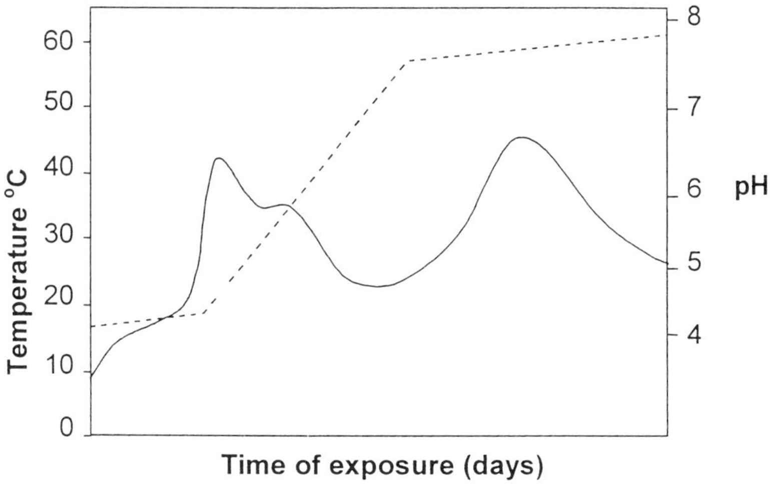

Fig. 8.5 Changes in silage pH (---) and temperature (—) during the aerobic spoilage of untreated grass silage (source: R. Jones *et al.*, previously unpublished data, 1999).

the silo has given rise to the concept of 'stable' and 'unstable' silages. A stable silage has a sufficiently high lactic acid concentration to prevent its deterioration by clostridia, and with a stable pH negligible amounts of butyric acid accumulate (Fig. 8.4a). The acidity of an unstable silage (Fig. 8.4b) will increase initially, but not sufficiently to inhibit clostridial growth. Subsequently, the quality of the silage will deteriorate as clostridia ferment lactic acid to butyric acid, with a consequential rise in pH. There is also evidence to suggest that some lactic acid bacteria can metabolize lactic acid to acetic acid under nutrient-limiting conditions (Lindgren *et al.* 1990), thus causing the pH of the ensiled herbage to increase.

8.6.3 The feed-out phase

Silage is generally stored in a clamp or bale under anaerobic conditions at ambient temperature for a minimum of three months, although it is sometimes fed earlier if there is a shortage of herbage for grazing. When a silo is opened and a clamp face exposed, oxygen has free access to the surface of the silage, promoting the development of an aerobic, spoilage microflora. The micro-organisms usually associated with the aerobic spoilage of silage include yeasts, moulds, *Bacillus* spp. and some acetic acid bacteria (Woolford 1990). In addition, bale silage appears to be particularly prone to contamination with *Listeria monocytogenes*, a pathogen that enters the silo via soil, livestock waste or animal vectors. This organism proliferates as a consequence of air ingress where penetration of the film wrap has occurred (Fenlon 1986). The aerobic spoilage organisms are successional in their growth pattern and dependent on temporal changes in silage composition and environmental conditions. A typical sequence of events during aerobic spoilage, in terms of the temperature and pH profiles, is shown in Fig. 8.5. Microbial activity during aerobic deterioration leads to a progressive increase in silage pH as the organic acids produced during ensilage are oxidized to carbon dioxide and water. This activity generates heat, and in the example shown in Fig. 8.5, two peaks in temperature were observed after exposing the silage to air. These peaks coincided with the proliferation of yeasts and acetic acid bacteria during the initial peak, and mould growth, which was more closely associated with development of the second temperature peak. Acetic acid bacteria may not always play a role in aerobic spoilage of grass silage but can precede yeast growth in maize silage (Spoelstra *et al.* 1988). Where moulds proliferate in the terminal stages of spoilage mycotoxins may be produced, and this has implications for the health of both animals and humans (Woolford 1990, Oldenburg 1991).

8.7 Evaluation of silage quality

Silages are generally described in terms of their pH values, the concentration of their fermentation end-products (lactic and volatile fatty acids) and their ammonia-N contents. They are also characterized in terms of their potential nutritive value via measurements of crude protein, crude fibre, metabolizable energy and digestibility. However, these chemical, biological and nutritional characteristics are not always representative of silage quality, particularly as they do not adequately describe the protein content of the silage in relation to that present in the original herbage. For example, of the nitrogen present in fresh herbage, 70–90% is typically in the form of foliar protein. This is in contrast to a well-preserved, low pH silage, where residual protein is considerably lower and most of the nitrogen is in the form of amino acids and soluble non-protein nitrogen compounds, including ammonia (Table 8.4). This observation is of significance, given the major impact that protein has on silage quality and animal performance (McDonald *et al.* 1991). The value for ammonia-N is often used as an indicator of the extent of protein breakdown in silage, but evidence of a poor correlation with residual protein content is accumulating (Cussen *et al.* 1995). This is perhaps not surprising, given that much of the protein degradation in silage is mediated by plant proteases, whereas the ammonia is produced through the microbial deamination of protein degradation products.

8.8 Silage additives

Many factors influence the silage-making process, and the nutritive value of the product hinges around the diversity and population size of the epiphytic microflora and the availability of a suitable substrate to fuel the fermentation.

Table 8.4 The pH and nitrogen fractions of ryegrass and the corresponding silage prepared without additive treatment (source: Heron *et al.* 1986).

	Herbage	Silage
pH	5.65	3.79
Protein-N	858	293
Ammonia-N	11	136
Amino acid-N	43	285
Amide-N	10	16
Nitrate-N	4	ND

ND, not detected.
All values except pH are expressed as $g\,kg^{-1}$ TN.

Limitations in microflora and substrate availability can be compensated for by the use of silage additives. However, different grass species and cultivars have different water-soluble carbohydrate contents and respond differently to environmental conditions. Some basic crop management principles (see Section 8.3) are therefore needed to maximize the benefits that may accrue through using additives.

8.8.1 Types of additives and their mode of action

A considerable amount of research has gone into both chemical and biological approaches to control and direct the silage fermentation, and a bewildering range of additives is available to the farmer. The general principle of chemical additives is to inhibit or completely stop fermentation or increase aerobic stability, whereas that of biological additives is to promote efficient homolactic fermentation (i.e. predominantly lactic acid production) and in some cases prevent aerobic spoilage as well. In this section we take a detailed look at the biological additives for silage (the reader is referred to the books by Woolford (1984) and McDonald *et al.* (1991) for a more comprehensive account of chemical additives).

Silage inoculants contain competitive strains of lactic acid bacteria that are applied usually at about 10^6 viable cells g^{-1} fresh herbage, and produce mainly lactic acid, making them more effective at reducing the pH of herbage than the epiphytic population. Other biological additives (used with or without inoculant bacteria) may be applied where the water-soluble carbohydrate content of the cut herbage is low. These include the direct addition of substrate (molasses for example) or the addition of enzymes capable of releasing soluble sugars from the grass cell walls. Absorbents can also be added to herbage at the time of ensiling to reduce effluent loss in low dry matter silages.

A total of 86 additives were available in the UK in 1998. Of these, 72% were inoculants (with or without enzymes), 21% were chemicals, and the remainder were enzyme alone, sugar substrates or effluent absorbents (Anon. 1997). This contrasts markedly with the situation 10 years earlier (Wilkinson & Stark 1988), when the total number of additives (108, excluding intake/palatability enhancers) was greater. Although the number of inoculants was higher (51), they represented a lower proportion of the total market (47%), as twice as many chemicals were available at that time. The decline in numbers of chemical additives for silage was due to opposition from farmers and silage contractors to using hazardous, polluting chemicals, and to the availability of better inoculants. These changes have been reflected in an increase in popularity of biological additives, particularly inoculants, which have replaced acids in most parts of the UK (Wilkinson *et al.* 1996).

Relatively few of the silage inoculants (<15%) contain a single species of lactic acid bacteria, usually *Lactobacillus plantarum*. Mixed-species inoculants are used because development of the lactic acid bacterial population in the silo is pH-dependent and successional. Thus, in mixed-species inoculants, one or more of *Pediococcus*, *Enterococcus* and *Lactococcus* spp. are included because they grow rapidly in herbage at *c.* pH 6.0, often initiating the fermentation. On the other hand, *Lactobacillus* species are included because they grow well at lower pH and are effective at consolidating the pH decline as it falls below 4.5. The effect of using a single- or mixed-strain inoculant, relative to no treatment, or formic acid treatment (at 3 l tonne^{-1} fresh matter) of grass herbage, on lactic acid production and pH decline is shown in Fig. 8.6. In comparison with the inoculant-treated silages, the fermentation was slow in the untreated (control) silage, whilst formic acid treatment reduced the herbage pH to 4.2 within 24 h without the production of lactic acid. Lactic acid production and consequent pH decline in the inoculated herbage was more pronounced than in the uninoculated control, with the *Lactobacillus plantarum–Pediococcus pentosaceous* mixture being more effective than the single species, *Lactobacillus plantarum*, inoculant, particularly in the first 2 days of the fermentation.

8.8.2 Effects of additives on silage fermentation and quality

In general, acid additives are applied to herbage to induce rapid pH decline, to prevent microbial activity and to preserve protein. They have, therefore, a distinct mode, quite different from that of the fermentation-enhancing biological inoculants. The effect of an acid additive on herbage is shown in Table 8.5. In this experiment, increasing concentrations of formic acid were added to a grass/clover mixture at the time of ensilage, resulting in an increasingly restricted fermentation with reduced lactic acid production, a reduction in ammonia and the enhanced preservation of protein. These are all desirable quality characteristics in silage, but they are set against a backdrop of problems with acid additives related to their corrosive and toxic nature, and the fact that

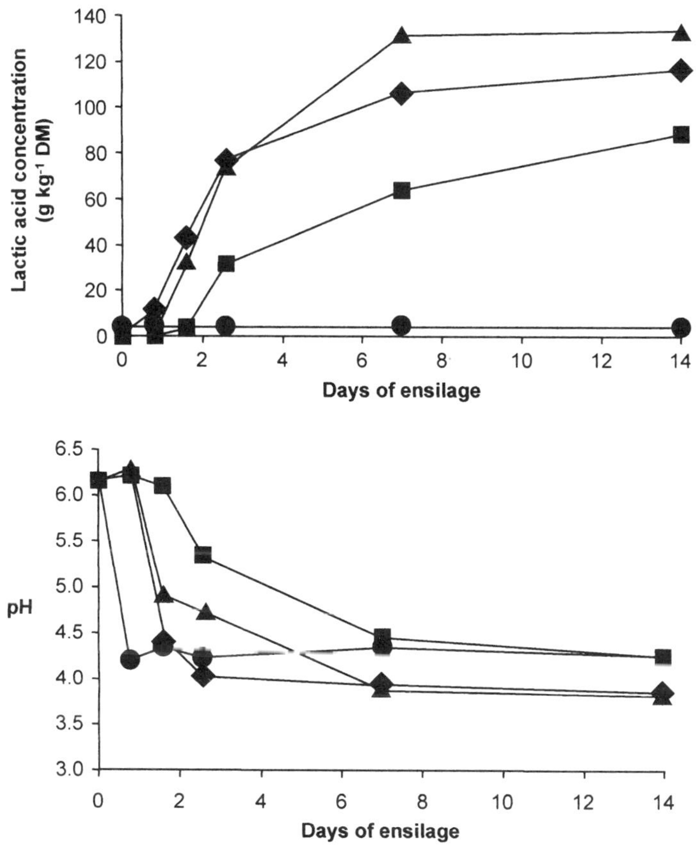

Fig. 8.6 Changes in lactic acid concentration and pH in untreated (■) grass silages, silages treated with formic acid (●), or a single strain (▲) or mixed strain (◆) inoculant (source: R.J. Merry *et al.*, previously unpublished data, 1999).
Single strain inoculant, *Lactobacillus plantarum*; mixed strain inoculant, *Lactobacillus plantarum* + *Pediococcus pentosaceous*.

they have related palatability/intake problems in livestock. However, it is important to note that even at a very high application rate of 91 tonne^{-1}, more than 30% of the foliar protein was degraded in the silo.

In 1990, the results of a joint collaborative European programme (Eurobac) on the use of silage inoculants were published (Lingvall & Lindgren 1990). This study confirmed previous research findings which indicated that under certain conditions, inoculant-treated silages were preserved better and gave a positive response in terms of animal production. It was concluded that the efficacy of a silage inoculant was dependent on the relationship between the dry matter and

Table 8.5 Chemical composition of herbage and grass/clover silages treated with different levels of formic acid (source: Carpintero *et al.* 1979).

	Herbage	Silage Acid application rate ($l\,t^{-1}$)				
		0	1.2	2.2	4.9	9.2
pH	5.85	3.87	3.67	3.81	3.88	3.80
WSC ($g\,kg^{-1}$ DM)	203	71	162	233	233	233
Total N ($g\,kg^{-1}$ DM)	19.3	18.7	17.0	17.8	18.1	17.6
Protein N ($g\,kg^{-1}$ TN)	819	338	376	496	523	550
Ammonia-N ($g\,kg^{-1}$ TN)	ND	95	59	46	12	12

Herbage DM = $199\,g\,kg^{-1}$. ND, not detected.

water-soluble carbohydrate content of the forage (Castle 1990). Thus, although some benefits could be been seen with inoculants where the water-soluble carbohydrate concentration in the fresh crop was as low as $15\,g\,kg^{-1}$, at that time they could not be recommended for direct-cut, unwilted herbage. However, with herbages rapidly wilted to 250–350 DM $g\,kg^{-1}$, inoculants markedly improved the fermentation and silage quality and were as effective as formic acid. Subsequently, results from studies on water-soluble carbohydrate requirements for ensiling unwilted herbage have indicated that the concentration should be at least $25\,g\,kg^{-1}$ in the fresh crop to obtain a silage of acceptable quality, whereas with inoculation this threshold value could be reduced to $20\,g\,kg^{-1}$ fresh herbage (Lunden Pettersson & Lindgren 1990).

The application rate of viable lactic acid bacteria to the crop is also critical to the success of inoculants, and to have a pronounced effect on silage quality it has been indicated that 10-fold greater numbers need to be added than those found naturally on the crop (Pahlow 1990). Thus, as the epiphytic population often does not exceed 10^5 bacteria g^{-1} of fresh herbage, most of the European inoculants are applied at 10^6 viable lactic acid bacteria g^{-1} of fresh herbage. Data illustrating the effect of level of inoculation on ammonia production and the preservation of grass protein after ensilage are summarized in Table 8.6. As the rate of inoculation was increased up to $10^6\,g^{-1}$ fresh herbage, the resultant silages had higher residual protein, and lower pH and ammonia concentrations. As with the acid additives, the rapidity of the initial rate of pH decline was probably responsible for reducing the degradation of protein.

Recognition of the importance of the initial rate of pH decline in relation to improving silage quality (Spoelstra 1991), particularly with respect to the preservation of protein in silage, led to the concept of using bacterial inoculants that had been freshly cultured (actively growing) instead of freeze-dried prior to their application to herbage (Nesbakken & Broch-Due 1991, Merry *et al.* 1995a). In a comparative study by Merry *et al.* (1995a), in which grass was inoculated with either freshly cultured or freeze-dried *Lactobacillus*

Table 8.6 The effect of level of inoculation with lactic acid bacteria on pH, ammonia production and protein in grass silage at different times during the fermentation (source: Heron *et al.* 1988).

Inoculation level	pH		Protein-N (% grass protein-N)		Ammonia-N ($g\,kg^{-1}$ TN)	
	24 h	76 d	24 h	76 d	24 h	76 d
0	5.75	3.75	78	33	35	114
$10^4\,g^{-1}$ FM	5.43	3.70	79	36	29	94
$10^6\,g^{-1}$ FM	4.54	3.57	81	45	23	60
$10^8\,g^{-1}$ FM	4.03	3.56	75	38	21	48

plantarum, the bacteria in the freshly cultured treatment had a reduced lag phase and increased growth rate, significantly reducing reduced the time taken to reach a pH of 4.2 in the silo from 70 h to 46 h. This research culminated in the on-farm culturing of silage inoculants (in an approach akin to home brewing), leading ultimately to the commercial availability of cheaper and more active inoculants. It has been confirmed subsequently, that the addition of such inoculants to low dry-matter grasses with high water-soluble carbohydrate content (>160 $g\,kg^{-1}$ DM) produced high quality, low pH silages with low acetic acid and ammonia-N concentrations and higher protein contents, compared with either untreated or formic-acid-treated silages (Cussen *et al.* 1995, Davies *et al.* 1998).

There is now a considerable body of scientific evidence to indicate that inoculants improve silage fermentation, and this is demonstrated in a summary of data from research published between 1985 and 1992 (Fig. 8.7). Based on the three fermentation characteristics of silage (pH, the lactic acid : acetic acid ratio, and ammonia concentration) 50–60% of the inoculants were successful in improving the fermentation. For further information on silage additive trials, the reader is referred to the reviews by McDonald *et al.* (1991), Spoelstra (1991), Henderson (1993) and Weinberg & Muck (1996).

8.8.3 Future developments in silage inoculant technology

Despite the generally positive findings with silage inoculants, they still have some shortcomings, and even the best inoculants are ineffective where the supply of water soluble carbohydrate is limiting or unavailable for microbial growth (Davies *et al.* 1998). This problem of substrate availability has been addressed in several ways, including the direct addition of sugars (molasses) or enzymes (cellulases and hemicellulases) to the herbage to provide extra substrates (McDonald *et al.* 1991). Genetically modified lactic acid bacteria with the ability to ferment cellulose have also been constructed and shown to survive

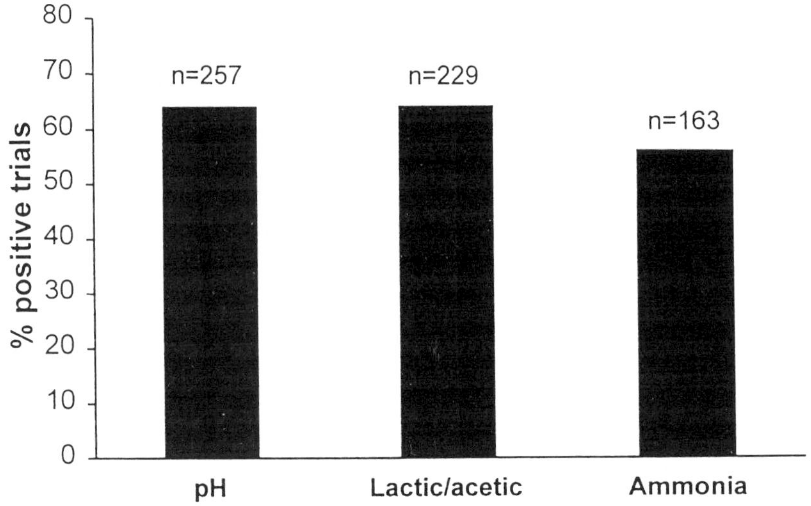

Fig. 8.7 A summary of trials results, for silages produced from a range of forages on different scales in trials in Europe and North America, showing the proportion of trials where inoculation had a positive effect on fermentation characteristics compared with untreated controls (*n*, number of trials per characteristic). Source: Weinberg & Muck (1996).

during ensilage, although further work is needed to demonstrate their ability to express cellulase activity in the silo (Sharp *et al.* 1992). Another novel approach to solve the problem of substrate limitation arose from an observation made by Mueller & Lier (1994) that less than 2% of 600 lactic acid bacteria isolated from grasses could ferment fructan, a polymer of fructose that is generally the major water-soluble carbohydrate component of temperate grasses. Thus, it may be possible to use these fructan-degrading bacteria as silage inoculants for the conservation of herbage low in water-soluble carbohydrates. The possibility of using fructan-degrading lactic acid bacteria as silage inoculants was investigated by Merry *et al.* (1995b). They inoculated sterile grass with a fructan-degrading strain of *Lactobacillus paracasei* subsp. *paracasei*, and found that although fructan degradation was fastest in the inoculated treatment, it also occurred in the sterile control, indicating a role for plant enzymes in the degradation process (Fig. 8.8). Thus, in addition to using selected bacteria, it may also be possible to select for grasses in plant-breeding programmes with enhanced fructan hydrolase activity during ensilage.

The removal of pathogens from silage is particularly important. Several serious outbreaks of poisoning have been associated with food hygiene, and there is a need to allay public concerns over meat quality and livestock feeding practices, particularly following the BSE crisis. Micro-organisms such as clostridia, listeria, enterobacteria and some of the aerobic spoilage yeasts and moulds are potentially harmful, pathogenic or spoilage organisms in silage.

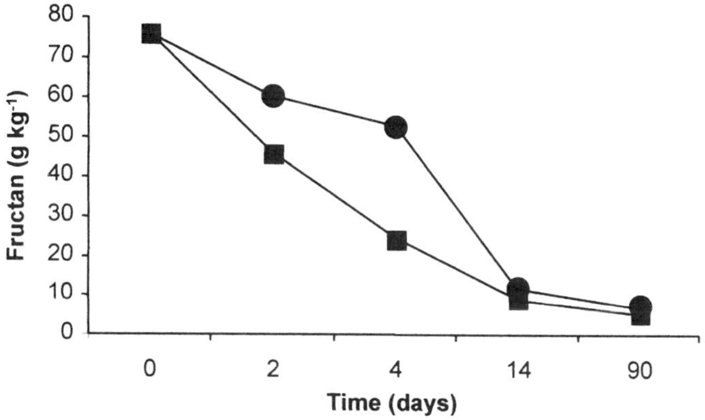

Fig. 8.8 Changes in the fructan content of sterile ryegrass during ensilage either untreated (●) or inoculated with *Lactobacillus paracasei* subsp. *paracasei* (■) (after Merry *et al.* 1995b).

Approximately 20% of the silage inoculants available in the UK contain clostridiaphages (Anon. 1997). These viruses destroy clostridia, although evidence of their efficacy as silage inoculants is equivocal and strain-specificity is reported to be a problem (Wight 1990). Aerobic spoilage by yeasts, moulds and acetic acid bacteria represents one of the most intractable problems with silage. Paradoxically, aerobic deterioration in silage often tends to be exacerbated by conventional inoculants, partly due to the development of a homolactic fermentation with consequential lower production of acetic acid and other minor products, which may act as preservatives or antimycotic compounds in silage (Merry & Davies 1999). Examples of bio-control inoculants for silage which are effective against aerobic spoilage micro-organisms include species of the genera *Propionibacteria*, *Bacillus* and *Serratia* (Moran *et al.* 1993, Bolsen *et al.* 1996). There is also the possibility of introducing species/strains of lactic acid bacteria that express anti-microbial agents, such as bacteriocins (Lindgren & Dobrogosz 1990) to inhibit spoilage and also pathogenic bacteria. An approach based on the antagonistic properties of lactic acid bacteria against *Listeria monocytogenes* is being sought, and several lactic acid bacteria that produce bacteriocins have been isolated from grasses and silages (Mayer *et al.* 1995). It was shown subsequently, that a strain of *Listeria monocytogenes* added to grass at 10^6 viable cells g^{-1} fresh herbage was completely eliminated after 2 days of ensilage when the grass had been co-inoculated with a bacteriocin-producing strain of *Lactobacillus curvatus*, whereas *Listeria* numbers surviving after this time in silage where the *Lactobacillus* had not been added were still in excess of $10^6\,g^{-1}$ fresh herbage (Mayer 1997). Lowes *et al.* (1997) discuss the possibility of using mycocins produced by some yeasts to prevent spoilage of silage by other yeasts. This approach shows promise, as a partially purified yeast mycocin extract has been shown to delay yeast growth and aerobic spoilage in maize silage. As the direct use of the extract may have economic disadvantages, the possibility of obtaining expression of the product in the silo by a genetically modified lactic acid bacterium is being investigated.

8.8.4 UKASTA silage additive registration schemes

Despite the move towards lower numbers of higher quality silage additives, farmers can find it difficult to make rational and well-informed decisions on their efficacy under their own particular conditions. The introduction of product registration schemes in several European countries will contribute toward alleviating this problem. The scheme in place in the UK (Weddell *et al.* 1996) is based on gaining approval for a product in a specified number of scientific trials designed to test the efficacy of the additive on silage, in comparison with no treatment, and in a number of categories. Category A considers animal performance benefits (improvements in liveweight gain and in milk production and quality). Category B comprises animal/feed effects (improvements in dry matter intake, silage digestibility and efficiency of protein and energy use). Category C includes effects related to ensiling (improvements in fermentation and aerobic stability and reduction in effluent losses). Although the scheme has its limitations, registration has at least provided the farmer with some information on which to base a choice.

8.9 Production response of animals offered additive-treated silages

As discussed in Section 8.7, during ensilage a large proportion of the true protein is hydrolysed by plant and microbial proteases to peptides and amino acids. The efficiency with which silage-N is converted into microbial protein depends on the rate of ammonia release and assimilation in the rumen. If more ammonia is produced than is assimilated, the excess will be absorbed in the blood supply and converted to urea in the liver, and then lost to the environment in the form of urine. Since a high proportion of silage nitrogenous components are in a soluble form they are rapidly degraded in the rumen, giving rise to extreme ammonia patterns which may affect feed intake. In a trial conducted at IGER, silages were either untreated, or treated with formic acid or an inoculant additive, and fed to 400-kg beef cattle for a period of 10 weeks (Table 8.7). All the silages in this trial were excellently preserved, with a pH value of < 4.1 and ammonia-N content of <50.0 $g\,kg^{-1}$ total N. The lower ammonia-N concentrations and free amino acid contents found in the treated silages would suggest that treatment with formic acid or the inoculant reduced proteolytic activity compared with the untreated silage. When these silages were fed to beef cattle, improvements in animal performance were observed, with an

Table 8.7 Effect of silage additive treatments on components of silage quality and on animal performance (source: R. Jones, unpublished data, 1999).

	Treatments		
	Control	Formic acid	Inoculant
Dry matter $g\,kg^{-1}$	213.2	225.3	215.2
pH	3.95	4.04	3.91
Composition of DM $g\,kg^{-1}$			
Crude protein	155.7	156.2	154.3
Ammonia-N $g\,kg^{-1}$ of TN	45.01	25.0	21.0
Free amino acids $g\,kg^{-1}$ of TN	127	109	69
Lactic acid	120.0	66.3	113.3
Acetic acid	25.64	26.85	10.33
ME ($MJ\,kg^{-1}$ DM)	10.56	10.63	10.67
Animal performance			
Live weight gain ($kg\,d^{-1}$)	0.66	0.94	0.89
DM intake ($kg\,d^{-1}$)	7.41	8.41	8.23
FCE kg intake per 1 kg gain	11.3	8.9	9.4

increased dry matter intake of 11%, and daily live-weight gains of 35 and 42% from inoculated and formic-treated silage, respectively, compared with untreated silage. The improvements in animal gain from the treated silages in this experiment may be explained partly by the increased dry matter intake and marginal improvements in the energy content of the silages. However, the lower free amino acid content and ammonia-N of the treated silages could have resulted in improved protein-N utilization in the rumen. Other workers have proposed this hypothesis for improvement in protein-N utilization of inoculated silages. Sharp *et al.* (1994) investigated the digestion and utilization of inoculated silages using 16-month-old Jersey heifers and reported a 33% improvement in the efficiency of microbial protein synthesis when compared with well-preserved untreated silage. Mayne & Steen (1990) observed relatively small effects of inoculation on fermentation characteristics, but positive responses were obtained in both silage digestibility and N-retention when fed to sheep. Keady & Steen (1995) evaluated a bacterial inoculant based on a single strain of *L. plantarum* as a silage additive, compared with formic acid or no treatment. The silages were made from low dry matter crops and were poorly preserved. Nevertheless, heifers fed inoculant-treated silage gave 11% higher liveweight gains, compared with formic-treated or untreated silages. In this particular experiment the true protein content of the inoculated silage as fed was similar to the untreated, while that of the formic-treated silage was significantly higher. N-retention, however, was higher (86%) in the inoculated silage compared with both formic treated or untreated silages.

Other trials conducted at IGER over a period of 5 years using sheep and beef steers fed inoculated silage have also shown improvements in animal

Table 8.8 Effect of inoculant additives on silage fermentation characteristics and performance of beef and sheep (source: R. Jones, unpublished data, 1999).

Experiment	pH		Ammonia-N ($g\,kg^{-1}$ TN)		N-retention ($g\,d^{-1}$)		Dry matter intake (as % of C)	Liveweight gain (as % of C)
	C	T	C	T	C	T		
1 Beef	4.6	3.9	68	64	na	na	103	138
2 Sheep	4.6	4.15	10.1	7.1	8.5	13.5	195	na
3 Sheep	4.3	4.39	58	31	9.4	12.7	192	na
4 Beef	4.5	3.9	104	40	na	na	102	120
5 Beef	4.1	3.9	64	59	na	na	102	133

na, not available; C, control (untreated); T, inoculant treated.
Experiments 1, 2, 3 and 4: precision-chop harvested and ensiled in clamps.
Experiment 5: ensiled as bales.
Beef, Charolais steers (400 kg liveweight) fed for 10 weeks.
Sheep, 10-month-old lambs fed for 21 days with detailed feed intake and outputs recorded during the last 7 days.

performance, often independent of silage preservation status (Table 8.8). When fed to beef steers, silages treated with a multi-strain inoculant containing lactobacilli, streptococci and pediococci (experiment 1) showed increases in daily weight gains, while daily dry matter intakes were similar to those obtained with untreated silage. Of particular interest were experiments 2 and 3, which used sheep housed in metabolism crates for 21 days and fed inoculated silages. These showed mean increases in whole-tract nitrogen retention between the two experiments of 59 and 35%, respectively, when compared with values for the untreated silage. Other inoculants based on a single strain of *Lactobacillus plantarum* applied to precision-chop harvested silage (experiment 4) or baled grass (experiment 5) led to the production of silages that gave increases of 20 and 33%, respectively, in daily liveweight gain of beef steers, compared with untreated silages. There is substantial evidence that treatment of herbages with effective bacterial inoculants improves animal performance (Weinberg & Muck 1996). Characterizing the product or products responsible for this improved utilization from inoculated silages is difficult. Evidence from the limited number of trials conducted at IGER would suggest the changes in free amino acid content of inoculated silages and its relationship with reduced proteolytic activity and improved N-utilization may be contributory factors.

8.10 Silage technology

The most popular type of harvester in the UK is the high-output metered-chop forage harvester. This machine produces short chop material that aids consolidation and the more rapid release of sugars for fermentation, which results in a more rapid rate of acidification. Evidence has shown that this type of silage is more palatable for ruminants than the longer-chop silage harvested by forage wagons (Dulphy & Demarquilly 1991).

Almost 75% of the silage made in the UK is stored in solid horizontal-wall clamp silos fitted with a concrete base and effluent drainage systems. There has been a trend towards using baled silage on smaller farms, and this provides a flexible feeding system in association with buffer grazing systems.

8.10.1 Big-bale silage

Big-bale silage was introduced to the UK in the late 1970s (Forster 1989), and by the late 1990s approximately 20% of the total silage made in the UK was in this form. In the early years of big-bale technology it was used primarily on small farms, and particularly on farms that had not made silage previously. The system allowed a changeover from hay making to silage without expensive capital investments in constructing clamp silos. Big-bale silage also provided a flexible low-cost system for existing silage makers who wished to increase their silage production, or required additional silage to supplement summer-grazed grass. During the 1990s baling technology developed considerably, with several different models of balers being produced. Most modern balers offer chopping facilities as well as electro-mechanical applicators to allow the even distribution of silage additives. The most popular multi-purpose baler is the fixed-chamber baler, which produces bales of a fixed diameter. Grass is picked up from the field windrow and is fed into the bale chamber and subjected to an increasing rolling action as it begins to come into contact with the fixed bale-forming mechanism on the periphery of the chamber. The other type of baler is the variable chamber, where bale diameter can be variable as the bale formation is subjected to continuous pressure by belts, and the grass is rolled from the centre core to the outside. Bales may vary in diameter between 0.9 m and 1.8 m.

Grass is normally wilted to DM contents of 250–400 g kg^{-1} for baled silage, and on certain farms, especially those feeding bales to horses or sheep, it is not uncommon to wilt the grass to 500–650 g kg^{-1}. In the latter case, the ensiled material is referred to as haylage. Bales prepared from lower dry matter crops can weigh between 600 and 800 kg. In 1986, machinery was developed for film-wrapping of bales, and more than 95% of bales are now covered using this technology. The wrapping system uses a thin (20 micron) polyethylene film that can be stretched to 55–70% of the original width to ensure an airtight seal, when four to six overlapping layers are used. Research summarized by Kennedy (1989) indicated that dry matter losses in baled silage (10%) were less than half of those encountered with conventional precision-chop harvested silage stored in silos. Data presented in Table 8.9 show the quality of grass silage prepared by different methods, as big bales, using a pick-up wagon without chopping,

Table 8.9 The effect of different methods of harvesting grass on silage quality (after Fychan & Jones 1996a).

	Bales	Wagon	Precision-chop forage harvester
Toluene DM ($g\,kg^{-1}$)	198.6	192.6	202.3
pH	4.23	4.10	3.93
Ammonia ($g\,kg^{-1}$ N)	99.9	102.5	85.3
Nitrogen ($g\,kg^{-1}$ DM)	18.2	18.9	18.3
ME ($MJ\,kg^{-1}$ DM)	10.53	10.64	10.60
Lactic acid ($g\,kg^{-1}$ DM)	72.0	96.4	114.5
Acetic acid ($g\,kg^{-1}$ DM)	15.2	24.0	44.2
DM loss ($g\,kg^{-1}$)	112	121	180

Bales, grass harvested using a fixed-chamber baler.
Wagon, harvested with a pick-up wagon unchopped.

or a precision-chop forage harvester (Fychan & Jones 1996a). A restricted fermentation was observed with the unchopped material harvested as round bales, or unchopped using a pick-up wagon and stored in a silo, with both pH and ammonia-N being higher and lactic acid content lower than for the precision-chopped material. The restricted fermentation in the unchopped silage material is likely to be due to the lower density of the silage and a slower acidification during the early part of the ensiling period. The lower dry matter losses with bales in this trial confirms the earlier findings of Kennedy (1989).

8.11 Silage losses

A survey conducted by ADAS (Bastiman & Altman 1985) of 200 commercial farms in the UK showed that losses in the silage clamp ranged between 25% and 45%. These losses were estimated as the difference in weight of silage entered into the clamp and the weight of silage removed from the clamp during feed-out. The main losses in the silage process can be summarized as follows: field (respiration DM losses), storage (fermentation DM losses), effluent and feed-out. Field losses can be 2–5% of the dry matter and will depend on weather conditions during harvesting, type of mechanical treatment used to wilt the crop, and grass species. Storage losses can range from 2% to 10% of the total dry matter depending on silo management, time of filling the clamp, consolidation and sealing. The other major area of loss occurs during feed-out, where as much as 10% of the total silage dry matter can be lost as oxygen. This is then available for yeast and mould activity, which can rapidly degrade the lactic acid and residual sugars, resulting in the release of carbon dioxide. The losses from aerobic deterioration appear to be higher in wilted grass silage. However, in unwilted or direct-cut silage, considerable losses can occur as silage effluent will be produced. As much as 500 litres of effluent per tonne of silage can be released from a crop of 18% DM.

8.11.1 Silage effluent

The increase in silage production in the UK, particularly since the mid-1980s, coupled with changes in agronomic and ensiling processes, has greatly increased the amount of effluent produced on farms (Offer *et al.* 1991). Silage effluent has a biochemical oxygen demand of up to 90 000 mg $O_2 l^{-1}$, which is almost 200 times more polluting than raw domestic sewage and is extremely polluting if released to water courses (see also Chapter 14). In the UK, very stringent penalties can be issued to polluters of water courses under the Control of Pollution Act 1992. The amount of effluent produced during ensiling varies from as much as 500 l t^{-1} for wet crops, to little or no effluent for crops wilted to a DM content of 300 g kg^{-1} or more. In general, research work has shown an imprecise relationship between crop DM and effluent production. Predictive equations for effluent production based solely on crop DM are extremely variable (Jones & Jones 1995). Effluent production is influenced by a number of factors additional to crop DM, notably fertilizer input, silage additives and grass species or varieties.

8.11.2 Effluent production from different grass varieties

Perennial and Italian ryegrass varieties were ensiled in small-scale silos (10 kg) at IGER. The data for grass and silage chemical composition and total effluent production are presented in Table 8.10. Crop DM content (*c.* 190 g kg^{-1}) was

Table 8.10 Chemical composition and effluent production from different varieties of perennial (PRG) and Italian (IRG) ryegrasses (source: R. Jones, previously unpublished data, 1999).

Chemical composition (g kg^{-1} DM)	Aberelan (PRG)	Ba11353 (PRG)	Fenemma (PRG)	Abercomo (IRG)	RVP (IRG)
Grass					
Dry matter (g kg^{-1})	192.5	198.5	189.0	199.0	192.2
DOMD	621.7	651.3	616.5	625.1	603.5
WSC	182.0	215.0	170.0	281.0	225.5
Silage					
Dry matter (g kg^{-1})	204.3	217.5	203.3	212.8	204.2
pH	4.30	4.33	4.23	4.16	4.25
Ammonia-N (% of N)	6.49	6.43	6.19	6.92	4.94
Crude protein	136.3	136.5	136.4	121.3	122.7
ME (MJ kg^{-1} DM)	10.74	11.15	10.82	10.74	10.95
Silage effluent (ml kg^{-1})	94.3	125.1	108	102	68.7

similar for grass varieties; however, effluent production ranged between 68 and 125 ml kg^{-1}. It is likely, therefore, that differences in the anatomical structures of the grass varieties may be a further contributing factor to the rate and extent of silage effluent release.

8.11.3 Nitrogen losses in silage effluent

Silage effluent dry matter content is usually low, and may range between 40 and 100 g l^{-1} depending on the crop. Detailed studies of the free amino acid composition of silage effluent (Winters *et al.* 1996) from a range of treated silages are summarized in Table 8.11. Treatment of silage with additives had a marked effect on lowering the free amino acid content in released effluent. The effluent released from a formic-acid-treated silage showed the lowest loss of total free amino acid. However, this may well be related to the increased volume of effluent, and hence dilution effect, on the free amino acid content compared with untreated and inoculant-treated silage. Some variation was observed between silage treatments in individual free amino acid content, e.g. the leucine

Table 8.11 Free amino acid content of silage effluent (g l^{-1})* released from grass silage either untreated or treated with formic acid (3 l t^{-1}) or with *Lactobacillus plantarum* inoculant (source: Winters *et al.* 1996).

	Control	Inoculated	Formic
Methionine	0.18	0.16	0.08
Threonine	0.45	0.33	0.23
Serine	0.36	0.28	0.30
Asparagine	0.05	0.13	0.24
Aspartic acid	0.51	0.40	0.35
Glutamic acid	0.37	0.35	0.22
Glutamine	0.03	0.06	0.30
Proline	0.37	0.31	0.27
Glycine	0.44	0.33	0.21
Alanine	1.04	0.84	0.51
Valine	0.66	0.50	0.28
Isoleucine	0.39	0.29	0.12
Leucine	0.93	0.81	0.41
Phenylalanine	0.54	0.47	0.35
Histidine	0.12	0.09	0.07
Tyrosine	0.14	0.23	0.15
Lysine	0.14	0.17	0.38
Arginine	0.03	0.34	0.26
Ornithine	0.27	0.08	0.01
λ-aminobutyric acid	1.41	1.06	0.66
Total amino acids	8.46	7.26	5.41

* Weighted means.

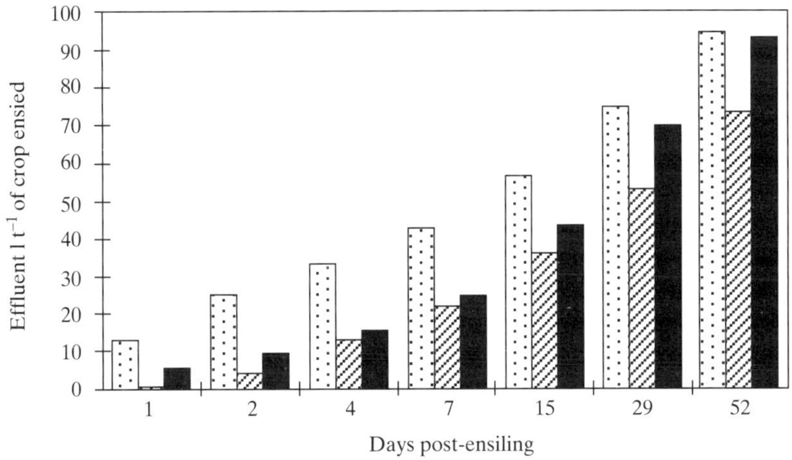

Fig. 8.9 Rate of effluent production from grass harvested as bales, ▤, unchopped pick-up wagon, ▨, or a precision-chop forager, ■ (source: Fychan & Jones 1996a).

and isoleucine contents of inoculant-treated silage were almost half that of the other treatments.

8.11.4 Effluent from baled silage

Effluent production from baled silage has been shown to equal that of forage-harvested clamp silage prepared from the same crop (Fig. 8.9). However, release of effluent appeared to be more rapid from baled silage, probably due to less compaction in the baled silage resulting in easier effluent drainage. As shown in Fig. 8.10, effluent release from baled silage can be reduced by as much as 38% by using six layers of a wide film wrap (750 mm width), compared with the conventional system of film wrapping with four layers of the same width, or with four layers of a narrow (500 mm) width film (Fychan & Jones 1996b). Although potential effluent release from low dry-matter baled silage can be extensive, in a farm situation the problem is almost non-existent as the majority of baled silage in the UK is made from crops of high DM (>300 g kg^{-1}), as reported by Haigh (1990).

8.12 Future trends in silage making

The problems of BSE and its association with animal-derived protein feed, together with the shrinking world supplies of protein concentrate feeds to supplement grass silage, have highlighted the value of forage resources in producing animal products economically. Apart from the advances in mechanical

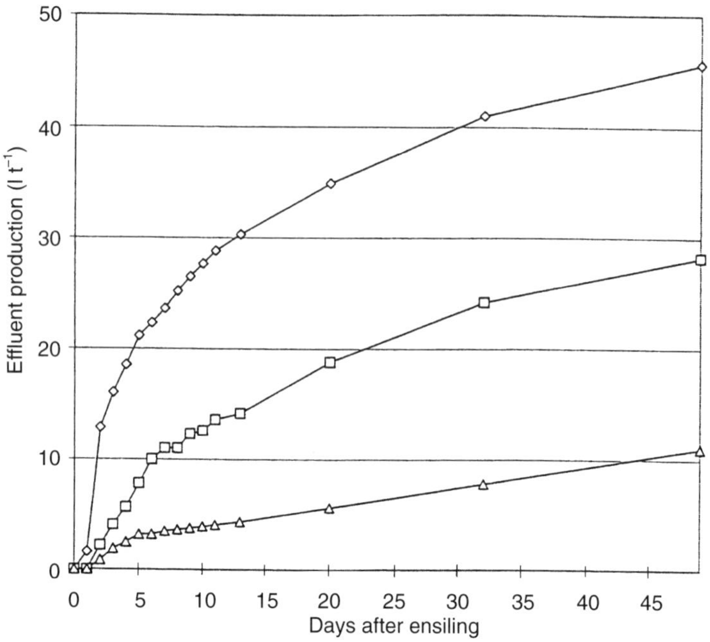

Fig. 8.10 Effluent production from baled silage film-wrapped with four layers of narrow (500 mm) film, ◇, four layers of wide (750 mm) film, □, or six layers of wide film, △. Source: Fychan & Jones (1996b).

technology and biotechnology already described, there is a concerted move towards high-protein alternative forages such as legumes and brassicas to reduce the reliance on imported concentrate feed supplements. To implement such changes in farming practice, conservation techniques are being developed to overcome the problems associated with ensiling legumes. It is anticipated that although the current levels of grass-silage production are likely to be maintained, a two-pronged approach should be considered. Firstly, an improvement in the varieties of grasses that we currently ensile, in particular targeted at reducing protein degradation and increasing the sugar content, and secondly, by combining grass and legume silages, preparing silages of bi-cropped materials. Research at IGER is being directed towards selection and breeding of forages to meet these requirements.

8.13 References

Anonymous (1997) Forage additives. *Farmers Weekly*, 21 November 1997, S1–S17.

Bastiman B. & Altman J.F.B. (1985) Silage losses on commercial farms. *Research and Development in Agriculture* **2**, 19–25.

Bolsen K.K., Bonilla D.R., Huck G.L., Young M.A. & Hart-Thakur R.A. (1996) Effect of propionic acid bacterial inoculant on fermentation and aerobic stability of wholecrop maize silage. In Jones D.I.H., Jones R., Dewhurst R., Merry R.J. & Haigh P.M. (eds) *Proceedings of 11th International Silage Conference University of Wales Aberystwyth*, 154–5. IGER, Aberystwyth.

Bosma A.H. (1987) Making wilted silage with low weather risk. *Proceedings of International Forage Conservation Conference, Nitra (CSSR), 1987*, 50–7.

Carpintero C.M., Henderson A.R. & McDonald P. (1979) The effect of some pre-treatments on proteolysis during the ensiling of herbage. *Grass and Forage Science* **34**, 311–15.

Castle M.E. (1990) Conclusions and future prospects. In Lingvall P. & Lindgren S. (eds) *Proceedings of the Eurobac Conference. Swedish University of Agricultural Sciences. Grass and Forage Reports* **3**, 184–8.

Cussen R.F., Merry R.J., Williams A.P. & Tweed J.K.S. (1995) The effect of additives on the ensilage of forage of differing perennial ryegrass and white clover content. *Grass and Forage Science* **50**, 249–58.

Davies D.R., Merry R.J. & Bakewell E.L. (1996) The effect of slurry application on the microflora of grass and changes occurring during the silage fermentation. *Grass and Forage Science* **51**, 42–51.

Davies D.R., Merry R.J., Williams A.P., Bakewell E.L., Leemans D.K. & Tweed J.K.S. (1998) Proteolysis during ensilage of forages varying in soluble sugar content. *Journal of Dairy Science* **51**, 1–10.

Dulphy J.P. & Demarquilly C. (1991) Digestibility and voluntary intake of conserved forage. In Pahlow G. & Honig H. (eds) *Forage Conservation Towards 2000. Landbauforschung Völkenrode* **123**, 140–60.

Fenlon D.R. (1986) Growth of naturally occurring *Listeria* spp. in silage: A comparative study of laboratory and farm-ensiled grass. *Grass and Forage Science* **41**, 375–8.

Forster L. (1989) Handling and storing of big bales. *Big Bale Silage. Proceedings of British Grassland Society Conference, February 1989*, 8.1–8.4. NAC, Stoneleigh.

Fychan R. & Jones R. (1996a) The effect of harvesting technology on silage quality and effluent production. In Jones D.I.H., Jones R., Dewhurst R., Merry R. & Haigh P.M. (eds) *Proceedings 11th International Silage Conference, University of Wales Aberystwyth, 1996*, 218–19. IGER, Aberystwyth.

Fychan R. & Jones R. (1996b) Effect of varying film wrap width and layering on effluent production from baled silage. In Jones D.I.H., Jones R., Dewhurst R., Merry R. & Haigh P.M. (eds) *Proceedings 11th International Silage Conference, University of Wales Aberystwyth, 1996*, 88–9. IGER, Aberystwyth.

Green J.O., Corrall A.J. & Terry R.A. (1971) Grass species and varieties: Relationships between stage of growth, yield and forage quality. *GRI Technical Report No. 8.* Grassland Research Institute, Hurley.

Haigh P.M. (1990) The effect of dry matter content on the preservation of big bale silages made during the autumn on commercial farms in South Wales 1983–87. *Grass and Forage Science* **45**, 29–34.

Harkess R.D. & Alexander R.A. (1969) The digestibility and productivity of selected herbage varieties. *Journal of the British Grassland Society* **24**, 282–9.

Henderson A.R. (1993) Silage additives. *Animal Feed Science and Technology* **45**, 35–56.

Henderson A.R., McDonald P. & Anderson D.H. (1982) The effect of silage additives containing formaldehyde on the fermentation of ryegrass ensiled at different dry-matter levels and on the nutritive value of direct-cut silage. *Animal Feed Science and Technology* **7**, 303–14.

Heron S.J.E., Edwards R.A. & McDonald P. (1986) Changes in the nitrogenous components of gamma-irradiated and inoculated ensiled ryegrass. *Journal of the Science of Food and Agriculture* **37**, 979–85.

Heron S.J.E., Edwards R.A. & McDonald P. (1988) The effects of inoculation, addition of glucose and mincing on fermentation and proteolysis in ryegrass ensiled in laboratory silos. *Animal Feed Science and Technology* **19**, 85–96.

Jones D.I.H. & Jones R. (1995) The effect of crop characteristics and ensiling methodology on grass silage effluent production. *Journal of Agricultural Engineering Research* **60**, 73–81.

Jones T.N. & Palmer L.O. (1932) Grass physiology. *Agricultural Engineering* **13**, 199–200.

Keady T.W.J. & Steen R.W.J. (1995) The effects of treating low dry-matter, low digestibility grass with a bacterial inoculant on the intake and performance of beef cattle, and studies on its mode of action. *Grass and Forage Science* **50**, 217–26.

Kennedy S.J. (1989) Methods of making and feeding silage. *Annual Report on Research and Technical Work of the Department of Agriculture for Northern Ireland*, p. 285.

Langston C.W., Irvin H., Gordon C.H., Bouma C., Wiseman H.G., Melin C.G., Moore L.A. & McCalmont J.R. (1958) Microbiology and chemistry of grass silage. *US Department of Agriculture Technical Bulletin*, No.1187.

Lindgren S., Axelsson L.T. & McFeeters (1990) Anaerobic L-lactate degradation by *Lactobacillus plantarum*. *FEMS Microbiological Letters* **66**, 209–14.

Lindgren S.E. & Dobrogosz W.J. (1990) Antagonistic activities of lactic acid bacteria in food and feed fermentations. *FEMS Microbiology Reviews* **87**, 149–64.

Lingvall P. & Lindgren S. (eds) (1990) *Proceedings of the Eurobac Conference. Swedish University of Agricultural Sciences. Grass and Forage Reports* **3**.

Lowes K.F., Shearman C.A.S., Gasson M.J. & Merry R.J. (1997) An experimental application of a yeast killer toxin to improve the aerobic stability of silage. *Supplement to the Journal of Applied Microbiology* **83**, 13.

Lunden Pettersson K. & Lindgren S. (1990) The influence of the carbohydrate fraction and additives on silage quality. *Grass and Forage Science* **45**, 223–33.

Mawdsley J.L., Bardgett R.D., Merry R.J., Pain B.F. & Theodorou M.K. (1995) Pathogens in livestock waste, their potential for movement through soil and environmental pollution. *Applied Soil Ecology* **2**, 1–15.

Mayer J.A. (1997) *Biocontrol of Listeria monocytogenes* in silage. Ph.D. Thesis, University of Manchester.

Mayer J.A., Merry R.J., Epton H.A.S. & Sigee D.C. (1995) Biological control of *Listeria monocytogenes* in silage. *Annales de Zootechnie* **44** (Suppl.1), 99.

Mayne C.S. & Steen R.W.J. (1990) Recent research on silage additives for milk and beef production. In Annual Report No. 63, 31–42. *Agricultural Research Institute, Northern Ireland (1989/90)*.

McDonald P., Henderson A.R. & McGregor A.W. (1968) Chemical changes and losses during the ensilage of wilted grass. *Journal of the Science of Food and Agriculture* **19**, 125–32.

McDonald P., Heron S.J.E. & Henderson A.R. (1991) *The Biochemistry of Silage*. Chalcombe Publications, Marlow.

Merry R.J. & Davies D.R. (1999) Review: Propionibacteria and their role in the biological control of aerobic spoilage in silage. *Lait* **79**, 1–16.

Merry R.J., Cussen-Mackenna R.F. & Jones R.F. (1993) Biological silage additives. *Ciencia E Investigacione Agraria* **20**, 372–401.

Merry R.J., Dhanoa M.S. & Theodorou M.K. (1995a) Use of freshly cultured lactic acid bacteria as silage inoculants. *Grass and Forage Science* **50**, 112–23.

Merry R.J., Winters A.L., Thomas P.I., Mueller M. & Mueller Th. (1995b) Degradation of fructans by epiphytic and inoculated lactic acid bacteria and plant enzymes during ensilage of normal and sterile hybrid ryegrass. *Journal of Applied Bacteriology* **79**, 583–91.

Merry R.J., Mawdsley J.L., Brooks A.E. & Davies D.R. (1997) Viability of *Cryptosporidium parvum* during ensilage of perennial ryegrass. *Journal of Applied Microbiology* **82**, 115–20.

Moran J.P., Pullar D. & Owen T.R. (1993) The development of a novel bacterial inoculant to reduce mould spoilage and the silage fermentation in big bale silage. In O'Kiely P., O'Connel M. & Murphy J. (eds) *Silage Research 1993, Proceedings of the 10th International Conference on Silage Research*, 85–6. Dublin City University, Ireland.

Mueller M. & Lier D. (1994) Fermentation of fructans by epiphytic lactic acid bacteria. *Journal of Applied Bacteriology* **76**, 406–11.

Murdoch J.C. (1989) The conservation of grass. In Holmes W. (ed) *Grass, its Production and Utilization*. 2nd edn. 173–213. Blackwell Scientific Publications, Oxford.

Nash M.J. (1985) *Crop Conservation and Storage in Cool Temperate Climates*, 2nd edn. 64–6. Pergamon, Oxford.

Nesbakken T. & Broch-Due M. (1991) Effects of a commercial inoculant of lactic acid bacteria on the composition of silages made from grasses of low dry matter content. *Journal of the Science of Food and Agriculture* **54**, 177–90.

Offer N.W., Chamberlain D.G. & Kelly M. (1991) Management of silage effluent. In Pahlow G. & Honig H. (eds) *Forage Conservation towards 2000, Landbauforschung Völkenrode* **123**, 129–36.

Oldenburg E. (1991) Mycotoxins in conserved forage. *Landbauforschung Volkenrode* **123**, 191–205.

Pahlow G. (1990) Microbiology of inoculants, crops and silages – small scale experiments. In Lingvall P. & Lindgren S. (eds) *Proceedings of the Eurobac Conference. Swedish University of Agricultural Sciences. Grass and Forage Reports* **3**, 45–59.

Patterson D.C., Yan T. & Gordon F.J. (1996). The effects of wilting of grass prior to ensiling on the response to bacterial inoculation. 2. Intake and performance by dairy cattle over three harvests. *Animal Science* **62**, 419–29.

Pell A.N. (1997) Manure and microbes: Public and animal health problem? *Journal of Dairy Science* **80**, 2673–81.

Playne M.J. & McDonald P. (1966) The buffering constituents of herbage and of silage. *Journal of the Science of Food and Agriculture* **17**, 264–8.

Rammer C., Ostling C., Lingvall P. & Lindgren S. (1994) Ensiling of manured crops – effects on fermentation. *Grass and Forage Science* **49**, 343–51.

Rooke J. (1990) The numbers of epiphytic bacteria on grass at ensilage on commercial farms. *Journal of the Science of Food and Agriculture* **51**, 525–33.

Seale D.R., Henderson A.R., Petterson K.O. & Lowe J.F. (1986) The effect of addition of sugar and inoculation with two commercial inoculants on the fermentation of lucerne silage in laboratory silos. *Grass and Forage Science* **41**, 62–70.

Sharp R., O'Donnel A.G., Gilbert H.G. & Hazlewood G.P. (1992) Growth and survival of genetically manipulated *Lactobacillus plantarum* in silage. *Applied and Environmental Microbiology* **58**, 2517–22.

Sharp R., Hooper P.G. & Armstrong D.G. (1994) The digestion of grass silages produced using inoculants of lactic-acid bacteria. *Grass and Forage Science* **49**, 42–53.

Spoelstra S.F. (1990) Comparison of the content of clostridial spores in wilted grass silage ensiled in either laboratory, pilot-scale or farm silos. *Journal of Agricultural Science* **38**, 423–34.

Spoelstra S.F. (1991) Chemical and biological additives in forage conservation. *Landbouforschung Volkenrode* **123**, 48–70.

Spoelstra S.F., Courtin M.G. & Van Beers J.A.C. (1988) Acetic acid bacteria can initiate aerobic deterioration of wholecrop maize silage. *Journal of Agricultural Science, Cambridge* **111**, 127–32.

Thomas C. & Fisher G. (1991) Forage conservation and winter feeding. In Thomas C., Reeve A. & Fisher G.E.J. (eds) *Milk from Grass*. ICI, SAC, IGER, 2nd edn. 27–51. British Grassland Society, Reading.

Watson S.J. & Nash M.J. (1960) *Conservation of Grass and Forage Crops*. Oliver & Boyd, Edinburgh.

Weddell J.R., Haigh P.M. & Steen R.W.J. (1996) The UK forage additive approval scheme – development and guidelines. In Jones D.I.H., Jones R., Dewhurst R. Merry R. & Haigh P.M. (eds) *Proceedings 11th International Silage Conference, University of Wales Aberystwyth, 1996*, 144–5. IGER, Aberystwyth.

Weinberg Z.G. & Muck R.E. (1996) New trends and opportunities in the development and use of inoculants for silage. *FEMS Microbiology Reviews* **19**, 53–68.

Wight P.N. (1990) An investigation of clostridiaphage additives as an aid to silage production. In *Proceedings of the British Grassland Society Fifth Research Conference, Newton Abbot*, 29–30.

Wilkinson J.M. & Stark B.A. (1988) *Silage UK.* Chalcombe, Marlow.

Wilkinson J.M. & Stark B.A. (1992) *Silage in Western Europe*, 2nd edn. Chalcombe Publications, Canterbury.

Wilkinson J.M., Wadephul F. & Hill J. (1996) *Silage in Europe*. Chalcombe, Lincoln.

Winters A.L., Fychan R. & Jones R. (1996) Nitrogen losses in silage effluent. In Jones D.I.H., Jones R., Dewhurst R., Merry R. & Haigh P.M. (eds) *Proceedings 11th International Silage Conference, University of Wales Aberystwyth, 1996*, 70–71. IGER, Aberystwyth.

Woolford M.K. (1984) *The Silage Fermentation.* Marcel Dekker, New York.

Woolford M.K. (1990) The detrimental effects of air on silage. *Journal of Applied Bacteriology* **68**, 101–16.

Zimmer E. & Wilkins R.J. (eds) (1984) Efficiency of silage systems: A comparison between unwilted and wilted silages. Results of a collaborative programme of European Research Institutes 1980–1983, *Landbauforschung Völkenrode*, Sonderheft 69, 88 pp.

Chapter 9
Principles of Foraging and Grazing Behaviour

A.J. Rook

9.1 Introduction

A sward and the animals grazing on it form a dynamic system, in which plants and animals interact in a complex way with each having effects upon the other. A grazing animal is confronted with a food source that is heterogeneous in vertical and horizontal distribution and may contain many different plant species. Within the constraints imposed by the sward and by its own anatomy, the animal must make behavioural choices, from the placement of individual bites to the choice of grazing area at a field or landscape scale. These choices will not only affect the animal's nutritional status, but will also feed back on the productive capacity of the sward via the resulting pattern of defoliation. An understanding of the fundamental processes of foraging and grazing behaviour is thus a prerequisite for the design of efficient grazing management systems.

9.2 Measuring grazing behaviour

Daily herbage intake may be summarized (Allden & Whittaker 1970) as

$$\begin{aligned}\text{Daily intake} &= \text{Instantaneous intake rate} \times \text{Grazing time}\\ &= (\text{Bite mass} \times \text{Biting rate})\\ &\quad \times (\text{Meal duration} \times \text{Number of meals}).\end{aligned}$$

Progress in understanding sward–animal interactions has been dependent on the development of methods for measuring these components of grazing behaviour.

Recording grazing time, whether by direct observation or using video recordings, is labour-intensive, limited by daylight, and may disturb the animals resulting in abnormal behaviour. These limitations have led to the development of automatic methods for recording grazing behaviour. Some systems record the position and acceleration of the animal's head, while others record jaw movements. The latter approach, coupled with sophisticated computer analysis, allows ruminating and biting and chewing while grazing to be distinguished. Recent technological developments allow these systems to record to a resolution of 20 Hz for 24 h, or alternatively to sample at a reduced rate for many days without the need to handle the animal (Rutter *et al.* 1997a). The systems have also been linked to a satellite global positioning system to obtain information on animal location, a crucial element in diet selection studies (Rutter *et al.* 1997b).

Direct recording of bite weight or of diet selection requires the use of oesophageally fistulated animals. This is expensive, compromises animal welfare and may result in abnormal behaviour. Therefore, bite weight is often calculated as the product of the number of bites and the total intake over a given period. In short-term experiments, turves or artificially constructed swards can be weighed before and after grazing to obtain a direct measurement of intake (e.g. Laca *et al.* 1992a). This method may not reflect normal foraging behaviour. The method has also been widely used in the study of diet selection (e.g. Newman *et al.* 1992). Herbage intake has often been estimated in the field by cutting herbage before and after grazing. However, this method is susceptible to bias due to plant growth and senescence, and selective grazing by the animal. It cannot be used to obtain the individual intakes of animals in groups. Intake can be estimated over a period of days by using indigestible markers. Many different markers have been used, but recent attention has focused on the use of natural alkanes in plant cuticular waxes (Dove & Mayes 1996). These have also been proposed as a means of indirectly estimating diet composition, but this is still a matter of debate. The use of long-term intakes to calculate bite weight assumes that bite weight is constant over the measurement period, and is therefore prone to bias. Intake can be estimated over a period of 1 h by weighing the animal before and after grazing. This method requires the collection of faeces and urine voided during this period, and also an estimate of insensible weight loss due to evaporation from the respiratory tract and other gaseous losses (Penning & Hooper 1985). For further discussion of methods for studying foraging behaviour see the review by Gordon (1995).

9.3 Factors affecting bite mass

Bite mass is affected by the constraints of the grazing animal's anatomy, including its mouth and body size, and by sward factors such as height and bulk density.

9.3.1 Constraints imposed by animal anatomy

Bite mass is the product of bite volume and the bulk density of the grazed horizon, with bite volume being the product of bite area and bite depth. Bite area is ultimately constrained by the dimensions of the animal's mouth. The breadth of the incisor arcade is proportional to body mass (*M*) to the power 0.36 (Illius & Gordon 1987). These authors also predicted that when sward height is non-limiting, bite mass will scale with $M^{0.75}$, but on very short swards it will scale with $M^{0.36}$ the same as incisor breadth, the only unrestricted bite dimension. The sward height at which this occurs is shorter for small animals, which can thus subsist on a shorter sward.

Allometric relationships constrain the digestion as well as the ingestion of food. Gut capacity increases faster than metabolic rate as body size increases. Thus, large animals have a relatively long retention time of food in the gut, allowing them to digest fibre more fully and to subsist on low-quality diets. Small animals, by contrast, are forced to feed selectively to obtain a higher quality diet (Demment & Van Soest 1985).

Bite mass may not always be constrained by mouth dimensions. For example, sheep sometimes insert their mouths sideways into swards in order to gather larger mouthfuls of herbage (Edwards *et al.* 1995). Conversely, animals may not always fully exploit their anatomical capabilities. For example, animals can increase bite mass (Chacon & Stobbs 1977, Newman *et al.* 1994) or biting rate (Greenwood & Demment 1988), and thus their intake rate after fasting. This suggests that non-fasted animals behave sub-optimally by not exploiting their maximum possible intake rate. However, an increased intake rate may lead to an increased need for rumination, highlighting the need to study optimal foraging in terms of absorbed rather than ingested nutrients (Greenwood & Demment 1988).

9.3.2 Constraints imposed by the sward

Sward height is a major constraint on bite mass in temperate grass swards, the effect being primarily on bite depth rather than on bite area. Dairy cows have been shown to consistently remove around 34% of the height of marked tillers, whatever the tiller height and whether or not the tillers have been grazed previously (Wade *et al.* 1989). It has been calculated that most of the energetic cost of grazing is due to chewing rather than biting, and that energy gain exceeds energy cost whatever the depth of a bite (Illius *et al.* 1995). Thus, energetics cannot explain why animals remove only the surface layer of the sward, suggesting that the greater force required to sever the sward at lower depth is a more likely constraint. This is supported by studies with perennial ryegrass swards continuously grazed by sheep, which have shown depth of grazing to be independent of sward height above a height of 6 cm, but apparently inhibited by the presence of the pseudo-stem horizon below this height (Barthram 1980, Barthram & Grant 1984). However, Illius *et al.* (1995) have suggested that canopy structure (number of tillers, etc.) is more important in constraining bite mass, by determining the force needed to sever a mouthful, than are the mechanical properties of the individual leaves.

Although sward height is an important determinant of bite depth, and hence bite mass, it cannot be used indiscriminately to compare swards of different structure. The regression of bite mass on sward height differs for swards of different structure. This has been shown both in theoretical models (e.g. Ungar & Noy-Meir 1988) and in experimental studies. For example, the relationship has been shown to differ between reproductive and vegetative swards (Prache & Petit 1995) and between continuously or rotationally grazed swards (Penning *et al.* 1994). However, the relationship with green leaf mass was similar in the different treatments in these experiments and closer than that with sward height. These different relationships with height result mainly from changes in the leaf:stem ratio. Leaf:stem ratio may be a more important determinant of bite mass than height for tropical grasses (Stobbs 1973a, b, Chacon & Stobbs 1976). Flores *et al.* (1993) found that in hand-constructed vegetative swards, pseudo-stems did not form a barrier to defoliation, but in reproductive swards, stems trimmed at ligule height did form a barrier, with steers grazing only the laminae. Bite area was also reduced by the presence of stems because the tillers bent at the ligule rather than at the base, thus restricting the number of tillers which could be gathered into the mouth by a sweep of the tongue. They suggest that to predict bite mass in reproductive swards, the height of the ligule and the relative length of lamina are needed in addition to sward height.

Bulk density of the grazed horizon is also important in determining bite mass. For example, bite mass can be greater for legumes than for grasses despite a shallower bite depth, as a result of the vertical distribution of bulk density; with bulk density greatest close to the ground for grasses but near the top of the sward for legumes (Gong *et al.* 1996). Because sward height and sward bulk density tend to covary, studies designed to separate their effects have mostly used artificial hand-constructed swards. Laca *et al.* (1994) used this approach and found that sward height and bulk density explained 44% and 27%, respectively, of the variation in bite mass. They concluded that bite mass is the driving variable in intake rate, not available biomass. This contrasts with the results of Black & Kenney (1984), who found that intake rate of sheep grazing artificial perennial ryegrass swards was related to sward height only at constant bulk density, and to bulk density only at similar sward heights, and that herbage mass per unit area was a better predictor of intake rate than either of these measurements. Burlinson *et al.* (1991) confined sheep in modified metabolism cages on swards of oats or grass manipulated to give independent variation in sward height and bulk density. They found that bite mass was directly related to sward height by the mechanism of bite depth, and that grazed stratum bulk density had a direct, independent and additive effect on bite mass.

Laca *et al.* (1992b) used artificial grass swards consisting of lamina only at four heights and three densities, and lucerne swards at three heights and three densities, and allowed cattle to take six bites from each sward. They found that animals obtained heavier bites on tall sparse swards than on short dense swards of equal mass per unit area. Thus, in contrast to Black & Kenney (1984), they concluded that recording of herbage mass is insufficient, and that both density and height are needed to predict bite mass even on homogeneous artificial swards. They also found that the animals removed a constant proportion of sward height, in line with the results of Wade *et al.* (1989). For both the grass and lucerne swards, bite area decreased linearly with density and increased quadratically with height to an asymptote of around $170\,cm^2$ determined by the sweep of the tongue. The animals appeared to adjust their tongue sweep to obtain as large a bite as possible given the stiffness of the sward. The reduction in bite area at a given height as density increases is consistent with the idea that bite area is limited by the force needed to sever the herbage. Average bite depth increased linearly with height, but the rate of increase was less in dense swards. Bite mass was less variable than bite dimensions. Mitchell *et al.* (1991) grew swards in trays, and then clipped out rows to vary the density and inserted a grid to control sward height. They found that the effect of a 100% change in sward height on bite depth was around 11 times greater than a 100% change in bulk density. Bite area only changed with height at low density. As a result, bite volume increased asymptotically with sward height but declined with bulk density. After correcting for the bulk density of different horizons, bite mass increased with sward height and bulk density but at a decreasing rate.

Ungar *et al.* (1991) found that in a heterogeneous hand-constructed sward, the residual height after grazing of long leaves was shorter than the height of leaves which had escaped grazing. Thus, the minimum grazeable height and residual height are not identical. They suggested that this is due to shorter leaves slipping out of the mouth.

There is some evidence that animals alter their response to sward state depending on their physiological state. For example, Jamieson & Hodgson (1979) found that calves strip-grazing perennial ryegrass swards initially took larger bites on shorter swards. Bite mass declined as the animals grazed down the sward, but the effect was more marked on short swards so that bite mass on these swards was lower after 19h on the pasture. They suggested that the high initial bite mass on shorter swards was due to greater hunger drive in these animals. Prache & Petit (1995) found between-animal differences in the regression of bite mass on green leaf mass, and suggested that ewes with high milk yields increase their bite mass more as green leaf mass increases.

9.4 Factors affecting biting rate

Biting rate is constrained by the time required to search for each bite and by the time required to process it, i.e. to sever, chew and swallow the food. A herbivore grazing dense, homogeneous swards characteristic of temperate grasslands can be assumed to require minimal searching time, as not only is the next bite readily available, but it can also be searched for while the current bite is being processed (Spalinger & Hobbs 1992, Farnsworth & Illius 1996). Conversely, processing time will only be limiting if the time to process a bite is longer than that required to find the next bite (Gross *et al.* 1993b). Animals foraging in spatially diverse environments may have to devote some time exclusively to searching. This leads to a reduction in biting rate, compared with foraging in homogeneous environments. In this situation the walking rate while searching and the standing density of the preferred food type become important additional considerations (Owen-Smith & Novellie 1982).

The time cost of severing a bite is relatively fixed, as it is determined by the time required to open and close the jaw. Chewing time, however, increases linearly with bite mass (Parsons *et al.* 1994). This leads to the inverse relationship between biting rate and bite mass observed in many experiments (e.g. Penning *et al.* 1991a). Small bites are handled less efficiently, as total handling time per unit mass increases exponentially as bite mass declines (Parsons *et al.* 1994). Thus, biting rate is largely constrained by bite mass, and the scope for active compensation for constraints on bite mass is small. As a result, bite mass is the main determinant of intake rate across a wide range of mammalian herbivores (Gross *et al.* 1993a,b). Animals grazing short swards usually have high biting rates which are inversely related to the low bite mass on these swards (e.g. Petit & Bechet 1995). However, the higher biting rate is often insufficient to maintain intake rate (e.g. Petit & Bechet 1995), as would be expected from the exponential increase in handling time per unit mass described above.

The total daily number of jaw movements in sheep is relatively fixed, but these can be apportioned between biting and chewing (Penning *et al.* 1991a). For cattle the situation is more complicated, as they have been shown to chew food simultaneously with biting (Laca *et al.* 1994). These authors distinguished these compound jaw movements from pure chews, and from what they termed manipulative movements concerned solely with gathering herbage into the mouth. They found that time per bite increased quadratically with bite mass because the number of exclusively manipulative movements decreased, with a converse increase in compound jaw movements and an increase in chews per bite. Total chews per bite increased linearly with increasing bite mass, but because of the positive intercept the number of chews per gram decreased. They speculated that this does not necessarily result in reduced comminution as the full molar area may not be used for smaller bites. They calculated that the use of compound jaw movements allowed cattle to increase intake rate by a factor of 1.3–1.4 for bite mass below 5 g. Modified models of the functional response of intake rate to bite mass have been proposed that account for an overlap between chewing and biting (Ginnett & Demment 1995, Farnsworth & Illius 1996).

The time required to sever a bite is independent of body mass (Shipley *et al.* 1994). However, chewing rate across a large number of mammalian species scales with body mass$^{-0.128}$ due to scaling with jaw length$^{0.312}$, suggesting that the energy cost of chewing per unit body mass is similar across species. The preferred rate seems to be chosen to optimize muscle stress at about one-third maximum (Druzinsky 1993). Smaller animals invest more time both in chewing while eating and in ruminating to compensate for their lower rumen residence time (Domingue *et al.* 1991, Gross *et al.* 1995). Illius (1989) found that herbage organic matter intake increases at a slower rate than bite weight as animals mature, suggesting that the number of bites per day declines with maturity.

Biting rate can be varied by the animal independently of bite mass. For example, Newman *et al.* (1994) showed that sheep fasted for 24 h had a higher biting rate than unfasted animals when first released back to the pasture, but this effect declined rapidly. Related effects have been seen in strip-grazing systems, where calves offered a lower allowance under the same sward conditions had a higher initial biting rate when offered their new daily allowance, but a lower biting rate later (Jamieson & Hodgson 1979).

9.5 Factors affecting grazing time

Grazing time is an important compensatory mechanism to counter the effect of variation in bite mass and intake rate, as the following examples demonstrate. Sheep graze for longer on grass than on white clover; this compensates for the lower intake rate of grass, resulting in similar daily intakes (Penning *et al.* 1991b). Heifers have been shown to graze for longer on reproductive swards than on vegetative swards of cocksfoot (D'Hour *et al.* 1995), and lactating ewes to graze for longer on short swards than on long swards (Petit & Bechet 1995). Heifers subject to winter food restriction achieve compensatory growth at pasture by increasing grazing time to compensate for their smaller bite mass (Ferrer Cazcarra & Petit 1995). Grazing time of set-stocked dairy cows decreases with increasing sward surface height, increases with higher initial milk yield, and decreases with increased concentrate supplementation (Pulido & Leaver 1995).

Animals are sometimes only partially able to compensate for low intake rate by increasing grazing time due to other constraints. An upper limit to grazing time is set by the need to undertake other activities such as ruminating, although lactating sheep and cows in negative energy balance have been shown to graze for up to 12 h per day when sward state is limiting intake rate (e.g. Penning *et al.* 1991b, Rook *et al.* 1994). Grazing time may also be constrained by available daylight. Many studies have shown that around 80% of grazing occurs during daylight, for both cattle and sheep, but the proportion of night time grazing increases as day length decreases (Penning *et al.* 1991b, Rook *et al.* 1994). Social facilitation and competition between animals also constrain grazing time. The grazing activity of sheep (Rook & Penning 1991) and cows (Rook & Huckle 1995) grazing in groups is more synchronized than would be expected by chance. The start of meals is also more synchronized than the end, suggesting that the start of meals is subject to social facilitation, while the end is under physiological control. Penning *et al.* (1993b) showed that grazing time is reduced when sheep are kept in groups of less than four, due to increased vigilance activity. Previous experience has also been shown to affect the ability of sheep to alter both their grazing time and intake rate in cross-over experiments between high and low stocking rate treatments (Curll & Davidson 1983).

Animals do not always use grazing time to compensate for low intake rates, although they appear able to do so. For example, Hendricksen & Minson (1980) found that cattle grazing a tropical forage legume reduced grazing time when they were forced to eat stem material, even though the intake rate was lower and their rumens were not distended. Similar results were seen by Chacon & Stobbs (1976). Rook *et al.* (1994) found that cows grazing a very short sward and offered a supplement decreased their grazing time compared with cows on a longer sward offered the same supplement, despite the lower herbage intake rate on the shorter sward. These results were obtained under continuous stocking systems. In short-term rotational systems animals may also graze for less time at a low herbage allowance in anticipation of a new allowance later (Jamieson & Hodgson 1979). These effects may be related to the low marginal energy gain of additional grazing time under very constrained conditions. Murray (1991), in a study of wild ruminants, found that the distance travelled per unit of food ingested increased exponentially as bite mass declined, resulting in increased energy costs.

9.6 Meal patterns

The number and duration of meals are important, as the factors controlling the start and end of meals provide a link between the behavioural mediation of intake and the underlying physiological control mechanisms; this is an area that is still poorly understood. The definition of meals is not straightforward. Animals eat in bouts punctuated by intervals of rest or other activities. Some of these can be regarded as intra-meal intervals, and others as inter-meal intervals. The criterion for allocating intervals to one of these groups is usually derived empirically from data using methods of varying degrees of sophistication (e.g. Slater & Lester 1982, Sibley *et al.* 1990). Penning *et al.* (1993a) have shown that cumulative intake by sheep declines as a meal progresses, due to an increase in the number and duration of intra-meal intervals rather than a decrease in instantaneous intake rate while actually eating. D'Hour *et al.* (1995) showed that biting rate gradually increased by around 8% over the first three-quarters of a meal, but then declined in the last quarter.

Many studies have shown that grazing ruminants take a large meal just before sunset, with the next largest meal at dawn. Environmental factors such as temperature and rainfall may modify this diurnal grazing pattern, but these effects are relatively small (Dudzinski & Arnold 1979, Champion *et al.* 1994). Variability in grazing time between animals is less during the evening meal than at other times (Phillips & Denne 1988, Rook & Huckle 1996) and this meal is more intense, i.e. more of it is actually spent eating (Rook & Huckle 1997). Sheep have been found to have higher intake rates during the evening meal as a result of greater bite mass (Orr *et al.* 1997), while the biting rate of both cows (Phillips & Leaver 1986, Phillips & Hecheimi 1989) and heifers (D'Hour *et al.* 1995) has been shown to be higher in the evening than in the morning. A possible explanation for these effects is that the animal is adopting an optimal foraging strategy by taking advantage of the higher concentration of dry matter and photosynthetic products in the leaves late in the day (Orr *et al.* 1997). An alternative suggestion is that it is a strategy to ensure a high rumen fill and hence sufficient substrate for rumen flora during the night when there is little grazing (Penning *et al.* 1991b). Meal patterns are also affected by diet. For example, sheep grazing clover have more but shorter meals than those grazing grass. This is probably related both to the higher intake rate for clover and to its faster breakdown in the rumen (Penning *et al.* 1991b, Penning & Boval 1997).

9.7 Foraging in heterogeneous environments

The choices facing the animal described in the previous sections largely concern the placement of bites so as to select different vertical components at a point in the sward. On a larger scale, choices are often complicated because of horizontal heterogeneity in the environment, for example, in mixed grass–clover swards, in species-rich permanent grasslands, or in monocultures with patches of high and low vegetation (Milne 1991). Selection of different species or sward types by the animal not only affects its diet, but also has a profound effect on the stability of the plant community. For example, Yarrow & Penning (1994) showed that the proportion of white clover in mixed swards is lower when grazed by sheep than when grazed by cattle, both of which are lower than under a cutting regime. These effects are due to the different ability of the species to select clover. The effects are transient, and can be changed rapidly by altering management.

It is important to distinguish between selection, that is the diet actually consumed by the animal, and preference, the diet the animal would select given the minimum of physical constraints (Hodgson 1979, Newman *et al.* 1995). It is also important to distinguish between active selection by the animal and passive selection due to the distribution of different species within the grazed horizon of the sward. Dove (1996) used the terms 'principal food', i.e. one forming a large part of the diet, and 'preferred food', i.e. one which is proportionately more common in the diet than in the environment. This raises the issue of how to define the environmental proportion, that is to what should the diet be compared. While there may be no selection in a particular sward horizon, the animal may have actively chosen that horizon, or alternatively have been constrained to that horizon. The definition of selection even when the correct sward horizon is used for comparison has been the subject of methodological debate. Thus, care is required in interpreting published results. For example, the conclusion of Clark & Harris (1985) that sheep grazing intimate mixtures of grass and clover did not actively select clover was challenged following a re-analysis of the data by Ridout & Robson (1991). There is also debate regarding the currency in which selection and preference are expressed. However, whether intake should be approximated by grazing time, number of bites, or number of visits to particular species etc., appears to matter little in grass swards as the ranking of preferences using the different methods is highly correlated (Ganskopp *et al.* 1997)

Sheep, goats and cattle (Newman *et al.* 1992, Penning *et al.* 1995a, b, respectively), when offered adjacent monocultures of grass and white clover, have all been shown to exhibit a partial preference for clover; i.e. they include 60–80% clover in the diet but continue to eat some grass. Further, the pattern of preference changes over the day, with more clover being eaten in the morning and more grass at night (Penning *et al.* 1995a), a result which has also been predicted from optimal foraging models (Newman *et al.* 1992). One proposed explanation for partial preference is that animals constantly need to resample the less preferred species. However, this argument has been seriously weakened by the work of Edwards *et al.* (1996), who showed that sheep can learn and remember the location of food patches within a few days and that this effect persists over a period of at least 3 days. Edwards *et al.* (1997) also showed that sheep could make flexible use of visual or olfactory cues (turves of grass or clover) to locate preferred food pellets, thus demonstrating that they do not need to sample grass and clover to distinguish between them. However, Distel *et al.* (1995) found that cattle entered patches of ryegrass of different density and height in a random search pattern, but increased their residence time and number of bites in patches which allowed the greater instantaneous intake rate. They concluded that the cattle did not base their choice between contrasting patches solely on the cues offered by height and density, but adjusted their patch residence time instantaneously on the basis of the intake rate achieved. Illius *et al.* (1992) found that sheep offered grass–white clover mixtures of different proportion and height appeared to make their initial patch choice on the basis of sward height, but frequently switched between patches, suggesting that they were using information gathered during grazing. Clover content did not affect initial patch choice, but did affect the ultimate distribution of bites between the patches. They also found a carry-over effect of previous experience. They concluded that pre-grazing cues are less important than information obtained during grazing, but that previous experience may initially distort the use of this information.

The cues used by sheep to locate food have been studied in detail by impairment of various senses (e.g. Arnold 1966a,b, Krueger *et al.* 1974). The relative importance of the different senses seems to change depending on the choice with which the animal is faced. Thus, taste is important in distinguishing between sweet and sour plants, while touch is more important in identifying succulents, and sight in distinguishing plants of different height. Horizontal location of food patches seems to be less affected by blindfolding, but blindfolded sheep tend to change position less frequently. Taken over a whole season, sense impairment did not affect the overall performance of the sheep, indicating that they were able to compensate for the loss of various senses.

Dietary preferences have been shown to be strongly influenced by learning. In a review of the role of learning in dietary preference of domestic ruminants, it was concluded that learning from the dam is more important than later learning from conspecifics, and that the food choices of adults are more stable: they accept new foods less readily and are better at avoiding foods with adverse post-ingestive consequences (Provenza & Balph 1987, 1988). For example, Key and MacIver (1980) compared cross-fostered and naturally reared lambs of either Clun or Welsh Mountain breeds. They found that the diet selected subsequently was related to that of the rearing dam rather than the breed of the lamb. However, Orr *et al.* (1995) did not find significant effects of rearing in a similar cross-fostering experiment with sheep and goats. Prior experience of a species in the diet affects selection (Flores *et al.* 1989a), and the effect appears to be related to development of the specific foraging skills required for that species as the effects are not apparent if cut herbage is fed (Flores *et al.* 1989b). Newman *et al.* (1992) found that the preference of sheep for grass or white clover was affected by which species they had been grazing immediately prior to the test period, with selection for novelty generally occurring. Effects of prior experience can be very long lasting; for example, it has been found that the diet selected by goats could be affected for up to 4 years by their experience in the first grazing season (Biquand & Biquand-Guyot 1992).

Sheep are generally more selective than cattle as a result of their different jaw morphology, with sheep having narrower muzzles (Grant *et al.* 1985). Sheep are able to graze lower into the sward, than cattle, giving them further opportunity for selection. This result would be predicted from allometric considerations. Larger ruminants have relatively large rumens compared to their metabolic requirements. Thus they have a longer retention time and are able to utilize a more fibrous diet. Smaller animals cannot utilize such diets efficiently and are thus forced to be more selective (Demment & Van Soest 1985). Larger animals also have relatively large foraging costs, and may therefore need to be less selective in order to contain these costs (Murray 1991). Sheep and cattle have been shown to follow the same sequence of species preference when grazing down a rangeland sward, but the sheep took longer to switch to the less preferred species as they were able to be more selective. Both species began regrazing the preferred species at around the same time as they began to include less preferred foods in the diet (O'Reagain & Grau 1995). Evidence for greater selectivity by small as opposed to large cattle has been given by Lazo and Soriguer (1993), who found that small animals took fewer bites per step when food was scarce, i.e. they were engaging in more searching for preferred foods, while larger animals did not alter their bite/step ratio as they were not foraging selectively.

Preference and selection can be affected by the animals' state. Newman *et al.* (1994) compared sheep fasted for 24h with unfasted controls when offered a choice between monocultures of grass and white clover. The fasted animals spent less time grazing on clover, i.e. the selection of their normally preferred species declined. This was probably related to the higher intake rate of fasted animals leading to less opportunity for selectivity.

Selection does not only take place between plant species, but also between areas of the sward of different structure within the same species. There is some disagreement in the literature as to the basis of such selection. Some authors have found that animals offered choices between swards of different height, density and species composition select those swards which allow them to maximize their intake rate (e.g. Black & Kenney 1984, Arnold 1987, Distel *et al.* 1995, Griffiths *et al.* 1996), while others have not found evidence to support this conclusion (Illius *et al.* 1992). Griffiths *et al.* (1996) offered lactating cows a choice between patches of different sward height and bulk density arranged along a linear race. They found that patch residence time and number of bites per patch were higher on taller patches, but that bulk density within sward height had little effect. Illius *et al.* (1992) offered sheep a choice of intimate grass–white clover mixtures (with different proportions of clover and different sward heights) in sward trays; the effects of height and proportion were found to be additive, with tall swards and those of intermediate clover content being preferred.

Theoretical models of selection have often focused on the spatial aspects of selection. The preferred food resource is assumed to occur in a number of discrete patches, and the residence time of the animals within each patch and its 'rules' for leaving the patch are studied. For example, Charnov (1976) proposed the 'marginal value theorem', which states that an animal will stay in a patch until the marginal rate of resource capture within the patch has fallen to the mean capture rate over the whole habitat. Edwards *et al.* (1994) offered sheep a simple choice of two different types of pellets mixed in different proportions in a number of bowls spread over a pen at different spatial aggregations. They found that sheep were able to select between pellets at a fine scale, and that selection of the preferred pellets was greatest at the largest scale of aggregation of patches due to better selection between, rather than within, patches.

Some workers have considered patch residence time in terms of 'feed station' residence time, where a feed station is defined as the area an animal can graze without moving its forefeet. Shipley *et al.* (1996) produced a conceptual model of foraging locomotion. They showed that there may be energetic advantages to taking many bites at a single feeding station, and that the feeding station may be the best functional unit for defining patch size for herbivores. Experimental results have shown that as sheep grazed down a reproductive cocksfoot sward, the mean residence time at each feed station decreased, but the number of stations visited per minute increased, resulting in little change in the overall time spent at feeding stations (Roguet & Prache 1995). The animals moved less with their heads up as the sward height decreased, indicating that the grazing was of a more continuous nature with shorter distances between feeding stations. Similar effects were seen by Ruyle & Dwyer (1985). Rogalski *et al.* (1996) found that weaned lambs not only took more bites per minute on short swards, but also more bites per step, and that they moved more slowly.

9.8 Conclusions

The research evidence presented in this chapter demonstrates the importance and complexity of animal behaviour in understanding the grazing process. Much work remains to be done, particularly in understanding the interaction between physiological and behavioural signals, and in understanding the effects of spatial and temporal scale. Such knowledge is needed to explain why animals sometimes appear to be acting sub-optimally, given the apparent environmental and anatomical constraints. Despite these gaps in our knowledge, the interplay between farm animal experimentation and ecological modelling described in this chapter already provides a firm basis for the design of practical grazing management systems, and the setting of realistic objectives for both plant and animal breeding.

9.9 References

Allden W.G. & Whittaker I.A.McD. (1970) The determinants of herbage intake by grazing sheep: The interrelationships of factors influencing herbage intake and availability. *Australian Journal of Agricultural Research* **21**, 755–66.

Arnold G.W. (1966a) The special senses in grazing animals. 1. Sight and dietary habits in sheep. *Australian Journal of Agricultural Research* **17**, 521–9.

Arnold G.W. (1966b) The special senses in grazing animals. 2. Smell, taste and touch and dietary habits of sheep. *Australian Journal of Agricultural Research* **17**, 531–42.

Arnold G.W. (1987) Influence of the biomass, botanical composition and sward height of annual pastures on foraging behaviour by sheep. *Journal of Applied Ecology* **24**, 759–72.

Barthram G.T. (1980) Sward structure and the depth of the grazed horizon. *Grass and Forage Science* **36**, 130–1.

Barthram G.T. & Grant S.A. (1984) Defoliation of ryegrass dominated swards by sheep. *Grass and Forage Science* **39**, 211–19.

Biquand S. & Biquand-Guyot V. (1992) The influence of peers, lineage and environment on food selection of the Criollo goat (*Capra hircus*). *Applied Animal Behaviour Science* **34**, 231–45.

Black J.L. & Kenney P.A. (1984) Factors affecting diet selection by sheep. *Australian Journal of Agricultural Research* **35**, 565–78.

Burlinson A.J., Hodgson J. & Illius A.W. (1991) Sward canopy structure and the bite dimensions and bite weight of grazing sheep. *Grass and Forage Science* **46**, 29–38.

Chacon E. & Stobbs T.H. (1976) Influence of progressive defoliation of a grass sward on the eating behaviour of cattle. *Australian Journal of Agricultural Research* **27**, 709–27.

Chacon E. & Stobbs T.H. (1977) The effects of fasting prior to sampling and diurnal variation on certain aspects of grazing behaviour in cattle. *Applied Animal Ethology* **3**, 163–71.
Champion R.A., Rutter S.M., Penning P.D. & Rook A.J. (1994) Temporal variation in grazing behaviour of sheep and the reliability of sampling periods. *Applied Animal Behaviour Science* **42**, 99–108.
Charnov E.L. (1976) Optimal foraging, the marginal value theorem. *Theoretical Population Biology*, **9**, 129–36.
Clark D.A. & Harris P.S. (1985) Composition of the diet of sheep grazing swards of differing white clover content and spatial distribution. *New Zealand Journal of Agricultural Research* **28**, 233–40.
Curll M.L. & Davidson J.L. (1983) Defoliation and productivity of a *Phalaris*–subterranean clover sward, and the influence of grazing experience on sheep intake. *Grass and Forage Science* **38**, 159–67.
Demment M.W. & Van Soest P.J. (1985) A nutritional explanation for body size patterns of ruminant and non-ruminant herbivores. *The American Naturalist* **125**, 641–72.
D'Hour P., Petit M. & Garel J.P. (1995) Components of grazing behaviour of three breeds of heifers. *Annales de Zootechnie* **44** suppl, 270.
Distel R.A., Laca E.A., Griggs T.C. & Demment M.W. (1995) Patch selection by cattle: Maximisation of intake rate in horizontally heterogeneous pastures. *Applied Animal Behaviour Science* **45**, 11–21.
Domingue B.M.F., Dellow D.W. & Barry T.N. (1991) The efficiency of chewing during eating and ruminating in goats and sheep. *British Journal of Nutrition* **65**, 355–63.
Dove H. (1996) Constraints to the modelling of diet selection and intake in the grazing ruminant. *Australian Journal of Agricultural Research* **47**, 257–75.
Dove H. & Mayes R.W. (1996) Plant wax components: A new approach to estimating intake and diet composition in herbivores. *Journal of Nutrition* **126**, 13–26.
Druzinsky R.E. (1993) The time allometry of mammalian chewing movements: Chewing frequency scales with body mass in mammals. *Journal of Theoretical Biology* **160**, 427–40.
Dudzinski M.L. & Arnold G.W. (1979) Factors influencing the grazing behaviour of sheep in a Mediterranean climate. *Applied Animal Ethology* **5**, 125–44.
Edwards G.R., Newman J.A., Parsons A.J. & Krebs J.R. (1994) Effects of the scale and spatial distribution of the food resource and animal state on diet selection: An example with sheep. *Journal of Animal Ecology* **63**, 816–26.
Edwards G.R., Parsons A.J., Penning P.D. & Newman J.A. (1995) Relationship between vegetation state and bite dimensions of sheep grazing contrasting plant species and its implications for intake rate and diet selection. *Grass and Forage Science* **50**, 378–88.
Edwards G.R., Newman J.A., Parsons A.J. & Krebs J.R. (1996) The use of spatial memory by grazing animals to locate food patches in spatially heterogeneous environments: An example with sheep. *Applied Animal Behaviour Science* **50**, 147–60.
Edwards G.R., Newman J.A., Parsons A.J. & Krebs J.R. (1997) Use of cues by grazing animals to locate food patches: An example with sheep. *Applied Animal Behaviour Science* **51**, 59–68.
Farnsworth K.D. & Illius A.W. (1996) Large grazers back in the fold: Generalizing the prey model to incorporate mammalian herbivores. *Functional Ecology* **10**, 678–80.
Ferrer Cazcarra R. & Petit M. (1995) The effect of winter feeding level on subsequent grazing behaviour and herbage intake of Charolais heifers. *Animal Science* **61**, 211–17.
Flores E.R., Provenza F.D. & Balph D.F. (1989a) Role of experience in the development of foraging skills of lambs browsing the shrub serviceberry. *Applied Animal Behaviour Science* **23**, 271–8.
Flores E.R., Provenza F.D. & Balph D.F. (1989b) The effect of experience on the foraging skill of lambs: Importance of plant form. *Applied Animal Behaviour Science* **23**, 285–91.
Flores E.R., Laca E.A., Griggs T.C. & Demment M.W. (1993) Sward height and vertical

morphological differentiation determine cattle bite dimensions. *Agronomy Journal* **85**, 527–32.

Ganskopp D., Cruz R. & Fajemisin B. (1997) Relationships among variables indexing selective grazing behaviour of goats. *Applied Animal Behaviour Science* **51**, 75–85.

Ginnett T.F. & Demment M.W. (1995) The functional response of herbivores: Analysis and test of a simple mechanistic model. *Functional Ecology* **9**, 376–84.

Gong Y., Hodgson J., Lambert M.G. & Gordon I.L. (1996) Short term ingestive behaviour of sheep and goats grazing grasses and legumes. 1. Comparison of bite weight, bite rate and bite dimensions for forages at two stages of maturity. *New Zealand Journal of Agricultural Research* **39**, 63–73.

Gordon I.J. (1995) Animal based techniques for grazing ecology research. *Small Ruminant Research* **16**, 203–14.

Grant S.A., Suckling D.E., Smith H.K., Torvell L., Forbes T.D.A. & Hodgson J. (1985) Comparative studies of diet selection by sheep and cattle: The hill grasslands. *Journal of Ecology* **73**, 987–1004.

Greenwood G.B. & Demment M.W. (1988) The effect of fasting on short term cattle grazing behaviour. *Grass and Forage Science* **43**, 377–86.

Griffiths W.M., Hodgson J., Holmes C.W. & Arnold G.C. (1996) The use of a novel approach to determine the influence of sward characteristics on the discriminatory grazing behaviour of dairy cows. *Proceedings of the New Zealand Society of Animal Production* **56**, 122–4.

Gross J.E., Hobbs N.T. & Wunder B.A. (1993a) Independent variables for predicting intake rate of mammalian herbivores: Biomass density, plant density or bite size? *Oikos* **68**, 75–81.

Gross J.E., Shipley L.A., Hobbs N.T., Spalinger D.A. & Wunder B.A. (1993b) Functional response of herbivores in feed-concentrated patches: Tests of a mechanistic model. *Ecology* **74**, 778–91.

Gross J.E., Demment M.W., Alkon P.U. & Kotzman M. (1995) Feeding and chewing behaviours of Nubian ibex: Compensation for sex-related differences in body size. *Functional Ecology* **9**, 385–93.

Hendricksen R. & Minson D.J. (1980) The feed intake and grazing behaviour of cattle grazing a crop of *Lablab purpureus* cv. Rongai. *Journal of Agricultural Science, Cambridge* **95**, 547–54.

Hodgson J. (1979) Nomenclature and definitions in grazing studies. *Grass and Forage Science* **34**, 11–18.

Illius A.W. (1989) Allometry of food intake and grazing behaviour with body size in cattle. *Journal of Agricultural Science, Cambridge* **113**, 259–66.

Illius A.W. & Gordon I.J. (1987) The allometry of food intake in grazing ruminants. *Journal of Animal Ecology* **56**, 989–99.

Illius A.W., Clark D.A. & Hodgson J. (1992) Discrimination and patch choice by sheep grazing grass–clover swards. *Journal of Animal Ecology* **61**, 183–94.

Illius A.W., Gordon I.J., Milne J.D. & Wright W. (1995) Costs and benefits of foraging on grasses varying in canopy structure and resistance to defoliation. *Functional Ecology* **9**, 894–903.

Jamieson W.S. & Hodgson J. (1979) The effect of daily herbage allowance and sward characteristics upon ingestive behaviour and herbage intake of calves under strip grazing management. *Grass and Forage Science* **34**, 261–71.

Key C. & MacIver R.M. (1980) The effects of maternal influences on sheep: Breed differences in grazing, resting and courtship behaviour. *Applied Animal Ethology* **6**, 33–48.

Krueger W.C., Laycock W.A. & Price D.A. (1974) Relationships of taste, smell, sight, and touch to forage selection. *Journal of Range Management* **27**, 258–62.

Laca E.A., Ungar E.D., Seligman N. & Demment M.W. (1992a) Effect of sward height and bulk density on bite dimensions of cattle grazing homogeneous swards. *Grass and Forage Science* **47**, 91–102.

Laca E.A., Ungar E.D., Seligman N.G., Ramey M.R. & Demment M.W. (1992b) An integrated methodology for studying short term grazing behaviour of cattle. *Grass and Forage Science* **47**, 81–90.

Laca E.A., Ungar E.D. & Demment M.W. (1994) Mechanisms of handling time and intake rate of a large mammalian grazer. *Applied Animal Behaviour Science* **39**, 3–19.

Lazo A. & Soriguer R.C. (1993) Size-biased foraging behaviour in feral cattle. *Applied Animal Behaviour Science* **36**, 99–110.

Milne J.A. (1991) Diet selection by grazing animals. *Proceedings of the Nutrition Society* **50**, 77–85.

Mitchell R.J., Hodgson J. & Clark D.A. (1991) The effect of varying leafy sward height and bulk density on the ingestive behaviour of young deer and sheep. *Proceedings of the New Zealand Society of Animal Production* **51**, 159–65.

Murray M.G. (1991) Maximizing energy retention in grazing ruminants. *Journal of Animal Ecology* **60**, 1029–45.

Newman J.A., Parsons A.J. & Harvey A. (1992) Not all sheep prefer clover: Diet selection revisited. *Journal of Agricultural Science, Cambridge* **119**, 275–83.

Newman J.A., Penning P.D., Parsons A.J., Harvey A. & Orr R.J. (1994) Fasting affects intake behaviour and diet preference of grazing sheep. *Animal Behaviour* **47**, 185–93.

Newman J.A., Parsons A.J., Thornley J.H.M., Penning P.D. & Krebs J.R. (1995) Optimal diet selection by a generalist grazing herbivore. *Functional Ecology* **9**, 255–68.

O'Reagain P.J. & Grau E.A. (1995) Sequence of species selection by cattle and sheep on South African sourveld. *Journal of Range Management* **48**, 314–21.

Orr R.J., Penning P.D., Parsons A.J., Harvey A. & Newman J.A. (1995) The role of learning and experience in the development of dietary choice by sheep and goats. *Annales de Zootechnie* **44** suppl, 111.

Orr R.J., Penning P.D., Harvey A. & Champion R.A. (1997) Diurnal patterns of intake rate by sheep grazing monocultures of ryegrass or white clover. *Applied Animal Behaviour Science* **52**, 65–77.

Owen-Smith N. & Novellie P. (1982) What should a clever ungulate eat? *The American Naturalist* **119**, 151–78.

Parsons A.J., Thornley J.H.M., Newman J. & Penning P.D. (1994) A mechanistic model of some physical determinants of intake rate and diet selection in a two-species temperate grassland sward. *Functional Ecology* **8**, 187–204.

Penning P.D. & Boval M. (1997) Effects of fasting on ingestive behaviour of sheep grazing grass or white clover monocultures. *18th International Grassland Congress, Winnipeg, Canada 1987*, paper 410.

Penning P.D. & Hooper G.E. (1985) An evaluation of short term weight changes in grazing sheep for estimating herbage intake. *Grass and Forage Science* **40**, 79–84.

Penning P.D., Parsons A.J., Orr R.J. & Treacher T.T. (1991a) Intake and behaviour responses by sheep to changes in sward characteristics under continuous grazing. *Grass and Forage Science* **46**, 15–28.

Penning P.D., Rook A.J. & Orr R.J (1991b) Patterns of ingestive behaviour of sheep continuously stocked on monocultures of ryegrass or white clover. *Applied Animal Behaviour Science* **31**, 237–50.

Penning P.D., Orr R.J., Parsons A.J. & Rutter S.M. (1993a) Factors controlling meal duration and herbage intake rate by grazing sheep. *Proceedings of the 27th International Congress of the International Society of Applied Ethology, Berlin.*

Penning P.D., Parsons A.J., Newman J.A., Orr R.J. & Harvey A. (1993b) The effect of group size on grazing time in sheep. *Applied Animal Behaviour Science* **37**, 101–109.

Penning P.D., Parsons A.J., Orr R.J. & Hooper G.E. (1994) Intake and behaviour responses by sheep to changes in sward characteristics under rotational grazing. *Grass and Forage Science* **49**, 476–86.

Penning P.D., Newman J.A., Parsons A.J., Harvey A. & Orr R.J. (1995a) The preference of adult sheep and goats grazing ryegrass and white clover. *Annales de Zootechnie* **44** suppl, 113.

Penning P.D., Parsons A.J., Orr R.J., Harvey A. & Yarrow N.H. (1995b) Dietary preference of heifers for grass or clover, with and without romensin slow-release anti-bloat boluses. *Animal Science* **60**, 550.

Petit M. & Bechet G. (1995) Grass intake and grazing behaviour of dry and suckling ewes according to sward height. *Annales de Zootechnie* **44** suppl, 250.

Phillips C.J.C. & Denne S.K.P.J. (1988) Variation in the grazing behaviour of dairy cows measured by a Vibracorder and bite count monitor. *Applied Animal Behaviour Science* **21**, 329–35.

Phillips C.J.C. & Hecheimi K. (1989) The effect of forage supplementation, herbage height and season on the ingestive behaviour of dairy cows. *Applied Animal Behaviour Science* **24**, 203–16.

Phillips C.J.C. & Leaver J.D. (1986) Seasonal and diurnal variation in the grazing behaviour of dairy cows. In Frame J. (ed) *Grazing*. British Grassland Society Occasional Symposium, No. 19, 98–104. BGS, Hurley.

Prache S. & Petit M. (1995) Influence of maturity of the sward on the bite mass of lactating ewes. *Annales de Zootechnie* **44** suppl, 108.

Provenza F.D. & Balph D.F. (1987) Diet learning by domestic ruminants: Theory, evidence and practical implications. *Applied Animal Behaviour Science* **18**, 211–32.

Provenza F.D. & Balph D.F. (1988) Development of dietary choice in livestock on rangelands and its implications for management. *Journal of Animal Science* **66**, 2356–68.

Pulido R. & Leaver J.D. (1995) Influence of initial milk yield, sward height and concentrate level on herbage intake and grazing behaviour of dairy cattle. *Annales de Zootechnie* **44** suppl, 129.

Ridout M.J. & Robson M.J. (1991) Composition of the diet of sheep grazing swards of differing white clover content and spatial distribution – a re-evaluation. *New Zealand Journal of Agricultural Research* **34**, 89–93.

Rogalski M.T., Krysak J., Keos J.M. & Stanejko M. (1996) Ingestive behaviour of weaned lambs in a continuous grazing system. *Proceedings 16th General Meeting, European Grassland Federation, Grado, Italy*, 583–5.

Roguet C. & Prache S. (1995) Influence of forage availability on feeding station behaviour of ewes. *Annales de Zootechnie* **44** suppl, 106.

Rook A.J. & Huckle C.A. (1995) Synchronization of ingestive behaviour by grazing dairy cows. *Animal Science* **60**, 25–30.

Rook A.J. & Huckle C.A. (1996) Sources of variation in the grazing behaviour of dairy cows. *Journal of Agricultural Science, Cambridge* **126**, 227–33.

Rook A.J. & Huckle C.A. (1997) Activity bout criteria for grazing dairy cows. *Applied Animal Behaviour Science* **54**, 89–96.

Rook A.J. & Penning P.D. (1991) Synchronization of eating, ruminating and idling activity by grazing sheep. *Applied Animal Behaviour Science* **32**, 157–66.

Rook A.J., Huckle C.A. & Penning P.D. (1994) Effects of sward height and concentrate supplementation on the ingestive behaviour of spring calving dairy cows grazing grass clover swards. *Applied Animal Behaviour Science* **40**, 101–12.

Rutter S.M., Beresford N.A. & Roberts G. (1997a) Use of GPS to identify the grazing areas of hill sheep. *Computers and Electronics in Agriculture* **17**, 177–88.

Rutter S.M., Champion R.A. & Penning P.D. (1997b) An automatic system to record foraging behaviour in free-ranging ruminants. *Applied Animal Behaviour Science* **54**, 185–95.

Ruyle G.B. & Dwyer D.D. (1985) Feeding stations of sheep as an indicator of diminished forage supply. *Journal of Animal Science* **61**, 349–53.

Shipley L.A., Gross J.E., Spalinger D.E., Hobbs N.T. & Wunder B.A. (1994) The scaling of intake rate in mammalian herbivores. *The American Naturalist* **143**, 1055–82.

Shipley L.A., Spalinger D.E., Gross J.E., Hobbs N.T. & Wunder B.A. (1996) The dynamics and scaling of foraging velocity and encounter rate in mammalian herbivores *Functional Ecology* **10**, 234–44.

Sibley R.M., Nott H.M.R. & Fletcher D.J. (1990) Splitting behaviour into bouts. *Animal Behaviour* **39**, 63–9.

Slater P.J.B. & Lester N.P. (1982) Minimizing errors in splitting behaviour into bouts. *Behaviour* **79**, 153–61.

Spalinger D.E. & Hobbs N.T. (1992) Mechanisms of foraging in mammalian herbivores: New models of functional response. *The American Naturalist* **140**, 325–48.

Stobbs T.H. (1973a) The effect of plant structure on the intake of tropical pastures. 1. Variation in the bite size of grazing cattle. *Australian Journal of Agricultural Research* **24**, 809–19.

Stobbs T.H. (1973b) The effect of plant structure on the intake of tropical pastures. 2. Differences in sward structure, nutritive value, and bite size of animals grazing *Setaria anceps* and *Chloris gayana* at various stages of growth. *Australian Journal of Agricultural Research* **24**, 821–9.

Ungar E.D. & Noy-Meir I. (1988) Herbage intake in relation to availability and sward structure: Grazing processes and optimal foraging. *Journal of Applied Ecology* **25**, 1045–62.

Ungar E.D., Genezi A. & Demment M.W. (1991) Bite dimensions and herbage intake by cattle grazing short hand-constructed swards. *Agronomy Journal* **83**, 973–8.

Wade M.H., Peyraud J.L., Lemaire G. & Comeron E.A. (1989) The dynamics of daily area and depth of grazing and herbage intake of cows in a five-day paddock system. *Proceedings 16th International Grassland Congress, Nice, France*, 1111–12.

Yarrow N.H. & Penning P.D. (1994) Managing grass/clover swards to produce differing clover proportions. *Grass and Forage Science* **49**, 496–501.

Chapter 10
Grassland Management under Grazing and Animal Response

C.S. Mayne, I.A. Wright and G.E.J. Fisher

10.1 Introduction

The primary objectives of grassland management under grazing are to supply high quality forage for as long as possible through the season, and to ensure efficient utilization by the grazing animal whilst maintaining acceptable levels of animal performance, i.e. efficient conversion of grass to animal product. There is also a need for these to be achieved within a sustainable farming system, whilst maintaining or improving the rural landscape. At the outset, it is important to highlight that utilization of grass through grazing requires much less support energy relative to other methods of grass utilization. Consequently, the relative energy costs of grazed grass are approximately half those required for conserved forage (as shown in Table 10.1). The costings in Table 10.1 are similar to those produced in the United States (Hlukik & Smith 1988), with estimated costs of grazed pasture of £26–40 t^{-1}, maize silage of £34–42 t^{-1} and alfalfa haylage of £50–70 t^{-1} DM. In addition, the intake characteristics and nutritive value of grazed grass are higher than those of conserved forage, resulting in potentially greater animal production from the former.

Given the wide range in both annual dry matter (DM) production and seasonal pattern of grass production in temperate grassland regions of the world, and major differences between grazing livestock species in terms of grass requirements through the season, there is no universal system of grassland management. Furthermore, it is important to recognize that there are many interactions between the sward, plant and animal (as reviewed in Chapter 9) which may influence the choice of grazing system. For example, improving the efficiency of conversion of grass to animal product implies the need for animals with high potential levels of performance, which, in the absence of supplementation, may conflict with the objective of maximizing grazing efficiency.

The principles and practices of grazing management for a range of animal production systems are outlined in this chapter. Consideration is also given to the role of supplementation at pasture and its potential impact on the overall efficiency of conversion of herbage to animal product. Furthermore, whilst grazing is the major source of feed for ruminant animals for 6–9 months of the year in most temperate grassland regions of the world, it cannot be considered in isolation from the need to ensure efficient provision of winter feed. Specific

Table 10.1 Use of support energy for grazed and conserved grass and total cost of grass production and conservation (after Wilkins 1990).

	Grazed grass	Grass silage		Field-cured hay
		Direct cut	Wilted	
Energy				
(a) Support energy for production, harvesting and storage (GJ ha^{-1})	34.9	48.0	39.8	33.6
(b) Feed metabolizable energy (ME) output (GJ ha^{-1})	118	110	104	86
Energy efficiency (b/a)	3.4	2.3	2.6	2.6
Feed costs				
£ ha^{-1}	213	398		322
£ (GJME)$^{-1}$	1.80	3.62	3.83	3.74

factors influencing animal performance from conserved forage are reviewed in Chapter 8.

10.2 Potential production from grazed pasture

Potential animal production from grazing systems on an annual basis depends on the quantity and quality of herbage produced, the efficiency of utilization of pasture and the potential productivity of the animal. Previous chapters have reviewed the major factors influencing grass productivity (Chapter 4) and nutritive value (Chapter 7). The potential annual yield of DM ranges from about 2000 to 25 000 kg ha^{-1}, and the metabolizable energy content from 6 to 13 MJ kg^{-1} DM. In general, the higher the quality of herbage offered, in terms of digestibility, the higher the intake, dependent of course on the availability. Data on the effect of annual grass DM yield and herbage quality on the levels of animal performance attainable from grazing are presented in Table 10.2. These data indicate that daily milk yields up to 30 kg cow^{-1}, and daily gains of up to 1.25 kg from cattle and 300 g from growing lambs are theoretically obtainable with high quality pasture. Assuming annual utilized grass yields of up to 12 t DM ha^{-1}, these levels of animal performance are potentially capable of producing 19.8 t of milk from dairy cows or 2.0 t of liveweight gain from beef cattle per ha per year.

The effects of changes in potential productivity of the animal on the efficiency of feed use and animal output from pasture, using the dairy cow as an example, are presented in Table 10.3. These data indicate that, providing animal feed requirements can be met from pasture, then increasing individual animal performance can improve the efficiency of pasture use and increase animal

Table 10.2 Levels of animal performance attainable from grazed grass assuming a range of annual DM yields and variation in grass quality.*

	Annual utilized DM yield ($t\,ha^{-1}$)					
	8.0	8.0	10.0	10.0	12.0	12.0
Metabolizable energy content ($MJ\,kg^{-1}$ DM)	10.5	12.0	10.5	12.0	10.5	12.0
Dairy cows (600 kg)						
Dry matter intake ($kg\,cow^{-1}\,day^{-1}$)	16.4	18.2	16.4	18.2	16.4	18.2
Milk yield ($kg\,day^{-1}$)	21.0	30.0	21.0	30.0	21.0	30.0
Milk yield ($kg\,ha^{-1}\,year^{-1}$)	10250	13190	12800	16480	15365	19780
Beef cattle (300 kg)						
Dry matter intake ($kg\,DM\,day^{-1}$)	7.04	7.41	7.04	7.41	7.04	7.41
Daily gain ($kg\,day^{-1}$)	0.87	1.25	0.87	1.25	0.87	1.25
Annual gain ($kg\,ha^{-1}\,year^{-1}$)	990	1350	1235	1690	1480	2025
Lambs (20 kg)						
Dry matter intake ($kg\,DM\,day^{-1}$)	0.90	1.00	0.90	1.00	0.90	1.00
Daily gain ($kg\,day^{-1}$)	0.22	0.30	0.22	0.30	0.22	0.30

* Based on ME requirements AFRC (1993) and herbage intake predicted for dairy cows using Equation 6a (Caird & Holmes, 1986).

Table 10.3 Effect of changes in level of individual animal performance on potential output in a grazing system, assuming an annual utilized grass DM yield of $12\,t\,ha^{-1}$ at $12\,MJ\,kg^{-1}$ DM (after Gordon 1996).

Milk yield per cow ($kg\,year^{-1}$)	5000	6000	7000
Forage energy utilized per ha (GJ ME)	144	144	144
Energy required per cow (GJ ME)	50	55	60
Stocking rate achievable ($cows\,ha^{-1}$)	2.88	2.62	2.40
Milk output per ha (t milk)	14.4	15.7	16.8
Efficiency of feed use (kg milk per kg pasture DM utilized)	1.20	1.31	1.40

production per hectare despite a reduction in stocking rate. However, it is important to note that a 40% increase in milk yield per cow only increases output per ha by 17%. This reflects the fact that a key assumption in the data presented in Table 10.3 is that grass supply is controlled according to animal needs, i.e. lower yielding cows are stocked at a higher level to ensure efficient pasture use.

Furthermore, the data presented in Table 10.3 assume that pasture can continue to supply sufficient nutrients to maintain individual animal performance as performance increases. At very high levels of individual animal performance, i.e. above those shown in Table 10.2, additional non-pasture nutrients may be required to sustain production and reproduction. If this involves replacement of grazed pasture with more expensive feeds, then attempting to improve the efficiency of conversion of feed into animal product through improving individual animal performance may have detrimental effects on the overall efficiency of pasture use (and financial performance) (see Section 10.6.2.).

10.3 Effects of stocking rate, herbage allowance and grazing severity on animal performance

One of the most important factors influencing both animal output per hectare and individual animal performance is stocking rate (normally expressed as number of animals per ha for a given time period). As stocking rate is increased, grazing pressure (number of animals per unit mass of herbage) increases, herbage allowance (weight of herbage DM per animal) decreases and the level of competition between animals increases. As a consequence of increasing competition between animals for available herbage, herbage intake and animal performance decreases, but the efficiency of herbage utilization (proportion of herbage removed relative to that available) increases. An example of the effect of altering grass allowance in the short term on herbage intake, grazing efficiency and animal performance is presented in Table 10.4. At low herbage allowances (high stocking rate) grass is efficiently grazed, but competition between animals reduces herbage intake and individual animal performance.

When more herbage is offered (by reducing stocking rate) the efficiency of utilization of herbage decreases, but intake and performance per animal increase. It is also important to note that the relationship between herbage allowance and herbage intake is asymptotic, i.e. intake declines at a progressively faster rate when the daily herbage allowance is reduced below 20 kg DM cow^{-1}. For example, decreasing herbage allowance from 24 to 20 kg DM cow^{-1} decreased herbage intake by 3.6%, whereas a further decrease in allowance from 20 to 16 kg DM cow^{-1} decreased intake by 7.8%. Consequently, a high proportion of the additional herbage offered at the high herbage

Table 10.4 Effect of increasing the quantity of grass offered (i.e. reducing stocking rate) on herbage intake and efficiency of utilization of grass at grazing (Stakelum 1996).

	Grass offered (kg DM cow^{-1} day^{-1} above 3.5 cm)		
	16	20	24
Intake (kg DM cow^{-1})	15.3	16.5	17.1
Efficiency of utilization (%)	95.6	82.5	71.2
Residual sward surface height (cm)	5.8	6.7	7.4
Milk yield (kg cow^{-1} day^{-1})	20.6	22.2	22.9

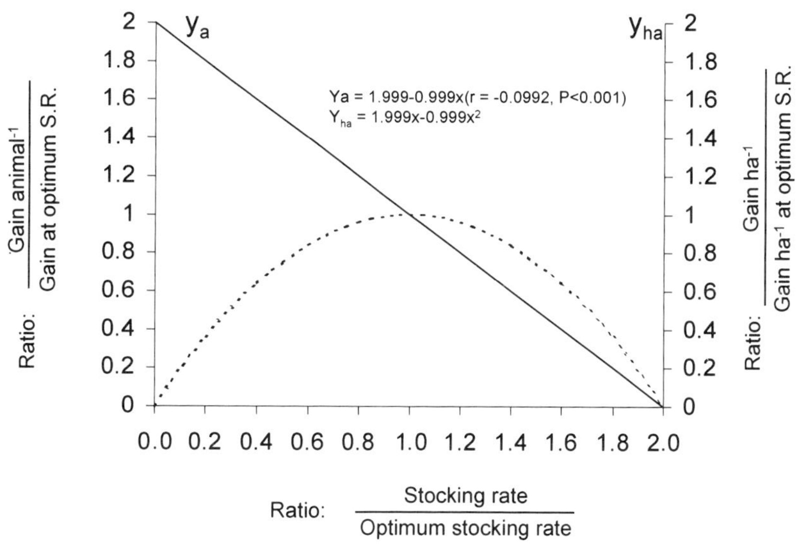

Fig. 10.1 The relation between stocking rate (SR) and both gain head^{-1} and gain ha^{-1} from grazing experiments conducted with a variety of pasture species in a wide range of environments (Jones and Sandland 1974).

allowance remains uneaten, and this can contribute to a subsequent deterioration in herbage composition and sward structure when the herbage re-grows.

Longer-term effects of changes in stocking rate on individual animal performance and output per ha have been examined in many investigations. For example, the data presented in Fig. 10.1 illustrate the effect of increasing stocking rate with growing cattle (350 kg), based on a review by Jones & Sandland (1974). Individual animal performance declined linearly with increasing stocking rate, reflecting reductions in herbage intake, whereas production per ha increased to a peak and then declined. Jones & Sandland (1974) concluded that with growing beef cattle and sheep, liveweight gain per ha was maximized at a stocking rate half of that which gave zero liveweight gain, and at this point individual animal performance was reduced by approximately 24% relative to the maximum achievable at low stocking rate.

10.3.1 Assessment of grazing severity

Most assessments of grazing severity involve either direct or indirect estimates of herbage mass or sward height. Herbage mass is normally assessed by clipping samples from the sward to a height lower than that likely to be grazed by the animals. For example, in grazing studies, with dairy cows, herbage yields are normally assessed either above ground level or above a cutting height of 4 cm, using long, narrow quadrats, e.g. 1.0 × 0.25 m. Direct estimation of herbage mass is a time consuming operation, and is normally undertaken mainly in research studies or as a means of cross checking visual assessments.

Sward height estimates involve a number of measurement techniques including measurement with a rule, or with a more sophisticated measuring device which provides a measure of the vertical height of an undisturbed sward, sward surface height (SSH), e.g. an HFRO sward stick (Bircham 1981), or alternatively by a rising plate meter (PM) which exerts pressure on the sward and enables the height of the compressed sward to be recorded (Holmes 1974, Castle 1976). There is considerable variation in height measurements within grazed swards, with coefficients of variation normally in the 20–25% range. Whilst some measures of height, e.g. rising plate meter, are highly correlated with herbage mass ($r^2 = 0.7$ or greater), the relationship changes with changes in sward type and over the duration of the grazing season. Similarly, there is no consistent relationship between the various sward height measurements, although approximate relationships for determining SSH from plate meter measurements are as given below.

Continuously stocked swards:

$$\text{SSH} = 1.64 + 1.15\ (\text{PM}) \qquad R^2 = 0.78 \qquad (1)$$

Rotationally grazed swards – post grazing residual:

$$\text{SSH} = 2.55 + 0.75\ (\text{PM}) \qquad R^2 = 0.47 \qquad (2)$$

PM represents plate meter height (30 cm × 30 cm square plate meter weighing 428 g). Equations 1 and 2 are from Davies (1986). Throughout this chapter, all references to sward height relate to SSH as measured by an HFRO sward stick.

10.3.2 Effects of stocking rate and herbage allowance on herbage intake

Whilst herbage allowance is one of the primary factors influencing daily herbage intake, the form in which herbage is presented to the cow, i.e. sward structural characteristics, can also have major effects. For example, recent studies (Peyraud *et al.* 1996) indicate the overriding effect of low herbage mass in reducing herbage intake at a constant herbage allowance (Fig. 10.2). These results indicate a marked reduction in herbage intake with animals grazing swards with less than 2.5 t OM ha^{-1} (i.e. short swards), even though herbage allowance was maintained on these swards by decreasing stocking density. This illustrates one of the disadvantages of relying solely on herbage mass (allowance) as a means of allocating herbage to grazing animals.

The key sward structural characteristics influencing herbage intake have been reviewed in Chapter 9. The primary reasons for the reduction in herbage intake with increased stocking rate, or reduced herbage allowance, relate to the effects of herbage allowance on intake per bite and grazing behaviour. It is now gen-

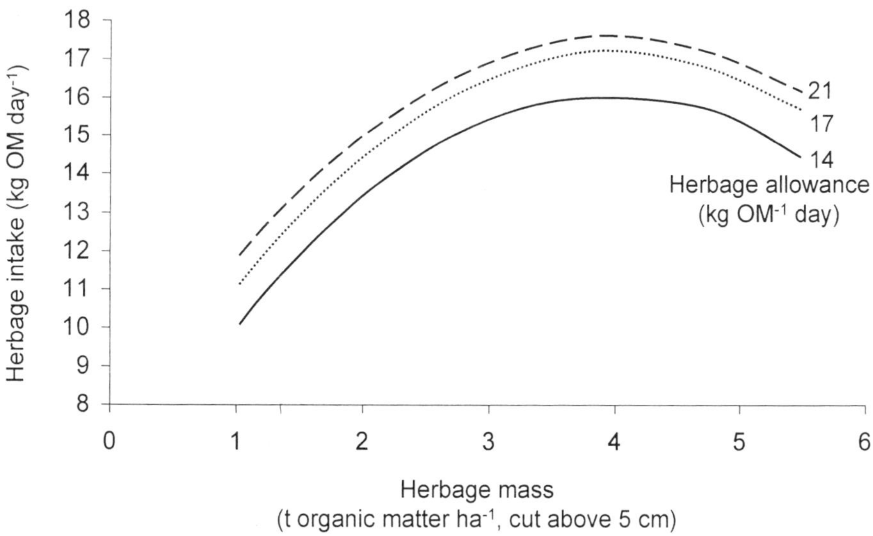

Fig. 10.2 Effect of herbage mass on herbage intake at differing herbage allowances (after Peyraud *et al.* 1996).

erally accepted that as swards are grazed to lower sward heights, intake per bite decreases. However, as intake per bite declines there is usually an increase in biting rate due to a change in the ratio of manipulative to biting jaw movements, i.e. fewer manipulative jaw movements are required to process a reduced bite size. Thus, as sward height declines initially, e.g. at sward heights greater than 10 cm, daily herbage intake of cows may be maintained through concurrent changes in intake per bite and biting rate. However, on very short swards, e.g. at sward heights less than 7.5 cm, the increase in biting rate is insufficient to offset the lower intake per bite, and consequently daily herbage intake decreases markedly (as shown in Table 10.4).

10.4 Grazing systems and integration with forage conservation

There have been a number of major developments in grazing systems since the 1950s, with a full cycle of management systems having been implemented on farms over this period. These began with continuous stocking systems, through controlled rotational grazing, back to continuous grazing and, more recently, an increasing uptake of management-intensive rotational grazing systems, particularly in Western Europe and parts of the USA. These changes have been part of an evolution in grassland management (Frame *et al.* 1995), and the use of controlled rotational grazing, coupled with progressively increased use of nitrogenous fertilizer from about 1960 to the mid-1980s, led to a fuller realization of the production potential of grazed grass.

The two major categories into which most grazing systems can be ascribed are either continuous stocking systems or rotational grazing systems. These two grazing systems have quite different effects on sward structure. Continuous grazing at high stocking rates encourages the development of a dense sward with a high tiller population (up to 40000 tillers m^{-2}), whereas rotational grazing, particularly with extended grazing intervals, results in the development of a much more open sward structure (10000–15000 tillers m^{-2}).

10.4.1 Continuous stocking

Under continuous grazing or set stocking, animals have access to the entire grazing area for the majority of the grazing season (Fig. 10.3a). In the purest sense, true continuous grazing applies only in very extensive grazing systems, for example in conventional hill and range grazing systems, where stocking rate is very low relative to pasture production. However, modifications of continuous grazing are the 'norm' for sheep and beef cattle grazing systems given the low capital and labour requirements required for the implementation of such systems. In practice, livestock are normally turned out on to the grassland area in spring, with a number of alternative procedures being used to balance grass supply and requirements. These include making additional areas available for grazing through the season by introducing silage aftermaths to compensate for lower grass growth rates and/or increased animal requirements. Alternatively, stocking rate adjustments may be achieved by the sale of fat stock or the removal of weaned lambs and calves to other areas on the farm.

Most continuous grazing systems involve close integration of grazing and forage conservation. This can be achieved either by 'buffer grazing' or by operating a 'full-graze' system. With buffer grazing, animals are turned out in spring at a relatively high stocking rate. As grass growth rates decrease and/or animal requirements increase through the season, the stocking rate is gradually reduced by increasing the area available for grazing (Fig. 10.3b). Herbage accumulation on the buffer area is controlled by conservation of excess forage as hay or silage, although careful management is required to avoid grass shortages in the immediate period following a conservation cut.

In the full-graze system (Fig. 10.3c) the grazing area is divided in two in early season, for example on a ratio of one-third : two-thirds. The larger area is normally grazed for the first few weeks in early season, with stock then moved to the smaller area during the period of peak grass growth in late spring/early summer. The larger area then becomes available for either silage or hay making in early summer. Following removal of hay or silage, stock return to the larger area for grazing in mid-summer. The smaller area is then closed up for a second cut of silage or a second hay crop, with the re-growths available for grazing in late summer. Consequently the area available for grazing increases in the ratio 1 : 2 : 3, and the system has been described as the '1,2,3' system.

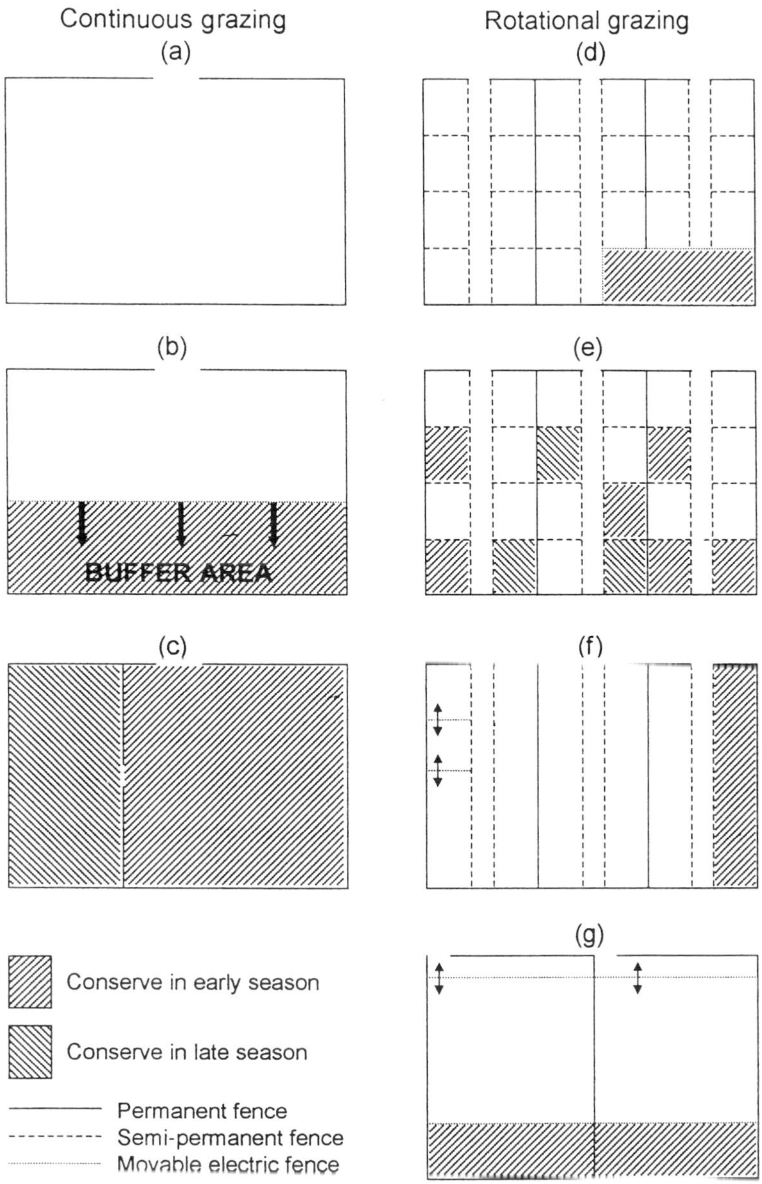

Fig. 10.3 Outline of grazing systems. (a) Continuous stocking; (b) buffer grazing, (c) full-graze system; (d) rigid rotational grazing; (e) flexible rotational grazing; (f) flexigraze rotational grazing; (g) strip-grazing. Adapted from Holmes (1989).

Both the buffer grazing and full-graze systems require minimal investment in fencing and water troughs. Furthermore, in the full-graze system, as each area is rested from grazing for between 5 and 7 weeks, the build-up of parasitic worms is reduced. Both grazing systems are particularly suitable for growing cattle and suckler cows, and can also be adopted for sheep grazing, for example in the full-graze system weaned lambs can be moved on to clean pasture on the re-growth areas. However, continuous grazing systems require a high level of management if they are to operate successfully. Regular assessment of sward height, herbage availability and/or animal performance is required in order to ensure efficient herbage utilization and to achieve target levels of animal performance.

10.4.2 Rotational grazing

With rotational grazing, the sward is defoliated at regular intervals following a period of re-growth. The grazing area is normally divided into a number of paddocks (Fig. 10.3d), with the animals being restricted to a paddock for a relatively short period, e.g. 1 day. As with continuous grazing, there are many variations of rotational grazing in practice, ranging from rigid rotational grazing through flexible systems involving integration with conservation and strip-grazing.

Rigid rotational grazing

Rigid rotational grazing systems operate a fixed rotation length throughout most of the grazing season, e.g. Fig. 10.3d. In this system, a fixed number of paddocks (e.g. 21) of similar size are established at the start of the season and stock move according to a pre-determined timetable (e.g. moved daily to a new paddock) throughout the grazing season. Whilst this system is extremely easy to operate in practice, fluctuations in grass growth over the season can result in a risk of understocking in early season when grass growth rates are high, and overstocking later in the season when growth rates decline. Consequently, in practice, rigid rotational grazing is only appropriate in those situations where the nutrient requirements of the stock are declining through the grazing season, e.g. autumn-calving dairy cows. Even in this situation, topping of pastures after grazing to control pasture quality may be required in early season, and in mid- and late season the stocking rate should be reduced, e.g. by removal of dry cows or introduction of silage aftermath areas.

Flexible rotational grazing

In order to better accommodate fluctuations in grass growth through the season, most rotational grazing systems incorporate some degree of flexibility (Fig. 10.3e). This can be achieved by closer integration of grazing and forage conservation, with paddocks set aside for silage making in early season, when grass growth rates are high, and then reintroduced into the grazing cycle as the season progresses. Big-bale silage is particularly useful in this context as it allows greater flexibility to conserve fodder from small areas, which may be surplus to grazing requirements, at various times throughout the grazing season. Furthermore, removal of surplus herbage from previously grazed paddocks facilitates improvements in sward quality later in the season, particularly if paddocks have been undergrazed in the first or second grazing cycles. When grass growth rates are high, shortening rotation length reduces both average daily growth rate and total net herbage accumulation over the re-growth period, whereas the opposite effect occurs when rotation lengths are increased. Successful operation of flexible rotational grazing requires effective grass budgeting, based on a knowledge of seasonal patterns of grass growth and changes in nutrient requirements of stock during the season (see Section 10.6.3).

Further flexibility in rotational grazing systems can be achieved by adopting either flexigraze or strip-grazing systems (see Fig. 10.3f, g). In these systems, the area allocated to stock is controlled on a daily basis and portable electric fencing is used to adjust the grazing area. With flexigraze systems, more investment in roadways and fencing is required, but animals move on to a fresh area of pasture each day and are not allowed to regraze. With strip-grazing, animals normally have access to previously grazed areas, and this can increase the risk of sward damage (particularly in wet conditions) and delay subsequent re-growth of pasture.

A further modification of rotational grazing involves the practice of leader/follower grazing. This involves offering stock with the highest nutrient requirements first access to the pasture, with a second group of animals with lower nutrient requirements following behind and grazing the sward down closely. In sheep and lamb systems the term forward creep grazing is used. Lambs preferentially graze paddocks ahead of ewes by accessing paddocks through specially constructed gaps in the fences. A further advantage of this approach is that lambs are less exposed to parasitic worm larvae, as they are not forced to graze into the base of the sward. Leader/follower grazing is also widely used in dairy heifer rearing systems, with younger heifers (6–12 months old) grazing ahead of the older (18–24 months old) cattle. Leaver (1970) showed that leader/follower grazing of heifers produced better growth rates, particularly in younger heifers, in addition to reducing the incidence of disease associated with parasitic worms. Other studies (Mayne *et al.* 1988) have shown that leader/follower grazing systems can increase the performance of lactating dairy cows, with the milk yield of high-yielding cows increased by 26%, and total milk output per ha increased by 8.2%, when high-yielding cows preferentially grazed as a leader group ahead of a lower-yielding group. However, in practice leader/follower grazing of dairy cows is difficult to implement, and in many situations non-lactating animals, e.g. dry cows, growing cattle etc., are used as the follower group. It is worth noting that the follower group may not be required after every grazing cycle, e.g. grazing after every second or third grazing may be sufficient to control pasture growth and maintain sward quality.

Length of rotation cycle

Numerous studies have shown that under cutting, the yield of herbage DM produced over the season as a whole is inversely related to the frequency of defoliation (Bartholomew & Chestnutt 1977, Reid 1986). The depression in growth rate with increased frequency of defoliation is greatest in early season when growth rates are highest (Binnie *et al.* 1997). However, in grazing studies there is no evidence of a depression in animal performance with shorter grazing cycles (Marsh *et al.* 1971), even when comparing 15- and 30-day rotations (McFeely *et al.* 1975). The absence of an effect on animal performance may be related to the fact that comparisons of rotation length have normally been undertaken at one stocking rate only, and inefficiencies in herbage utilization may mask treatment effects on grass production. Nevertheless, current recommendations suggest a minimum rotation length of 18 days in early season, increasing up to 50–70 days for late autumn grazing.

Duration of grazing period

For a given herbage allowance, herbage intakes and animal performance are similar in rotational grazing systems in which fresh pasture is allocated on a daily basis, or over several days (Peyraud *et al.* 1989). In the latter system, however, there is a marked cyclic variation in nutrient intake and daily milk output, as described by Hoden *et al.* (1991). More recent studies with grazing dairy cows suggest there were no significant differences in herbage intake or milk production with animals offered a fresh herbage allowance either once daily, following the morning milking, or in six equal allocations throughout the day (Dalley 1998b). Other studies (Orr *et al.* 1998) indicate that in 24-h paddock grazing systems, total nutrient intake and milk yield were increased with cows given a new allocation of pasture following afternoon rather than morning milking. The increased nutrient intake associated with an afternoon move was attributed to cows consuming a greater proportion of their daily intake when the DM and sugar contents of pasture were highest, i.e. in the evening.

10.4.3 Comparison of continuous and rotational grazing systems

The majority of experimental evidence indicates that the effects of grazing system on animal output per ha, or on individual animal performance, are small in comparison with those due to varying the stocking rate. For example, Ernst *et al.* (1980), in a review of published data from western Europe, concluded that milk production was similar from rotational and continuous grazing systems operated at identical stocking rates. In more recent studies, Pulido (1997) also observed similar animal performance with continuous and rotational grazing, even with high yielding dairy cows producing up to 32.5 l milk day^{-1} at the commencement of the trial. However, there is evidence that rotational grazing systems are superior to continuous grazing at high stocking rates, with increases in output of milk solids per cow of 4% and 16% being obtained at low and high stocking rates, respectively (McMeekan & Walshe 1963). There is also a more uniform pattern of herbage intake and animal performance through the season (Evans 1981).

Whilst experimental studies have failed to demonstrate a consistent production benefit from rotational grazing over continuous grazing, rotational grazing systems offer a number of other advantages. For example, they facilitate identification of grass surpluses and deficits much more readily than with continuous grazing. Rotational grazing systems also offer greater flexibility to adjust grass supply, with paddocks being taken out of the system during periods of surplus growth and then returned to the grazing cycle when grass growth rates decline. However, one of the most important characteristics of rotational grazing systems is that they facilitate presentation of herbage to the animal in an 'optimum' form for prehension, i.e. tall, dense leafy swards. Finally, within rotational grazing systems several options are available for controlling the high residual herbage mass remaining following grazing by high producing cattle, thereby maintaining sward quality in mid- and late season. For example, swards can be mechanically topped following grazing to control stem extension, or alternatively leader/follower grazing can be implemented to utilize residual herbage. Other options include a period of severe grazing by lower-producing stock in late June, early July to remove stem material (Chestnutt 1994), or a system of alternate grazing and cutting of swards for conservation.

10.4.4 Zero grazing

One of the limitations of grazing is that utilization of herbage per ha is reduced relative to that obtained under cutting systems. This reflects the effects of more frequent defoliation and the impact of treading and fouling of herbage during grazing. At similar levels of nitrogen application, Richards (1977) estimated that 30% more herbage was utilized under cutting compared with grazing. Similarly, in a review of pasture utilization across 34 farms in the United Kingdom, Walshe (1982) concluded that utilized metabolizable energy (UME) output was 15% greater on average with cut swards, even when allowing for nutrient losses during ensiling. Consequently, there is potential to increase animal output from pasture by adopting zero-grazing systems in which herbage is harvested and fed fresh to stock. Marsh (1976) observed an improvement of 20% in liveweight gain per hectare with zero-grazing in a review of comparisons of zero grazed and conventionally grazed beef cattle. This suggests that grassland productivity can be improved by zero-grazing, and this approach may be particularly useful on fragmented farms where access is difficult. However, there are many drawbacks with the system, such as the increased labour and machinery costs, the need for improved housing and slurry disposal systems, and the difficulty in maintaining a supply of herbage of high nutritive value throughout the season. Nonetheless, zero-grazing may have a role for short periods in early or late season, when grazing conditions are often difficult, or during temporary periods of grass shortage on the core grazing area.

10.4.5 Extending the grazing season

Given the superior feeding value and reduced cost of grazed grass relative to conserved forage, there is currently considerable interest in opportunities to extend the grazing season. The potential to produce grass over the winter period (i.e. November to March in the British Isles) varies widely, with the grass growing season varying from less than 200 days up to 365 days per year between different regions. There is also great variation in grass growth, even within relatively small areas such as the British Isles. Brereton *et al.* (1995) estimated that mean grass growth rates in winter (1 November–31 March) varied from 3 kg DM ha^{-1} day^{-1} in Aberdeen (northern Scotland) to 11 kg DM ha^{-1} day^{-1} in Kerry (southwest Ireland). Nonetheless, herbage availability for late autumn/early winter grazing can be increased by deferred grazing of the herbage produced in late August/September. For example, in Northern Ireland, Laidlaw & Mayne (2000) have shown that with swards closed on 5 September, herbage DM yield increased to 2.0 t ha^{-1} by 21 October, and 2.7 t ha^{-1} by 17 November, but then declined to 2.0 t ha^{-1} by 15 December. Similarly, the pattern of grass growth in early spring can be manipulated by choice of grass species or variety, timing of last grazing in the previous autumn/winter and timing of N fertilizer application in spring. In the 1950s and 1960s, tall fescue (*Festuca arundinacea*), and to a lesser extent cocksfoot (*Dactylis glomerata*), were identified as having potential for winter growth (Baker *et al.* 1965). Italian (*Lolium multiflorum*) and hybrid ryegrass, and some cultivars of perennial ryegrass (*L. perenne*), are also capable of rapid growth in early spring. For example, in Northern Ireland the early *L. perenne* variety Moy produces 2.15 t DM ha^{-1} between 1 March and 14 April, whereas the late variety Carrick produces only 0.7 t DM ha^{-1} during the same period (D. Johnston, personal communication, 1999).

The major challenge in extending the grazing season is to utilize grass efficiently, whilst minimizing sward damage. This requires flexible grazing management strategies involving the use of short (2–6 h day^{-1}) grazing periods and block-grazing techniques, with animals confined to a limited area each day and prevented from regrazing. Nonetheless, research has shown that there is considerable potential to extend the grazing season in the British Isles (Dillon & Crosse 1994, Mayne & Laidlaw 1995), allowing substantial savings in silage requirements and improving animal performance relative to systems based on indoor housing.

10.4.6 Grass/white clover systems

Grass/white clover (*Trifolium repens*) swards are particularly important in extensive grazing systems, given that biological fixation of nitrogen from the atmosphere can contribute up to 280 kg N ha^{-1} $year^{-1}$ (Cowling 1982). Grass/clover swards without fertilizer N can produce similar yields to those of pure grass swards receiving around 200 kg N ha^{-1} $year^{-1}$ (Davies & Hopkins 1996) (see also Chapter 4). Furthermore, incorporation of clover in the sward can markedly improve the nutritive value of pasture, with increases in voluntary food intake and nutrient supply, particularly protein (Ulyatt *et al.* 1980). A major factor limiting the widespread adoption of white clover-based systems is their apparent unpredictability (Newton & Davies 1987), and consequently effort has been focused on opportunities to improve the persistence and reliability of white clover in mixed swards. Research indicates that factors such as choice of companion grass (e.g. use of tetraploid rather than diploid perennial ryegrass varieties), tight grazing in late autumn and early winter, and avoidance of undergrazing during the main grazing period are all important factors in maintaining a high clover content in the sward. Effects of grazing on clover morphology vary between studies, with some suggesting benefits from rotational grazing, whilst other studies suggest that clover content can be maintained under continuous grazing, although with the latter system a rest period, achieved by integrating grazing with a conservation cut, can also contribute to increased clover content.

Grazing studies have reported significant increases in liveweight gain of lambs and growing cattle, and in milk production of dairy cows, associated with inclusion of white clover in the sward. For example, Clark & Jans (1995) concluded that milk yield increases of 0.30–0.45 kg cow^{-1} day^{-1} could be expected for each 10% increase in white clover content of the pasture. Further examples are given in Davies & Hopkins (1996).

10.5 Principles of supplementation at pasture

The provision of supplementary feeds to animals at pasture is normally undertaken either to maintain performance during periods of herbage shortage, or to improve animal performance over and above that which can be produced from pasture alone. Improvements in the production potential of livestock in recent years have resulted from improvements in genetic merit or changes in production system (e.g. summer calving suckler and dairy herds). Consequently, potential levels of individual animal performance commonly exceed those which can be sustained from pasture only (see Table 10.2), and thus provision of supplementary feeds is often required to realize this potential and to maintain acceptable levels of reproduction (see Chapter 7 for further discussion of this subject).

One of the major difficulties in supplementary feeding at pasture is estimating herbage DM intake on a daily basis, and hence in determining the level of supplementation required to sustain a given level of performance. Furthermore, individual animal response to supplementation is largely a function of the effect of supplementation on herbage intake, i.e. the substitution rate. This is influenced by a wide range of sward and animal factors. Nevertheless, most grazing studies indicate that one of the key factors influencing herbage substitution rate is herbage availability. For example, the data presented in Table 10.5 indicate that at a high herbage allowance, supplementation results in a large substitution effect with a relatively small increase in total DM, and hence ME, intake. In contrast, when herbage availability is reduced, substitution rate is lower and the response in daily ME intake is increased. A logical interpretation of these data is that in most situations supplementation at pasture is unlikely to produce a reasonable response in total nutrient intake unless pasture is very tightly grazed, i.e. at a very low herbage allowance.

10.5.1 Supplementation during pasture shortage

In periods of grass shortage, for example as a result of mid-summer drought, provision of supplementary feeds results in increased total DM intake and improved animal performance. A range of alternative feeds have been used in this situation, including forage supplements such as grass or maize silage, or concentrate supplements based on cereals or by-product feeds. A summary of published responses with forage supplementation of grazing dairy cows under conditions of low herbage availability is presented in Table 10.6.

These results indicate that buffer feeding, i.e. offering forage supplements once or twice per day after milking, increased milk yield relative to pasture only when grass availability was restricted. A further advantage of the buffer feeding approach, under conditions of grass shortage, is that it reduces the use of available herbage, preventing the spiral into acute grass shortage. Consideration should also be given to zero-grazing as a means of providing additional grass during periods of grass shortage on the core grazing area. A key feature of

Table 10.5 Effect of supplementation of dairy cows at different herbage allowances on substitution rate and daily ME intake (after Meijs & Hoekstra 1984).

Herbage allowance (kg OM cow^{-1} day^{-1})	Substitution rate (kg herbage OM (kg concentrate OM)$^{-1}$)	Increase in ME intake* (MJ day^{-1})
15	0.105	11.6
20	0.30	9.1
25	0.50	6.5
30	0.69	4.0

* Assumes ME content of herbage and concentrate of 13.0 MJ (kg OM)$^{-1}$.

Table 10.6 Effect of a range of supplements on herbage intake and animal performance of dairy cows offered a restricted herbage allowance.

	Supplement type		
	Grass silage*	Hay*	Brewers grains/treated straw**
Supplement intake (kg DM day^{-1})	4.94	4.35	3.2
Substitution rate kg herbage (kg supplement)$^{-1}$	0.28	0.34	NA
Milk yield response (kg day^{-1})	+1.6	+2.6	+1.9
Fat + protein yield response (g day^{-1})	+119	+208	+124

* Phillips (1988).
** Leaver & Campling (1990).

supplementation during periods of grass shortage is the need to monitor grass supply frequently and to remove the supplementary feed when grass supply returns to target (see Section 10.6.3 on grass budgeting).

10.5.2 Supplementation when adequate grass is available

The primary objective of effective grazing management, which can only be achieved by grass budgeting through the season, is to provide a high allowance of high quality grass right through the season. In this situation, feeding of supplements reduces the animal's motivation to graze and consequently grass intake is reduced. A number of studies have indicated that the extent of the depression in intake is related to supplement type and the production potential of the animal. In general, greater depressions in grass intake are obtained with forage supplements (e.g. grass or maize silage) reflecting the greater rumen-fill effect compared with concentrates, and depressions are also greater with low- compared with high-producing animals.

Consequently, decisions on supplementation at pasture when pasture is available *ad libitum* primarily involves establishing the levels of performance which can be sustained from a range of pasture allowances, relative to the production potential of the animal. One example of this approach, reported by Leaver (1982), involved using grass height measurements under continuous grazing as an indicator of the need for supplementary feeding. In this study, concentrates were offered when the sward surface height declined below either 11.0, 9.0 or 6.9 cm, at the rate of 1 kg cow day^{-1} for each 0.2 cm decline below these threshold levels. Leaver (1982) concluded that sward height monitoring could be used as an indicator of when supplementary feeding was necessary, although he also suggested that the critical height below which milk yield was reduced may vary according to the milk yield of the cow (see Section 10.6.1).

In summary, provision of supplementary feeds at pasture can be a useful management tool, either to maintain animal performance during periods of grass shortage, or as a means of improving individual animal performance above that which can be sustained from pasture alone. In situations of grass shortage, herbage substitution rates are generally low and a wide range of supplements can be used, with the key factor being value for money in terms of energy cost. In contrast, where the objective is to improve animal performance when grass supply is adequate, choice of supplement becomes much more critical, given the major variation between supplements in substitution rate (see subsequent sections for management guidelines for individual animal species).

10.6 Practical grazing management – dairy cows

Data presented previously in Table 10.1 indicate that in intensive grassland systems, the relative cost of energy as grazed grass is approximately half that of conserved forage (either silage or hay). In addition, a number of studies have shown that the intake characteristics and nutritive value of grazed grass are higher than those of conserved forage, resulting in increased production from forage with the former. For example, Mayne & Cushnahan (1995), in a review of 13 studies which collectively summarized 432 individual grass/silage comparisons, noted that, on average, DM intake by ruminants was reduced proportionately by 27% following ensiling, with a range from 1% to 64%. These data suggest that in temperate grassland regions of the world, with good grass-growing conditions, maximum reliance on grazed grass offers the best opportunity to produce milk cost-effectively. Furthermore, milk production systems based on grazing require fewer resource inputs, are more sensitive to the welfare needs of animals than indoor-housed systems and, when appropriately managed, can contribute to an enhanced rural landscape. However, one of the disadvantages of grazing relative to indoor systems of milk production is the greater management effort and expertise required, given seasonal fluctuations in grass growth and chemical composition, and the major interactions between sward and animal factors on herbage intake. Increases in genetic merit, and hence production potential, of dairy herds worldwide in the last 20 years has also increased the difficulty of balancing both individual animal performance and efficient pasture utilization.

10.6.1 Factors influencing herbage intake and milk yield

Feed intake in the grazing dairy cow is primarily determined by how effective the cow is in harvesting grass in the field, and this is influenced by a large number of both sward and animal factors (see Chapter 8). Dairy cows normally spend between 420 and 720 min day^{-1} grazing, with an average of about 510 min

Table 10.7 Effect of changes in intake per bite on potential daily DM intake with high-yielding dairy cows.

	Small bite size	Large bite size
Intake per bite (g)	0.6	0.8
×		
Biting rate (bites per minute)	55	50
×		
Grazing time (minutes per day)	450	450
↓		
Daily grass intake (kg DM)	14.8	18.0

(Arnold 1981). Biting rate generally varies in the range 45–65 bites min^{-1} (Chacon & Stobbs 1976), but is influenced by intake per bite, tending to increase with reductions in bite size. Phillips & Leaver (1986) suggested that dairy cows grazing temperate swards are constrained to a maximum of approximately 40000 bites per day, which effectively restricts the ability of the cow to compensate for reductions in bite size. This is borne out in the results of recent research (A. Cushnahan and C.S. Mayne *et al.*, unpublished results, 1999), which have shown that the most important parameter influencing daily herbage intake is intake per bite, as illustrated in Table 10.7.

An increase in intake per bite of 0.2 g DM has the potential to increase daily herbage intake by up to 22%. Intake per bite is largely controlled by sward factors, principally sward height, although sward density and the proportion of green leaf in the sward are also important. The effects of changes in sward height and bulk density on intake per bite under rotational grazing are illustrated in Fig. 10.4. These data indicate that intake per bite declines more quickly with decreasing sward height in low-density swards. For example, in order to achieve an intake per bite of 0.7 g DM, sward heights of 8.4, 11.8 and 12.9 cm were required with dense, intermediate and open swards, respectively. Similarly, Le Du *et al.* (1981) have shown that sward canopy height has a major impact on intake per bite and daily herbage intake in continuous grazing systems, with intake being reduced when sward surface heights decline below 7–8 cm. Given the effects of reductions in sward height on herbage intake, it is possible to formulate sward height guidelines for cows of differing yield potential through the season in both continuous and rotational grazing systems, as shown in Table 10.8.

10.6.2 Nutrient requirements of grazing dairy cows

Increases in milk yield of grazing dairy cows have major implications for energy requirements, as shown in Table 10.9. Assuming similar liveweight loss, herbage intakes of 15.0 and 18.7 kg DM day^{-1} are required to meet the energy

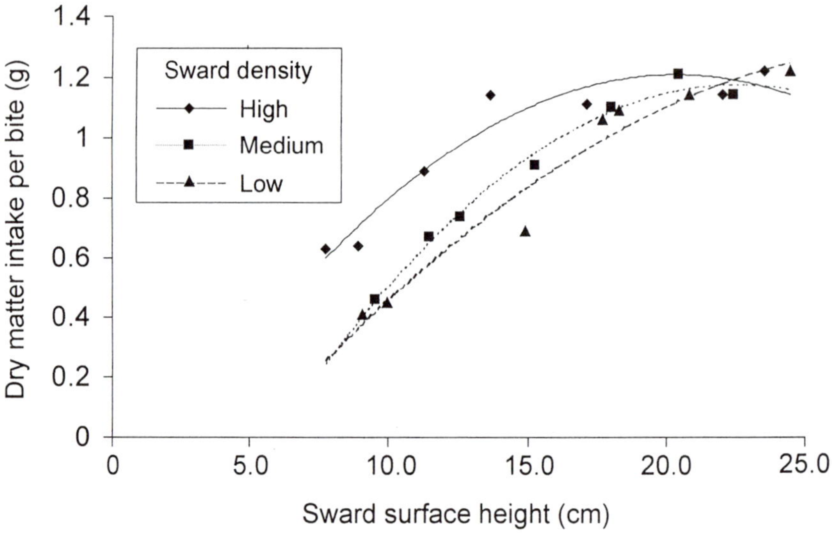

Fig. 10.4 Effect of sward surface height on DM intake per bite at different sward densities (D.M. Patterson, C.S. Mayne, D.G. McGilloway and A.S. Laidlaw, personal communication, 1999).

Table 10.8 Suggested sward surface height targets (cm) for high- and low-yielding cows through the season.

	Continuous grazing		Rotational grazing (residual sward height with dense swards)	
	High yielders	Low yielders	High yielders*	Low yielders
April–June	7–8	5–6	8–10	6–8
July–August	8–10	6–8	8–10	6–8
September–October	10–12	8–9	8–10	6–8
November	7	7	6	6

* Topping, alternate cutting and grazing or leader/follower grazing will be required to maintain sward quality.

requirements of cows producing 25 or 32.5 kg milk day^{-1}, respectively. Studies in The Netherlands (Meijs & Hoekstra 1984) have suggested that under ideal grazing conditions there is an upper limit to grass intake of 16.9 kg DM day^{-1}. However, levels of herbage intake of up to 20.7 kg DM day^{-1} have been reported, with high-yielding cows offered a herbage allowance of 27 kg DM cow^{-1} day^{-1} assessed above a height of 4 cm (Buckley & Dillon 1998). These results suggest that under good grassland management, herbage intake at pasture has the potential to meet the needs of cows yielding up to approximately 30 kg milk

Table 10.9 Theoretical energy and intake requirements for moderate- and high-yielding dairy cows.

Yield potential	Moderate	High
Liveweight (kg)	550	650
Liveweight loss (kg day^{-1})	0.5	0.5
Milk yield (kg day^{-1})*	25.0	32.5
Metabolizable energy requirements (MJ day^{-1})		
Maintenance**	55.4	62.5
Total requirement	180	225
Dry matter intake required (kg day^{-1}), assuming ME of 12.0 MJ kg^{-1} DM	15.0	18.7

* Assuming milk composition of 39.4 g kg^{-1} butterfat, 31.9 g kg^{-1} protein and 44.2 g kg^{-1} lactose.
** Maintenance energy requirements estimated by AFRC (1993).

day^{-1}. However, in order to achieve such high levels of individual animal performance, milk output per ha and the efficiency of grassland utilization will be compromised, given the need to offer a high herbage allowance (i.e. to maintain a high sward height).

Furthermore, reduced grazing severity, with residual sward heights above 9 cm in rotational grazing, or sward canopy heights in excess of 8 cm in continuous grazing systems, will result in a deterioration in sward quality, particularly during the spring/early summer period. For example, a number of studies (Mayne *et al.* 1987, Stakelum 1993) have shown increases in the proportion of stem and dead material and reductions in herbage digestibility following lax grazing in early season. The implication is that in grazing systems designed to maximize individual animal performance, utilization per hectare will be poor, and the major challenge in grazing management is to develop alternative strategies to utilize residual herbage to a degree commensurate with maintaining high quality swards through the grazing season. Whilst this is possible with rotational grazing systems, with options such as leader/follower grazing or alternate grazing and cutting of paddocks, control of sward quality under lax grazing is much more difficult with continuous grazing systems. An alternative option to sustain high levels of individual animal performance, whilst maintaining efficient herbage utilization, could involve the use of high stocking rates coupled with provision of supplementary feeds (see Section 10.6.4).

However, attaining high levels of individual animal performance may not always be the most appropriate objective in dairy cow grazing systems. In situations where land is expensive, but fixed costs per cow are relatively low, lower individual animal performance coupled with higher output per hectare may be more appropriate. The data presented in Table 10.10 indicate that when fixed costs per cow and milk price are relatively low, increasing milk yield per cow produces considerably less benefit in the margin remaining to cover variable

Table 10.10 Effect of milk yield and level of fixed costs per cow on economics of milk production at two milk price levels (UK currency).

	Milk price (pence litre^{-1})			
	10		20	
	£ cow^{-1}	p litre^{-1}	£ cow^{-1}	p litre^{-1}
Fixed costs	85		425	
4000-l cow				
Milk sales	400	10	800	20
Margin to cover variable costs + profit	315	7.8	375	9.4
6000-l cow				
Milk sales	600	10	1200	20
Margin to cover variable costs + profit	515	8.6	775	12.9
Increase in margin available to cover variable costs + profit by increasing yield cow^{-1}	200	0.8	400	3.5

costs of production and profit, compared with the situation where fixed costs per cow and milk price are relatively high. In the low-cost, low-price situation, it is only profitable to increase milk yield per cow from 4000 to 6000 l if this can be achieved without increasing the variable costs of production by more than £200 per cow, or 0.8 p l^{-1}. In contrast, in the high fixed cost, high price situation, profitability per cow could be increased by increasing milk yield per cow providing less than £400 per cow, or 3.5 p l^{-1}, are incurred in the increased variable costs of the additional production. These data highlight the need to assess production costs at the individual farm level and to consider implications for the production system used on the individual farm.

10.6.3 Pasture budgeting

Grass production varies considerably depending upon local soil and climatic conditions, in addition to annual and seasonal variation within a given location. In order to balance grass supply against the changing nutrient requirements of the dairy herd through the season, effective grass budgeting is required. Grass budgeting can be considered in three phases: (a) an annual feed budget; (b) an intermediate feed budget; (c) a short-term feed budget. Grass budgeting can be used with both continuous and rotational grazing systems. In both cases, effective budgeting requires close integration of grazing and conservation management, which provides the flexibility to remove grass surpluses at times of peak grass growth or alternatively to prevent grass shortage during periods of poor growth.

Annual feed budget (time scale = 1 year)

The annual feed budget involves setting targets for areas required for grazing through the season. This involves assessment of overall grass production for the season, taking account of local soil and climatic conditions and the level of nutrients which will be applied. Under typical weather conditions prevailing in western regions of the UK and assuming applied N levels up to 350 kg N ha^{-1}, it should be possible to produce 13 t DM ha^{-1}, with 75% of this being utilized, i.e. 9.75 t DM ha^{-1}. Assuming an average grass intake of 16 kg DM cow^{-1} day^{-1} from 15 April to 1 October, this equates to a total grass requirement of 2.69 t DM cow^{-1} (168 days × 16 kg DM cow^{-1}). Consequently, the average grazing stocking rate in this example will be (9.75/2.69) = 3.6 cows ha^{-1}. For example, for 100 cows the average grazing area required over the season will be 28 ha, i.e. (100/3.6).

Intermediate feed budget (time scale = 1 month)

The two major objectives of the intermediate feed budget are firstly, to adjust the stocking rate during the season to match changes in grass growth rate with changes in demand by the cow, and secondly, to conserve surplus grass for winter feeding. An example of an intermediate feed budget for 100 spring-calving dairy cows at three times during the season is shown in Table 10.11. Assuming a fixed paddock size of 0.8 ha, the rotation length required in May can be calculated as follows:

Grass consumed = 17.1 kg DM cow^{-1} × 100 cows = 1710 kg DM
Assuming 70% utilization of grass growth, require (1710/0.70) = 2443 kg grass DM available
Grass growth rate = 120 kg DM ha^{-1} day^{-1}
Hence require (2443/120) = 20.3 days regrowth

Similarly, the annual stocking rate equivalent in May can be calculated as follows:
100 cows grazing 20, 0.8 ha paddocks (16 ha)
Mean stocking rate = (100/16) = 6.25 cows ha^{-1}

Table 10.11 Example of an intermediate feed budget for a 100-cow spring-calving dairy herd.

	May	July	September
Grass growth rate (kg DM ha^{-1} day^{-1})	120	90	50
Milk yield cow^{-1} day^{-1} (kg)	30	25	20
Grass intake required (kg DM cow^{-1} day^{-1})*	17.1	15.0	13.1
Rotation length (days)	20	24	37
Area required day^{-1} for 100 cows (ha)**	0.80	0.80	0.80
Stocking rate (annual equivalent, cows ha^{-1})	6.25	5.20	3.38

* Assumes 550 kg dairy cow producing milk of 39.4 g kg^{-1} butterfat, 31.9 g kg^{-1} protein and 44.2 g kg^{-1} lactose, with ME content of herbage of 12.0 MJ kg^{-1} DM. Energy requirements based on AFRC (1993).
** Assumes 70% of the grass available is utilized at each grazing.

The data in Table 10.11 indicate that, assuming average grass growth rates of 120, 90 and 50 kg DM ha^{-1} day^{-1} in May, July and September, respectively, and a fixed paddock size of 0.8 ha, then 20 paddocks are required in May, increasing to 24 in July and 37 in September. In summary, the intermediate feed budget enables calculation of the grazing areas likely to be required at various points through the season, allowing for typical grass growth rates and changes in grass requirements of the cow. In the example given in Table 10.11, the increase in grazing area required through the season could be facilitated by bringing in 4 × 0.8 ha temporary paddocks from first-cut silage aftermaths in mid-June, with a further 15 × 0.8 ha temporary paddocks being brought in from second-cut aftermaths in August. (The grass budget shown in Table 10.11 could equally be used to calculate likely additional areas needed through the season in a continuous grazing system.)

Short-term feed budget (time scale = days or weeks)

The objective of short-term feed budgeting is to supply an adequate quantity of high quality grass to the cow whilst maintaining sward quality. This is best achieved by regular monitoring of the grazing area. Monitoring can involve assessment of grass yields and/or sward heights before and after grazing, and basically involves comparing actual grass availability against target. For example, weekly assessment of residual sward height in a rotational grazing system, or sward canopy height under continuous grazing, facilitates early assessment of grass surpluses or deficits. Techniques which can be used in short-term pasture budgeting include altering rotation length in rotational grazing, e.g. closing up surplus paddocks for big-bale silage, use of buffer grazing in continuous grazing systems and/or the use of feed supplements during periods of pasture shortage. In essence, short-term feed budgeting involves fine tuning of the grazing management to take account of fluctuations in grass growth from 'average' growth patterns. Target sward height guidelines for short-term feed budgeting are presented in Table 10.8.

10.6.4 Supplementation of grazing dairy cows

The principles of supplementation at pasture have been considered in Section 10.5 and also in Chapter 7. Responses to supplementation of dairy cows at pasture are largely dependent upon effects on herbage intake, i.e. substitution rate, and this varies with grazing conditions, production potential of the cow, supplement type and level of feeding. This leads to considerable difficulty at the farm level in deciding when supplements should be fed and the type of supplement to offer. A key priority for future grazing research must be to deliver reliable decision support systems, which will enable accurate assessment of herbage intake for particular milk yield levels and sward conditions and, in addition, will predict effects of specific supplement types on substitution rate and animal response.

One of the primary factors influencing substitution rate is the level of available herbage. Grainger & Mathews (1989), in a review of a series of Australian studies undertaken with high quality herbage, demonstrated a significant positive correlation between substitution rate and level of pasture intake when pasture was the sole feed (Fig. 10.5). Consequently, in periods of grass shortage, pasture intake as the sole feed will be reduced and provision of supplementary feeds will result in low substitution rates and hence increase nutrient intake and milk production. In these situations, supplement type has relatively little effect on substitution rate, and the key factor in this scenario is choosing a type of supplement which offers best value for money in terms of energy cost. (In extreme grass shortage situations, where very high levels of supplementation are required, care must also be taken to balance the diet in terms of fibre and protein concentrations.)

In contrast, under conditions of high herbage allowance, substitution rates are generally much higher, normally around 0.6 kg herbage per kg additional supplement, and there is considerable variation between different supplement types. For example, substitution rates with forage supplements, such as grass silage or hay, are normally much higher than with concentrate supplements. This may be related to effects on grazing time, with greater depressions in grazing time reported with grass silage (up to 37 min day^{-1} kg^{-1} silage DM; Phillips & Leaver 1986) than with concentrates (15–22 min day^{-1} kg^{-1} supplement DM; Sarker & Holmes 1974, Cowan *et al.* 1975).

Consequently, in situations of high herbage allowance, supplementation with

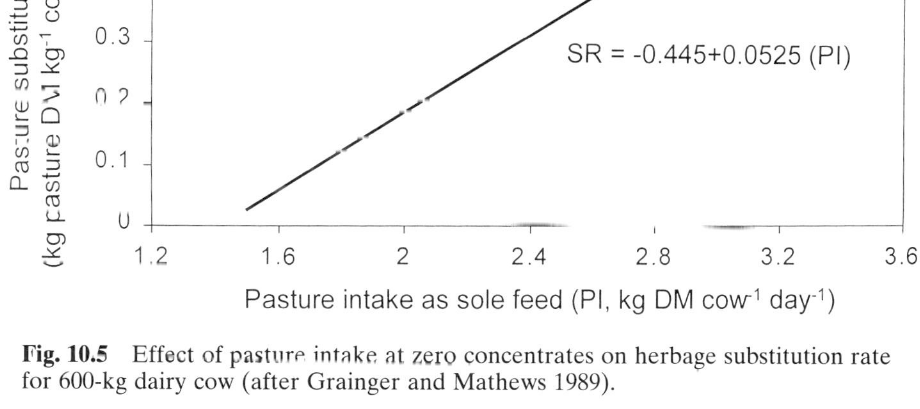

Fig. 10.5 Effect of pasture intake at zero concentrates on herbage substitution rate for 600-kg dairy cow (after Grainger and Mathews 1989).

hay or grass silage has resulted in substitution rates of 0.84–1.02 kg herbage DM per kg supplement DM and reductions in both milk and milk protein yield. Similarly with maize silage, Bryant & Donnelly (1974) observed a reduction in animal performance relative to grass only, whereas more recent work in the USA (Holden *et al.* 1995) reported no effect on animal performance of feeding 2.3 kg maize silage DM per day to high-yielding (29.0 kg milk per day) grazing dairy cows.

The lower substitution rates with concentrate supplements when adequate grass is available generally result in higher milk production responses compared with forage supplements. Furthermore, work in France (Hoden *et al.* 1991) suggests that responses to concentrate supplementation are influenced by the milk yield potential of the cow, with daily milk yield responses of 0.55, 0.77 and 0.84 kg milk kg^{-1} concentrates for cows yielding 25, 30 or 35 kg milk day^{-1}, respectively, at turnout in spring. The greater responses with high-yielding cows may reflect behavioural constraints on biting rate and grazing time. A number of studies (Stakelum 1993, Peyraud *et al.* 1996) indicate that as milk yield increases, the incremental increase in herbage DM intake tends to decrease, and consequently the incremental increase in intake provides only approximately half to two-thirds of the net energy requirement per kg of additional milk produced for high-yielding cows.

These data suggest that a combination of moderate herbage allowance (20 kg DM cow^{-1} day^{-1} assessed above 3.5 cm) presented in tall, dense, leafy swards, coupled with strategic use of concentrate supplements fed on a 'feed to yield' basis, offers the most effective basis for managing high-yielding dairy cows at pasture. Given that high quality grazed grass has the potential to support milk yields up to 30.0 kg day^{-1} in spring, declining to 25.0 kg day^{-1} in late summer (Table 10.2), suggested concentrate feeding levels for cows of differing yield potential are presented in Table 10.12. These estimates suggest that cows producing up to 40 kg milk day^{-1} in early season may require up to 7.0 kg concentrates day^{-1}, whereas in late season up to 8.5 kg concentrates day^{-1} may be required for cows yielding 35 kg milk day^{-1}.

Concentrate composition

A wide range of concentrate supplements are fed to dairy cows at pasture, including cereal grains and by-product feeds such as sugar beet pulp, citrus pulp, brewers grains, etc. Given the increases in milk production potential of the dairy herd and the higher levels of supplementation being used with these herds, the effects of changes in concentrate composition on herbage intake and milk production are becoming increasingly important.

Most modelling studies which have examined nutrient flow with dairy cows offered high quality pasture indicate that energy, rather than protein supply, is likely to be the first limiting nutrient for milk production (Kolver *et al.* 1998). Recent studies (Hongerholt & Muller 1998) appear to confirm this observation, supplementation of undegradable protein having little effect on milk yield from

Table 10.12 Suggested concentrate feeding levels for high-yielding dairy cows in early and late season offered a moderate herbage allowance (20 kg DM cow^{-1} day^{-1} assessed above 3.5 cm, equating to a residual sward height in rotational grazing of 8 cm).

	Early season target milk yield (kg cow^{-1} day^{-1})			Late season target milk yield (kg cow^{-1} day^{-1})	
	25.0	35.0	40.0	25.0	35.0
Potential milk yield from grass (kg day^{-1})*	27.0	29.4	30.9	20.0	24.5
ME required from supplement (MJ day^{-1})**	0	28.0	45.5	25.0	52.5
Supplement feed level required (kg cow^{-1} day^{-1})†	0	4.5	7.0	4.0	8.5

* Assumes increase in herbage intake of 0.125 kg DM kg^{-1} milk.
** Assumes ME required for milk production of 5.0 MJ kg^{-1} milk.
† Assumes substitution rate of 0.4 kg herbage DM kg^{-1} supplement DM, and ME concentration of herbage and supplement of 12.0 MJ kg^{-1} DM.

cows yielding 34.2 kg day^{-1} (fat-corrected milk) and offered high quality pasture (256 g CP kg^{-1} DM) with 8.9 kg day^{-1} of a cereal-based concentrate containing 147 g crude protein kg^{-1} DM. Similarly, Kellaway & Porta (1993), in a review of the effects of protein supplementation on dairy cow performance with pasture-based diets, suggested that responses to additional protein were only obtained when pasture quality was poor, with cows in early lactation, or when high levels of grain supplementation were being used. Nonetheless, it is important to note that a high proportion of the rumen-degradable protein in fresh herbage is absorbed from the rumen, with less than 30% of the ingested grass protein reaching the duodenum (Beever *et al.* 1986). Consequently, supplementation strategies should be designed to increase utilization of rumen-degradable nitrogen by improving the synchronization of energy and protein supply to the rumen. Possible options include feeding supplements with high levels of starch or sugars, or alternatively varying the timing of supplementary feeding during the day. With respect to the latter, the results of research are somewhat conflicting. Data from the USA (Buckmaster *et al.* 1995) suggest no effect of increased frequency of concentrate feeding on animal performance, whereas data from Australia (Dalley 1998a) indicate increases in milk yield of up to 2.0 l day^{-1} with 8.0 kg concentrates day^{-1} fed over four rather than two meals per day.

Effects of concentrate energy source on substitution rate and animal performance of dairy cows at grass are also inconsistent. For example, Meijs (1986) observed higher substitution rates and lower animal performance responses with high starch (350 g starch + sugars kg^{-1} DM) relative to high fibre (100 g starch + sugars kg^{-1} DM) concentrates. In contrast, Van Vuuren *et al.* (1993)

observed no effect of concentrate energy source on herbage intake, although organic matter digestion in the rumen was reduced and duodenal amino acid flow increased with high starch (487 g starch + sugars kg^{-1} DM) relative to high fibre (94 g starch + sugar kg^{-1} DM) concentrates. Van Vuuren *et al.* (1993) concluded that supplementation of herbage with high starch concentrates increased the efficiency of microbial protein synthesis through better synchronization of nutrient supply to the rumen.

The apparent conflict between these studies relating to energy source may be partly attributable to changes in the chemical composition of herbage through the season. Further research is required on the effect of concentrate energy source on rumen fermentation characteristics, herbage substitution rate and animal performance with a range of herbage types.

10.6.5 Opportunities to extend the grazing season with dairy cows

As reviewed in Section 10.4.5, the major challenges in extending the grazing season are to produce grass of high quality during periods of low grass growth rate, and to utilize the grass efficiently whilst minimizing sward damage. With dairy cows, the latter can be achieved by using short (2–6 h day^{-1}) grazing periods and block-grazing techniques, with animals confined to a limited area each day and prevented from re-grazing. Using these techniques, Mayne & Laidlaw (1995) observed that autumn grazing (29 October to 26 November) with cows grazed for 3 h day^{-1} reduced silage intake by 4.2 kg DM day^{-1} whilst increasing milk yield by 2.1 kg day^{-1}, when compared with cows maintained indoors on similar silage and concentrate feed levels. Similarly, with early spring grazing (7 March to 17 April), Sayers & Mayne (1998) observed increases in milk yield and milk composition following turnout for 2 h day^{-1}, as shown in Table 10.13. Animals with access to pasture for 2 h day^{-1} had significantly lower

Table 10.13 Effect of early turnout in spring for 2 h day^{-1} on dairy cow performance (Sayers & Mayne 1998).

	Treatment		
	Fully housed	Grazed for 2 h day^{-1} at residual sward height (cm)	
		5.5	6.5
Silage intake (kg DM day^{-1})	10.3	7.3	6.6**
Milk yield (kg day^{-1})	25.9	28.0	28.8**
Milk composition (g kg^{-1})			
Milk fat	39.9	41.4	40.2
Milk protein	28.6	29.5	30.3

silage intakes, whilst milk yield was increased and there was also a trend towards improved milk composition, even when cows grazed to a residual sward height of 5.5 cm. These results illustrate the potential contribution of grazed grass in late autumn and early spring towards reducing requirements for conserved forage, whilst improving animal performance.

10.7 Practical grazing management – beef cattle

10.7.1 Herbage intake and animal performance

The level of liveweight gain achieved by grazing beef cattle is primarily a consequence of their level of herbage intake. Efficient grazing systems for beef cattle should therefore aim to maximize the intake of grass by grazing cattle. As with dairy cows (see Section 10.6.1), the major determinant of daily herbage intake is intake per bite, which is primarily a consequence of sward height (Fig. 10.4). Sward height is therefore a major determinant of daily herbage intake. Figure 10.6 shows the relationship between herbage intake and grass height when suckler cows grazed continuously on a perennial ryegrass sward. At heights of below about 10–11 cm the intake is progressively reduced (Wright 1988), and this is reflected in the liveweight gain of the animals. Maximum liveweight gain of continuously grazing beef cows and growing/finishing cattle occurs at sward heights of 8–10 cm. Above that height, liveweight gain decreases because of the reduction in the digestibility of the herbage, especially in late summer (Swift *et al.* 1989, Wright & Whyte 1989). This research led to recommendations on the optimum sward height for beef cattle shown in Table 10.14.

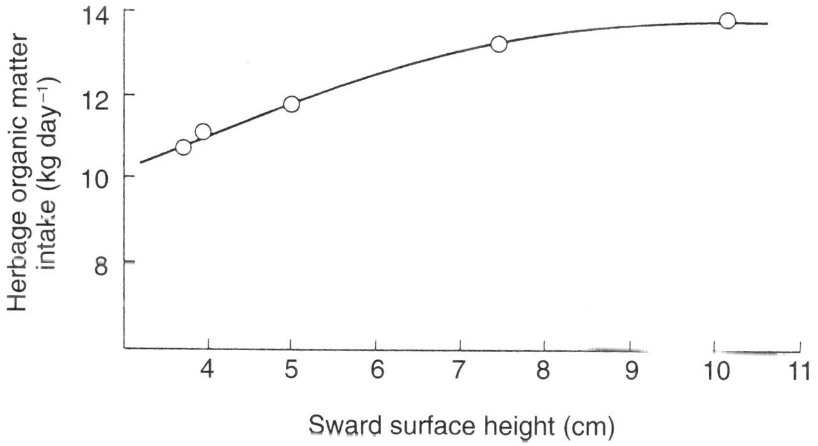

Fig. 10.6 The effect of sward height on herbage intake by continuously grazing beef cows (from Wright 1988).

Table 10.14 Recommended sward heights for beef cattle (cm).

	Continuous stocking		Rotational grazing (stubble height)
	Early summer	Late summer	
Growing/finishing cattle	7–8	10–12	7–10
Beef cows and calves	7–8	10–12	7–10

In spring, beef cattle should be turned out to pasture before sward height reaches 8 cm. If turnout is delayed until the sward height is over 8 cm it can be difficult to control sward height thereafter. In spring and early summer it is important to keep the sward surface height of continuously grazed swards to 7–8 cm. Above that height there is increased development of flower heads in grassland, and this leads to a reduction in the digestibility and feeding value of the grass. Once the risk of seedhead development has passed, then sward height can be allowed to increase to 10–12 cm in late summer. Grass should be grazed down to 4 cm in autumn as tall grass can suffer from winter kill, in which a large number of tillers die off. The same heights are recommended for mixed grass/white clover swards, although the clover component may benefit from a rest period followed by a conservation cut (Section 10.4.6).

By adhering to these sward height guidelines, the pasture is used efficiently. When sward heights are too high, cattle start to concentrate their grazing on certain parts of the field, keeping the grass short, and undergraze other areas, which become tall with a high proportion of reproductive tillers. Gibb & Ridout (1986, 1988) demonstrated that swards grazed by cattle comprise patches of frequently grazed and infrequently grazed areas. Figure 10.7 shows the proportion of a field which was infrequently grazed when beef cows and calves continuously grazed at a sward height of 8 or 12 cm. The 8-cm sward quickly reached a state when only 20% of the area was infrequently grazed, and for much of the season only 10% of the sward was infrequently grazed – mainly around dung pats. In contrast, the sward grazed at 12 cm had over 40% of the area infrequently grazed for much of the season.

Since the growth rate of grass varies seasonally, a way of changing stocking rate needs to be found if sward height is to be controlled. In practice, there are two ways of achieving this, either by changing the number of animals on a fixed area, or by changing the area of grassland available to the herd. The first can be achieved in some circumstances. For example, in a grass-based finishing system, the decline in cattle numbers in late summer and autumn as cattle are sold for slaughter coincides with the decline in grass growth rate. In the case of a breeding herd there is little opportunity to change animal numbers. However, since most grazing beef systems also require winter fodder to be produced there is a need to make silage or hay, and the integration of grazing with fodder

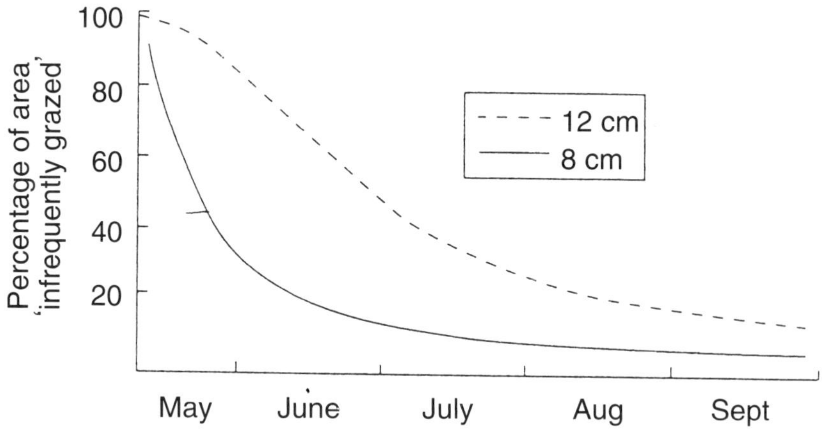

Fig. 10.7 The percentage of area 'infrequently grazed' when sward heights of 8 and 12 cm are maintained by continuously grazing cattle (from Wright 1988).

conservation provides an ideal method of controlling sward height (see Section 10.4).

10.7.2 Supplementation of beef cattle at pasture

If grass is managed properly, and kept at the heights suggested in the previous section, there should be little need to supplement beef cattle at pasture during most of the grazing season. However, in spring, immediately following turnout and before sward heights reach 8 cm, there may be a need to provide some supplementary feeding to growing beef cattle, and 1–2 kg of a cereal-based supplement is adequate. Beef cows are susceptible to grass staggers (hypomagnesaemia) at this time, and the provision of 1 kg of a high-magnesium (Mg) supplement is an effective means of providing Mg. Supplements should be reduced gradually over a period of 1–2 weeks when grass height is sufficient for growing beef cattle or when the risk of staggers is reduced.

During most of the season, grass maintained at the recommended height will have an organic matter digestibility of over 70%, and there will be no need to provide supplements. Supplements will act as a substitute for herbage, and the cattle will consume less grass. However, in autumn the quality of grass declines, and it can also be difficult to maintain the recommended sward heights. If sward height falls below 8 cm in autumn, then suckled calves should be offered 1–2 kg of supplementary feed per day, but when sward height is above 8 cm there is no benefit in providing supplements (Wright 1992). To maintain liveweight gain, growing/finishing cattle also need to be offered supplementary feeding in autumn, to compensate for the decline in both the quality and quantity of grass.

10.8 Practical grazing management – sheep

10.8.1 Herbage intake and animal performance

As with other types of livestock, sward height is a major determinant of the intake, and therefore performance, of grazing sheep. Figure 10.8 shows the herbage intake of ewes suckling twin lambs when sward height was

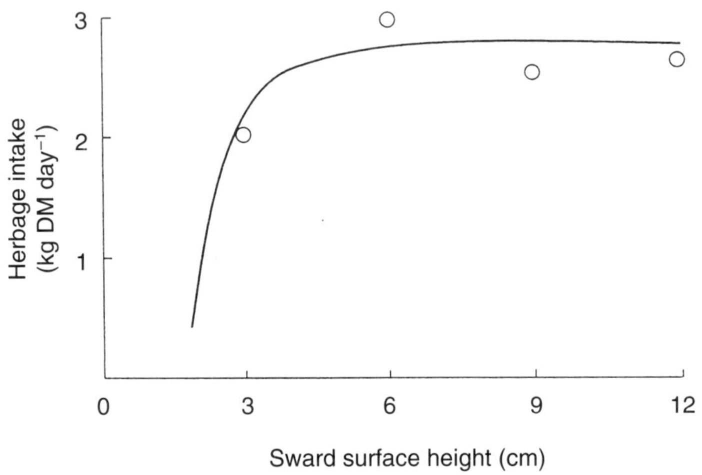

Fig. 10.8 The herbage intake of ewes suckling twins grazed on swards of different heights (from Penning *et al.* 1991).

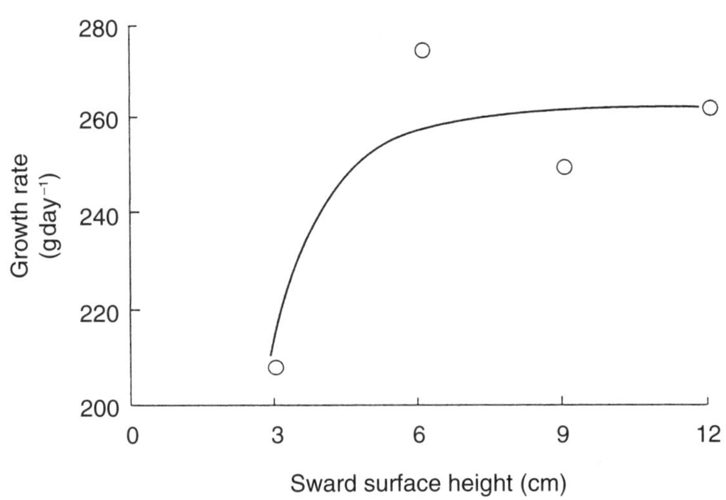

Fig. 10.9 The growth rate of twin lambs kept on swards of differing heights (from Penning *et al.* 1991).

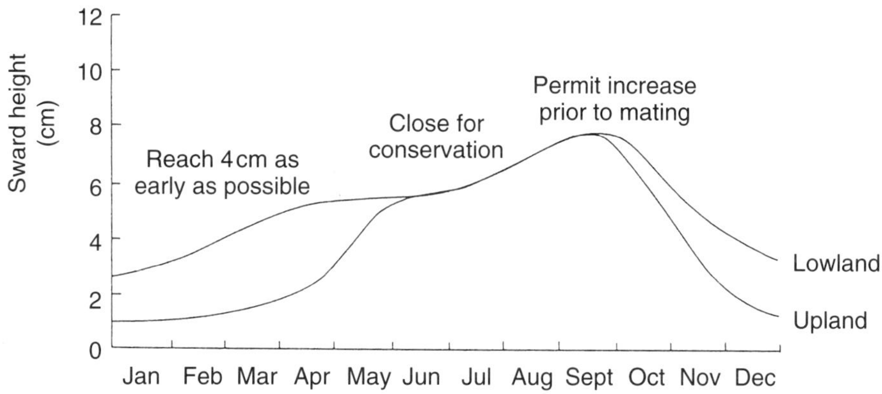

Fig. 10.10 The recommended annual profile of sward height for lowland and upland sheep systems (from Maxwell & Treacher 1987).

maintained at 3, 6, 9 or 12 cm. Maximum intake occurred when sward height was 6 cm, with no increase at higher heights. The growth rate of the lambs showed a similar response (Fig. 10.9), with reduced lamb liveweight gain on 3-cm swards and no increase in performance at sward heights above 6 cm.

10.8.2 Optimum sward height profile

The optimum sward height profiles for lowland and upland systems are shown in Fig. 10.10. The guidelines are for the five main periods in the grazing year (Maxwell & Treacher, 1987).

Winter grazing (late pregnancy)

Grazing of the main spring and summer area over winter will depress spring growth, and so ideally the flock should be housed or graze another area of grassland.

Early spring (post-lambing)

A sward height of 4 cm should be achieved as quickly as possible. The time at which this is achieved will vary. It will be later on upland farms than on lowland farms.

Mid-lactation

To maintain high levels of ewe and lamb performance, the sward height should be kept between 4 and 6 cm. Below 4 cm, liveweight gain will be depressed, while above 6 cm herbage utilization will be low, and a poor structure will develop in the sward, with high numbers of reproductive tillers and a decline in digestibility.

Weaning and ewe-recovery phase

Dry ewes will maintain high intakes on swards above 4 cm. However, it is desirable at this stage to allow sward height to increase to about 8 cm to build up a reserve of good quality herbage to provide a high level of nutrition to ewes in the mating period. For weaned lambs, the sward height should be maintained at around 6 cm and should not be allowed to decline. The performance of weaned lambs is severely depressed when they graze on a sward that is declining rapidly in height, as they are then forced to eat stem.

Pre- and post-mating phase

Rapid decline in sward height should be avoided during mating and immediately afterwards, as the intake of ewes may fall below maintenance levels. Then before winter, the swards should be grazed down to 3 cm to allow pastures to enter winter in a good condition to avoid winter kill.

As with other types of livestock, the optimum sward height profile can be maintained by the integration of grazing management and fodder conservation, and by varying the area closed for fodder conservation to maintain sward height.

10.8.3 Supplementation of ewes and lambs at pasture

Supplements may be needed for ewes in spring and autumn. Milne *et al.* (1981) suggested that there is no advantage in feeding supplements when sward height is over 4 cm. When sward height is below 4 cm, ewe milk yield and liveweight gain will be depressed. In this case 500–600 g of supplement should be fed to lactating ewes. When ewes are lambed later in the spring (e.g. in May) there is usually no need to feed supplements prior to lambing. Grass height should be kept at around 4 cm in the few weeks before lambing (by keeping stocking rates high), and ewes put on a sward height of 6 cm after lambing.

Provided swards reach 8 cm by early autumn (mid-September), it is usually possible to ensure that sward height does not fall to below 3 cm before the end of the mating period. However, if this is not possible and sward height is less than 3 cm during mating, supplements should be fed at a rate of 300 g day^{-1}.

10.9 Mixed and sequential grazing

Different species of livestock can graze together in the same field (mixed grazed) or in sequence (sequential grazing).

10.9.1 Mixed grazing

Many advantages have been claimed for mixed grazing, including increased animal performance and output per hectare (Nolan & Connolly 1989), better matching of animals' seasonal energy requirements to herbage production, reduced gastro-intestinal parasite burdens, diversification of animal production and manipulation of botanical composition (Lambert & Guerin 1989). In Ireland, Nolan and Connolly (1989) compared mixed sheep and cattle grazing with grazing by sheep only or cattle only. They suggested that the animal output per hectare from perennial ryegrass pasture could be increased by about 10% under mixed grazing compared with sheep or cattle alone. The increased output may partly be explained by a higher level of utilization of the pasture. In particular, sheep graze much closer to cattle dung pats than cattle, and so utilize herbage that is not normally grazed by cattle. However, not all research has shown such an advantage, and the circumstances under which there are advantages for mixed grazing are unclear.

10.9.2 Sequential grazing

Sequential grazing occurs when one species of livestock follows another. For example, on dairy farms in autumn when grass heights and pasture quality are too low for cows, sheep can be used to graze down swards to ensure that they are in a suitable state to enter winter and avoid winter kill.

On mixed grass/white clover swards cattle tend to select less clover than sheep, and so a higher clover content develops under cattle grazing. Thus, when weaned lambs follow cattle in these circumstances, they have liveweight gains of about 30 g day^{-1} higher than when they follow sheep, because of the higher clover content in their diet.

10.10 Hill and upland grazing systems

Hill and upland grazing land accounts for approximately 8 million hectares in the UK. In these areas pasture growth is limited by climatic conditions and soil fertility. Annual dry matter production decreases by about 2% for each 30 m rise in altitude, but the effect of altitude does depend on season. In spring, each 30-m rise in altitude is associated with a decrease in plant production by about 5%, while in autumn the decrease is only 1.8%. In many hill areas soils are acidic because of the combined effects of the parent material and leaching of anions due to high rainfall. Low pH can be a major cause of slow rates of decomposition of organic matter in soils and hence may limit the cycling of nutrients.

With a short growing season, stocking rates of sheep were traditionally determined by the ability of the hills to support sheep through the winter, with

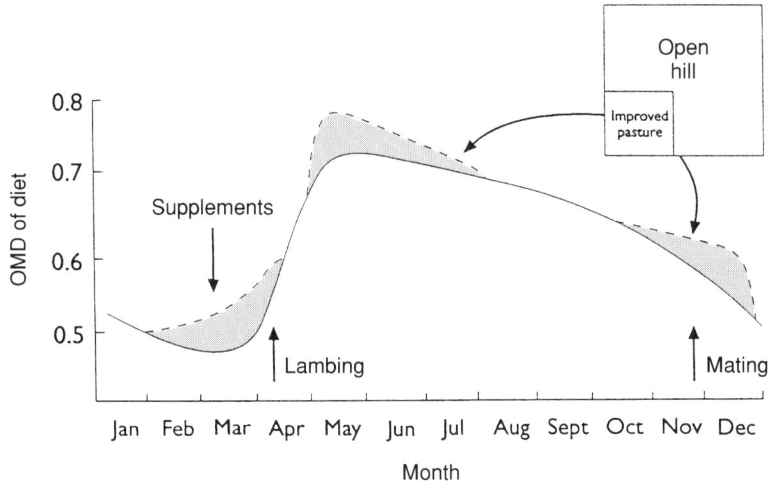

Supplements	Improved pastures	
Increase lamb birthweight	Increase milk yield	Increase ewe liveweight and condition
Reduce ewe and lamb mortality	Increase lamb growth rate	Increase conception rate

Fig. 10.11 Improving diet quality of hill sheep by use of the two-pasture system (from Cunningham & Russel 1979).

only little supplementary feeding. This led to under-utilization of the summer growth because of the low stocking rates in summer. This under-utilization in early summer led to a large proportion of mature, low-digestibility herbage, which diluted the feeding value of the pasture. Ewes could not attain good body condition at mating, and therefore ovulation rates were low. Ewes were also underfed in winter during pregnancy and could not provide a high enough milk yield to support high lamb growth rates. Weaning rates were typically 60–90% and the weight of lambs at weaning was low.

Identification of these constraints led to the development of the Hill Farming Research Organization (HFRO) 'two-pasture system'. In this system, a small area of accessible, well-drained land was fenced and improved by liming and fertilization, and possibly reseeding. This improved area is used to provide improved nutrition at key times in the annual production system, in particular around mating and for twin-bearing ewes during lactation. In addition, supplementary feed is given during the last 6 weeks of pregnancy (Fig. 10.11). The two-pasture system has allowed an increase in individual animal performance and an increase in stocking rate (Table 10.15).

Table 10.15 The increase in production following introduction of the HFRO two-pasture system (Armstrong *et al.* 1984).

	Traditional system	Two-pasture system
Flock size	387	669
Weaning rate	91%	111%
Total weight of lambs weaned (kg)	7900	19170
Total weight of wool produced per year (kg)	870	1720

10.10.1 Utilization of heather

Heather (*Calluna vulgaris*) is an important component of many hill grazings. It can contribute to the nutrition of sheep and also red deer (*Cervus elaphus*) and grouse (*Lagopus scoticus*). Although heather cannot maintain sheep on its own, in conjunction with grass it provides a valuable source of feed for sheep. It is a woody shrub, but young growing shoots can have a digestibility of 60%, although when mature the digestibility declines to 45%. The young shoots can be grazed provided no more than 40% of each season's growth is removed. Levels of utilization above 40% lead to a reduction in the vigour of the plants, and eventually they are killed and replaced by grass species (Grant *et al.* 1978). Heather becomes woody with age and needs to be burned at 7–15-year intervals depending on the level of grazing. The important role of heather in both the nutrition of sheep and as a wildlife habitat has stimulated the development of computer models (Armstrong *et al.* 1997) to help manage heather so that an appropriate balance between agriculture and conservation objectives can be achieved.

10.11 Animal health on grazed pasture

The major issues relating to animal health of grazing animals concern mineral and trace element nutrition, and parasitic diseases caused by both gastro-intestinal and lung worms.

10.11.1 Mineral and trace element nutrition

The concentration in forage of many important minerals varies considerably between different locations and soil types. It can be influenced further by management and botanical composition, for instance between extensively managed permanent swards and N-fertilized ryegrass (Hopkins *et al.* 1994), and chemical analysis has also indicated that a number of herb species have higher concentrations of some mineral elements than *L. perenne* (see Chapter 14). The role of mineral elements in relation to the feeding value of grass has also been discussed in Chapter 7. The following additional comments relate more specifically to mineral nutrition and grazing management.

Magnesium

One of the major animal health problems associated with grazing is hypomagnesaemia, which arises from low absorption of magnesium (Mg) from the diet, or low availability of Mg in herbage, and is most apparent in lactating ruminants, i.e. in both spring and autumn. Herbage Mg is lower in spring and autumn, and this is often compounded by high levels of potassium (K) and a low ratio of sodium (Na) to K. High levels of K depress Mg absorption in the rumen (Tomas & Potter 1976). Stress is also an important factor in the occurrence of acute hypomagnesaemia (grass staggers). Rapid changes in diet from conserved forage to grazing in spring, the reverse in autumn, and the onset of inclement weather are factors that can predispose animals to suffer from staggers. Gradual changes in diet and provision of additional Mg can help to avoid acute symptoms.

A wide range of strategies has been used to prevent Mg deficiency. Binnie *et al.* (1995) demonstrated that the higher herbage-Mg of swards established from varieties bred for a high Mg content was reflected in the magnesium metabolism of sheep. The Mg content of pasture can also be improved by application of fertilizers containing magnesium, e.g. dolomite. However, in most situations, Mg deficiency is prevented by supplementation (discussed in Section 7.9.2) or by pasture dusting. Todd (1962) first used pasture dusting in Northern Ireland and showed that application of 31 kg Mg ha^{-1}, applied as finely ground calcined magnesite just prior to grazing, maintained high serum magnesium levels in grazing dairy cows.

Sodium

In areas where soil Na or K levels are low, grass growth may be limited by Na deficiency, and there is also evidence that up to 25% of herbage grown in Britain is deficient in Na for lactating dairy cows (Blaxter 1968). Permanent pasture swards have also been shown to have lower Na levels than sown *L. perenne* on the same sites (Hopkins *et al.* 1994). Deficiencies of Na in herbage can result in depressed appetite, reduced growth rates in cattle and sheep and depressed milk yield in dairy cattle. Chiy & Phillips (1991) observed increased herbage growth rates following application of 50 kg sodium chloride per hectare to natrophilic herbage, with associated increases in milk yield and liveweight gain of grazing dairy cows. In other studies, Chiy *et al.* (1994) observed that application of sodium fertilizer to pasture improved the mineral status of grazing sheep in addition to increasing fibre digestibility and dry matter intake.

Copper

Copper (Cu) is an essential element for all ruminants and Cu deficiency in grazing animals is relatively common worldwide. This reflects the considerable variation in Cu content of pasture, with Minson (1990) reporting a range from 2.5 to 20.0 mg kg^{-1} DM in *L. perenne*. Copper deficiency results in depressed animal performance and a range of clinical symptoms (see Section 7.9.3, where methods to prevent Cu deficiency are discussed).

Selenium

Selenium (Se) and vitamin E are essential nutrients for a range of metabolic processes (McMurray *et al.* 1983), and deficiency can result in reduced food intake, weak calf syndrome, retained placenta, nutritional muscular dystrophy, and depressions in milk production. Selenium concentration in herbage is extremely variable, reflecting levels of available Se in soil. Control measures to prevent Se deficiency include inclusion of Se in mineral supplements fed at pasture, and oral or subcutaneous administration (but see comments on substitution of Vitamin E in Section 7.9.1).

Iodine

Iodine is an essential element in the hormones thyroxine (T4) and tricodothyronine (T3) which are involved in a wide range of metabolic processes. Typical iodine levels in *L. perenne* range from 0.16 to 0.52 (Johnson & Butler 1957), with values tending to decrease with increased plant maturity. Symptoms and control methods are discussed in Section 7.9.3.

Cobalt

Cobalt (Co) is required for the synthesis of vitamin B_{12} by rumen bacteria, and anaemia and loss of appetite (e.g. pine fever) normally characterize Co-deficiency in ruminants. There is considerable variation in Co concentration of herbage, mainly associated with differences in soil-Co levels, and there are numerous clearly defined Co-deficient areas worldwide. Control methods commonly used to prevent Co and iodine deficiencies are discussed in Section 7.9.3.

10.11.2 Pastures and parasitic diseases

Parasitic worms cause a range of disease conditions in grazing animals, as reviewed by Cawthorne (1986). An understanding of the biology of the parasites enables appropriate management and control methods to be applied.

***Liverfluke* (Fasciola hepatica)**

Liverfluke is a common disease in grazing livestock, particularly those grazing in warm, moist conditions. The disease results in liver damage with subsequent ill thrift, and in some instances death of both sheep and cattle. The life history of the fluke involves the snail *Limnaea truncatula*, which acts as an intermediate host. Control measures involve treatment of infected animals and elimination of the snail. The latter can be achieved by draining or fencing off wet areas, or by application of molluscicides. The risk of fluke infestations can now be accurately forecast, based on an analysis of weather conditions, and disease risk assessments are normally widely publicized.

Lungworm (Dictyocaulus viviparus)

Husk or parasitic bronchitis is a common disease in calves during their first summer at grass, but is also occasionally observed in older cattle. Symptoms include severe bronchitis, which can result in permanent lung damage and, in severe cases, be fatal. Control measures include use of clean pastures for young calves, avoiding mixing young stock with older cattle, vaccination with irradiated larvae 6 weeks and 2 weeks prior to turnout to grass, or regular treatment with ivermectin during the grazing season. Animals normally acquire active immunity during their first grazing season, although immunity may not be complete for all cattle in a group, and adult cattle may become infected in their second season after a severe challenge.

Parasitic gastroenteritis (stomach worms)

A number of organisms are implicated in parasitic gastroenteritis, the symptoms of which are scouring and reduced animal performance. The major round worms are *Ostertagia*, *Trichostrongylus* and *Haemonchus contortus*, and in addition *Nematodirus* species cause nematodiarisis in lambs. Young calves and lambs are particularly susceptible to roundworm infection. Control measures include grazing clean pastures and the use of anthelmintics in spring and mid-summer.

Protozoa

Coccidiosis is a common disease in young calves and lambs, resulting in ill thrift and in severe cases causing heavy losses through death, particularly in young lambs. The disease results from severe infection with *Coccidia*, an organism commonly found in faeces samples, but which can multiply rapidly under dense stocking and favourable weather conditions. Control measures include grazing on clean pasture, particularly with newborn lambs, and in some cases use of coccidiostats may be required.

10.12 References

AFRC (Agricultural and Food Research Council) (1993) *Energy and Protein Requirements of Ruminants*. An advisory manual prepared by the AFRC Technical Committee on Responses to Nutrients. CAB International, Wallingford.

Armstrong H.M., Gordon I.J., Hutchings N.J., Illius A.W., Milne J.A. & Sibbald A.R. (1997) A model of the grazing of hill vegetation by sheep in the UK. II. The prediction of offtake by sheep. *Journal of Applied Ecology* **34**, 187–207.

Armstrong R.H., Eadie J. & Maxwell T.J. (1984) Hill sheep production: A modified management system in practice. In O'Toole M.A. (ed) *Hill Land Symposium, Galway*, 230–47. An Foras Taluntais, Irelands.

Arnold G.W. (1981) Grazing behaviour. In Morley F.H.W. (ed) *World Animal Science. 1. Grazing Animals*, 79–104. Elsevier, New York.

Baker H.K., Chard J.R.A. & Hughes W.E. (1965) A comparison of cocksfoot and tall fescue-dominant swards for out-of-season production. *Journal of the British Grassland Society* **20**, 84–90.

Bartholomew P.W. & Chestnutt D.M.B. (1977) The effect of a wide range of fertilizer nitrogen application rates and defoliation intervals on the dry matter production, seasonal response to nitrogen, persistence and aspects of chemical composition of perennial ryegrass. *Journal of Agricultural Science, Cambridge* **88**, 711–21.

Beever D.E., Losada H.R., Cammell S.B., Evans R.T. & Haines M.J. (1986) Effect of forage species and season on nutrient digestion and supply in grazing cattle. *British Journal of Nutrition* **56**, 209–25.

Binnie R.C., Johnston D.T. & Chestnutt D.M.B. (1995) An evaluation of a high magnesium perennial ryegrass variety. In Pollott G.E. (ed) *Grassland into the 21st Century*. British Grassland Society Occasional Symposium No. 29, 286–8. BGS, Reading.

Binnie R.C., Chestnutt D.M.B. & Kilpatrick D.J. (1997) The effect of altering the length of the regrowth interval in the early, mid and late season on the productivity of grass swards. *Journal of Agricultural Science, Cambridge* **128**, 303–9.

Bircham J.S. (1981) *Herbage Growth and Utilization under Continuous Stocking Management*. Ph.D. Thesis, University of Edinburgh.

Blaxter K.L. (1968) Fertilizers and animal production. *Proceedings of the Fertilizer Society* **105**, 1–46.

Brereton A.J., Danielov S.A. & Scott D. (1995) *Agrometeorology of Grasslands for Middle Latitudes*, World Meteorology Organization, Technical Note. WMO, Geneva.

Bryant A.M. & Donnelly P.E. (1974) Yield and composition of milk from cows fed pasture herbage supplemented with maize and pasture silages. *New Zealand Journal of Agricultural Research* **27**, 491–3.

Buckley F. & Dillon P. (1998) The effect of genotype and feeding system on the performance of Holstein Friesian cows at pasture. *Proceedings Irish Grassland Association Dairy Conference, Limerick*, 17 April 1998, 1–15.

Buckmaster D., Muller L., Hongerholt D. & Gardner M. (1995) Mobile computer controlled concentrate feeder for lactating cows. *Summaries of Pasture Research and Extension Activities*, Pennsylvania State University, University Park, PA.

Caird L. & Holmes W. (1986) The prediction of voluntary intake of grazing dairy cows. *Journal of Agricultural Science, Cambridge* **107**, 43–54.

Castle M.E. (1976) A simple disc instrument for estimating herbage yield. *Journal of the British Grassland Society* **31**, 37–40.

Cawthorne R.J.G. (1986) Management for control of parasites. In Frame J. (ed) *Grazing*. British Grassland Society Occasional Symposium No. 19, 89–97. BGS, Hurley.

Chacon E. & Stobbs T.H. (1976) Influence of progressive defoliation of a grass sward on the eating behaviour of cattle. *Australian Journal of Agricultural Research* **27**, 709–18.

Chestnutt D.M.B. (1994) Effect of early season sward management on sward quality and liveweight gain during autumn. *Grass and Forage Science* **49**, 405–13.

Chiy P.C. & Phillips C.J.C. (1991) The effects of sodium chloride application to pasture, or its direct supplementation, on dairy cow production and grazing preference. *Grass and Forage Science* **46**, 325–31.

Chiy P.C., Phillips C.J.C. & Ajele C.L. (1994) Sodium fertilizer application to pasture. 5. Effects on herbage digestibility and mineral availability in sheep. *Grass and Forage Science* **49**, 25–33.

Clark D.A. & Jans F. (1995) High forage use in sustainable dairy systems. In Journet M., Grenet E., Farce M.H., Theriez M. & Demarquilly C. (eds) *Recent Developments in the Nutrition of Herbivores, Proceedings of the IVth International Symposium on the Nutrition of Herbivores*, 497–526. INRA Editions, Paris.

Cowan R.T., Byford I.J.R. & Stobbs T.H. (1975) Effects of stocking rate and energy supplementation on milk production from tropical grass–legume pasture. *Australian Journal of Experimental and Animal Husbandry* **15**, 740–6.

Cowling D.W. (1982) Biological nitrogen fixation and grassland production in the United Kingdom. *Philosophical Transactions of the Royal Society of London, Series B* **296**, 397–404.

Cunningham J.M.M. & Russel A.J.F. (1979) The technical development of sheep production from hill land in Great Britain. *Livestock Production Science* **6**, 379–85.

Dalley D. (1998a) Synchrony of protein and energy supply during early lactation. In *Annual Report 1997–98, Agriculture Victoria*, 50–1. Ellinbank, Australia.

Dalley D. (1998b) Is more frequent feeding better? *Proceedings, Open Day 30 April 1998, Agriculture Victoria*. Ellinbank, Australia.

Davies D.A. & Hopkins A. (1996) Production benefits of legumes in grassland. In Younie D. (ed) *Legumes in Sustainable Farming Systems.* British Grassland Society Occasional Symposium No. 30, 234–46. BGS, Reading.

Davies O.D. (1986) Comparisons between measurements obtained from a falling plate meter and sward stick on cattle swards under various management systems. *ADAS Internal Report No. 86/12, Trawsgoed Experimental Husbandry Farm.*

Dillon P. & Crosse S. (1994) Effect of early season grazing on dairy cow performance. *Proceedings, Irish Grassland Association Dairy Conference, March 1994*, 45–64.

Ernst P., Le Du Y.L.P. & Carlier L. (1980) Animal and sward production under rotational and continuous grazing management – a critical review. *Proceedings of the International Symposium on the Role of Nitrogen in Intensive Grassland Production, Wageningen, 1980*, 119–26.

Evans B. (1981) Production from swards grazed by dairy cows. *Grass and Forage Science* **36**, 132–4.

Frame J., Baker R.D. & Henderson A.R. (1995) Advances in grassland technology over the past fifty years. In Pollot G.E. (ed) *Grassland into the 21st Century*. British Grassland Society Occasional Symposium No. 29, 31–63. BGS, Reading.

Gibb M.J. & Ridout M.S. (1986) The fitting of frequency distributions to height measurements on grazed swards. *Grass and Forage Science* **41**, 247–9.

Gibb M.J. & Ridout M.S. (1988) Application of double normal frequency distributions fitted to measurements of sward height. *Grass and Forage Science* **43**, 131–6.

Gordon F.J. (1996) Potential milk production from grass and limiting factors. *Proceedings, Irish Grassland Association Dairy Conference, April 1996*, 45–64.

Grainger C. & Mathews G.L. (1989) Positive relation between substitution rate and pasture allowance for cows receiving concentrates. *Australian Journal of Experimental Agriculture* **29**, 355–60.

Grant S.A., Barthram G.T., Lamb W.I.C. & Milne J.A. (1978) Effects of season and level of grazing on the utilization of heather by sheep. 1. Response of the sward. *Journal of the British Grassland Society* **33**, 289–300.

Hlukik J.G. & Smith T.R. (1988) Economics of pasture systems. *Proceedings: Pasture in the Northeast Region of the United States, NRAES-36*. Cornell University, Ithaca, NY.

Hoden A., Peyraud J.L., Muller A., Delaby L. & Faverdin P. (1991) Simplified rotational grazing management of dairy cows: Effects of rates of stocking and concentrate. *Journal of Agricultural Science, Cambridge* **116**, 417–28.

Holden L.A., Mulle L.D., Lykos T. & Cassidy T.W. (1995) Effect of corn silage supplementation on intake and milk production in cows grazing grass pasture. *Journal of Dairy Science* **78**, 154–60.

Holmes C.W. (1974) The Massey grass meter. In *Dairy Farming Annual*, 26–30. Ruakura, New Zealand.

Holmes W. (1989) Grazing management. In Holmes W. (ed) *Grass, its Production and Utilization*, 2nd edn, 130–72. Blackwell Scientific Publications, Oxford.

Hongerholt D.D. & Muller L.D. (1998) Supplementation of rumen undegradable protein to the diets of early lactation Holstein cows on grass pasture. *Journal of Dairy Science* **81**, 2204–14.

Hopkins A., Adamson A.H. & Bowling P.J. (1994) Response of permanent and reseeded grassland to fertilizer nitrogen. 2. Effects on concentrations of Ca, Mg, K, Na, S, P, Mn, Zn, Cu, Co and Mo in herbage at a range of sites. *Grass and Forage Science* **49**, 9–20.

Johnson J.M. & Butler G.W. (1957) Iodine content of pasture plants. 1. Method of determination and preliminary investigation of species and strain differences. *Physiologia Plantarum* **10**, 100–11.

Jones R.J. & Sandland R.L. (1974) The relation between animal gain and stocking rate. Derivation of the relation from the results of grazing trials. *Journal of Agricultural Science* **83**, 335–42.

Kellaway R. & Porta S. (1993) Feeding concentrates: Supplements for dairy cows. In Hopkins R (ed) *Dairy Research and Development Corporation*. Victoria, Australia.

Kolver E.S., Muller L.D., Barry M.C. & Penno J.W. (1998) Evaluation and application of the Cornell Net Carbohydrate and Protein System for dairy cows fed diets based on pasture. *Journal of Dairy Science* **81**, 2029–39.

Laidlaw A.S. & Mayne C.S. (1999) Setting management limits for the production and utilization of herbage for out of season grazing. *Grass and Forage Science* **55**, 14–25.

Lambert M.G. & Guerin H. (1989) Competitive and complementary effects with different species of herbivore in their utilization of pastures. In *Proceedings of the XVI International Grassland Congress, Nice*, 1785–9.

Leaver J.D. (1970) A comparison of grazing systems for dairy herd replacements. *Journal of Agricultural Science* **75**, 265–72.

Leaver J.D. (1982) Grass height as an indicator for supplementary feeding of continuously stocked dairy cows. *Grass and Forage Science* **37**, 285–90.

Leaver J.D. & Campling R.C. (1990) Supplementation of grazing dairy cows with a brewers grains/treated straw mixture. *British Society of Animal Production, Winter Meeting 1990*, Paper 46.

Le Du Y.L.P., Baker R.D. & Newberry R.D. (1981) Herbage intake and milk production by grazing dairy cows. 3. The effect of grazing severity under continuous stocking. *Grass and Forage Science* **36**, 307–18.

Marsh R. (1976) Systems of grazing management for beef cattle. In Hodgson J. & Jackson D.K. (eds) *Grazing: Sward Production and Livestock Output*. British Grassland Society Occasional Symposium No. 8, 119–28. BGS, Hurley.

Marsh R., Campling R.C. & Holmes W. (1971) A further study of a rigid grazing management system for dairy cows. *Animal Production* **13**, 441–8.

Maxwell T.J. & Treacher T.T. (1987) Decision rules for grassland management. In Pollot G.E. (ed) *Efficient Sheep Production from Grassland*. British Grassland Society Occasional Symposium No. 21, 67–78. BGS, Hurley.

Mayne C.S. & Cushnahan A. (1995) The effects of ensilage on animal performance from the grass crop. *68th Annual Report, Agricultural Research Institute of Northern Ireland*, 30–41.

Mayne C.S. & Laidlaw A.S. (1995) Extending the grazing season – a research review. In *Extending the Grazing Season. Proceedings British Grassland Society Discussion Meeting, Reaseheath College, Cheshire, 27 April 1995* (unpaginated). BGS, Reading.

Mayne C.S., Newberry R.D., Woodcock S.C.F. & Wilkins R.J. (1987) Effect of grazing severity on grass utilization and milk production of rotationally grazed dairy cows. *Grass and Forage Science* **42**, 59–72.

Mayne C.S., Newberry R.D. & Woodcock S.C.F. (1988) The effects of a flexible grazing management strategy and leader/follower grazing on the milk production of grazing dairy cows and on sward characteristics. *Grass and Forage Science* **43**, 137–50.

McFeely P.C., Browne D. & Carty O. (1975) Effect of grazing interval and stocking rate on milk production and pasture yield. *Irish Journal of Agricultural Research* **14**, 309–19.

McMeekan C.P. & Walshe M.J. (1963) The inter-relationships of grazing method and stocking rate on the efficiency of pasture utilization by dairy cattle. *Journal of Agricultural Science, Cambridge* **61**, 147–66.

McMurray C.H., Rice D.A. & Kennedy S. (1983) Nutritional myopathy in cattle: From a clinical problem to studying selenium, vitamin E and polyunsaturated fatty acid interactions. In Suttle N.F., Gunn R.G., Allen W.M., Linklater K.A. & Wiener G. (eds) *Trace Elements in Animal Production and Veterinary Practice*. British Society of Animal Production Occasional Publication No. 7, 61–73. BSAP, Edinburgh.
Meijs J.A.C. (1986) Concentrate supplementation of grazing dairy cows. 2. Effect of concentrate composition on herbage intake and milk production. *Grass and Forage Science* **41**, 229–35.
Meijs J.A.C. & Hoekstra J.A. (1984) Concentrate supplementation of grazing dairy cows. 1. Effect of concentrate intake and herbage allowance on herbage intakes. *Grass and Forage Science* **61**, 147–66.
Milne J.A., Maxwell T.J. & Souter W. (1981) Effect of supplementary feeding and herbage mass on the intake and performance of grazing ewes in early lactation. *Animal Production* **32**, 185–95.
Minson D.J. (1990) *Forage in Ruminant Nutrition*, 483 pp. Academic Press, London.
Newton J.E. & Davies D.A. (1987) White clover and sheep production. In Pollott G.E. (ed) *Efficient Sheep Production from Grass*, British Grassland Occasional Symposium No. 21, 79–87. BGS, Hurley.
Nolan T. & Connolly J. (1989) Mixed versus mono grazing of steers and sheep. *Animal Production* **48**, 519–33.
Orr R.J., Rutter S.M., Penning P.D., Yarrow N.H., Atkinson L.D. & Champion R.A. (1998) Matching grass supply to grazing patterns for dairy cows under strip-grazing management. *Proceedings of the British Society of Animal Science, 1998*, 49.
Penning P.D., Parsons A.J., Orr R.J. & Treacher T.T. (1991) Intake and behaviour responses by sheep to changes in sward characteristics under continuous grazing. *Grass and Forage Science* **46**, 15–28.
Peyraud J.L., Comeron E.A. & Wade M.H. (1989) Some factors affecting herbage intake of high yielding dairy cows at grazing. *Proceedings XVI International Grassland Congress, Nice, France*, 1151–2.
Peyraud J.L., Comeron E.A., Wade M.H. & Lemaire G. (1996) The effect of daily herbage allowance, herbage mass and animal factors upon herbage intake by grazing dairy cows. *Annales de Zootechnie* **45**, 201–17.
Phillips C.J.C. (1988) The use of conserved forage as a supplement for grazing dairy cows. *Grass and Forage Science* **43**, 215–30.
Phillips C.J.C. & Leaver J.D. (1986) The effect of forage supplementation on the behaviour of grazing dairy cows. *Applied Animal Behaviour Science* **16**, 233–47.
Pulido R. (1997) *Interaction of Pasture Conditions, Concentrate Supplementation and Milk Yield Level in Relation to Dairy Cow Performance and Behaviour.* Ph.D. Thesis, Wye College, University of London.
Reid D. (1986) The effects of frequency of cutting and nitrogen application rates on the yields from perennial ryegrass plus white clover swards. *Journal of Agricultural Science, Cambridge* **107**, 687–96.
Richards I.R. (1977) Influence of soil and sward characteristics on the response to nitrogen. In Gilsenan B. (ed) *Animal Production from Temperate Grassland*, 45–9. Irish Grassland and Animal Production Association, Dublin.
Sarker H.B. & Holmes W. (1974) The influence of supplementary feeding on the herbage intake and grazing behaviour of dry cows. *Journal of the British Grassland Society* **29**, 141–3.
Sayers H.J. & Mayne C.S. (1998) The effect of extending the grazing period in spring on dairy cow performance. *Proceedings of the British Society of Animal Science, Winter Meeting, 1998*, 50.
Stakelum G. (1993) Achieving high performance from dairy cows on grazed pastures. *Irish Grassland and Animal Production Association Journal* **27**, 9–18.
Stakelum G. (1996) Practical grazing management for dairy cows. *Irish Grassland and Animal Production Association Journal* **30**, 33–45.

Swift G., Lowman B.G., Scott N.A., Peebles K., Nelson D.R. & Hunter E.A. (1989) Control of sward surface height and the growth of set-stocked finishing cattle. *Research and Development in Agriculture* **6**, 91–7.

Todd J.R. (1962) Magnesium in forage plants. III. Magnesium distribution in pastures of low magnesium content. *Journal of Agricultural Science, Cambridge* **58**, 277–9.

Tomas F.M. & Potter B.J. (1976) The effect of site of action of potassium upon magnesium absorption in sheep. *Australian Journal of Agricultural Research* **27**, 873–80.

Ulyatt M.J., Beever D.E., Thomson D.J., Evans R.T. & Haines M.J. (1980) Measurement of nutrient supply at pasture. *Proceedings of the Nutrition Society* **39**, 67A.

Van Vuuren A.M., Van der Koclen C.J. & Vroons-de Bruin J. (1993) Ryegrass versus corn starch or beet pulp fibre diet effects on digestion and intestinal amino acids in dairy cows. *Journal of Dairy Science* **76**, 2692–700.

Walshe A. (1982) *The Contribution of Grass to Profitable Milk Production. An On-Farm Study Based on the Results from 34 Herds*. Rex Patterson Trust Memorial Study, 44 pp. British Grassland Society, Reading.

Wilkins R.J. (1990) Grass utilization for animal production. In *Soil – Grassland – Animal Relationships. Proceedings 13th General Meeting of the European Grassland Federation, Banska Bystrica, June 1980, Vol. 1*, 22–34.

Wright I.A. (1988) Suckler beef production. In Frame J. (ed) *Efficient Beef Production from Grassland*. British Grassland Society Occasional Symposium No. 22, 51–64. BGS, Hurley.

Wright I.A. (1992) The response of spring-born suckled calves to the provision of supplementary feeding when grazing two sward heights in autumn. *Animal Production* **54**, 197–202.

Wright I.A. & Whyte T.K. (1989) Effects of sward surface height on the performance of continuously stocked spring-calving beef cows and their calves. *Grass and Forage Science* **44**, 259–66.

Chapter 11
Grassland Management for Natural Landscapes and Wildlife

R.G. Jefferson and H.J. Robertson

11.1 Introduction

Successful conservation management of wildlife in British grassland landscapes depends on one guiding principle: *use traditional agricultural practices or suitable low-intensity substitutes.* The need for low-intensity methods to achieve non-agricultural objectives contrasts with the aims of productivity and resource efficiency, which underlie techniques described elsewhere in this book. Many of the operations required for intensive grass production are often totally inappropriate for nature conservation purposes, although low-intensity and high-intensity grassland management can co-exist on the same farm. The maintenance of low-intensity farming to conserve wildlife has been recognized as critical in the British Isles and mainland Europe (Bignal *et al.* 1994). This chapter outlines the variety, scarcity and conservation importance of native grassland flora and fauna, and the key management practices employed to conserve this biodiversity, and demonstrates how conservation of grassland biodiversity and traditional landscapes can be integrated with sustainable low-intensity agriculture. The final section covers the process of reaching decisions about the most appropriate way to manage particular grasslands, and the role of incentive schemes and designations.

The chapter focuses on, and draws its examples from, the enclosed lowland grasslands (typically below 300 m) of the British Isles, although many of the principles are equally applicable to temperate semi-natural grasslands elsewhere in Europe, including those in upland areas.

11.2 Value of grasslands for wildlife and landscape conservation

11.2.1 Origin and history in relation to present value and the impact of recent changes

Lowland grasslands in the British Isles have been created by forest clearance and subsequent management by grazing and cutting. Forest clearance largely occurred hundreds, and in some areas thousands, of years ago, so many grasslands are ancient in historical terms (Greig 1988). The native plant species favoured by forest clearance were those that had been part of the flora in Late-Glacial times, before forest cover spread across the British Isles and thus restricted open areas to gaps such as eroding river banks, cliffs or woodland clearings (Godwin 1975, Greig 1988). British grasslands composed of native plant species which are characteristic of open ground in turn support native fauna. Such grasslands are termed 'semi-natural', to reflect their native origin and the human management on which their survival depends.

Over most of the historical period, management of semi-natural grasslands has been of low intensity; grazing or cutting for hay were the main activities, and inputs were mainly in the form of farmyard manure and sometimes lime or marl (Pitt 1813, Shaw 1994, Simpson & Jefferson 1996). In places, for instance the Somerset Levels, drainage of fens, swamps and bogs produced semi-natural grasslands composed of native plants that were more tolerant of seasonally dry conditions than the species found in these wetland habitats. Traditional management, superimposed on the British Isles' varied climate, geology, soils and topography, produced a very wide variety of grassland habitat types and maintained the high biodiversity of these habitats. Semi-natural grasslands were also an integral part of characteristic landscapes such as the sweeping chalk downland of Wiltshire or the pattern of small hay-meadows of the Pennine Dales, enclosed within an intricate network of stone walls.

In more recent times, especially since the Second World War, great changes have taken place. The introduction of chemical fertilizers and herbicides, reseeding with high-yielding grass varieties, the change from hay-making to silage, and greater land drainage capability have all contributed to a substantial reduction in the area of semi-natural grasslands by eliminating most of their component plant and animal species (Hopkins & Hopkins 1994). Conversion to arable land has also had a major impact, both directly and indirectly. Neglect of relict fragments of semi-natural grasslands in arable landscapes where livestock farming has ceased has led to scrub invasion and eventual loss of open grassland. Fuller (1987) estimated that in 1984, unimproved grassland occupied only 3% of the area it had covered in 1930 in England and Wales. Recent surveys by English Nature and others have shown that loss is continuing, e.g. a resurvey of Berkshire neutral grassland sites in 1995, previously surveyed in 1984 or 1987, showed that 50% (60% by area) had been damaged or destroyed (Redgrave 1995). Remaining grassland has also become more fragmented: mapping of the decline in the extent of unimproved chalk grassland on Chanctonbury Hill in the South Downs between 1947 and 1991 provides a clear illustration of this trend (Steven 1992). In the Lincolnshire Wolds, of 55 sites surveyed by the Nature Conservancy Council in 1988, none was found to have more than 10 ha of chalk grassland, and over 75% comprised grassland fragments of less than 1 ha (Keymer & Leach 1990). Agriculturally improved or partially improved grassland can sometimes retain some conservation interest, e.g. as winter feeding areas for brent geese or, where allowed to flood, as habitat for wildfowl and waders, but in general the biodiversity associated with lowland grassland has been much reduced.

11.2.2 Plant communities of semi-natural grasslands

Lowland semi-natural grasslands can be divided into broad groups, three of which are based on soil reaction (acidic, calcareous and neutral) and one group (wet grassland) related to moisture status. All of these groups are composed of characteristic assemblages of plants, called plant communities. The British National Vegetation Classification (NVC) (Rodwell 1991, 1992, 1995) provides a more detailed breakdown (see Chapter 1). NVC types of botanical interest are listed in Table 11.1 and include some types in the mires and swamps sections of the NVC as these communities are commonly described and managed as 'grassland'. Table 11.1 also includes estimates for the remaining extent of semi-natural grasslands (Crofts & Jefferson 1999, UK Steering Group 1995). While comprehensive NVC surveys have not been done, in all cases types are very restricted compared with the area of improved 'permanent' grassland (over 5 years old), which covers over 5 million hectares in the UK (MAFF *et al.* 1995).

Calaminarian grassland containing metallophyte species are also of conservation interest, and this is described in Rodwell (1999).

The plant communities that make up the semi-natural grasslands of Britain are often colourful and attractive and form an intrinsic part of grassland landscapes. For example, acid U1, calcareous CG, and inland sand-dune SD10 and SD11 collectively make up the 'grass-heaths' of Breckland in eastern England. Northern hay meadows, MG3, are a feature of the dales landscapes of the Pennines and Lake District, while MG13 'silver meadows', and swamp types are characteristic of the washlands of the Fens.

11.2.3 Ecological relationships and dynamics

The traditional managements associated with each community type are summarized in Table 11.2. The relatively low nutrient conditions that are found in soils of semi-natural grasslands are critical to their conservation, and differ from the high nutrient status of soils of improved grasslands that have been treated with chemical fertilizers or high rates of organic manures. The characteristic

Table 11.1 Estimated areas of lowland semi-natural grassland in the UK.

Group and estimated area in the UK	NVC types
Acid grasslands (<30 000 ha)	U1–U6, *U20, SD10, SD11
Calcareous grasslands (<50 000 ha)	CG1–CG10
Dry neutral grasslands (<15 000 ha)	MG2, 3, 5
Wet grasslands (<65 000 ha)	MG4, 8, MG11–13, M13, M16, M22–28, S5, S22, S28

* Where grassland plants are a significant proportion of the sward.

Table 11.2 'Traditional' management treatments of semi-natural lowland grassland communities.

NVC Grassland Community	1	2	3	4	5	6
MG1 c, d, e					✓	
MG2					✓	
MG3			✓			
MG4		✓				
MG5	✓	✓	✓			
MG8	✓	✓	✓			
MG11	✓	✓				
MG13	✓	✓				
CG1	✓				✓	
CG2	✓					
CG3	✓				✓	
CG4	✓				✓	
CG5	✓			✓	✓	
CG6	✓				✓	
CG7	✓				✓	✓
CG8	✓					
CG9	✓					
CG10	✓					
U1	✓					✓
U2	✓					
U3	✓			✓		
U4	✓			✓		
M22	✓	✓				
M23	✓	✓				
M24	✓	✓		✓		
M25	✓			✓		
M26	✓		✓			
M27					✓	
OV37					✓	

1, livestock grazing.
2, mowing and aftermath grazing.
3, spring grazing, mowing and aftermath grazing.
4, burning and grazing.
5, unmanaged/sporadic management by grazing and/or mowing.
6, sporadic cultivation/ploughing/disturbance.

environmental conditions and management associated with these communities are described more fully in Rodwell (1991, 1992, 1995, 1999).

11.2.4 Plant species diversity and scarcity

Some semi-natural plant communities can be exceptionally species-rich on a unit area basis. The richness of chalk grassland has long been recognized, and totals of 30–40 species per m^2 have been recorded (Wells 1975). It is now clear that equivalent diversity can occur in other types. A recent study of MG5 (Gibson 1997) found almost 40 species per m^2, while NVC samples recorded a maximum of 62 species per $4\,m^2$ in U4 grassland (Rodwell 1992).

Semi-natural grasslands are also of significance for their rare and scarce plants. Crofts and Jefferson (1999) list nearly 90 nationally rare and scarce plants in lowland grassland (rare being defined as recorded in 15 or fewer 10 × 10-km squares, and nationally scarce as recorded in 16–100 10 × 10-km squares in Britain). Two nationally scarce species, perennial flax (*Linum perenne* subsp. *anglicum*) and early gentian (*Gentianella anglica*), are endemic to Britain. Most plant communities have their rare members, e.g. snake's-head fritillary (*Fritillaria meleagris*) in MG4, Jacob's ladder (*Polemonium caeruleum*) in MG2, dark red helleborine (*Epipactis atrorubens*) in CG8 and CG9, and four orchid species in CG2: late spider-orchid (*Ophrys fuciflora*), early spider-orchid (*Ophrys sphegodes*), monkey orchid (*Orchis simia*) and burnt orchid (*Orchis ustulata*). Some grassland categories have more than others, in particular calcareous grassland with around 70 rare and scarce species (Crofts & Jefferson 1999), and acid grassland with over 30 species (Sanderson 1998).

Lower plants (bryophytes and lichens) are not as well documented, but certain grassland habitats have been recognized as important: chalk downland, chalk cliffs and lowland grass-heaths for bryophytes and lichens, and metallophyte communities for lichens (Gilbert 1993, Palmer 1994).

11.2.5 Grassland fauna

Invertebrates

Lowland semi-natural grasslands are of considerable value for invertebrate conservation, although grassland types vary in the richness of the fauna they support and the number of rare and scarce species they contain. Calcareous grasslands generally have richer faunas than acid or neutral grasslands. Kirby (1994) lists 153 'Red Data Book' invertebrate species in grassland and 350 notable species. Table 11.3 gives the conservation status of one group, rare butterflies. Further information can be found in Kirby (1994), Shirt (1987) and Bratton (1991).

Vertebrates

Wet grassland, which is often a composite habitat of grassland, swamp and mire communities, is very important for birds, particularly wildfowl and waders (Fuller 1982). Twenty-four 'Red-listed' and 'Amber-listed' species of conservation concern are wholly or partly dependent on the habitat, including breeding species such as snipe (*Gallinago gallinago*), redshank (*Tringa totanus*) and shoveler (*Anas clypeata*), and wintering species like wigeon (*Anas penelope*), golden plover (*Pluvialis apricana*) and Bewick's swan (*Cygnus bewickii*). Dry grasslands generally have a smaller bird fauna, but this includes 'Red-listed' species

Table 11.3 Rare and scarce butterflies associated with lowland semi-natural grassland.

Scientific name	English name	Red Data Book category (see key)	Red list (R = Red A = Amber) (after Warren *et al.* 1997)	Wildlife & Countryside Act 1981: Schedule 5	Nationally rare (1–15 10 × 10 km squares	Nationally scarce (16–100 10 × 10 km squares)	List of globally threatened/ declining species	Grassland type
Aricia artaxerxes	Northern Brown Argus	—	R	✓ Section 9 (5) only	—	✓	✓ LL	Calcareous
Cupido minimus	Small Blue	—	A	✓ Section 9 (5) only	—	✓	✓ LL	Calcareous
Eurodryas aurinia	Marsh Fritillary	—	R	✓	—	✓	✓ P	Calcareous, fen meadow, rush pasture
Hamearis lucina	Duke of Burgundy	—	—	✓ Section 9 (5) only	—	✓	✓ LL	Scrub margins on calcareous grassland
Hesperia comma	Silver-spotted Skipper	3	R	✓ Section 9 (5) only	—	✓	✓ P	Calcareous
Lysandra bellargus	Adonis Blue	—	A	✓ Section 9 (5) only	—	✓	✓ P	Calcareous
L. coridon	Chalkhill Blue	—	—	✓ Section 9 (5) only	—	✓	✓ LL	Calcareous
*Maculinea arion**	Large Blue	1	R	✓	✓		✓ P	Calcareous
Plebejus argus	Silver-studded Blue	—	A	✓ Section 9 (5) only	—	✓	✓ P	Calcareous, acid
Thymelicus acteon	Lulworth Skipper	—	A	✓ Section 9 (5) only	—	✓	✓ LL	Calcareous, neutral
Boloria euphrosyne	Pearl-bordered Fritillary	—	R	✓ Section 9 (5) only	—	✓	✓ P	Bracken slopes and rough grassland

Key: * Became extinct 1979; Swedish sub-species subsequently re-introduced. P, priority species (formerly Short and Middle list species) (see The UK Steering Group 1995).
RDB categories: 1, endangered; 2, vulnerable; 3, rare; 4, out of danger.

such as stone curlew (*Burhinus oedicnemus*) and woodlark (*Lullula arborea*) (Gibbons *et al.* 1996, Jefferson & Robertson 1996).

Semi-natural grasslands sometimes form parts of the ranges of mammals and reptiles, although usually these animals do not depend wholly on these habitats. An example of a rare mammal which uses a habitat mosaic including grassland is the greater horse-shoe bat (*Rhinolophus ferrumequinum*), which searches for prey over permanent pasture (Ransome 1996).

11.3 Grazing management

Grazing by sheep, cattle and sometimes by horses is essential for maintaining the wildlife value of many grassland communities. Most of the predominantly perennial grasses and dicotyledonous species (herbs) occurring in semi-natural swards are adapted to frequent defoliation, treading and manuring (Duffey *et al.* 1974). In the absence of grazing (or some other means of defoliation such as mowing), the botanical composition of most grasslands will change through the process of succession, which leads to invasion of woody species and the eventual transformation of grassland to woodland (Harper 1977). Maintenance of the nature conservation interest of many semi-natural grasslands requires the removal of a high proportion of the annual herbage production.

11.3.1 Livestock systems

Semi-natural grasslands are less productive than intensively managed grassland (Kirkham & Tallowin 1995, Tallowin 1997). However, the output in terms of dry matter yield and utilizable metabolizable energy will vary according to the type of grassland: neutral grasslands on deeper soils are more productive than calcareous grasslands on thin nutrient-poor limestone soils. Where such grasslands remain as part of farm businesses, extensive sheep and beef enterprises will thus generally be more compatible with their utilization. These will include beef-suckler and store enterprises, and lamb-rearing systems where finishing and periods of higher nutritional requirements (e.g. at lambing time) are met by the use of more productive grassland or supplementary feeding. Semi-natural grasslands can also be grazed by barren ewes or dry cows, and for rearing ewe replacements.

11.3.2 Stock type and breed

Sheep and cattle graze in different ways (see Chapter 9) and they differ in their diet selection (Harper 1977, Crawley 1983, Putnam 1986). The grazing impact of both types of livestock will also vary between breed and age of animal.

However, in general terms, both sheep and cattle will be suitable for managing semi-natural grasslands, although the choice of species will depend on the specific nature conservation objectives set for a site. These are determined by the key grassland habitats, the species present and their ecological requirements, as well as practical considerations such as stock availability, land tenure and farm system, geographical location and the physical nature of the site (topography, hydrology, etc.). For example, beef cattle will be better suited than sheep to grazing wet fen meadows and rush pastures, but sheep may be better on steeper slopes as they are less likely to cause excessive poaching of the sward.

Although little research has been carried out on the subject, anecdotal observation and practical experience suggests that certain breeds of sheep and cattle, particularly primitive and hill/mountain sheep breeds (e.g. Hebridean, Herdwick, Swaledale) and traditional cattle breeds (e.g. Highland, Aberdeen Angus, Galloway, Hereford) may be better suited to grazing semi-natural swards. This is thought to be due to their resilience to harsh climatic conditions, greater resistance to common diseases, ability to select and utilize nutritionally poor, coarse herbage often avoided by other breeds and reduced care requirements. Primitive sheep breeds may also show more propensity for browsing shrubs or tree saplings, the presence of which may be undesirable on semi-natural grasslands (see Section 11.7.4).

In contrast, some Down and Closewool sheep breeds, continental cattle such as Charolais, Simental and Limousin, and most dairy cattle have attributes that make them less suitable, in being less hardy, with higher nutritional requirements and less likely to eat coarse vegetation. The problem in utilizing some of the former livestock types, particularly the primitive sheep breeds, is that they may prove difficult to use in lowland commercial farming enterprises, although they can be used on nature reserves where profit and output are less important.

11.3.3 Intensity and timing of grazing

The impact of grazing intensity on grassland structure is of fundamental importance. Different grassland species and species assemblages will have different structural requirements, which include sward height, density and amount of bare ground and litter (Hopkins 1991, Dolman 1992, Kirby 1992). For example, butterfly species of semi-natural grasslands have different sward height requirements (BUTT 1986), and the rare wart-biter cricket (*Decticus verrucivorus*) needs short turf for larval development and taller tussocks where adults can hide and hunt (Cherrill & Brown 1990). The amount and location of bare ground can also be important. The adonis blue butterfly (*Lysandra bellargus*) requires its food plant, horseshoe vetch (*Hippocrepis comosa*), to be in short turf of less than 3 cm in height, preferably where the plant is growing in tiny, sheltered sunny spots such as hoof-prints or scrapes (BUTT 1986). In contrast, the Duke of Burgundy (*Hamearis lucina*) favours cowslip (*Primula veris*) growing in a sward of 10–20 cm in height (BUTT 1986). Plants are also influenced by structure, e.g. several annual and biennial plants of chalk grassland are dependent on patches of bare ground for germination and establishment (Grubb 1976). The best way of adjusting grazing intensity is to keep a check on grassland structure rather than relying on a rigid stocking rate. The overall average annual stocking densities which are likely to be required for all types of semi-natural grasslands allowing for different objectives range from *c.* 0.2 to 0.8 livestock units ha^{-1} $year^{-1}$.

Short periods of intense grazing, equivalent in rate to longer periods of extensive grazing, may not achieve the appropriate conditions for wildlife and sometimes may be damaging to nature conservation interests, particularly for invertebrates (Kirby 1992).

The sward structure requirements and appropriate grazing intensity for some species may need to be met at particular times of the year, although for plant community maintenance there is probably more flexibility in terms of the timing and periodicity of grazing. Lowland wet grassland pastures supporting breeding wading birds, for example, will require lower grazing intensities during the period April to early July to minimize nest trampling. Green (1986) has described a useful scale to assess the likely impact of different stocking densities on breeding wader success rates (reproduced in Crofts & Jefferson 1999). Winter grazing is generally thought to be better for invertebrate communities (Kirby 1992). Some rare plant species, such as the orchids associated with calcareous grassland, may benefit from no grazing during the flowering period which will allow seed maturation and dispersal.

Timing of grazing may also be important, e.g. in the control of competitive species in restoration of unmanaged sites. An example in limestone grasslands is tor-grass (*Brachypodium pinnatum*), which is palatable to livestock only in spring and early summer. This is also the period when its vigour can be most effectively suppressed (Crofts & Jefferson 1999).

Winter grazing can be desirable for some conservation objectives but may prove difficult to achieve, as this is a period of limited herbage biomass and quality, and an unattractive management option for farmers. It may also increase the need for supplementary feeding, which in itself may compromise nature conservation interests.

11.3.4 Supplementary feeding and use of mineral licks

Supplementary feeding

Outdoor feeding of imported hay, straw and other feedstuffs to livestock on semi-natural grasslands has disadvantages: it may compromise the objective of removal of a substantial amount of the annual herbage production and also lead to changes in the botanical composition and structure of the sward. The latter is caused by nutrient enrichment from the feedstuff, and the creation of bare ground by smothering and the trampling actions of feeding livestock. Thus it is generally undesirable, but it may be necessary in particular circumstances so that the basic nutritional maintenance requirements of livestock are met. If additional feeding with hay is undertaken, then the feeding point should be located in an area of low conservation value.

In contrast, the historic practice of winter feeding of hay on upland valley meadows was probably not a damaging practice from a nature conservation perspective, and may have been a factor in maintaining the botanical composition by dispersal and introduction of seed and the creation of germination niches by the hoof action of livestock. In such circumstances the hay was usually from species-rich semi-natural swards, and was not fed repeatedly in the same area but spread thinly over the ground.

Provision of mineral licks and blocks

The use of mineral and roughage supplementation licks and blocks can be useful on grasslands where there are likely to be mineral deficiencies. Supplements in the form of urea can also aid the digestibility and intake of poor quality herbage by livestock. Localized poaching may occur around these sites, so they should be sited in areas of low conservation value.

11.3.5 The role of horses

Horses played an important part both in farming and transport before the advent of the internal combustion engine, and thus would probably have been common grazing animals on semi-natural grasslands in historical times. In the early twentieth century their numbers declined sharply, but in recent decades numbers have steadily increased again as riding for leisure has become popular. As grazing animals with which to manage semi-natural grassland they differ from other livestock in some important respects. The objectives of horse owners are not the same as those of farmers who want to produce meat and milk. Horses are generally kept to do work, and heavily fertilized grassland is generally not favoured as grazing land because of horse health problems such as laminitis. In the best circumstances, semi-natural grassland provides sustainable forage (pasture and hay) that enables a horse to do work and remain in good condition. Sometimes supplementary feed, given off-site, is required where energy demands are heavy, e.g. for competition horses. Unfortunately, circumstances are often less than ideal. The available grazing area is often too small and over-grazing results, with supplementary feed, given on-site, making up the deficiency in forage.

Over-grazing brings a series of problems. Bare ground increases, especially where supplementary feeding is carried out, and problem weeds such as ragwort (*Senecio jacobaea*), thistles (*Cirsium* spp.) and docks (*Rumex* spp.) readily invade. In heavily grazed fields taller 'latrine areas', composed of a few coarse plant species, form where horses dung but do not graze. The net result of these problems is often termed a 'horse-sick pasture'. A study of the effect of horse grazing on neutral MG5 grassland (Gibson 1996, 1997) showed that heavy grazing also reduces plant species diversity in the sward, resulting in a grassland more akin to agriculturally improved MG6 grassland. However, this study also showed that where grazing was light or moderate (sward height at least 5 cm at any time), horses could maintain MG5 grassland in a satisfactory condition from a conservation point of view. Interestingly, the study also showed that heavy grazing by cattle had equally deleterious effects on plant species richness.

Many horses are kept on neutral grasslands in fields of a few hectares or less in size. However, it is becoming increasingly clear that horses, in particular hardy native pony breeds, can play a useful role in conservation management in more extensive systems, such as downland or unenclosed commons. Ponies are used by several conservation organizations to manage a variety of grasslands, including chalk, limestone, acid and wet grasslands. For instance, Exmoor ponies have been used successfully in restoration grazing to reduce the dominance of coarse *Brachypodium pinnatum* (Oates 1993). Ponies can also help control scrub by browsing, and in the New Forest they will even browse spiny gorse (*Ulex europaeus*) (Putnam 1986).

11.4 Mowing for hay

Management of grassland to produce hay for winter feed for domesticated livestock followed by grazing of the subsequent regrowth is a long-established practice, and archaeological evidence indicates that hay has been made since at least Roman times (Greig 1988). The remaining semi-natural meadows have retained their botanical richness under a traditional meadow management regime which normally involves summer hay making followed by late-summer and autumn grazing of the regrowth (or 'aftermath') coupled with the application of well-rotted farmyard manure and occasional applications of lime. In upland valley meadows, the management regime normally also includes a period of spring grazing, with the meadow being closed for hay in May (Crofts & Jefferson 1999, Younger & Smith 1994).

The main semi-natural grassland communities which are regularly managed as meadows, and where this management may be partly responsible for maintenance of their distinctive floristic composition, are MG3, MG4 and MG5. Other types of semi-natural grassland are occasionally mown for hay, including wet neutral grasslands (MG8, MG11, MG13) and fen meadows and rush pastures (M22, M23, M24, M26, M27) (see Table 11.2), with forage from the latter being described in the past as bog hay (Stapledon & Hanley 1927).

The following sub-sections outline the management techniques which are likely to sustain the botanical richness and nature conservation value of semi-natural hay meadows.

11.4.1 Hay making methods

Modern hay making methods are highly mechanized, and hay is normally cut with tractor-mounted drum, disc or horizontal rotary mowers, and either driven by the PTO or trailed. Reciprocating-blade, cutter and finger bar mowers were once widely used but are now virtually obsolete. However, on ridge and furrow fields, the modern cutters do not perform very well as they tend to cut into the ridges and undercut the furrows. In these situations finger bar cutters are better. Side-mounted cutters are best used at right angles to the ridge crests, whereas back-mounted cutters are better running along the line of the ridges and furrows. Nonetheless, modern haymaking methods in general appear capable of sustaining species-rich swards.

Since the Second World War there has been a reduction in the duration of hay making and earlier finish dates due to increased mechanization (Smith & Jones 1991). This has resulted in some meadows, particularly those at higher altitudes, being cut earlier in most years than previously, with consequences for their botanical composition.

11.4.2 Timing of cutting

Hay cutting dates, nature conservation objectives apart, are principally dictated by the growth stage of the grass crop, the prevailing summer weather conditions and, on farms with a number of meadows, the sequence of cutting.

Effect of time of hay cut on vegetation

Maintenance of the botanical richness of semi-natural meadows is best achieved by mimicking the historical management which was responsible for creating and maintaining them. In most years the hay should be cut at a suitable stage in terms of its biomass and quality, but ensuring that this does not take place before breeding birds have hatched or populations of characteristic annual, biennial or short-lived perennial species which depend on seed production have set seed, e.g. yellow rattle (*Rhinanthus minor*). Sustained early hay cutting reduces species richness in meadows (Smith 1994). The dates will vary according to location and the nature of the wildlife interest, but in practice will range from late June to late July in most years. Where there is reliable historical data on the sequence and timing of cutting of fields, this could be used as a guide (Smith & Jones 1991)

Examination of data on the phenology of species characteristic of neutral meadows from the botanical literature (e.g. Grime *et al.* 1988) suggests that, even with the above traditional cutting times, a number of vascular plant species will still be cut before they have set seed. Most of these species are perennials, and in the absence of long-term life-history data it is unclear what the impact of repeated cutting before seed maturation might be over a long period. Maintenance of populations of these species may only require intermittent seed production, particularly if vegetative regeneration is an important means of reproduction.

However, there is some evidence that selection in response to hay meadow management over a long period has led to populations of plant species in meadows having earlier flowering and seed maturation dates than their counterparts in other habitats (Lack 1982). In addition, some species, e.g. devil's bit scabious (*Succisa pratensis*), may flower and set seed after the hay is cut provided the flowering shoots are not removed during aftermath grazing. In some cases uncut field margins will be a source of seeds. In the absence of data, the strategy adopted by many management prescriptions for semi-natural hay meadows is to ensure an occasional late cut in August or September. Occasional late hay cuts would have occurred in the past in years with cold, wet weather in spring and summer (Smith 1997).

Effect of time of hay cut on breeding birds

On lowland wet meadows which support breeding waders, hay cutting should normally be delayed until after 1 July to allow chicks to hatch. Hay meadows in the upland dales of the Pennines are particularly favoured as nesting sites by the yellow wagtail (*Motacilla flava*) (Askew 1995), making the species potentially vulnerable to changes in the timing of hay making operations. In most years hay cutting during or after the middle of July allows most chicks to fledge and thus minimizes mortality of young birds (Askew 1995).

The corncrake (*Crex crex*), a UK 'Red-list' species (Gibbons *et al.* 1996), is now largely restricted in the British Isles to the Outer Hebrides, the Argyll Islands and western Ireland (Mayes & Stowe 1989, Gibbons *et al.* 1993). It uses hay meadows and other tall vegetation communities for breeding in spring and summer. Changes in agricultural practice, notably the earlier cutting dates associated with a shift from hay to silage, have contributed to its decline (Green & Stowe 1993). Survival of this species is likely to depend on maintaining and increasing traditional agricultural practices, including hay making regimes which involve, where practicable, cutting no earlier than the end of July, and avoiding mowing from the outside inwards which traps adults and chicks in the centre of the field (Andrews & Rebane 1994).

Effect of time of hay cut on invertebrates

Grasslands which have had a long history of hay management are generally not considered to be of great importance for invertebrates. Meadows may support large numbers of invertebrates, but these usually comprise a few common species (Kirby 1992). The species present will be those whose life cycles fit the existing long-established management regime, including species associated with other habitats which use meadows as a source of nectar or prey.

Although the invertebrate fauna of meadows could probably be diversified by a change in management such as leaving uncut areas, this is not generally recommended as it may prejudice the conservation of the botanical community.

11.4.3 Spring and autumn grazing

Meadow management involves a period of grazing which may occur in late summer and autumn (aftermath grazing) and in spring prior to the closure of a field for growing the hay crop. As semi-natural meadows have become increasingly marginal for modern farming systems there has been a decline in their use, particularly for aftermath grazing in the lowlands. Cessation of grazing in spring or late summer results in a decrease in species richness in meadows, and in the long term may result in the development of more species-poor grassland such as MG1 grassland (Smith & Rushton 1994). This loss of richness is thought to be due to a reduction in bare ground created by the treading action of livestock and hence regeneration niches for shorter-lived species. Grazing may also check the growth and abundance of competitive herbaceous dicotyledons and grasses which may increase at the expense of less competitive species. Thus, if the nature conservation value of meadows is to be maintained it is clearly important to maintain grazing.

11.4.4 Changes in management

Any proposed changes to a long-established traditional meadow management regime should be carefully evaluated using available knowledge of species and community ecology. If there is any doubt, the precautionary approach of avoiding changes in long-established management should be adopted where the latter is currently fulfilling nature conservation objectives.

Conversion from meadow to permanent pasture is likely to result in changes in plant species composition (Baker 1937, Rodwell 1992, Gibson 1997). Early flowering species which rely on seed production for maintenance of populations will be reduced or eliminated by such a change. Examples of vulnerable species include *Rhinanthus minor*, which occurs in a variety of neutral meadow types, and *Fritillaria meleagris*, which is largely confined to MG4 flood meadows.

Making silage occasionally at hay-time from fields normally managed in a traditional manner for hay is unlikely to change plant species composition, provided the hay making process is mimicked, e.g. by ensuring that the swath is well tedded prior to baling to allow seeds to be shed. Silage making is normally a quicker operation and may reduce the quantity of seed that is returned to the soil, which in turn may result in population decrease for species reliant on regeneration from seed (Smith *et al.* 1996).

11.5 Use of artificial fertilizers, farmyard manure and lime

11.5.1 Farmyard manure

The use of farmyard manure on hay meadows on neutral soils is a long-established agricultural practice, and it would appear that the botanical richness of semi-natural meadows has been maintained by periodic applications of well-rotted farmyard manure (Simpson & Jefferson 1996). Whilst there has been no experimental work undertaken which can help to provide guidance on rates, observations from existing sites suggest that rates up to 20 tonnes ha^{-1} every 3–5 years are consistent with the maintenance of botanical diversity (Simpson & Jefferson 1996). The slow release of nutrients by organic mineralization, in contrast to the rapid release with other types of fertilizer, are thought to account for differences in the impact on species richness, although the mechanism is unknown. However, excessive use of farmyard manure can have similar adverse effects to those of inorganic fertilizers.

11.5.2 Inorganic fertilizers

Research has shown that applications of inorganic fertilizers (N, P, K; singly or in compound form) to semi-natural grassland reduces the species-richness and diversity of swards, with a consequent loss of nature conservation value (Smith 1994, Tallowin 1996). Fertilizer addition encourages the growth of competitive species with concomitant extinction or reduction in the abundance of slower growing, less competitive species characteristic of less fertile soils. Even very low rates of N fertilizer application to neutral grassland on the Somerset Levels (25 kg ha^{-1} $year^{-1}$) resulted in a decrease in plant species richness (Tallowin 1996). Application normally converts semi-natural types to semi-improved grassland such as MG6 or MG7 grassland. Other materials which provide an immediately available source of N, P or K, such as animal slurry and other organic fertilizers, should not be applied to semi-natural grassland.

11.5.3 Lime

Liming is also a traditional treatment associated with meadows and pastures carried out to offset losses of calcium by leaching and cropping (see Chapter 4) especially on base-poor soils in areas of high rainfall. As part of traditional management, it may have an important influence on species composition of semi-natural grassland (Rodwell 1992). The occasional application of lime to neutral grasslands is thus normally acceptable for nature conservation management. Knowledge of historic practice is limited, but up to 3 tonnes ha^{-1} every 5–10 years is likely to mimic traditional application rates (Crofts & Jefferson 1999).

11.6 Use of herbicides, pesticides and veterinary products

11.6.1 Herbicides

Blanket-spraying of semi-natural grassland with herbicide clearly damages or destroys its botanical interest. However, there are a number of grassland 'weeds' which, when present in quantity, can hamper farming and reduce conservation interest (see Chapter 6). These species include ragworts (*Senecio jacobaea*, *S. aquaticus*), docks (*Rumex crispus*, *R. obtusifolius*), thistles (*Cirsium arvense*, *C. vulgare*), nettle (*Urtica dioica*), rushes (*Juncus* spp.) and bracken (*Pteridium aquilinum*). Control by knapsack spraying or weed-wiping may be necessary, but care must be taken not to damage other species in the sward.

Mechanical control methods such as cutting, pulling, spudding, swiping and targeted grazing are often preferable. It is important to limit weed infestations by following good sward management practice, e.g. not over-grazing or allowing trampling around stock-feeding areas.

11.6.2 Pest control

The rabbit is the key species which is likely to require control in relation to semi-natural swards. High numbers can cause degradation of semi-natural swards through intensive grazing and sward destruction through burrowing and scratching, and may reduce the forage available for livestock. However, in situations where livestock grazing or mowing are not possible and rabbit numbers are not too high, it can help to maintain grasslands of nature conservation importance (Sumption & Flowerdew 1985).

Other species which are viewed as pests from an agricultural perspective include moles, and various invertebrates such as soil-dwelling larvae of beetles and flies, molluscs, aphids and fungi. Use of pesticides against invertebrate pests is best avoided on semi-natural grasslands as the compounds may have deleterious effects on a range of non-target beneficial species of nature conservation significance, or may adversely affect ecosystem functioning.

11.6.3 Veterinary products

A range of chemicals is used to control internal and external parasites of livestock. These are targeted against internal parasitic worms and liver fluke, and ectoparasites such as biting flies, mites and lice. The majority of these compounds are not thought to have any long-term adverse effects on species of nature conservation importance provided they are correctly administered. However, one group of anthelmintic, the Avermectins, leave residues in dung which remain active for varying periods depending on the method of administration of the drug and can adversely affect dung-dwelling invertebrates (Strong 1992). The concerns over residues include the reduction in the number and variety of dung insects, which could reduce the food supply for insectivorous birds and mammals, and the impact on rare and scarce invertebrates, particularly those associated with dung, and on aquatic crustacea which are known to be killed by avermectins.

The use of avermectins should be avoided where possible and alternative anthelmintics used where it is not possible to prevent an overlap between the occurrence of residues in dung and the presence of the wildlife interest.

11.7 Other management practices

11.7.1 Burning

This practice is not commonly employed to manage lowland semi-natural grasslands, except for the purple moor-grass (*Molinia caerulea*) communities (Culm grasslands) of Devon and Cornwall (Wolton 1991) and the limestone grasslands of the Cotswolds. Burning is generally not an effective management tool, particularly if used alone rather than in conjunction with grazing. Indeed, burning can exacerbate problems such as the spread of *Brachypodium pinnatum* on chalk grasslands (Baxter & Farmer 1993).

Burning can relatively easily and cheaply remove dead accumulated plant litter and provide a flush of more palatable herbage for livestock. Thus, it can form the first stage in a programme of restoration management. However, burning when poorly planned and controlled can kill plants, invertebrates and other animals. Very hot fires cause the most damage, and originate when burning is done in areas with a great deal of dead plant material, or when burning is carried out in very dry conditions, or when a burn is made against the wind (back-burning) (Crofts & Jefferson 1999).

Winter burning reduces adverse effects on fauna and flora, and in fact burning is usually legally permitted only in winter and early spring in the lowlands. Various legal requirements govern the activity, for instance a licence is required for any proposed burn outside the permitted period (MAFF 1992).

11.7.2 Chain harrowing and rolling

These activities are usually carried out for agricultural rather than conservation purposes. They are generally acceptable from a conservation point of view, but restrictions may be required in some situations. No rolling or harrowing should be done from late winter to about mid-July where ground-nesting birds or early-flowering plants such as wild daffodil (*Narcissus pseudonarcissus*), snake's-head fritillary (*Fritillaria meleagris*) and green-winged orchid (*Orchis morio*) occur. Sites which have ant-hills should not be rolled or harrowed, and damp ground should be avoided where damage through rutting and compaction would result. Excessive harrowing can also lead to the spread of problem weeds such as *Senecio jacobaea*, *Rumex* spp. and *Cirsium* spp., and so should be carried out with care (Crofts & Jefferson 1999).

11.7.3 Sub-soiling and rotovation

Sub-soiling to improve soil drainage is likely to be deleterious in damp semi-natural grasslands and it is thus advisable not to carry it out. Rotovation or surface disturbance is also generally damaging except in limited circumstances. In the Brecklands, it has been used on a small scale to create open conditions to favour particular bird habitats, e.g. for stone curlew, and rare annual plants (Crofts & Jefferson 1999). These species can also benefit where close rabbit grazing of grassland has been encouraged by rotovation (Dolman & Sutherland 1994).

11.7.4 Scrub management

Without management, semi-natural lowland grasslands disappear through the process of succession, as trees and scrub invade and eventually woodland develops. It is therefore important to control the spread of woody species, although sometimes scrub, or more often the scrub–grassland margin, can have nature conservation significance. For instance, bloody cranesbill (*Geranium sanguineum*), fly orchid (*Ophrys insectifera*) and Jacob's-ladder (*Polemonium caeruleum*) are scarce plant species found on scrub margins. Equally, grassland glades in scrub are often favoured by butterflies seeking shelter and warmth. In these types of situation it is important to maintain a mix of grassland and scrub.

When considering scrub removal it is usually most beneficial to concentrate on areas that have not yet reached the closed canopy stage, as the associated soil enrichment that will have occurred in parallel with scrub development will mitigate against the re-establishment of species-rich grassland (Grubb & Key 1975). A range of scrub removal tools is available, including large machines such as tractor-mounted flails and swipes, sickle-bar mowers and forage harvesters. However, care must be taken when using these machines on slopes and wet ground, and they are not effective for creeping shrubs like privet (*Ligustrum vulgare*) and dogwood (*Cornus sanguinea*). Follow-up treatment of cut stumps with herbicide is recommended. Cuttings should always be removed to avoid smothering grassland and further enriching the soil (Crofts & Jefferson 1999).

Grazing animals can form part of a scrub control regime but, with the possible exception of goats, they are rarely effective alone where the aim is to reduce established scrub cover (Tutton 1994).

11.7.5 Drainage features and water levels

Several semi-natural grassland types are associated with low-intensity drainage systems, which generally comprise ditches, and less usually sub-surface drains (examples of these are found in MG3 meadows in the wet northwest of England). Such systems need careful maintenance to conserve the nature conservation of both the grassland and the ditches. Ditches should not be deepened, but maintained by rotational dredging, ideally from one side only. Spoil should be placed on existing, enriched spoil banks, not on semi-natural grassland. Late summer or early autumn are the best times for ditch management to avoid disturbance to breeding or wintering birds (Newbold *et al.* 1989).

The water-level regime is of great significance for wet grassland and even small changes can have major impacts. For example, monitoring of the Somerset Levels has shown that raising water levels to benefit breeding birds has caused species-rich MG8-related grassland to change towards species-poor swamp and inundation communities (Leach & Cox 1995, Prosser & Wallace 1996). Management objectives will depend on the nature conservation interest of the individual site, but a range of ways of controlling water levels and improving the condition of wet grasslands, particularly for birds, is available and include sluices, bunds, dams and pumps (Treweek *et al.* 1997).

11.8 Grassland management decision-making and management mechanisms

It is important to take a structured approach to making decisions about the management of lowland grassland for nature conservation. It is critical that nature conservation objectives are set for the particular grassland to be managed.

First, adequate baseline ecological information is necessary. This would normally include a map showing the distribution of plant communities and data on the distribution and abundance of priority plant and animal species.

Next, the nature conservation objectives must be defined. These will reflect the particular priorities for conservation, which in turn will be informed by criteria such as the relative rarity of particular grassland communities and associated species. It may not always be possible to manage a site to optimize the extent or numbers of all communities and species present. For example, raising water levels to benefit breeding and wintering birds on wet grassland may adversely affect the maintenance of a particular grassland community, as described in Section 11.7.5.

In order to understand what management will be necessary to sustain the priority conservation features, some knowledge of the ecological requirements of communities and species is necessary. This information would cover soil type and nutrient status, microclimate, hydrology, structure (sward height, amount of bare ground) and life history characteristics of priority species. Knowledge of past and current management may also help in elucidating the factors that are responsible for sustaining particular species and communities.

Subsequently, it should be possible to define the range of management techniques which could maintain the features of interest. Which option is ultimately selected will depend on three factors. The first factor is the technical feasibility of the regime, which will be influenced by the physical characteristics of the site (slope, water supply, location). The second factor is the resources available, and the third factor is the type of farming practised locally, which will determine the availability of appropriate grazing animals, machinery and so on.

It is important to remember that there will be a number of factors which cannot be directly controlled by management, and which may potentially prejudice the maintenance of communities and species on sites. These include climate, external inputs of nutrients and pollutants, and lowered water tables. Being aware of the existence of these and their potential impact is clearly desirable, as many of the problems caused by them can sometimes be overcome by adjusting management and by appropriate advocacy.

Regular monitoring of the condition of grassland sites against the initial baseline or target is essential to ensure that the key nature conservation features are being maintained under the management regime adopted.

In response to the documented losses of semi-natural lowland grassland over the last 50 years (Fuller 1987), a range of protected areas and incentive schemes has been introduced in the UK and elsewhere in Europe to protect, conserve and ensure positive management of habitats and species, including grasslands (IUCN 1992, Hopkins & Hopkins 1994, Baldock & Lowe 1996). These schemes range from statutory nature conservation designations such as Sites of Special Scientific Interest (SSSI) in Great Britain, designated under the 1981 Wildlife and Countryside Act, to voluntary incentive schemes such as the Environmentally Sensitive Areas (ESA) operated by the Agriculture Departments in the UK (similar arrangements to which exist elsewhere in Europe; Baldock & Lowe 1996). In addition, sites have been acquired by both government nature conservation agencies and charitable wildlife organizations such as Wildlife Trusts and the Royal Society for the Protection of Birds, and managed as nature reserves (Evans 1997). However, the vast majority of sites are not managed solely for nature conservation, but form part of farming systems. The same is true of much of the grasslands in high quality designated landscapes such as the

Yorkshire Dales National Park and the West Wiltshire Area of Outstanding Natural Beauty.

The various designations and schemes have slightly different objectives. SSSIs provide protection against damaging activities which would not normally be covered by planning controls, but allow for the conclusion of agreements with landowners which ensure that favourable management for wildlife is maintained. Incentive schemes such as ESA and the Countryside Stewardship Scheme make payments to landowners against specified management prescriptions, and can be aimed at maintaining or enhancing either landscape or wildlife, usually by ensuring the continuation of traditional farming practices (Coates 1997, Swash 1997). The latter schemes also have provisions for increasing public enjoyment of landscape and wildlife through improved access. Since the beginning of the 1990s, British grasslands have received international recognition through the European Union Habitats and Species Directive (CEC 1992) and the UK Biodiversity Action Plan (UK Steering Group 1995). These measures should also lead to improved conservation of semi-natural grasslands. However, whilst all of these measures have undeniably contributed to the conservation of lowland semi-natural grassland, it is widely believed that further integration of conservation objectives into European agricultural policy would produce still greater benefits (Baldock 1994).

11.9 References

Andrews J. & Rebane M. (1994) *Farming and Wildlife: a Practical Handbook for the Management, Restoration and Creation of Wildlife Habitats on Farmland*. Royal Society for the Protection of Birds, Sandy.

Askew D. (1995) The impact of meadow management on nesting birds in the Pennines. In Pollott G. (ed) *Grassland into the 21st Century: Challenges and Opportunities.* British Grassland Society Occasional Symposium No. 29, 163–5. BGS, Reading.

Baker H. (1937) Alluvial meadows: A comparative study of grazed and mown meadows. *Journal of Ecology* **25**, 408–20.

Baldock D. (1994) Possible policy options and their implications for conservation. In Haggar R.J. & Peel S. (eds) *Grassland Management and Nature Conservation*. British Grassland Society Occasional Symposium No. 28, 167–76. BGS, Reading.

Baldock D. & Lowe P. (1996) The development of European agri-environment policy. In Whitby M. (ed) *The European Environment and CAP Reform. Policies and Prospects for Conservation.* CAB International, Wallingford.

Baxter D. & Farmer A. (1993) The control of *Brachypodium pinnatum* in chalk grasslands: Influence of management and nutrients. *English Nature Research Reports No. 100*, Peterborough.

Bignal E.M., McCracken D.I. & Curtis D.J. (1994) *Nature Conservation and Pastoralism in Europe*. Joint Nature Conservation Committee, Peterborough.

Bratton J.H. (ed) (1991) *British Red Data Books. 3. Invertebrates other than Insects*. Joint Nature Conservation Committee, Peterborough.

BUTT (Butterflies Under Threat Team) (1986) *The Management of Chalk Grassland for Butterflies*. Nature Conservancy Council, Peterborough.

CEC (Council of the European Communities) (1992) Council Directive 92/43/EEC of

21 May 1992 on the conservation of natural habitats and of wild fauna and flora. *Official Journal of the European Communities* No. L.206, 7–50.

Cherrill A.J. & Brown V.K. (1990) The habitat requirements of adults of the wart-biter *Decticus verrucivorus* (L.) (Orthoptera: Tettigoniidae) in southern England. *Biological Conservation* **53**, 145–57.

Coates D. (1997) UK policy for the ESAs. In Sheldrick R.D. (ed) *Grassland Management in Environmentally Sensitive Areas.* British Grassland Society Occasional Symposium No. 32, 5–11. BGS, Reading.

Crawley M.J. (1983) *Herbivory. The Dynamics of Animal–Plant Interactions. Studies in Ecology Vol. 10.* Blackwell Scientific Publications, Oxford.

Crofts A. & Jefferson R.G. (1999) *The Lowland Grassland Management Handbook*, Second edition. English Nature/The Wildlife Trusts, Peterborough.

Dolman P. (1992) A review of lowland dry grassland birds in Britain: Their status, ecological requirements and priorities for conservation. *Joint Nature Conservation Committee Reports No. 125.* Joint Nature Conservation Committee, Peterborough.

Dolman P.M. & Sutherland W.J. (1994) The use of soil disturbance in the management of Breckland grass heaths for nature conservation. *Journal of Environmental Management* **41**, 123–40.

Duffey E., Morris M.G., Sheail J., Ward L.K., Wells D.A. & Wells T.C.E. (1974) *Grassland Ecology and Wildlife Management.* Chapman & Hall, London.

Evans D. (1997) *A History of Nature Conservation in Britain*, 2nd edn. Routledge, London.

Fuller R.J. (1982) *Bird Habitats in Britain.* T. & A.D. Poyser, Calton.

Fuller R.M. (1987) The changing extent and conservation interest of lowland grasslands in England and Wales: A review of grassland surveys 1930–1984. *Biological Conservation* **40**, 281–300.

Gibbons D.W., Reid J.B. & Chapman R.A. (1993) *The New Atlas of Breeding Birds in Britain and Ireland: 1988–1991.* British Trust for Ornithology/Scottish Ornithologist's Club/Irish Wildbird Conservancy. T. & A.D. Poyser, London.

Gibbons D.W., Avery M., Baillie S., Gregory R., Kirby J., Porter R., Tucker G. & Williams G. (1996) Bird species of conservation concern in the United Kingdom, Channel Islands and Isle of Man: Revising the Red Data List. *RSPB Conservation Review* **10**, 7–18.

Gibson C.W.D. (1996) The effects of horse grazing on species-rich grasslands. *English Nature Research Reports No. 164*, Peterborough.

Gibson C.W.D. (1997) The effects of horse and cattle grazing on English species-rich grasslands. *English Nature Research Reports No. 210*, Peterborough.

Gilbert O. (1993) The lichens of chalk grassland. *Lichenologist* **25**, 379–414.

Godwin H. (1975) *The History of the British Flora.* 2nd edn. Cambridge University Press, Cambridge.

Green R.E. (1986) The management of lowland wet grassland for breeding waders. *CSD Report No. 626*, Nature Conservancy Council, Peterborough.

Green R.E. & Stowe T.J. (1993) The decline of the corncrake *Crex crex* in Britain and Ireland in relation to habitat change. *Journal of Applied Ecology* **30**, 689–95.

Greig J. (1988) Some evidence for the development of grassland plant communities. In Jones M. (ed) *Archaeology and the Flora of the British Isles.* Botanical Society of the British Isles, Conference Report No. 19, 39–54.

Grime J.P., Hodgson J.G. & Hunt R. (1988) *Comparative Plant Ecology: A Functional Approach to Common British Species.* Unwin Hyman, London.

Grubb P.J. (1976) A theoretical background to the conservation of ecologically distinct groups of annuals and biennials in the chalk grassland ecosystem. *Biological Conservation* **10**, 53–76.

Grubb P.J. & Key B.A. (1975) Clearance of scrub and re-establishment of chalk grassland on the Devil's Dyke. *Nature in Cambridgeshire* **18**, 18–22.

Harper J. (1977) *Population Biology of Plants.* Academic Press, London.

Hopkins A. & Hopkins J.J. (1994) UK grasslands now: Agricultural production and nature conservation. In Haggar R.J. & Peel S. (eds) *Grassland Management and Nature Conservation*. British Grassland Society Occasional Symposium No. 28, 10–19. BGS, Reading.

Hopkins J.J. (1991) Vegetation structure and the conservation of wild plants and animals. In Curtis D.J. Bignall E.M. & Curtis M.A. (eds) *Birds and Pastoral Agriculture in Europe*, 12–17. Scottish Chough Study Group/Joint Nature Conservation Committee, Peterborough.

IUCN (International Union for the Conservation of Nature) (1992) *Protected Areas of the World: A Review of National Systems. Vol. 2. Paleartic*. IUCN, Gland, Switzerland, and Cambridge.

Jefferson R.G. & Robertson H.J. (1996) Lowland grassland: Wildlife value and conservation status. *English Nature Research Reports No. 169*, Peterborough.

Keymer R.J. & Leach S.J. (1990) Calcareous grassland – a limited resource in Britain. In Hillier S.H. Walton D.W.H. & Wells D.A. (eds) *Calcareous Grasslands – Ecology and Management. Proceedings of British Ecological Society/Nature Conservancy Council Symposium, 1987*, 11–17. Bluntisham, Huntingdon.

Kirby P. (1992) *Habitat Management for Invertebrates: A Practical Handbook*. Royal Society for the Protection of Birds, Sandy.

Kirby P. (1994). Habitat fragmentation – species at risk. Invertebrate Group Information. *English Nature Research Reports No. 89*, Peterborough.

Kirkham F.W. & Tallowin J.R.B. (1995) The influence of cutting date and previous fertilizer treatment on the productivity and botanical composition of species-rich hay meadows on the Somerset Levels. *Grass and Forage Science* **50**, 365–77.

Lack A.J. (1982) Competition for pollinators in the ecology of *Centaurea scabiosa* L. and *Centaurea nigra* L. I. Variation in flowering time. *New Phytologist* **91**, 297–308.

Leach S.J. & Cox J.H.S. (1995) *Effects of raised water-levels on grasslands at Southlake Moor SSSI. First Progress Report, 1994*. English Nature, Taunton.

MAFF (Ministry of Agriculture, Fisheries and Food) (1992) *The Heather and Grass Burning Code*. HMSO, London.

MAFF (Ministry of Agriculture, Fisheries and Food), SOAFD, DANI & WO (1995) *The Digest of Agricultural Census Statistics. United Kingdom 1995*. The Stationery Office, London.

Mayes E. & Stowe T.J. (1989) The status and distribution of the corncrake in Ireland in 1988. *Irish Birds* **4**, 1–12.

Newbold C., Honnor J. & Buckley K. (1989) *Nature Conservation and the Management of Drainage Channels*. Nature Conservancy Council, Peterborough (reprinted 1997).

Oates M. (1993) The management of Southern Limestone grasslands. *British Wildlife* **5**, 73–82.

Palmer M. (1994) *A UK Plant Conservation Strategy: A Strategic Framework for the Conservation of the Native Flora of Great Britain and Northern Ireland*. Joint Nature Conservation Committee, Peterborough.

Pitt W. (1813) *General View of the Agriculture of the County of Worcester, With Observations on the Means of its Improvement*. Sherwood Neely and Jones, London. Facsimile reprinted 1969, David and Charles, Newton Abbot.

Prosser M.V. & Wallace H.L. (1996) *National Vegetation Classification Survey of West Sedgemoor*. Royal Society for the Protection of Birds, Sandy.

Putnam R.J. (1986) *Grazing in Temperate Ecosystems: Large Herbivores and the Ecology of the New Forest*. Croom Helm, London.

Ransome R.D. (1996) The management of feeding areas for greater horseshoe bats. *English Nature Research Reports No. 174*, Peterborough.

Redgrave L.J. (1995) *Berkshire Unimproved Neutral Grassland Survey*. English Nature, Newbury.

Rodwell J.S. (ed) (1991) *British Plant Communities. 2. Mires and Heaths*. Cambridge University Press, Cambridge.

Rodwell J.S. (ed) (1992) *British Plant Communities. 3. Grassland and Montane Communities*. Cambridge University Press, Cambridge.

Rodwell J.S. (ed) (1995) *British Plant Communities. 4. Aquatic Communities, Swamps and Tall-Herb Fens*. Cambridge University Press, Cambridge.

Rodwell J.S. (ed) (1999) *British Plant Communities. 5. Maritime Communities and the Vegetation of Open Habitats.* Cambridge University Press, Cambridge.

Sanderson N.A. (1998) A review of the extent, conservation interest and management of lowland acid grassland in England. *English Nature Research Reports No. 259*, Peterborough.

Shaw J. (1994) Manuring and fertilizing the lowlands 1650–1850. In Foster S. & Smout T.C. (eds) *History of Soils and Field Systems*, 111–18. Scottish Cultural Press, Aberdeen.

Shirt D.B. (ed) (1987) *British Red Data Books. 2. Insects*. Nature Conservancy Council, Peterborough.

Simpson N.A. & Jefferson R.G. (1996) Use of farmyard manure on semi-natural (meadow) grassland. *English Nature Research Reports No. 150*, Peterborough.

Smith R.S. (1994) Effects of fertilizers on plant species composition and conservation interest of UK grassland. In Haggar R.J & Peel S. (eds) *Grassland Management and Nature Conservation.* British Grassland Society Occasional Symposium No. 28, 64–73. BGS, Reading.

Smith R.S. (1997) Upland meadow grasslands in the Pennine Dales Environmentally Sensitive Area. In Sheldrick R.D. (ed) *Grassland Management in Environmentally Sensitive Areas.* British Grassland Society Occasional Symposium No. 32, 80–90. BGS, Reading.

Smith R.S. & Jones L. (1991) The phenology of mesotrophic grassland in the Pennine Dales, northern England: Historic hay cutting dates, vegetation variation and plant species phenologies. *Journal of Applied Ecology* **28**, 42–59.

Smith R.S. & Rushton S.P. (1994) The effects of grazing management on the vegetation of mesotrophic (meadow) grassland in northern England. *Journal of Applied Ecology* **31**, 13–24.

Smith R.S., Pullan S. & Sheil R.S. (1996) Seed shed in the making of hay from mesotrophic grassland in a field in northern England: Effects of hay cut date, grazing and fertilizer in a split-split-plot experiment. *Journal of Applied Ecology* **33**, 833–41.

Stapledon R.G. & Hanley J.A. (1927) *Grassland: Its Management and Improvement.* Clarendon Press, Oxford.

Steven G. 1992. *A Botanical Survey of Unimproved Grassland on the South Downs in West Sussex.* English Nature, Lewes.

Strong L. (1992) Ivermectins – a review of their impact on insects of cattle dung. *Bulletin of Entomological Research* **82**, 265–74.

Sumption K.J. & Flowerdew J.R. (1985) The ecological effects of the decline in rabbits *Oryctolagus cuniculus* L. due to myxomatosis. *Mammal Review* **15**, 151–86.

Swash A. (1997) Environmentally Sensitive Areas in the UK and their grassland resource. In Sheldrick R.D. (ed) *Grassland Management in Environmentally Sensitive Areas.* British Grassland Society Occasional Symposium No. 32, 34–43. BGS, Reading.

Tallowin J.R.B. (1996) Effects of inorganic fertilizers on flower-rich hay meadows: A review using a case study on the Somerset Levels, UK. *Grasslands and Forage Abstracts* **66**, 147–52.

Tallowin J.R.B. (1997) The agricultural productivity of lowland semi-natural grassland: A review. *English Nature Research Reports No. 233*, Peterborough.

Treweek J., Jose P. & Benstead P. (eds) (1997) *The Wet Grassland Guide: Managing Floodplain and Coastal Wet Grasslands for Wildlife.* Royal Society for the Protection of Birds, Sandy.

Tutton A. (1994) Goats versus holm oak. *Enact* **2**, 8–9.

UK Steering Group (1995) *Biodiversity: The UK Steering Group Report*. HMSO, London.

Warren M.S., Barnett L.K., Gibbons D.W. & Avery M.I. (1997) Assessing national conservation priorities: An improved red list of British butterflies. *Biological Conservation* **82**, 317–28.

Wells T.C.E. (1975) The floristic composition of chalk grassland in Wiltshire. In Stern L.F. (ed) *Supplement to the Flora of Wiltshire*, 99–125. Natural History Section of Wiltshire Archaeological and Natural History Society, Devizes.

Wolton R.J. (1991) *Wildlife Enhancement Scheme: Management Guidelines for Culm Grassland (Rhos Pasture) in South-West England*. English Nature, Peterborough.

Younger A. & Smith R.S. (1994) Hay meadow management in the Pennine Dales, northern England. In Haggar R.J. & Peel S. (eds) *Grassland Management and Nature Conservation*. British Grassland Society Occasional Symposium No. 28, 137–43. BGS, Reading.

Chapter 12
Amenity Grassland

D. Thorogood

12.1 Introduction

Amenity grassland in the widest sense is defined as 'all grass with recreational, functional or aesthetic value and of which agricultural productivity is not the primary aim' (NERC 1977). Amenity grassland has been classified into (1) intensively managed areas used mainly for sport and recreation, (2) trampled open spaces, both man-made (e.g. parks and lawns) and semi-natural (road verges, country parks etc.), and (3) untrampled open spaces such as airfields and railway and motorway embankments. In the UK these categories collectively account for about 850 000 ha (NERC 1977), equivalent to about 4% of the land area and 12% of total grassland. Semi-natural and untrampled spaces are subject to minimal management and are not considered in any detail in this chapter. There are additional categories of grassland that can be regarded as having an amenity function, but this may in some cases overlap with, or be subsidiary to, agricultural use. These include land for the pasturing of horses kept primarily for sporting or recreational purposes, estimated at 200 000 ha in the UK, as well as agricultural common land and other farmland with public access where agricultural use is sometimes shared with, and possibly affected by, the activities of the public. Grasslands, particularly in downland, coastal heath and mountain areas, are highly valued for their short, springy turf and open spaces, and rambling and informal countryside recreational pursuits have become among the most popular outdoor leisure activities in Britain.

This chapter deals primarily with amenity turfgrass, i.e. grassland that is used for sport and recreation, and which is subject to human trampling and a consistent mowing regime. These grasslands cover some 345 000 ha in the UK (Symes 1987) (Table 12.1).

12.2 The importance of amenity turfgrass

The area occupied by amenity turfgrass is only one indicator of its economic importance; another, because maintenance and creation may often require reseeding, is the sale of grass seed. Perennial ryegrass (*Lolium perenne*) is not only the most extensively used grass in UK agriculture (see Chapter 2), but is

Table 12.1 Areas of amenity turfgrass in the UK (adapted from Symes 1987).

ha	Category
125 000	Intensively managed sports turf
153 000	Domestic lawns
64 000	Urban parks
3 000	Turf farms
345 000	Total

also widely used for amenity sowings. Statistics of estimated annual seed sales (MAFF 1998, and other years) distinguish between amenity varieties of ryegrass and those used for agriculture. During the 20 years after 1980 (when varieties bred specifically for turf use became freely available) the demand for turf-type perennial ryegrass increased ten-fold, whilst that of perennial ryegrass sown for forage remained fairly constant (Fig. 12.1). Seed of red fescue (*Festuca rubra*), which is also used primarily for amenity purposes, shows continued availability over the same period (Fig. 12.2).

Turfgrass is taken for granted by some, and seen by others as requiring financial input, but with little financial gain. The benefits of turfgrass are numerous, although often not easy to quantify because it is a resource that is *maintained*, albeit often to an exacting standard, rather than one resulting in a tangible end product.

12.2.1 Amenity turfgrass for sport

Participation in sport and leisure activities, both as spectator and player, is predicted to increase. It is estimated that in Britain 4.8% and 4.7% of adults play soccer and golf, respectively, at least once a month (Anon. 1996). There are approximately 1.65 million members of football clubs, 1.22 million members of golf clubs, 0.44 million members of bowls clubs and 0.28 million members of rugby union clubs in England alone (Anon. 1993). Active participation in sport provides both physical and mental health benefits (Steptoe & Butler 1996). Many of the most popular sports (soccer, rugby, cricket, golf, tennis, bowls and horse racing) are played on grass, so there is a need for natural grass surfaces that meet the demands of sport and its participants. A grass surface has many attributes, and provides an ideal playing surface with predictable ball bounce and roll. It is also a cushioned comfortable surface which provides an appropriate degree of friction to allow for rapid running, turning and jumping movements to maximize the skill components of sport, whilst providing enough 'give' to reduce the risk of impact and motion-control injuries.

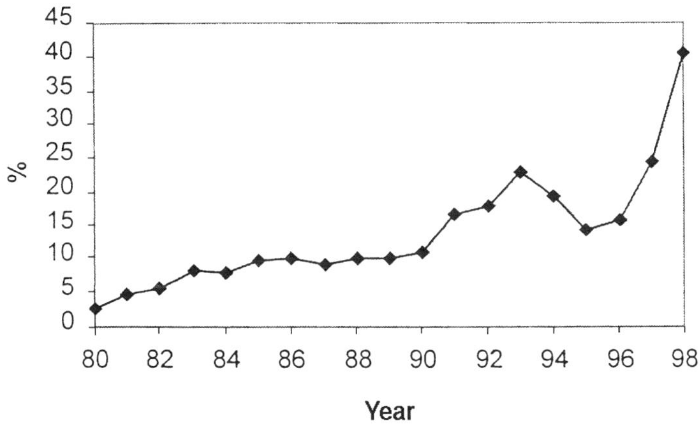

Fig. 12.1 Amenity perennial ryegrass seed delivered for use in the UK as a percentage of total (i.e. including forage) perennial ryegrass, 1980–1996.

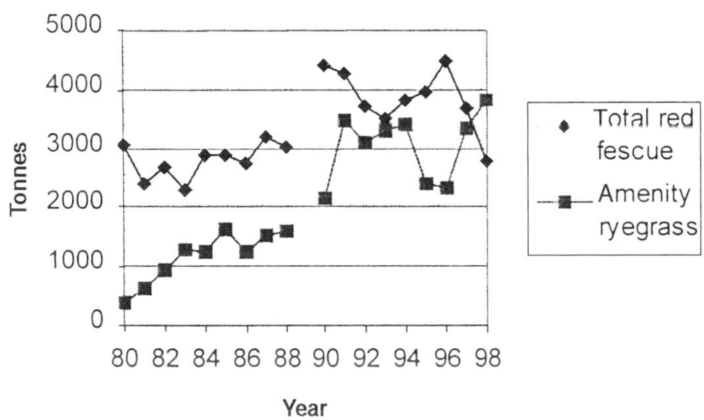

Fig. 12.2 Tonnes of amenity ryegrass and red fescue seed delivered for use in the UK, 1980–1996.

12.2.2 Aesthetic and functional values of amenity turfgrass

Whilst a high proportion of amenity turfgrass is used primarily for sport and other functional uses, the aesthetic value of turfgrass is probably of no less importance. In a survey of a representative sample of United States citizens, peoples' overwhelming preferred landscape was for short grass, and this was particularly so for children (Falk, cited by Bormann *et al.* 1993). Lawns and park grassland act as the framework around which both architectural and botanical features are enhanced. However, they are not only for looking at; they can be walked upon without causing major damage and act as informal seating. Lawns increase residential, commercial and industrial property values and appeal.

They also prevent water and wind erosion of soil, provide sound-proofing, reduce glare, act as dust, pollen and pollutant traps, and act as fire breaks, emergency stopping areas on roadside verges, and as bird deterrents at airports (Beard & Green 1994)

12.3 The evolution of amenity grassland and the demands of modern turfgrass

In areas where grassland is not the climax community, as in the British Isles, it needs to be maintained by frequent management, usually by cutting or with grazing animals. Bormann *et al.* (1993) provide an excellent summary of the historical development of the lawn. The non-agricultural use of grass began in earnest in France in the seventeenth century, when grass beds were shaped for viewing by the owners of large houses. This was developed in England in the eighteenth century, when landscape architects such as William Kent and Lancelot Capability Brown brought the lawn to its prominence. However, lawns were largely for the rich, as their management was labour-intensive and expensive. The village green provided the setting for recreation and sport for the general public, and the turf was kept short by the action of grazing animals. Edwin Budding, the inventor of the cylinder lawnmower in 1830, is responsible for bringing the benefits of turf to the masses as his machine made maintenance of turf for sport and leisure easier and cheaper.

The grass plant is adapted to cutting at extremely low levels (less than 5 mm for some species) because its vegetative meristems are at or near the soil surface. This allows for new shoots to develop even under the most extreme cutting regimes. However, continuous cutting puts the plants under a range of stresses. Continuous removal of vegetative matter removes useful nutrients, which would otherwise be recycled through breakdown by microbial action in the soil. The cut surfaces of leaves increase transpiration rates and make plants more susceptible to infection by disease organisms. Furthermore, these stresses will tend to make the grass plant less competitive, and the grass sward will become thin and ingress of undesirable plant species will occur. So not only is continuous cutting required, but chemical inputs (fertilizers, pesticides and herbicides), water and new grass seed/turf may also be needed to maintain a short, green and even surface.

Modern turf has become an intensively managed ecosystem with a restricted species composition. It is now supported by a multi-million pound chemical and seed industry, which meets the exacting requirements of modern sport and leisure. In the mid-1980s (the most recently available statistics), UK consumer expenditure on domestic lawn care products and services was estimated at £185 million, and total maintenance costs in the commercial sector at £150 million (Symes 1987). Mowing is still the major cost in terms of labour, machinery and fuel. The environmental costs are difficult to quantify, but also need to be considered. The future aim of turfgrass managers and scientists must be an integrated approach to turf maintenance whereby inputs are reduced or used efficiently, without detriment to the qualities sought by the end-user. This can be achieved by improving the nature of the turf surface (the grass itself and the growing medium or root zone) so that the inputs needed to maintain this surface are reduced, and also by developing more efficient methods of using the required resources.

12.4 Species of importance in amenity turfgrass

Intensively managed turfgrass is planted from a limited range of species, although the range of species that occur in amenity turfgrass is quite extensive. In order of seed quantities sold each year, they are perennial ryegrass, red fescues (*Festuca rubra* spp.), bentgrasses (*Agrostis* spp.) and smooth-stalked meadow grass (*Poa pratensis*). Detailed vegetative and reproductive morphological descriptions are given in Hubbard (1984). All of these perennial species have attributes which make them particularly suitable for one or more amenity situations, and plant breeders have actively exploited existing variation and created new variation to produce grasses adapted to their intended use. Varieties produced by breeders in the UK, mainland Europe and USA are trialled independently at the Sports Turf Research Institute (STRI), Bingley, resulting in a ranking list of varieties within each species according to a range of different attributes (Anon 1999).

12.4.1 Perennial ryegrass

Ryegrass is a loosely to densely tufted perennial, which responds to high fertility and is widely distributed and used extensively in agriculture. For amenity use, the main attribute of perennial ryegrass is its high tolerance to heavy wear and trampling, thus making it the major, and often sole, component of turf for football, rugby and tennis, and for general purpose lawns where the prime purpose is not ornamental. Although inter-species differences in wear tolerance have been attributed to anatomical features of individual leaves and stems (Shearman & Beard 1975a,b), inter-varietal differences in perennial ryegrass are largely due to differences in turf dry matter below the cutting height (verdure) before wear, and recuperative potential during and after wear (Ellis 1981). Ryegrass is not recommended for high quality lawns and sports surfaces which receive little wear, such as bowling and golf greens, because of its coarseness, low tolerance of close mowing and fast vertical growth. Modern varieties have improved these attributes considerably.

12.4.2 Red fescues

Three sub-species of red fescue are generally recognized as being extensively used in amenity turfgrass, although others have been classified and described (see Hubbard 1984). All are characterized by their bristle-like leaves. Chewings fescue (*F. rubra* subsp. *commutata*) is a densely tufted perennial. Slender creeping red fescue (*F. rubra* subsp. *litoralis*) is a densely tufted or mat-forming perennial with short, slender rhizomes. Strong creeping red fescue (*F. rubra* subsp. *rubra*) does not produce such a dense turf and has relatively longer scaly rhizomes. Its leaf sheaths are quite hairy or downy. There are many other fine-leaved fescues such as sheeps fescue (*F. ovina*), hard fescue (*F. longifolia*) and fine-leaved fescue (*F. tenuifolia*) which superficially resemble Chewings fescue, but which are not used so extensively for producing amenity turf. They are well adapted to UK conditions, are quite common in natural short turf and are used in low-fertility areas where low maintenance is required.

Red fescue is the major component by weight of seed mixtures for fine, closely mown (5 mm or less) and frequently mown lawns, and where a smooth even texture is a priority (e.g. on golf and bowling greens) and the turf is not subject to heavy wear and trampling. The species can also be used in mixtures which are less intensively managed (e.g. on golf fairways and for landscaping). Whereas ryegrass generally has a high optimum fertilizer requirement, red fescue is much more tolerant of low fertility, and varieties have also been bred which are tolerant of abiotic stresses such as drought, low temperatures, salt and heavy metal contamination.

Chewings and slender creeping red fescues are very much equal in shoot density, and in maintaining a short growth habit under a range of management conditions (although within each sub-species there is a range of variation shown by different varieties), and both tolerate close mowing. Chewings fescues tend to produce a lighter more vibrant green turf than the slender fescues. Strong creeping red fescue is poorer in all respects and is not suitable for turf cut below 13 mm, but its superior seed-yielding ability makes it a low-cost option for turfgrass mixtures.

12.4.3 Bentgrasses

The bentgrasses (*Agrostis* spp.) are rhizomatous or stoloniferous perennials. They have broader leaves than the fine-leaved fescues, which are characterized by their sharply pointed tips. They occur in a wide range of habitats and are often dominant or co-dominant with red fescue in unimproved permanent pasture. Browntop bent (*Agrostis capillaris*) is a commonly used turf grass species, usually in mixtures with red fescue. It spreads by short rhizomes or sometimes stolons. Creeping bent (*A. stolonifera*) is less commonly used, but it has been successfully used in the UK on golf greens, sown as a single species.

Velvet bent (*A. canina* subsp. *canina*) is still less commonly used, but has the potential to produce the most attractive turf when closely and frequently mown, with a texture that its name aptly describes. Highland bent (*A. castellana*) is a commonly used rhizomatous bent component in seed mixtures where, because of its ready availability and cheapness, it may replace a proportion of the browntop. It is inferior to the other bents in most respects, having broader leaves, a faster growth rate and lower shoot density. It does, however, have a darker colour, which it maintains in winter, and its seed-yielding ability is superior.

The bentgrasses are used either on their own or in mixtures with red fescue for fine turf situations and landscaping. When mixed with red fescue, bentgrass seed generally comprises only around 10–15% of the mixture by weight, because of its smaller seed size. Browntop bent has generally been regarded as being the better species for fine turf, and it produces better shoot density, shorter growth and a finer leaf. Creeping and velvet bent are susceptible to thatch (dead organic matter) build-up (see Section 12.6.5), and their extensive stolon production can make for an uneven playing surface. However, new varieties are now being produced which have similar shoot density and short growth characteristics to the better browntop varieties. When managed appropriately, by regular brushing to encourage vertical leaf growth rather than extensive stolon growth, and cultivation to discourage thatch build-up, golf greens based on these species can be maintained to a very high standard.

12.4.4 Other species

Smooth-stalked meadow grass (*Poa pratensis*) is not commonly used in the UK, despite the fact that it has the potential to produce an attractive, dense, dark green sward with similar wear tolerance to ryegrass if given adequate fertility and a neutral pH. It is, however, notoriously difficult to establish and will not tolerate wet conditions. Other species that have a minor role in amenity turf situations include crested dogstail (*Cynosurus cristatus*), rough-stalked meadow grass (*Poa trivialis*) and timothy grasses (*Phleum pratense* and *P. bertolonii*).

12.5 Establishment of turfgrass

12.5.1 Root zone construction methods

In most applications, the seedbed is prepared from the existing soil profile. The area is first cleared and ideally left fallow to allow weed seeds to germinate. Prior to sowing, the area is cleared of weeds with a broad spectrum herbicide, and the surface broken up to a depth of 10–15 cm. Large stones are removed and the area is raked until a reasonably fine tilth is created. This will help to

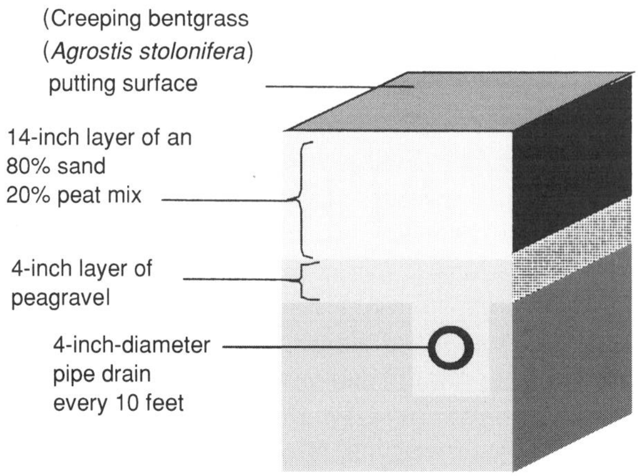

Fig. 12.3 Profile of a typical USGA specification putting green.

produce a quick establishing, even-textured turf. Grass will grow healthily at a soil pH of 5.5–7.5, the optimum being slightly below 7.0. Liming may be necessary if the pH is particularly low, and $1.2\,t\,ha^{-1}$ of lime (calcium carbonate) incorporated into the top 100 mm of the soil will increase pH by about 0.5 depending on the buffering capacity of the soil. Soils with low organic matter or clay content require less lime. The ground is then left to settle to create a firm surface.

For specialist sport applications, where there are exacting demands on the pitch in terms of turf surface quality and durability under play, the natural grass pitch will be produced on a specially constructed root zone with appropriate drainage (see Adams & Gibbs 1994). There are many different constructions (e.g. see the US Golf Association website http://www.usga.org/green/) although they share common features (Fig. 12.3). Typically there is a 100–250-mm layer of mainly sand-based particle size-graded topsoil above a blinding layer of fine material to provide a smooth level surface for the topsoil, with a lower layer of free-draining graded stone enabling water to drain into pipes positioned in trenches at the base. This whole profile may be laid on top of a robust but water-permeable membrane. Drainage may rely on gravity, or may be actively controlled by the use of pumps.

12.5.2 Seeding

Seed is sown at the recommended rate using a proprietary seed spreader, passes being made at right angles to each other to ensure an even distribution. Seed can be sown by hand-broadcasting, in which case marking and sowing smaller squares within the whole area will ensure a more even spread of seed. Soil samples should be taken for analysis, so that any nutrient deficiencies can be identified. In addition to any lime requirement to correct for low pH, inputs of other nutrients may be needed if analyses show concentrations below the following levels (all expressed in mg kg^{-1}): potassium 40, magnesium 40, calcium 10 and phosphorus 20. A compound seedbed fertilizer can then be applied, and a pre-emergence broad-spectrum herbicide such as paraquat may also be applied 2–3 days after sowing, before the grass seeds germinate.

Unless irrigation is readily available, seeding needs to be done in the autumn when moisture levels are adequate and competition with annual weeds, which will die out in the winter, is reduced. In most areas of the British Isles, a September sowing provides enough time for reasonably sized plants to establish before the shorter days of the winter months. Spring sowing is the second best, as soil temperatures will not be as high, irrigation may be required as water deficits may lead to poor establishment, and actively growing weeds will also compete with the grass.

Seed rates by weight will be dependent on seed size, which varies considerably between species (Table 12.2).

These rates seem to be confirmed in smooth-stalked meadow grass/perennial ryegrass mixtures by Brede & Duich (1984), who suggest an optimum seed rate of 196 viable seeds dm^{-2} regardless of mixture proportion to obtain maximum leaf area index (LAI). The rates for perennial ryegrass are nearly ten times higher than agricultural sowing rates (see Chapter 2). This reflects the requirement for rapid establishment of a dense weed-free sward. Rates can be reduced by as much as 75% without detriment to the final sward structure, but it will take longer to establish a turf of the same density as one produced by the higher seeding rate.

There are nearly 100 ryegrass, 70 red fescue, and approximately 20 bentgrass and 25 smooth-stalked meadow grass varieties currently listed by the STRI as available for commercial use (Anon. 1999). The potential number of combinations of varieties and species is therefore huge. In practice, however, as seed is

Table 12.2 Relative seed characteristics and sowing rates of the major cool season turfgrasses (adapted from Beard 1973).

Species	Number of seeds ($\times 10^6$) per kg	Seeding rate (kg ha^{-1})	Number of seeds ($\times 10^3$) per m^2
Perennial ryegrass	0.5	300	15
Chewings fescue	1.2	150	18
Red fescue	1.2	150	18
Browntop bent	19	20	38
Creeping bent	18	20	36
Velvet bent	26	20	52
Smooth-stalked meadow grass	4.8	40	19

sold through various breeders' agents, marketing considerations are paramount, and it may not be possible to choose exactly the varieties required for a particular use. In fact most seed houses prefer to produce their own mixtures, which are formulated for specific uses, and contain a good proportion of some of the best varieties available on the wholesale market. There is little control of mixture formulations of seed for the retail market. Grass seed to be sold for turf purposes does not have to have been bred for turf use, and so the market has long been regarded as an alternative one where surplus agricultural seed can be sold off in expensive packaging. However, some seed houses in the retail market, to their credit, sell carton mixtures containing STRI-listed varieties.

The newly sown turf will not be functional for two or three months, during which time the grass is mown regularly. The first cut of a newly established lawn is best made in spring with a rotary mower, as the grass will be fairly tall. Also, a cylinder mower would tend to pull out newly established grass plants from the seedbed. A cylinder mower can be used for subsequent cuts. Cutting frequency should be increased in step with grass growth rates, which will increase in late spring, and height of cut should be reduced gradually to the final desired height. Ensuring that no more than a third of the vegetation is removed will enable the turf to maintain net photosynthetic gain, not only for further leaf and tiller production, but also to maintain an adequate root system.

12.5.3 Turf laying

Where an instant surface is required, turf laying is preferable to seeding, and turf can be laid at any time except during frost. The quality of turf available varies considerably from ordinary 'meadow' turf harvested from agricultural land where quality and composition cannot be guaranteed, to specialist nursery turf grown from seed of registered varieties of grass in mixtures appropriate for their final function. The latter is usually harvested 12–24 months after sowing and meets guaranteed specification standards for varieties of grass, mowing height, shoot density, proportion of native grasses and broad-leaved weeds, soil depth and type. These standards may be self-tested by the turf growers or independently tested by an outside body such as the STRI. Many turf producers are members of Turf Producers International (http://www.turfgrasssod.org/), through which turf quality standards are set. Soil depth is kept to a minimum (1.5 cm), most of which will consist of grass roots. Turf nursery areas are often located on free-draining sandy loam soils. Turf is often harvested in rolls, which are easy to transport and lay. Turf washed of most soil can be produced if there is a concern for possible incompatibility with soil type or root-zone construction. Root establishment and depth may also be improved by using washed turf. Some turf producers also provide pallet-size slabs of turf with a specialist root

Table 12.3 The relative advantages and disadvantages of establishing from seed or turf.

Seed	Turf
Relatively cheap	Relatively expensive
Large choice of variety and species combinations	Choice restricted by the producer
Ideally sown in autumn (or spring)	Can be established most times of the year
No imported soil	Potential root-zone incompatibility
Establishment period of at least 2–4 months	Immediate establishment/ready for use
Weed seeds in seed bed	Weed seeds if poor quality turf is used

zone up to a depth of 100 mm. These are suitable where regular replacement of hard-wearing areas, such as football goal mouths, is necessary. Table 12.3 summarizes the relative merits of turfing and seeding.

12.5.4 Turfgrass reinforcement

Reinforcement materials have been developed, for which there are claims of increased durability of turf (see Baker 1997). They are mainly plastic-based materials, which are either horizontally placed at or near the turf surface, or are incorporated into the root-zone layer. They may have advantages in terms of increased wearability, surface stability and drainage, reduced divoting and soil compaction, but have disadvantages in terms of practical maintenance and surface hardness. Other systems are being developed which combine the user-friendliness of natural grass with the durability of artificial materials. Grass roots grow down through a synthetic backing to create a mat. The grass stems and leaves grow between fibrillated synthetic strands that, it is claimed, protect the crowns of the grass plants from wear.

12.5.5 Artificial pitches and stadia

Pure artificial surfaces are common for many sports, especially in the USA where huge stadium complexes have been developed able to seat many thousands of spectators for an array of different sporting and entertainment activities. Natural grass is the preferred surface for most sports and provides an excellent surface in open playing situations. However, top-level sport is often played in highly developed, often at least partially enclosed, stadia, where environmental conditions, especially light levels, are sub-optimal, and the performance of an actively photosynthesizing playing surface suffers. This problem has been addressed with expensive schemes with opening roof structures, temporarily removing or raising the pitch away from the depths of the stadium bowl and using supplementary artificial lighting systems (John & Sheard 1994). It may be argued that the solution to the problem lies in a reappraisal of the basic stadium design, which historically has not addressed the problems of growing grass in the stressful micro-climate created.

12.5.6 Cricket pitches

The unique requirements of cricket pitches necessitate a specific surface created by the complex interaction of rootzone and grass roots. A pitch consists of a surface layer, mainly of clay (50–60%) with coarse sand (10%), which overlays a compacted stone/gravel bed. This provides a hard, smooth, flat surface. If the clay content is too high, this may lead to cracking. Perennial ryegrass is used in temperate climates because of its ability to withstand wear and its low thatch build-up, and also for its ability to grow well in heavy clay conditions. The newest varieties are also able to withstand close mowing and scalping, and are more likely to recover from the very low cutting height needed for the playing surface, which generally leaves only the crowns of the grass plants remaining. A strong vigorous rooting ability is essential to help maintain the integrity of the pitch. Detailed descriptions are given by Adams & Gibbs (1994) and Tainton *et al.* (1998).

12.6 Maintenance of turfgrass

12.6.1 General principles

In maintaining an appropriate sward structure that will tolerate frequent cutting, will wear well and will provide a smooth playing surface that is also aesthetically pleasing, consideration should be given to factors which determine sward structure.

The –3/2 self-thinning law (White & Harper 1970) describes the relationship between biomass and density of plants. It becomes the –1/2 power law when describing the relationship between biomass per unit area and plant density (Fig. 12.4), and can be applied to grass tillers in a sward (Matthew *et al.* 1995), including turf (Lush & Rogers 1992). It describes a limiting relationship between tiller density and biomass. The limit is determined by the maximum area of leaf per unit area of ground (the leaf area index or LAI) and the amount of supporting tissue. Only factors that increase these components of the sward can raise the limit. Below the limiting line, any combination of tiller density and biomass can occur. Once the limiting line is reached, increased biomass will be associated with a reduction in tiller density (self-thinning).

High biomass is considered to be important for wear tolerance (Ellis 1981, Shildrick & Peel 1984, Lush 1990). The simplest way of increasing biomass, and thus increasing wear tolerance, would be to increase the cutting height of a sward (Youngner 1961, Lush & Rogers 1992), but there are clearly

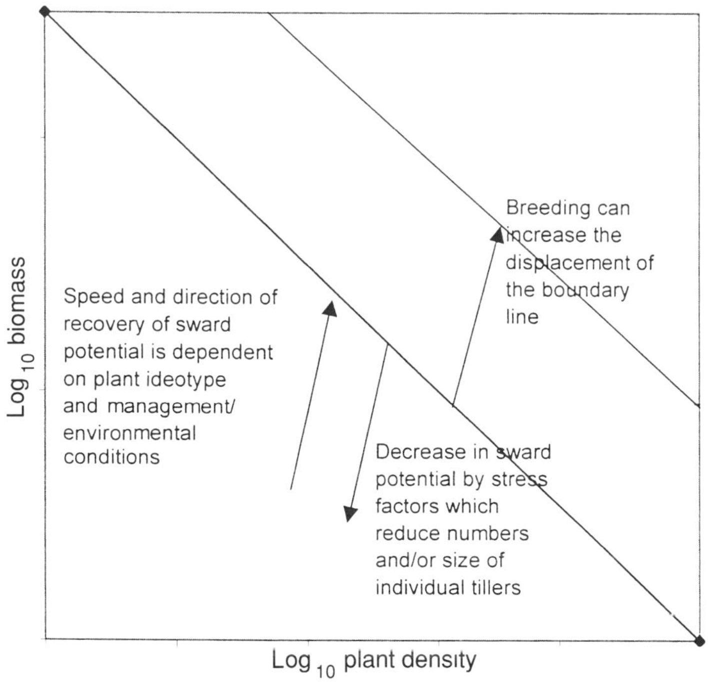

Fig. 12.4 The −1/2 self-thinning law. As biomass increases for a given sward, the density of plants decreases along the boundary line.

functional limitations to sward height in most turf applications. Additionally, an increase in biomass will be associated with a reduction in shoot density, further reducing the sward's usefulness and attractiveness. Plant breeders have made major improvements in sward structure, combining high shoot densities with high biomass and effectively raising the boundary line. This has been done by selecting high-tillering genotypes with small erect leaves with a relatively high LAI, and also by increasing the amount of plant tissue required to support that LAI.

Significant improvements in turfgrass varieties have been made, particularly in perennial ryegrass. Large-tillered ryegrass varieties that were used for forage were previously also used for amenity turfgrass. Late-flowering varieties such as Aberystwyth S.23 were preferred, because they tended to produce fewer unattractive flowering stems in the turf, and less growth in spring and early summer. The US-bred variety 'Manhattan' was the first to be produced specifically for turf use in the mid-1970s. By 1999, the Sports Turf Research Institute in the UK listed about 75 ryegrass varieties (Anon. 1999). Modern varieties sown as single-species mixtures make very acceptable quality lawns when mown as low as 8mm, and are suitable for landscaping and low-maintenance areas, as well as for sports pitches for winter (football, rugby, hockey) and summer (tennis, cricket) use. The cricket pitch, as well as the outfield, often consists of a mixture of turf ryegrass varieties.

The final boundary line (Fig. 12.4) remains constant for a given population and cannot be increased by any management practice apart from that which results in an increase in light intensity from a level that may have previously limited photosynthesis. However, the turf sward is a dynamic population, and the sward will drop below the boundary line through removal of plant material by mowing, wear, drought, pests and diseases, or other stress factors, all of which may result in a combination of reduced biomass and tiller density. Both management and breeding of improved varieties can help firstly to reduce the drop from the boundary line, and then optimize conditions for return. Once the sward has reached the boundary line it is important to maintain an appropriate biomass/tiller density relationship to optimize wear or playing quality/aesthetics, or a compromise between the two. This is simply done by relaxing or increasing the mowing frequency and cutting height.

12.6.2 Mowing

As a general rule, in order to maintain turf health, whatever the desired turf height, no more than a third of the grass height should be removed at any one time, otherwise carbohydrate reserves become limiting, especially for root growth. Cutting height will depend on the species used, the use to which the turf is put and the physiological condition of the turf. The coarser species (perennial ryegrass and smooth-stalked meadow grass) will not withstand the low cutting heights that finer-leaved species will. The lowest cutting heights are used for sports turf, where a true even rolling surface is required. On fescue/bentgrass bowling greens and putting surfaces, this is as low as 5 mm. For this height to be maintained, the surface will need to be cut three times weekly during the active growing season. Often during competition periods, twice-daily mowing would not be unusual. The newer, high-tillering perennial ryegrass varieties can be mown to 8 mm and still maintain very good ground cover and recovery growth. They are used for tennis and cricket pitches. They are able to maintain high biomass at high tiller densities, and therefore remain fairly wear-tolerant at these low cutting heights.

Regrowth after cutting will involve existing tiller growth if small quantities of leaf are removed, but will involve new tiller production if larger quantities are removed and there is a marked reduction in the maximum LAI (Madison 1960, 1962). Speed and method of recovery after cutting will be a function of environment and plant form.

In order to maintain a sward with a constant shoot density and biomass, mowing frequency is important. It is determined by the grass growth rate, so mowing frequency is greatest during the late spring and late summer/early autumn growing periods. Throughout most of the British Isles grass grows almost all the year round except in the coldest winters, so there will be a need to occasionally 'top' the grass to maintain shoot density if soil, weather and climatic conditions allow.

Irrespective of the mowing height and frequency, leaf material is continually being removed along with the nutrients they contain. These should be replaced by returning clippings or by fertilization. When clippings are returned, consideration needs to be given to the effects on the playability of a sports surface, of possible light exclusion from the growing grass, and of the creation of a microclimate conducive to the development of diseases such as *Fusarium* (see Section 12.7.1)

Mowing is by far the most costly maintenance procedure in terms of labour and equipment (Tranter & Tranter 1981), and one of the major challenges to turfgrass maintenance is to reduce mowing costs. A number of growth retardants that block gibberellin biosynthesis, which normally stimulates shoot growth, have been tested on turf. The initiation of reproductive meristems and subsequent stem elongation in the spring, induced by increased day length and higher temperatures after a winter vernalization period of short days and low temperatures, is the main cause of rapid vertical growth in turf. After this initial flush, growth is not so rapid. Some varieties do not have a marked period of rapid vertical growth because later flowering occurs naturally (compare the ryegrass variety 'Gator', which flowers in mid-May, with 'AberElf', which flowers in mid-June; (Fig. 12.5). Reproductive growth is still detrimental to turf

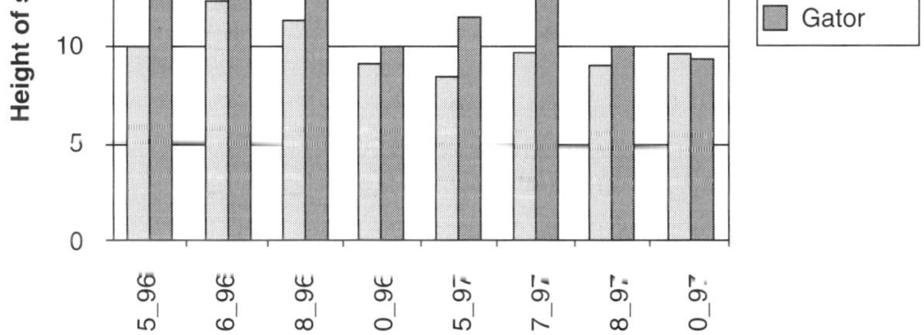

Fig. 12.5 Height of turf perennial ryegrass varieties prior to cutting (Sports Turf Research Institute low-maintenance trial 1995). Plots were cut at 5 cm.

performance in terms of vertical growth and lower tillering capacity, and grass stems affect both visual appearance and surface playing quality; hence continual cutting is required to maintain a short dense sward. The genes that control the transition of the vegetative grass apex to a reproductive one have been identified in *Lolium* and at least part-sequenced (Gocal 1997). By switching off such genes, it is possible to block flowering without altering vegetative growth. Reversal of the process would be necessary to enable seed production of such material, and the technology to switch genes on by external chemical stimulation is now a reality (King *et al.* 1997).

12.6.3 Fertility

Nitrogen (N) is the major nutrient required for maintaining adequate root and shoot growth essential for the production of good quality turf. It is a particularly important nutrient for intensively managed turf because, apart from being readily leached from soils, mowing removes a large proportion of the plant-tissue N normally available through internal plant remobilization to newly growing parts of the turf plant (Bakken *et al.* 1998). High N levels are important for increasing shoot growth rates in swards recovering from damage caused by wear or environmental stress, when both shoot density and overall biomass need to be returned quickly to their optimal level. In ornamental lawns not subject to extreme stresses and damage, N need only be supplied in small but frequently applied quantities, to maintain shoot density and colour. Excessive applications of N encourage shoot growth, so requiring more frequent mowing. Furthermore, if the mowing frequency were not increased under such conditions, shoot competition would lead to self-thinning, thus reducing sward quality.

A minimum of one fertilizer application in spring (usually made before the spring flush of growth) and one in autumn should be made each year. If the turf quality in spring is already good, fertilizer application can be delayed until just after the spring growth flush. Additional small applications can be made in the summer as long as there is enough soil moisture, either from rainfall or irrigation. This will rapidly increase the quality of the turf as long as mowing frequency is adjusted to the increased growth rates. Autumn fertilizer application should be timed so as to avoid new growth when there is a risk of low temperature and frost kill.

Fertilizer-N application will not increase turf quality when plant growth is minimal, such as during periods of drought or prolonged cold in the winter. Under these conditions turf appearance, in terms of colour and shoot density, will deteriorate, and improvements must be sought either by alleviating the stress condition, for example by irrigation (see Section 12.6.4), or by exploiting the genetic variation for stress tolerance that exists in turfgrass species (Aronson *et al.* 1987). This is an area where recent developments in plant breeding, for example the development of drought- and cold-tolerance in grasses

Table 12.4 Main plant nutrients and their requirements for turf growth.

Nitrogen	Vital constituent of amino acids, nucleic acids, vitamins and chlorophyll. Essential for all aspects of plant growth and development
Phosphorous	Establishment, rooting, tillering and reproduction. Important role in cell division and constituent of ATP used for energy transformation
Potassium	Rooting and disease resistance, winter-, drought- and heat-hardiness and wear tolerance due to its importance in carbohydrate synthesis and translocation, and regulation of respiration rate
Magnesium	Essential for maintaining greenness and growth; a constituent of chlorophyll and a co-factor for many enzymes
Calcium	Essential component of cell walls; role in cell division
Sulphur	Component of certain amino acids; if deficient can affect plant growth

(Humphreys *et al.* 1997), and the development of breeding lines containing a gene which stabilizes chlorophyll breakdown and so enhances colour stability (Thomas & Smart 1993, Thorogood 1996), have potential applications for improving turf performance under stress. Fungal endophyte associations (see Chapter 6, Section 6.5) of grass with *Neotyphodium* spp. are also known to be of some benefit during drought (Latch 1994).

Nitrogen fertilizers are described as quick-release (ammonium or potassium nitrate, ammonium sulphate and urea), which need to be added in small but regular amounts to avoid growth flushes, leaf burning and leaching, or as slow-release (ureaformaldehyde, methylene ureaisobutylidine diurea). Slow-release forms are more expensive, but avoid the problems associated with quick-release materials, and the N is in an insoluble organic form and its release is dependent on microbial decomposition. Alternatively, N release can be regulated by an inert coating material (e.g. sulphur-coated urea).

Although the nitrogen requirements for turf will vary depending on the species, or even variety, used and the intensity of management, an average of $50\,kg\,N\,ha^{-1}$ is required per growing month. The other major plant nutrients are not required in such high concentrations as N, but they are just as essential for plant growth. Plant deficiencies for the other major nutrients should not occur as long as soil levels do not fall below those stated in Section 12.5.2. The roles of the major nutrients are summarized in Table 12.4.

12.6.4 Irrigation and drainage

Irrigation of small areas is a simple procedure, but when large areas need irrigating, such as on golf courses and sports pitches, it often involves the use of expensive automatic systems of pipes, pumps and sprinklers (Shiels 1992). Drought rarely kills grass in the British Isles, but may result in temporary browning of the sward. Irrigation is necessary only when the turf is used for sport, where susceptibility of droughted tissues to wear is greatest. The simplest procedure for detecting drought at an early stage is the footprint test (Beard 1973). If the grass is water-deficient it will tend to remain squashed when trodden on, but if fully turgid it will spring back to its original position. Irrigation is most likely to be necessary in low rainfall areas where soil moisture deficits are common, e.g. in southeast England. Climate change models (Conway 1998) predict that reduced summer rainfall is likely in the future, and irrigation-need may therefore become more widespread. Where irrigation is applied, the frequency and timing of applications are important. By watering in the early morning or late evening, water has the opportunity to soak down to the roots rather than evaporate at the surface. Small, frequent quantities of water encourage shallow rooting and greater susceptibility to drought stress. Over-watering may cause additional problems, including increased leaching of N and susceptibility to disease, and may encourage weeds like annual meadow grass (*Poa annua*) which thrive in a moist environment.

Certain cultural practices will relieve the effect of low moisture levels on turf performance. Evapotranspiration can be reduced by raising the cutting height of the grass and by returning grass clippings. There may be opportunities for choosing grass varieties that are adapted to low moisture levels, and the development of drought-tolerant grasses is an area where there is potential for the plant breeders. For example, genes for drought tolerance have been transferred from tall fescue into ryegrass (Humphreys *et al.* 1997), providing useful germplasm for the creation of drought-tolerant turf ryegrass varieties. Drought-tolerance genes have also been combined with the meadow fescue 'stay green' gene in a ryegrass background (M. Humphreys, personal communication, 1999). Tropical turfgrasses such as Bermuda grass (*Cynodon dactylon*) may be considered as drought-tolerant alternative species that can also tolerate UK winter conditions by becoming dormant.

Drainage may also be required to alleviate problems caused by excess water, especially in the winter months and on sports surfaces subjected to soil compaction. On high-grade sports surfaces these problems have been alleviated by the development of sand-based root zones with pipe drainage systems (Adams & Gibbs 1994), although the high rates of water throughput may lead to an irrigation requirement during low rainfall periods. Some systems recycle drainage water to be used for irrigation. Where costs preclude the installation of drains, cultural practices that maintain soil structure and ensure a continuum between the turf surface and the soil (see Section 12.6.5) will help prevent the build-up of surface water.

12.6.5 Cultivation treatments

Resowing an established turf is often not an option, because of the expense and time required for re-establishing an appropriate functional or aesthetically pleasing surface. However, turf quality often deteriorates, even when management practices are ideal, and the effects of traffic, resulting in soil compaction and associated poor grass growth, increase deterioration further. There are several methods of cultivation, such as coring, grooving, slicing, forking and spiking (Beard 1973), which serve to aerate the soil and so allow gas and water exchange between soil and air. Cultivations are usually done in the spring or autumn when the soil is moist, grass growth is active and recovery from the inevitable damage and opening up of the sward is quickest. As cultivations expose some mineral soil, they are often done in conjunction with over-seeding. New seed can then come into contact with the soil to ensure rapid germination and thus improve sward composition, perhaps by introducing improved grass varieties. Topdressing (see Beard 1973) with a sand/soil mixture, which may also include a proportion of peat to increase the organic matter content of the soil, may also be considered. This may improve soil composition, control thatch and even-out small imperfections in the turf surface, particularly where a level playing surface is necessary. Nutrients will also be provided to encourage new shoot growth.

Turfgrass thatch (undecomposed organic matter) can be advantageous, as it provides cushioning, which reduces ball bounce on golf greens, and contributes to wear tolerance on sports pitches by increasing supporting tissue at the maximum LAI (effectively increasing the boundary line, Fig. 12.4). However, thatch deeper than about 1 cm can cause problems, including increased susceptibility to fungal diseases (caused by the favourable moist micro-environment). A deep thatch may also repel water, causing dry spots. It may also cause rooting and rhizome production above the soil surface, and may raise meristems where they become exposed to wear damage and drought stress. Cultivations and topdressing will encourage the breakdown of thatch by aerobic micro-organisms, as will pH increases following lime application. It may be necessary to physically remove thatch by scarification or vertical cutting, which involves pulling wire tines or metal blades through the turf.

12.7 Turfgrass stresses

12.7.1 Diseases, pests and weeds

The subject of diseases, pests and weeds of agricultural grassland is reviewed in Chapter 6. Whilst much of that information is also relevant to amenity grassland, this section considers further those aspects that relate specifically to the turfgrass situation. Turf grasses are attacked by a range of organisms that not only affect persistency and productivity (in terms of shoot density and biomass), but also affect the appearance of the turf. The first method of control is prevention, by the use of appropriate cultural practices and resistant varieties. Chemical controls should be used only when necessary.

Tani & Beard (1997) reviewed the diseases of turfgrass, the most common of which are caused by fungi. York (1998) summarizes the main fungal diseases of turfgrass in the UK. Meyer & Belanger (1997) have reviewed the target diseases for which resistant varieties need to be bred. Of the cool-season grasses, the bentgrasses are the most susceptible to a range of diseases. The turfgrass micro-climate is often ideal for the development of many fungal diseases. Any cultural practices that produce a high quality, actively growing turf, for example high fertilizer levels and optimum temperature, water and light, will tend to encourage obligate parasites. Mowing also provides ideal entry points at the cut ends of leaves for the introduction of plant pathogens. The moist micro-climate associated with surface thatch may encourage the build-up of saprophytic fungi that become pathogenic. Fungicides are available, but appropriate cultural practices can often prevent disease development.

Fusarium patch disease, caused by *Michrodochium nivale*, and *Laetisaria fuciformis*, the causal organism of red thread, were found to be common problems in a survey of winter games pitches in the UK (Raikes *et al.* 1994a, b). *Michrodochium* is a soil-borne pathogen causing a patch disease mainly, but not exclusively, on *Agrostis* spp., and is disseminated by mycelium spread by mechanical means. Occurring mainly in the winter and early spring, it is encouraged by mowing, high nitrogen and moisture levels, and excessive thatch build-up, and can be prevented if cultural practices avoid these conditions. In contrast, red thread disease caused by the fungus *L. fuciformis* is common on low-fertility turfgrass. It is often found in association with pink patch caused by a related organism, *Limonomyces roseipellis* (Baldwin 1987). This disease complex also causes patch symptoms, characterized by red threads or pink flocks of fused mycelium in the turf. Increasing the fertilizer N level prevents the disease by maintaining adequate shoot growth rates.

Seedling diseases (see Chapter 6, Section 6.4.1) can also affect the establishment of turfgrass. Leaf spot diseases such as *Drechslera* and the rusts (*Puccinia* spp.) are also common. They are normally seen on long grass, but rarely present a problem in close-mown turf as the inoculum is continually being removed by cutting. They can cause significant discoloration when they occur, but rarely affect the sward structure. The cereal disease take-all patch, caused by *Gaeumannomyces graminis*, can be found on bentgrass turf. It can cause severe damage, killing large areas of turf in rings up to one metre in diameter. It has become prevalent on sand-based golf greens, where there is a high pH and the organic matter content is low, with subsequently low numbers of antagonistic fungi that would otherwise compete with the take-all organism.

Resistance to fungal diseases is an objective of plant-breeding programmes. However, there is little totally effective resistance to many turfgrass diseases, and it can be broken down easily (Wilkins 1978, Meyer & Belanger 1997). Crown rust (*P. coronata*) resistance has been bred for in perennial ryegrass (Wilkins 1975), but despite high heritability, it is still difficult to achieve durable resistance. New sources of resistance in perennial ryegrass from meadow fescue (*F. pratensis*) are being incorporated into breeding programmes and can be characterized by being temperature-insensitive, whereas resistance in ryegrass at 10°C breaks down at 25°C (Adomako *et al.* 1997). Advances in biotechnology are helping to identify genes in other organisms that give general disease resistance, and turfgrass germplasm is being transformed with these genes (Meyer & Belanger 1997).

Insect pests are not as great a problem in the British Isles as they are in many other countries. Leatherjackets (*Tipula* spp.), which prefer cool moist conditions, are regarded as the most important pests that affect winter games pitches and they are prevalent in a range of turf and other grassland conditions. Chemical control methods are available (see Chapter 6, Section 6.3.3). The fungal endophyte *Neotyphodium* spp. (Chapter 6, Section 6.5) has a role to play in the control of stem-feeding insect pests in turfgrass in the USA (Funk *et al.* 1993). There is a range of approaches to insect resistance using transgenic plants (Gatehouse *et al.* 1993) which could be applied to turfgrasses.

Earthworms have a beneficial role in aerating soil and incorporating organic matter, but are regarded by many turf users as pests, especially on turf surfaces where an even surface is critical. Wormicides are used, although regular switching of the casts should prevent any serious problem.

Dicotyledonous weeds are considered a problem in quality turf, where there is a premium on an even-textured surface for both playing quality and appearance. Although broad-spectrum herbicides are available, cultural control can be achieved by ensuring optimum grass growth, which should compete with weeds intolerant of close mowing. White clover can become a problem in close-mown turf, but is discouraged by high fertilizer-N levels and, although tolerant to close mowing, it is susceptible to damage from wear, as its stolons can easily lose contact with the soil. Grass weeds can be more difficult to control. Annual meadow grass (*Poa annua*) is the most successful unsown grass species in turf. It competes with the sown species by virtue of its ability to flower and seed quickly, even in closely mown turf and, especially under moist conditions, can quickly become the dominant sward component. Although it has wear-tolerance and can form an acceptable turf in its own right, it flowers continuously, causing unacceptable turf quality. It is also susceptible to drought, due to its shallow rooting, and its yellow-green colour is also regarded as unattractive. There are active breeding programmes that aim to remove the unattractive features and poor adaptability to drought and cold conditions of this, undeniably successful, turfgrass species (Huff 1999).

12.7.2 Abiotic stress: wear and environmental effects

Wear

This is an unavoidable stress that amenity grassland has to endure because of its function as a leisure resource. Turfgrass wear is a complex and variable phenomenon describing the removal of grass material by abrasive forces, which vary both in frequency and intensity. They can result in (a) removal or damage to individual leaves or stems, for example, when the forces applied are mainly vertical and horizontal forces are relatively small (such as running and trampling), or (b) removal of complete areas of turf, including roots, where the major force is horizontal (skidding or divoting). Wear-tolerance is dependent on a number of factors. The inherent turf structure is important, and species differences in wear-tolerance have been associated partly with anatomical differences (Shearman & Beard 1975a, b). Perennial ryegrass and smooth-stalked meadow grass are the most wear-tolerant species. However, anatomical differences are unlikely to play a major part when damage is severe enough to remove whole plants and tillers. Wear is dependent on two major factors: the amount of biomass present before wear (Ellis 1981), and recovery from wear. Biomass can be maximized by choice of variety and by optimizing growth by appropriate management, i.e. by ensuring good drainage, fertility, light and temperature in a pest- and disease-free environment. Cutting height can be increased, as long as shoot density is not compromised. Wear-tolerance may be increased with turf age as the amount of supporting plant material (thatch and the less-degraded pseudo-thatch) to live leaf material is increased. However, excessive thatch will have detrimental effects on wear (Section 12.6.5). Recovery will be a function of growth rate and new tiller recruitment, which in turn will be a function of plant type, environmental conditions, management practices, and the nature of the damage caused by wear.

Apart from the effects of wear on the canopy structure, there will be major compaction effects on the soil profile, which in turn will affect turf recovery (Douglas & Crawford 1991, Sun & Liddle 1993, Frame & Merrilees 1996). The degree of compaction will depend on soil type and can be rectified by cultivation practices (Section 12.6.5). Modern sand-based constructions will not suffer from compaction but are prone to instability when large proportions of the turf have been removed by wear. Plastic reinforcement materials have been designed to stabilize sand-based constructions under wear (reviewed by Baker 1997). Although they have been found not to have a great impact on actual wear tolerance (Baker 1997), they do tend to reduce divot size (Adams 1997).

Drought

Severe drought is not a major problem in the UK, and irrigation is only essential on droughted turf that would otherwise suffer unacceptable treading and wear damage (Section 12.6.4). Otherwise, cultural practices and the use of drought-tolerant selections and species can decrease the temporary loss in turf quality.

Low temperatures and frost

Most temperate turf grasses are tolerant of sub-zero temperatures, as long as they have been appropriately hardened by progressively lower temperatures and reasonable levels of light. Sub-zero temperatures often preclude them being used for their intended purpose, and this is particularly so in frosty conditions where frozen tissues fracture when impacted upon. Where feasible, the use of mulches or artificial soil heating can prevent frost damage; this will also help maintain green colour during the winter. Grass plants, when subjected to low temperatures, will respond in much the same way as to drought by becoming dormant. Warm periods during winter will encourage new soft leaf growth, which will be susceptible to damage on return to cold conditions. Most damage will not be permanent, and turf plants will recover when temperatures become favourable as long as the meristematic regions are not affected. There are also other factors involved in cold stress, such as waterlogging, desiccation, ice encasement, and the fungal pathogens that operate at cold temperatures such as *Microdochium nivale*. Several cultural practices can increase cold hardiness, including improvements in soil drainage, ensuring that autumn fertilizer N applications are not excessive, and ensuring that there is not too much thatch which can expose meristems to air temperature extremes.

Photo-oxidative stress

Of relevance to plant performance, including turf, is the effect of low temperature/high light conditions, UV–B radiation and, particularly in urban environments, increasing levels of pollutants such as sulphur dioxide and ozone. These all cause photo-oxidative stress and the release of tissue-damaging free oxygen radicals that can affect plant performance. Introducing antioxidants as foliar applications in turf can reduce this damage and can indirectly increase drought tolerance and disease resistance (Schmidt 1999). Plant-breeding programmes are beginning to address the problem of photo-oxidative stress by exploiting genetic variation in the various anti-oxidant metabolic pathways present in plants (reviewed by Foyer *et al.* 1994).

Competition within and between sown species and mowing practices

It is one of the paradoxes of turf culture that the major stress on plants within a turf is competition from neighbouring plants. The better the conditions for growth, the greater the stress. Once ceiling LAI is reached, further growth will result in self-thinning, increased biomass and reduced shoot density. This is why frequent mowing is essential to maintain the equilibrium between shoot density and size of tillers and to produce an attractive, but wear-tolerant, turf. Mowing which exposes cut leaf surfaces, although essential, induces some stress in terms of increased susceptibility to pathogen infection and increased transpiration rates.

12.8 References

Adams W.A. (1997) The effect of fibremaster fibres on the stability and other properties of sand root zones. *International Turfgrass Society Research Journal* **8**, 15–26.

Adams W.A. & Gibbs R.J. (1994) *Natural Turf for Sport and Amenity: Science and Practice.* CAB International, Wallingford.
Adomako B., Thorogood D. & Clifford B.C. (1997) Plant reaction types to crown rust (*Puccinia coronata* Corda) in meadow fescue (*Festuca pratensis* L.), perennial ryegrass (*Lolium perenne* L. and *Lolium perenne*). 1. Introgression lines. *International Turfgrass Society Research Journal* **8**, 823–31.
Anonymous (1993) *General Household Survey, 1993.* Office of Population Censuses and Surveys, London.
Anonymous (1996) *Living in Britain, 1996.* Office of Population Censuses and Surveys, London.
Anonymous (1999) *Turfgrass Seed, 1999*. Sports Turf Research Institute, Bingley.
Aronson L.J., Gold A.J. & Hull R.J. (1987) Cool-season turf responses to drought stress. *Crop Science* **27**, 1261–6.
Baker S.W. (1997) The reinforcement of turfgrass areas using plastics and other synthetic materials: A review. *International Turfgrass Society Research Journal* **8**, 3–14.
Bakken A.K., Macduff J.H. & Collison M. (1998) Dynamics of nitrogen remobilization in defoliated *Phleum pratense* and *Festuca pratensis* under short and long photoperiods. *Physiologia Plantarum* **103**, 426–36.
Baldwin N.A. (1987) *Turfgrass Diseases.* Sports Turf Research Institute, Bingley.
Beard J.B. (1973) *Turfgrass: Science and Culture*, 658 pp. Prentice Hall, Englewood Cliffs, NJ.
Beard J.B. & Green R.L. (1994) The role of turfgrasses in environmental protection and their benefits to humans. *Journal of Environmental Quality* **23**, 452–60.
Bormann F.H., Balmori D. & Geballe G.T. (1993) *Redesigning the American Lawn: A Search for Environmental Harmony.* Yale University Press, New Haven, CT.
Brede A.D. & Duich J.M. (1984) Establishment characteristics of Kentucky Bluegrass–perennial ryegrass turf mixtures as affected by seeding rate and ratio. *Agronomy Journal* **76**, 875–9.
Conway D. (1998) Recent climate variability and future climate change scenarios for Great Britain. *Progress in Physical Geography* **22**, 350–74.
Douglas J.T. & Crawford C.E. (1991) Wheel-induced soil compaction effects on ryegrass production and nitrogen uptake. *Grass and Forage Science* **46**, 405–16.
Ellis C.J. (1981) *An Experimental Approach to Wear Tolerance in Lolium perenne.* PhD Thesis, University of Liverpool.
Foyer C.H., Lelandais M. & Kunert K.J. (1994) Photo-oxidative stress in plants. *Physiologia Plantarum* **92**, 696–717.
Frame J. & Merrilees D.W. (1996) The effect of tractor wheel passes on herbage production from diploid and tetraploid ryegrass swards. *Grass and Forage Science* **51**, 13–20.
Funk C.R., White R.H. & Breen J.P. (1993) Importance of *Acremonium* endophytes in turf-grass breeding and management. *Agriculture, Ecosystems and Environment* **44**, 215–32.
Gatehouse A.M.R., Shi Y., Powell K.S., Brough C., Hilder V.A., Hamilton W.D.O., Newell C.A., Merryweather A., Boulter D. & Gatehouse J.A. (1993) Approaches to insect resistance using transgenic plants. *Philosophical Transactions of the Royal Soiety Series B* **342**, 279–86.
Gocal G.F.W. (1997) *Molecular Biology of Floral Evocation in Lolium temulentum*. PhD Thesis, Australian National University, Canberra.
Hubbard C.E. (1984) *Grasses*, 3rd edn, 476 pp. Penguin, Harmondsworth.
Huff D.R. (1999). For richer, For *Poa*. Cultivar development of greens-type *Poa annua*. http://www.usga.org/green/index.html Green Section Record, January/February 1999.
Humphreys M., Thomas H.M., Harper J., Morgan G., James A., Ghamari-Zare A. & Thomas H. (1997) Dissecting drought- and cold-tolerance traits in the *Lolium–Festuca* complex by introgression mapping. *New Phytologist* **137**, 55–60.

John G. & Sheard R. (1994) *Stadia: A Design and Development Guide.* Butterworth Architecture, London.

King R.W., Gocal G.F.W. & Heide O.M. (1997) Regulation of leaf growth and flowering of cool season turf grasses. *International Turfgrass Society Research Journal* **8**, 565–76.

Latch G.C.M. (1994) Influence of *Acremonium* endophytes on perennial ryegrass improvement. *New Zealand Journal of Agricultural Research* **37**, 311–18.

Lush W.M. (1990) Evaluation of turf growth and performance based on turf biomass and tiller density. *Agronomy Journal* **83**, 800–803.

Lush W.M. & Rogers M.E. (1992) Cutting height and the biomass and tiller density of *Lolium perenne* amenity turfs. *Journal of Applied Ecology* **29**, 611–18.

Madison J.H. (1960) The mowing of turfgrass. I. The effect of season, interval, and height of mowing on the growth of 'Seaside' bentgrass turf. *Agronomy Journal* **52**, 449–52.

Madison J.H. (1962) Turfgrass ecology. Effects of mowing, irrigation and nitrogen treatments of *Agrostis palustris* Huds., 'Seaside' and *Agrostis tenuis* Sibth., 'Highland' on population yield, rooting and cover. *Agronomy Journal* **54**, 407–12.

MAFF (Ministry of Agriculture, Fisheries and Food) (1998) *Seed Traders' Annual Return Summary, Year Ended 30 June 1998.*

Matthew C., Lemaire G., Sackville-Hamilton N.R. & Hernandez-Garay A. (1995) A modified self-thinning equation to describe size/density relationships for defoliated swards. *Annals of Botany* **76**, 579–87.

Meyer W.A. & Belanger F.C. (1997) The role of conventional breeding and biotechnical approaches to improve disease resistance in cool-season turfgrasses. *International Turfgrass Society Research Journal* **8**, 767–76.

NERC (Natural Environment Research Council) (1977) *Amenity Grasslands – The Needs for Research.* NERC Publications Series C, No. 19. NERC, London.

Raikes C., Lepp N.W. & Canaway P.M. (1994a) Major diseases, pests and weeds of winter sports turf. 1. Results of a questionnaire survey of professional football clubs. *Journal of Sports Turf Research Institute* **70**, 55–75.

Raikes C., Lepp N.W. & Canaway P.M. (1994b) Major diseases, pests and weeds of winter sports turf. 2. Results of a questionnaire survey of local authorities. *Journal of Sports Turf Research Institute* **70**, 83–99.

Schmidt R.E. (1999) Benefits associated with increased antioxidant levels in turf. Abstract from a plenary paper given at the Eighth Annual Rutgers Turfgrass Symposium, p. 21.

Shearman R.C. & Beard J.B. (1975a) Turfgrass wear tolerance mechanisms. II. Effects of cell wall constituents on wear tolerance. *Agronomy Journal* **67**, 211–14.

Shearman R.C. & Beard J.B. (1975b) Turfgrass wear tolerance mechanisms. III. Physiological, morphological and anatomical characteristics associated with turfgrass wear tolerance. *Agronomy Journal* **67**, 215–21.

Shiels G. (1992) Thirst quenchers. *Turf Management*, November, 15–17.

Shildrick J.P. & Peel C.H. (1984) Shoot numbers, biomass and shear strength in smooth-stalked meadow grass (*Poa pratensis*). *Journal of the Sports Turf Research Institute* **60**, 66–72.

Steptoe A. & Butler N. (1996) Sports participation and emotional wellbeing in adolescents. *Lancet* **347**, 1789–92.

Sun D. & Liddle M.J. (1993) Trampling resistance, stem flexibility and leaf strength in nine Australian grasses. *Biological Conservation* **1**, 35–41.

Symes B.M.C. (1987) *A Research Study and Review of Intensively Managed Amenity Turfgrass in the UK.* National Turfgrass Council Special Report No. 2.

Tainton N.M., Klug J.R., Edmondson D., Campbell R.K., van Deventer P.W. & de Beer M.J. (1998) About cricket. Principles and practice of pitch preparation. University of Natal and Potchefstroom University, 33 pp (available through http://www-uk.cricket.org/link_to_database/ABOUT_CRICKET/PITCHES/PREP_OF_PITCHES.html)

Tani T. & Beard J.B. (1997) *Colour Atlas of Turfgrass Diseases*. Ann Arbor Press, Ann Arbor, MI.
Thomas H. & Smart C.M. (1993) Crops that stay green. *Annals of Applied Biology* **123**, 193–219.
Thorogood D. (1996) Varietal colour of *Lolium perenne* L. turfgrass and its interaction with environmental conditions. *Plant Varieties and Seeds* **9**, 15–20.
Tranter H.E. & Tranter R.B. (1981) The demand for amenity grassland. In Jollans J.L. (ed) *Grassland in the British Economy.* University of Reading, CAS Paper 10, 179–207.
White J. & Harper J.L. (1970) Correlated changes in plant size and number in plant populations. *Journal of Ecology* **58**, 467–85.
Wilkins P.W. (1975) Implications of host–pathogen variation for resistance breeding in the grass crop. *Annals of Applied Biology* **81**, 257–62.
Wilkins P.W. (1978) Specialization of crown rust on highly and moderately resistant plants of perennial ryegrass. *Annals of Applied Biology* **88**, 179–84.
York C.A. (1998) *Turfgrass Diseases and Associated Disorders*. Sports Turf Research Institute, Bingley.
Youngner V.B. (1961) Accelerated wear tests on turfgrasses. *Agronomy Journal* **53**, 217–21.

12.9 Further information

The following web sites are additional sources of information.

Turf Producers International: http://www.turfgrasssod.org/

United States Golf Association: http://www.usga.org/green/)

Pennsylvania State University Turfgrass Programme: http://www.cas.psu.edu/docs/casdept/agronomy/sts90.htm

University of Guelph Turfgrass Institute (with turfgrass links): http://www.uoguelph.ca/GTI/

Rutgers University Center for Turfgrass Science: http://www-cook.rutgers.edu/~turf/index.html

Michigan State University Turfgrass Science: http://www.msu.edu/user/turf/page2.html

Turf links from Michigan State University: http://www.msu.edu/user/karcherd/turflnk.html

Information on publications on many aspects of turfgrass management from the Sports Turf Research Institute: http://www.stri.org.uk/

Chapter 13
Control and Utilization of Livestock Manures

B.F. Pain

13.1 Introduction

Faeces and urine from housed livestock are commonly managed as semi-liquid slurries, e.g. from dairy cattle housed with cubicles but with minimal bedding, or solid manures, e.g. farmyard manure (FYM), from beef cattle loose-housed on straw. In this chapter, manure is used as a generic term to describe both slurries and more solid wastes. The quantities involved are huge. In the UK, for example, it is estimated that over 90 million tonnes of manure are produced annually by housed stock, with about 70 million tonnes arising on cattle enterprises. Nearly 80% of this is spread on to grassland together with some pig and poultry manures.

Livestock wastes are complex materials varying in physical, chemical and biological composition with class of stock and management. They all contain water and a wide range of carbon compounds, together with appreciable quantities of plant nutrients and trace elements. Ample supplies of readily biodegradable carbon, especially in slurries, provide excellent media for the multiplication of bacteria. Risks to human and animal health may arise from pathogenic organisms, and frequently there are complaints made from the public about offensive odours.

The perception of livestock wastes has changed over the years. Before the advent of the commercial fertilizer industry, wastes from livestock were used to make a very significant contribution to soil fertility. As mineral fertilizers became widely available and relatively inexpensive, management of wastes became to be viewed as a problem of disposal by the cheapest and most convenient means, a problem exacerbated by the rapid move towards much more intensive livestock production in the latter half of the twentieth century.

Increased awareness and concern about our environment, together with the recognition that livestock wastes represent significant sources of pollution, as well as having a possible role in disease transfer, have steered attitudes away from disposal towards more careful management. The risk of pollution from wastes is two-fold. Firstly, a point-source of watercourses may result from, for example, burst or overflowing slurry stores. These and similar occurrences are generally well documented and have tended to decrease over the years. Secondly, diffuse pollution can occur through leakages, especially of plant nutrients, to both the aqueous environment and the atmosphere. The main concerns in Europe are leaching of nitrates to water, emissions of ammonia (NH_3) and greenhouse gases to the atmosphere. The latter include nitrous oxide (N_2O) and methane (CH_4). Offensive odours cause nuisance to the public, but are more commonly associated with pigs and poultry than grassland-based cattle farming. In addition to European directives, such as the EU Nitrate Directive which aims to control nitrate leaching by a common upper permissible limit of nitrogen (N) loading to soils, there are national or regional recommendations and regulations for controlling pollution within the EU. These may include licensing requirements for housed animals, minimum periods of storage for manure, and prohibited periods and methods for application to land. There are differences between countries depending on the characteristics of the agricultural industry, and the extent and importance of different types of pollution. A common theme is the utilization of plant nutrients in manures for herbage and crop production. By ensuring that, wherever feasible, plant nutrients, especially N and phosphorus (P), excreted by livestock are recycled through the animal–plant–soil system, so that losses to the wider environment are minimized. The higher level of manure management required to achieve this aim can often go some way to lowering the risk of potential problems from pathogenic organisms and odours. Although this chapter focuses on the application of manure to land, it is important to bear in mind that losses of nutrients and pollution, e.g. of N as ammonia, also arise from animal buildings and manure stores.

13.2 Plant nutrients in manures

Large proportions of the major plant nutrients ingested by livestock in their diet are subsequently excreted. Manures therefore contain appreciable amounts of N, P and potassium (K), and so have value as fertilizers. Typical values are given in Table 13.1. Only a proportion of the total amounts are in forms available for uptake by plants. Whereas about 90% of K is 'plant available', the availability of N and P in slurries normally ranges between 30% and 60%. For N, this proportion is mostly as ammonium (NH_4^+), or uric acid in the case of poultry manure, which is readily lost from the soil, so the amount available for plant uptake is greatly influenced by the time of application. Solid manures, such as FYM, often contain very little plant-available N.

13.2.1 Nitrogen

Nitrogen is present in manures in both inorganic and organic forms, the former being readily available for plant uptake, whilst the latter must first undergo mineralization in the soil. The transformation and loss processes involving N are mediated by a wide range of biological, chemical, environmental and

Table 13.1 Typical total plant nutrient content of some manures.

	Dry matter (%)	N	P (kg/t)	K
Slurry				
Dairy cattle	6	3.0 (30)	0.5 (50)	3.0 (90)
Beef cattle	6	2.3 (30)	0.5 (50)	2.3 (90)
Pig	6	5.0 (30)	1.3 (50)	2.5 (90)
FYM				
Cattle	25	6.0 (20)	1.5 (60)	6.7 (60)
Pig	25	7.0 (20)	3.0 (60)	4.2 (60)

Numbers in parentheses indicate the additional total percentage available in the season of application (spring-applied only for N).

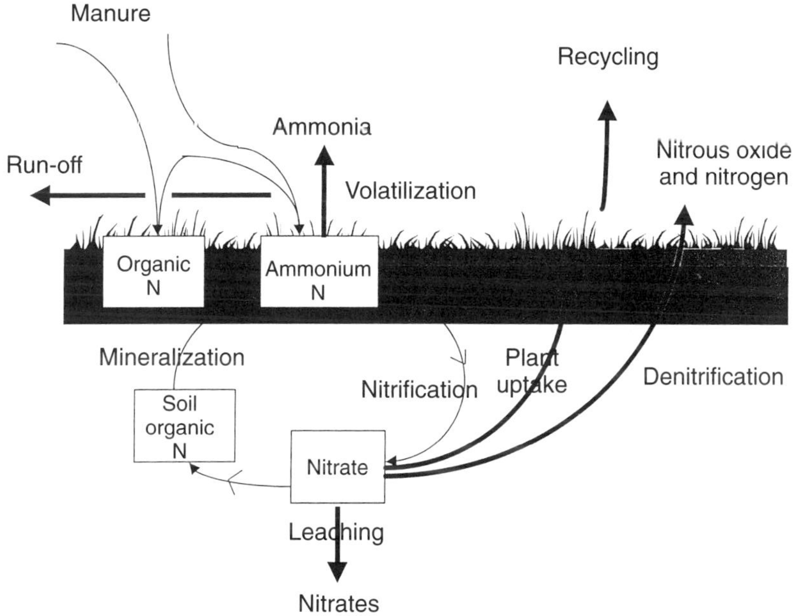

Fig. 13.1 Transformations, losses and plant uptake of N following the application of manure to land.

management factors, which make for a complex situation. The major processes are illustrated in simplified form in Fig. 13.1. A considerable amount of research has been directed towards quantifying N losses, identifying the factors controlling the processes, and developing strategies for limiting loss so that a high proportion of applied N is recovered by a crop. Such strategies not only enable fuller exploitation of the fertilizer value of manures, but also minimize the risk of pollution.

13.2.2 Ammonia volatilization

In most circumstances, NH_3 volatilization is the major pathway of N loss from manures. The process is relatively simple in chemical terms, and involves the hydrolysis of excreted urea to ammonium-N (NH_4^+-N), mediated by the enzyme urease, and the subsequent release of NH_3. It is common for over 50% of the NH_4^+-N, or plant-available N, to be lost as NH_3 following the spreading of cattle slurry on grassland. Much of this loss occurs within the first few hours after spreading (Pain *et al.* 1989), as illustrated in Fig. 13.2, the actual extent and rate of loss being influenced by many factors. Meteorological conditions, including wind speed, temperature and rainfall, are each important, together with soil type and condition, the presence of a crop and manure composition. In particular, there is a strong relationship between slurry dry matter content and NH_3 loss (Fig. 13.3). Losses are lower from more dilute slurries because infiltration into the soil is more rapid, assuming the soil is free-draining and not compacted or waterlogged, so that NH_4^+-N becomes bound and NH_3 volatilization ceases. Recent research has indicated that although losses from spreading FYM to land are lower than for slurry, losses during composting can be high.

Livestock production is the major source of ammonia emissions to the atmosphere. It is estimated that 74% of the 4016kt of NH_3-N emitted in western Europe comes from farm animals and their wastes (ECETOC 1994). A recent inventory for the UK (Pain *et al.* 1998) estimated an emission of 200kt NH_3-N per year from agriculture, which probably represents about 90% of the total national emissions (Sutton *et al.* 1995). The relative contributions to ammonia emissions are given in Table 13.2, and these emphasize the importance of cattle farming and the spreading of manures on to the land.

Although the emission of NH_3 is well recognized as a significant loss of N from livestock production systems, thereby decreasing the efficiency of N utilization, interest since the mid-1980s has been driven primarily by environmental concerns. Ammonia is the most important alkaline gas in the atmosphere, and as such is readily dissolved into water droplets, where it is transformed into the ionic form (NH_4^+) and reacts with acidic compounds to form ammonium sulphate or nitrate. These compounds can become dispersed in solution or in particulate forms and transported over long distances prior to deposition, so the problem is international in scale. Deposition to agricultural land is of little consequence and, indeed, may be a significant input of N in some areas, but deposition to natural or semi-natural ecosystems can have severe effects (Roelofs & Houdijk 1990): effects on the diversity of plant species can follow from enrichment of heathland with N, for example. Alternatively, the deposition and subsequent nitrification of ammonium salts

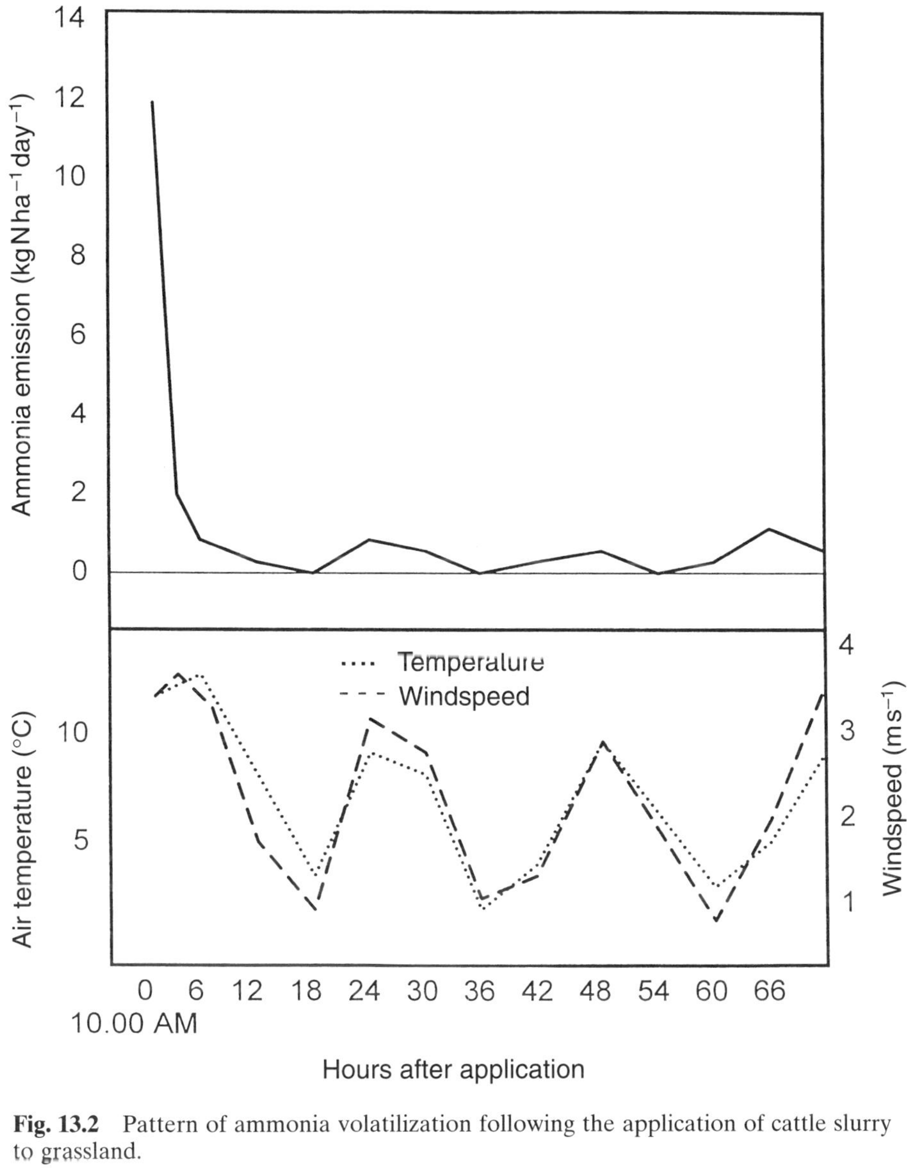

Fig. 13.2 Pattern of ammonia volatilization following the application of cattle slurry to grassland.

on poorly buffered soils can increase soil acidity, increasing the availability of toxic ions such as aluminium, but decreasing that of essential ions such as calcium (Ca) and magnesium (Mg). Adjacent to strong point-sources, such as animal housing, there may be direct toxic effects on plants. Although slurry spreading on grassland is known to be a strong source of ammonia, on-going research suggests that, under certain weather and sward conditions,

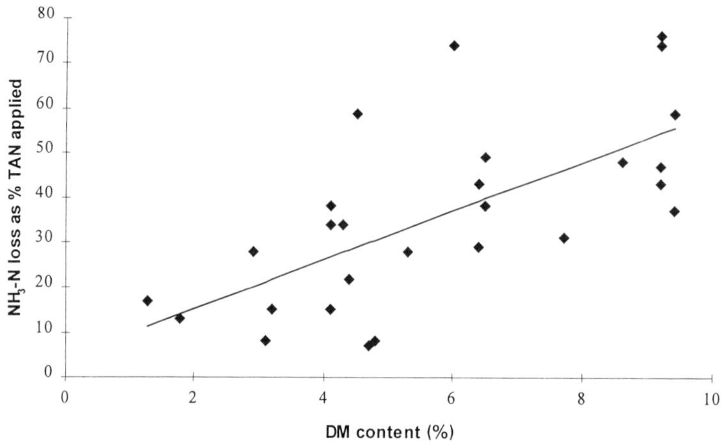

Fig. 13.3 Relationship between slurry dry matter (DM) content and ammonia loss following application to land.

Table 13.2 Contributions to ammonia emission from UK agriculture in 1996.

	% total emission
Livestock and N fertilizer	
Cattle	55
Poultry	19
Pigs	11
Sheep	6
Fertilizer	9
Management	
Manure spread on land	30
Animal housing	40
Animals outdoors	12
Fertilizers	9
Manure storage	9

NB. Total annual emission for UK agriculture = 230 kilotonnes NH_3-N.

a high proportion may be redeposited within 50 m or so (R. Hill, personal communication, 1998) on to adjacent grassland, so lowering the total emissions from the farm.

13.2.3 Nitrification and denitrification

Nitrification is the microbial transformation of NH_4^+-N to nitrite and then to nitrate (NO_3^-), denitrification is the microbial reduction of NO_3^--N eventually to nitrous oxide and dinitrogen gas (N_2). The former process requires oxygen and, through intermediate processes, can also yield N_2O. Denitrification is performed by a diverse and widely distributed group of bacteria in the soil under anaerobic conditions. In addition to N_2O, other oxides of N, often denoted as NO_x (nitric oxide and nitrogen dioxide (NO_2)) may also be released as intermediary products. The process is dependent upon, amongst other factors, a supply of freely available carbon (C). Hence, adding organic manures to soils can enhance denitrification rates. There is evidence for coupling of nitrification and denitrification in that both processes can occur simultaneously in adjacent soil pores of different aerobicity. The release of N_2O and N_2 not only represents an uncontrolled loss of N from agricultural soils, but also raises environmental concerns. Although N_2 is environmentally benign, it has been estimated that N_2O may account for 5–10% of the greenhouse effect, and it is known that atmospheric concentrations are continuing to increase, mainly due to the increased use of N fertilizers and to combustion processes.

When slurry is spread on the surface of grassland, between <0.1% and nearly 30% of the NH_4^+-N applied can be lost through denitrification. Higher losses occur in autumn than in spring. At the latter time, uptake by the actively growing crop provides a stronger sink for NO_3-N than denitrification. Nevertheless, denitrification losses are generally higher per kg N applied as slurry than as inorganic fertilizer, and increase with increasing application rate. There is also evidence that, through the provision of a C source, slurry application may increase denitrification of N applied in preceding or following applications of inorganic fertilizer N. Such interactions may have an insignificant effect on the response of herbage to applied N, but may be important to total emissions of N_2O.

There are few studies on the impact of manure application on emissions of N_2O from grassland soils, but Chadwick (1997) showed these to range from 0.03% to 0.50% of the NH_4^+-N applied, and to vary with time, rate and method of application.

13.2.4 Nitrate leaching

Interest in nitrate leaching in Europe has been driven by the 1980 EC Directive on the Quality of Water Intended for Human Consumption, which sets a maximum allowable concentration of 50 mg l^{-1} nitrate (11.3 mg l^{-1} NO_3^--N) in drinking water. The EC Nitrate Directive 1991 required member states to monitor water, set up nitrate vulnerable zones, and produce a Code of Good Agricultural Practice. Nitrates are also implicated in eutrophication, or nutrient enrichment of surface waters, which is associated with algal bloom and disturbance to ecological balances.

Although it is rare for nitrate to be present in manures, it is rapidly released into the soil from nitrification of NH_4^+-N under most circumstances. Spreading manures on to land can increase the risk of nitrate leaching on two accounts, both of which are associated with the lack of synchrony between release of nitrates into the soil and uptake by plants. Leaching can occur during the winter from application of manures in the autumn following mineralization of readily degradable organic matter and nitrification of NH_4^+-N. Prolonged mineralization of residual organic matter can lead to leaching in subsequent winters, irrespective of time of manure application.

The extent and rate of nitrate leaching from manures is influenced by a wide range of factors and, in general, the risk is greater from arable land than grassland. The time of application is of obvious importance, and in western Europe application before December generally results in a greater risk than applications made from January onwards (Chambers & Smith 1992). Interception and uptake of nitrates by grass roots lowers the risk of deep leaching to groundwater as the time between manure application and the onset of active plant growth decreases. High rates of manure application providing N in excess of the amounts recovered by the crop can lead to large accumulation in the soil in autumn, which will greatly increase the risk of leaching over the winter. These problems are especially relevant to forage maize because, since it is often left fallow over winter and the crop is very tolerant of heavy applications of manure, large quantities are sometimes disposed of on maize land. Apart from the immediate risks following winter application, surplus N may be present in the soil in leachable form after harvest. Mineralization of organic N in summer may lead to additional accumulation due to the relatively early cessation of N uptake by maize. It is also possible that the wide spacing of maize plants and a limited horizontal root extension can result in incomplete recovery of N from between the rows.

Manure composition, soil type and rainfall also influence nitrate leaching. The latter two factors are of general importance because the amount of rainfall passing through the soil has a large impact on the concentration of nitrate in groundwater. Thus, higher concentrations occur in drier areas, such as eastern England.

Increase in soil mineral N content, and hence leachable N, is closely related to the soluble NH_4^+-N content of manures. On non-retentive soils, the immediate risk is therefore greater for slurries where soluble N represents 40–60% of the total than for solid manures, such as FYM, which may contain as little as 10% as soluble material.

Application of manures with a relatively high C:N ratio (i.e. >20:1) are also likely to result in net immobilization of N in the soil. However, it is generally the case that prolonged mineralization of organic N will increase the risk of leaching.

13.2.5 Phosphorus

The form and content of P in manures varies widely, but it is present in both inorganic and organic compounds, with the former being most dominant. A proportion is in forms that can be taken up immediately by plants, whereas the remainder must first be mineralized or dissolved in soil water. The chemistry of P in soils is highly complex, mainly because it can occur in many different forms, the traditional view being that much is immobilized in the upper soil layers by adsorption on to the surface of clay minerals, and slowly released in forms available to plants. Soil organisms and microbes have a key role in immobilization/mineralization of organic P. Thus, application of manures has helped to maintain the long-term P status of soil with little risk of losses or of water pollution. Phosphorus makes a significant contribution to surface-water quality through its impact on eutrophication and associated algal blooms but, until recently, the emphasis was on P derived from waste industrial effluents. These sources are now decreasing, and there is more focus on diffuse leakages from agriculture (Haygarth 1997). The main pathway of transfer from agricultural land to surface waters is through erosion and run-off. Although this is associated mainly with arable land, grassland systems can provide high loads of P. Manure application to land is important in this respect because there is evidence for direct transfer of P through run-off (Harris *et al.* 1995), and for leaching of soluble organic forms down through the soil profile (Chardon *et al.* 1997). Heavy or repeated applications of manures can result in the retentive capacity of the soil being exceeded and 'breakthrough' of P to the aqueous environment. Large areas of farm land in The Netherlands are subject to this problem.

13.2.6 Potassium

Most of the potassium in manures is in forms that are readily available to plants. Although water soluble, K is retained well by heavier soils but to a lesser extent by sandy soils, where there is some risk of leaching. There is little evidence for adverse effects on surface and groundwater quality. Cattle manures contain large amounts of K relative to N and P, so there is a risk of applying excessive amounts of K. Potassium can be taken up by grass in large amounts at the expense of Mg, resulting in imbalance in the mineral composition of herbage and the risk of hypomagnesaemia in grazing stock (McAllister 1971, Minson 1990).

13.2.7 Other nutrients and heavy metals

In addition to N, P and K, manures contain a wide range of other plant nutrients and trace elements and so can play an important role in maintaining long-term soil fertility. These include sodium (Na), calcium (Ca), magnesium (Mg), manganese (Mn) and sulphur (S). The latter may be of particular significance because of the increasing evidence for sulphur deficiencies in grasslands. They also contain heavy metals which could potentially affect soil microbial function or be toxic to grazing animals. The risks are greater with manures from pigs because copper (Cu) and zinc (Zn) are added to their diet to increase food conversion.

13.3 Biochemical oxygen demand

Livestock wastes contain large quantities of readily biodegradable carbon and are excellent media for the multiplication of bacteria, which results in the consumption of oxygen. The potential for oxygen (O_2) removal in this way is often measured as biochemical oxygen demand (BOD). This is defined as the mass of dissolved oxygen required by a specified volume of liquid for the process of biochemical oxidation, under prescribed conditions over 5 days at 20°C. The BOD of livestock wastes is very high, and may range from 90 000 mg l^{-1} O_2 for silage effluent to 20 000–30 000 mg l^{-1} O_2 for slurry, and typically from 1000 to 3000 mg l^{-1} for 'dirty water' derived from washing milking parlours or floors, and rainfall run-off from concrete areas dirtied with silage or used by stock. These values compare with 300–400 mg l^{-1} O_2 for raw municipal sewage. Contamination of water courses, etc. by livestock wastes can therefore result in the rapid depletion of O_2 levels and the asphyxiation of fish and other aquatic life. Apart from point-source pollution, the main risk is from percolation or run-off following application of manures or dirty water to land. However, percolation vertically through soil is effective in lowering BOD (Chadwick & Pain 1998), and this principle has formed the basis of an experimental treatment system (Martinez 1997). The main danger is from application to land with field drains, when there is a high risk of rapid percolation through the disturbed soil above the drains and subsequent pollution of ditches and streams. Run-off across the surface and interflow in the upper soil layers commonly occurs on heavier sloping land, and for this reason it is recommended that manures and dirty water should not be spread within a 10 m wide buffer strip around watercourses. Recent research suggests that this may sometimes be inadequate on heavy soils under wet conditions.

13.4 Pathogens

Livestock wastes contain large numbers of micro-organisms, including many potential pathogens. The types of bacterial, viral and protozoon/parasitic pathogens encountered in livestock wastes were reviewed by Mawdsley *et al.* (1995), and potential risk to human health was discussed. Dissemination to water supplies can occur from direct leakages from wastes in buildings or stores, or following spreading on land. Much permanent grassland is on impermeable soils, and the most likely transport pathway is through surface or sub-surface run-off. Vertical transport to land drains may occur on lighter, freely drained soils or on cracking clays. Low-rate irrigation systems, commonly used for the disposal of dirty water, raise the probability of the dispersion of aerosols containing pathogens. Despite the potential risk, outbreaks of human disease proven conclusively to be related to livestock wastes are rare. This is no doubt due, at least in part, to effective water treatment facilities in most developed countries.

Mawdsley *et al.* (1995) highlighted the range of gastroenteritis-type diseases which may be transmitted from farm animals to humans by discussing selected examples of bacterial, protozoon and viral pathogens. Of the bacteria, *Escherichia coli* and *Salmonella* spp. are of most concern. *E. coli* is frequently used as an indicator of faecal contamination of water, and some types are responsible for diarrhoea outbreaks in man and stock. The organism can remain viable in slurry for several months (Burrows & Rankin 1970), but for a few weeks at most in soils (Linton & Hinton 1984). Salmonellae can survive for longer periods in slurry and soil (Jones *et al.* 1977); hence there is a real risk of transferring disease to healthy stock through grazing slurry-treated pasture. The occurrence of *Salmonella* in rivers has often been linked to contamination by cattle or their wastes (Clegg *et al.* 1983).

Research has highlighted the importance of the protozoa *Cryptosporidium* as an important pathogen of stock, especially cattle, and humans. Human symptoms include diarrhoea, vomiting, fever and flu-like illness. Direct links between contamination of water by infected livestock waste and outbreak of disease in the human population have also been documented. The problem arises because *Cryptosporidium* oocysts excreted by infected animals are not always totally removed by current water-treatment practices because the oocyst is resistant to the levels of chlorine used routinely in water-treatment plants. Furthermore, oocysts are excreted by infected animals in huge numbers, but ingesting as few as ten can cause disease.

Another protozoa, *Giardia*, is also responsible for gastroenteritis disease in mammals (including man), birds and reptiles in many parts of the world. Like *Cryptosporidium*, the cysts are highly resistant and ingestion of fewer than ten can cause disease. Giardiasis is the most common cause of water-borne epidemic diarrhoea in the United States (Deng & Cliver 1992), and levels of contamination in rivers have been shown to be especially high in areas of intensive livestock production (Rose *et al.* 1988).

Rotaviruses, a relatively recently discovered group of viruses, are recognized as one of the major causes of non-bacterial diarrhoea in small children. The number of recorded incidents of rotavirus infection in the UK almost doubled between the mid-1980s and mid-1990s, and it is a frequent cause of diarrhoea in calves.

Slurry may also contain large numbers of helminths and other pathogens, but the role of slurry in their epidemiology is probably minor in most circumstances.

13.5 Strategies for controlling losses and utilizing plant nutrients

13.5.1 Time and rate of application to land

There are considerable opportunities for reducing losses of plant nutrients from manures applied to land and for more fully exploiting their value as fertilizers. The key to effective utilization of manures is to apply at times and rates matched to the demands of the crop. It is evident from the discussions earlier in this chapter that losses of N through denitrification and leaching are much more likely from applications made in autumn than in spring or summer for a first or second silage cut. There is good evidence that slurry N is more efficient, in terms of increasing herbage yield and N uptake, when applied in spring than at other times of the year (Gracey 1983, Pain *et al.* 1986, Unwin *et al.* 1986). The effect of time of application on the efficiency of cattle slurry N is highlighted by the data in Table 13.3, which summarizes results of field experiments conducted in recent years. Restricting time of application to spring and early summer requires long-term storage of manure, and it is unlikely that the additional costs will be recovered as savings on N fertilizer use.

Applications that greatly exceed the amounts of N or P removed in the herbage can result in accumulations that may be lost (e.g. through leaching of NO_3-N or soluble P) during the following autumn. Manures do not contain N, P and K in the proportions appropriate for grass production, so it is important to adjust rates so that none of the major plant nutrients is applied in excessively high amounts, and then the balance can be supplied as inorganic fertilizer. On mixed swards of grass and white clover, slurry can provide a good source of P and K, but there is a risk of depressing clover from both smothering and the provision of N. It has been shown, however, that depression of the proportion

Table 13.3 Effects of time of application on per cent efficiency of N in cattle slurry compared with N fertilizer applied in spring.

	Application time		
	Winter	Spring	Summer
Herbage DM yield			
Mean	15	31	21
Range	8–25	15–46	13–35
N recovery			
Mean	9	22	11
Range	0–15	14–31	9–17

of white clover in the sward and in the yield of fixed N is lower for slurry than for fertilizer when similar amounts of N are applied (Nesheim *et al.* 1990). To calculate application rates, estimates of the available NPK content of manure can be obtained from published data (e.g. MAFF 1994) or by laboratory analysis of a representative sample. A range of rapid 'on farm' kits is also on the market for determining the available N content. Some of these have performed very well in comparison with laboratory analysis.

Application to steeply sloping land should be avoided, especially under wet conditions, because of the risk of surface run-off. Although the proportion of applied N and P lost in this way may be low, the concentrations of these nutrients and of BOD in run-off water can be very high (Misselbrook *et al.* 1995).

13.5.2 Other techniques for reducing nitrate leaching

Nitrification inhibitors (e.g. dicyandiamide marketed as 'Didin' or 'eNrich') temporarily curtail the activity of soil bacteria involved in the transformation of NH_4-N to nitrate, so their use has potential for delaying the release of nitrate until required for plant uptake (Pain *et al.* 1994). In practice, these products have proved to be costly and often unreliable when used with manures.

Cover crops can be useful in recovering potentially leachable N remaining in the soil in autumn, after maize harvest, for example. However, although there is evidence that autumn-sown cover crops such as rye (*Secale cereale*) or Italian ryegrass (*Lolium multiflorum*) may recover over $50\,kg\,N\,ha^{-1}$, this may be insufficient to lower nitrate concentrations in groundwater below the EC limit of $50\,mg\,l^{-1}$ (Schröder *et al.* 1992).

13.5.3 Techniques for reducing ammonia losses

Driven by the need to decrease the environmental impact of NH_3 in several parts of Europe, there is considerable interest in methods for controlling emissions from animal buildings and from the storage and application of manures to land. It is important to consider each stage of management. Nitrogen conserved by controlling NH_3 emission from buildings and stores can be lost through enhanced emissions from spreading manure on land. Controlling emissions from naturally ventilated cattle buildings is technically difficult and costly, but practical methods are under development (Monteny & Kant 1997). Covering slurry stores is an effective means of lowering emissions (Sommer *et al.* 1993) of both NH_3 and odour, and this is obligatory in some countries.

13.5.4 Alternative methods of slurry application

Paying attention to the time and rate of manure application to land does not reduce NH_3 losses; in fact these may be higher under warm spring or summer conditions than in winter. Alternatives to broadcasting on the surface, e.g. with a tanker fitted with a discharge nozzle and splashplate, are becoming increasingly commonplace in several European countries. Such types of machinery may be grouped according to their mode of operation as band spreaders, trailing shoes, injection with open slot, and injection with closed slots. Band spreaders usually comprise a series of hanging or trailing hoses designed to place slurry in even bands (Huijsmans *et al.* 1997). They aim to lower NH_3 volatilization by decreasing the surface area of the slurry exposed to the air, but in practice have achieved mixed results on grassland. Several workers have measured decreases of between 25% and 46% compared with broadcast application (Huijsmans *et al.* 1997; Lorenz & Steffans 1997). Thompson *et al.* (1990) reported that although initial rates of loss were much smaller than from broadcast applications, after 5 days the total losses from the former were 83% of the latter. Although very variable, results from a multi-site experiment in Ireland showed that band spreading improved the utilization of NH_4^+-N by conserved grass (Carton *et al.* 1994).

Trailing shoe applicators comprise a series of blades, or shoes, designed to part the grass leaves and place slurry in narrow bands on the soil beneath the canopy. Compared with broadcast application, decreases in NH_3 volatilization of 70% have been recorded (Huijsmans *et al.* 1997, Lorenz & Steffans 1997). To date, there is limited information on the effects on herbage yield, but the method has the potential advantage of decreasing problems associated with herbage contamination such as rejection by grazing stock and effects on silage fermentation.

Open slot injectors are designed to operate on shorter grass, i.e. in early spring or immediately after cutting for silage, than trailing shoes. V-shaped slots, up to 50–60 cm deep, are filled with slurry which is left exposed to the air. Rates of application are restricted to ensure slurry does not spill on to the surface. Reductions in NH_3 volatilization of between 50% and over 80% are achievable, depending on soil type and conditions and the weather. Utilization efficiencies of the NH_4^+-N in slurry, relative to inorganic N, of up to 90%, 40% and 20% for first, second and third grass cuts, respectively, were achieved in Ireland (Kiely 1987, Carton *et al.* 1994). In The Netherlands, apparent N recovery was increased by 23% for broadcast slurry compared with up to 49% for injected slurry (Meer, van der *et al.* 1987). Results from the UK showed no significant differences in total herbage yields from broadcast or shallow injected slurry, but with evidence for lower yields from slurry injected in early summer for a second silage cut (Misselbrook *et al.* 1996).

Closed-slot injection may either be shallow, where a 50–60 cm deep slot is closed by press wheels, or deep, in which winged tines aid lateral dispersion of slurry to 150 cm depth. Although very effective in lowering ammonia loss, both types are of limited applicability for grassland. The effectiveness of slot closure in the former is very dependent on suitable soil conditions and there is risk of squeezing slurry on to the surface even at moderate rates of application. Deep injection requires high tractor power, and mechanical damage can reduce herbage yields, especially at the first cut after injection (Rees *et al.* 1993).

13.5.5 Other methods for reducing NH_3 losses

The data illustrated in Fig. 13.3 show that there is potential for lowering NH_3 loss by diluting slurry with water, assuming soil type and conditions favour rapid infiltration. Ammonia volatilization can be lowered by up to 50% depending on the original total solids content of the slurry, the degree of dilution and the soil moisture content (Sommer & Olesen 1991, Stevens *et al.* 1992, Frost 1994). There are positive effects on herbage yield from slurry dilution, but the method has the practical disadvantages of requiring a readily accessible water supply and the need to transport and spread larger volumes.

The equilibrium between NH_4^+ and NH_3 in solution is very dependent upon pH, with higher values favouring release of NH_3. There has been considerable interest in adding sulphuric or nitric acid to slurries to lower pH, and hence NH_3 volatilization (Frost *et al.* 1990, Pain *et al.* 1990, Bussink *et al.* 1994). Using nitric acid was considered the more attractive option because of the potential savings on N fertilizer. Although effective in lowering NH_3 losses and improving the efficiency of N utilization, the method has not been put into practice. This is primarily due to the hazards involved in using concentrated acids on farms and the associated costs of equipment, and because of evidence that lowering ammonia loss in this way greatly increases N losses through denitrification and emissions of nitrous oxide.

Mechanical separators are sometimes used as an aid to slurry management in that they convert a semi-liquid material into a free-flowing pumpable liquid and a stackable solid. Frost *et al.* (1990) reported that losses could be decreased by about 60% compared with unseparated slurry, with corresponding increases in herbage yields.

Incorporation of manure into the soil by ploughing reduces NH_3 loss, and this is the only practical technique available for FYM, but its use is clearly limited to cultivated land. Incorporation should be done as soon as possible after spreading, preferably within a few hours (Pain *et al.* 1991).

13.6 Contamination of herbage

Although recommendations to store manures throughout the winter and to restrict spreading on land to spring and summer are sound in both environmental and agronomic terms, practical problems can arise. In some circumstances application of manures can result in smothering or scorch of the leaf area, causing a reduction in photosynthesis and herbage yield (Smith & Chambers 1993). The extent of the problem is influenced by the dry matter content of the slurry, the prevailing weather conditions (e.g. whether or not slurry is washed off the sward by rainfall) and the evenness of the spreading pattern (Prins & Snijders 1987).

Delaying application of slurry to spring or summer may greatly shorten the time elapsed between spreading and cutting for silage or grazing with livestock, so there is a greater risk of residual contamination. Late spring or summer applications of slurry can lead to poor silage fermentation (Tunney *et al.* 1980). Research with 'mini-silos' has shown there to be small but significant effects on silage composition, including increases in pH and changes in the volatile fatty acid profile, and more rapid deterioration which may affect animal intake. There is also a potential risk to animal health from pathogenic micro-organisms.

The rejection by cattle of herbage around dung deposited on pasture and the effects on herbage utilization are well documented (e.g. Martin & Donker 1966). Cattle may also reject herbage when grazing a pasture which has received slurry. When offered a choice between clean and slurry-treated swards, dairy heifers selectively grazed from the former and distinguished between different rates of slurry application (Pain *et al.* 1974, Broom *et al.* 1975). Results of experiments with dairy cows suggested that in a 'no choice' situation, they could detect slurry for 9 weeks after spreading. There were effects on grazing behaviour and a reduction in herbage intake (Pain & Broom 1978). More detailed studies have been conducted, using an indoor sward-box technique, on the factors influencing ingestive behaviour and diet selection. Method of slurry application (e.g. surface *v.* injection), time between application and grazing, and rainfall were important (Laws 1996). Recent field experiments with dairy cows have examined the impact of time and rate of slurry application. These have demonstrated important effects not only on herbage intake, but also on milk yield and composition (Danby *et al.* 1997). Although, surprisingly, cows were willing to graze swards only 1 week after slurry spreading, a period of more than 5 weeks between spreading and grazing was required to avoid reduction of herbage intake and milk yield. In normal farm practice, it is sensible to allow at least 4 weeks before grazing by adult cattle to avoid the risk of disease transfer, and several months in the case of more susceptible young stock.

13.7 Farm nutrient budgets

The amounts of N and P exported from farms in animal products is usually low compared with the amounts imported in fertilizers and animal feed, so manures may contain greater quantities of plant nutrients than can be effectively utilized on the farm for crop production. Farm nutrient budgets can help identify such surpluses, which pose a risk of long-term pollution. Obvious strategies for reduction include decrease in mineral fertilizer usage or, in more extreme cases, reduction in animal numbers or transport of manure to areas with fewer livestock. Modifying the protein nutrition of livestock has potential for decreasing the amount of N excreted and subsequent losses to air or water (Kay & Lee 1997, Smits *et al.* 1997, Misselbrook *et al.* 1998). More sustainable systems for cattle production, based on keeping nutrient budgets in balance through the introduction of a range of measures to reduce surpluses and losses, can be evaluated from models (e.g. Jarvis *et al.* 1996) and in practice (e.g. Mannetje 1994, Peel *et al.* 1997). They offer great potential for reducing surpluses of N and P, and for improved feeding strategies for cattle.

13.8 Treatment of wastes

Treatment normally involves harnessing a physical or biological process to improve the handling of wastes (e.g. making them easier to mix and pump), to make them more stable and less odorous, or to yield a useful product. The techniques are reviewed in Burton (1997). Separation of slurries using mechanical screening yields a free-flowing liquid and a stackable solid which composts readily (Pain *et al.* 1978). Both materials are easier to manage than slurry, with advantages for subsequent storage and utilization.

Biological treatment can be aerobic or anaerobic. The former involves the use of an aerator to dissolve enough oxygen into liquid manure to oxidize organic matter through the activities of aerobic bacteria. This has the effect of lowering BOD and decreasing odours. The process must be carefully controlled to avoid excessive loss of NH_3. Capital and operating costs can be high, but the technique may have a role in controlling odours from piggery wastes or in treating dirty water. Aerobic treatment, or composting, of FYM can yield a stable, deodorized material with value as a soil conditioner and source of plant nutrients. Other advantages include a reduction in the weight, volume and water content of the manure and, due to the high temperature reached during the process, the death of pathogens and weed seeds. The main factors controlling this microbial process are the level of aeration, C:N ratio, and the temperature and moisture content of the composting material (Stentiford 1996). Frequent turning, or preferably some form of forced aeration, is needed to ensure oxygen levels remain sufficiently high. The C:N ratio should be about 25:1, which may require the addition of an extra available carbon source to cattle manures, and the moisture content between 50% and 60%. To achieve high rates of decomposition, the composting temperature should be 45–55°C (Keener *et al.* 1993), although periods of higher temperature may be needed to ensure kill of pathogens. Although in many ways composting is environmentally attractive, high emissions of ammonia (Martin & Dewes 1992), and possibly also of N_2O and NO_x, occur during the process.

Anaerobic treatment normally refers to the microbial decomposition of slurries in the absence of oxygen. Slurry is retained in a sealed tank, or digester, stirred and maintained at a temperature of 35°C. The process has the benefits of reducing odour, achieving some degree of pathogen and weed seed kill, increasing the plant-available N content and lowering BOD. As for aerobic treatment, the reduction of BOD in slurries is insufficient to allow the treated material to be discharged direct to a watercourse. The process also yields biogas, a mixture of methane and carbon dioxide (CO_2). This can be used as fuel for a boiler for water heating or for an engine/generator for producing electricity. The 'energy crisis' in the 1970s gave great impetus to the development and installation of farm digesters, but the capital and operating cost, together with the technical complexity of the plant, has precluded their widespread uptake. In some countries there is interest in centralized plants which receive and treat organic wastes from farms and other sources.

13.9 References

Broom D.M., Pain B.F. & Leaver J.D. (1975) The effects of slurry on the acceptability of sward to grazing cattle. *Journal of Agricultural Science, Cambridge* **85**, 331–6.

Burrows M.R. & Rankin J.D. (1970) A further examination of the survival of pathogenic bacteria in cattle slurry. *British Veterinary Journal* **126**, 32–4.

Burton C.H. (ed) (1997) *Manure Management. Treatment Strategies for Sustainable Agriculture*. Silsoe Research Institute, Silsoe.

Bussink D.W., Huijsmans J.F.M. & Ketelaars J.J.M.H. (1994) Ammonia volatilization from nitric acid treated cattle slurry surface applied to grassland. *Netherlands Journal of Agricultural Science* **42**, 293–309.

Carton O.T., Cuddihy A. & Lenehan J.J. (1994) The effects of slurry spreading techniques on silage dry matter yields. In Hall J.E. (ed) *Proceedings of the 7th Technical Consultation on Animal Waste Management, Bad Zwischenahn, Germany*, 231–6. FAO, Rome.

Chadwick D. (1997) Nitrous oxide and ammonia emissions from grassland following applications of slurry: Potential abatement practices. In Jarvis S.C. & Pain B.F. (eds) *Gaseous Nitrogen Emissions from Grasslands*, 257–64. CAB International, Wallingford.

Chadwick D.R. & Pain B.F. (1998) Transport of nutrients and organic material in soils following application of animal wastes. In Matsunaka T. (ed) *Environmentally Friendly Management of Farm Animal Wastes*, 103–107. Rakuno Gakuen University, Sapporo.

Chambers B.J. & Smith K.A. (1992) Soil mineral nitrogen arising from organic manure applications. In Archer J.R. (ed) *Nitrates and Farming Systems. Aspects of Applied Biology* **30**, 135–43.

Chardon W.J., Oenema O., del Castilho P., Vriesema R., Japenga J. & Blaauw D. (1997) Organic phosphorus solutions and leachates from soils treated with animal slurries. *Journal of Environmental Quality* **26**, 372–8.

Clegg F.G., Chiejina S.N., Duncan A.L., Kay R.N. & Wray C. (1983) Outbreaks of *Salmonella* newport infection in dairy herds and their relationship to management and contamination of the environment. *Veterinary Record* **112**, 580–4.

Danby S., Penning P.D., Pain B.F., Owen E. & Laws J.A. (1997) Effect timing of slurry spreading on production in dairy cows. In *Proceedings of 5th British Grassland Society Research Conference, University of Plymouth*, 71–2. BGS, Reading.

Deng M.Y. & Cliver D.O. (1992) Degradation of *Giardia lamblia* cysts in mixed human and swine wastes. *Applied Environmental Microbiology* **58**, 2368–74.

ECETOC (European Centre for Ecotoxicology and Toxicology of Chemicals) (1994) *Ammonia Emissions to Air in Western Europe*. Technical Report No 62. ECETOC, Brussels.

Frost J.P. (1994) Effect of spreading method, application rate and dilution on ammonia volatization from cattle slurry. *Grass and Forage Science* **49**, 57–69.

Frost J.P., Stevens R.J. & Laughlin R.J. (1990) Effect of separation and acidification of cattle slurry on ammonia volatilization and on the efficiency of slurry nitrogen for herbage production. *Journal of Agricultural Science, Cambridge* **115**, 49–56.

Gracey H.I. (1983) Efficiency of slurry nitrogen as affected by the time and rate of slurry application and rate of inorganic nitrogen. In Corrall A.J. (ed) *Efficient Grassland Farming*. British Grassland Society Occasional Symposium No. 14, 179–84. BGS, Hurley.

Harris R.A., Heathwaite A.L. & Haygarth P.M. (1995) High temporal resolution sampling of P exported from grassland soil during a storm and the impact of slurry addition. In *Proceedings of International Workshop on Phosphorus Loss to Water from Agriculture*, 25–6. TEAGASC, Wexford.

Haygarth P.M. (1997) Agriculture as a source of phosphorus: Sources and pathways. *Scientific Committee on Phosphates in Europe (SCOPE) Newsletter No 21*. European Communications Unit, Paris.

Huijsmans J.F.M., Hol J.M.G. & Bussink D.W. (1997) Reduction of ammonia emission by new slurry application techniques on grassland. In Jarvis S.C. & Pain B.F. (eds) *Gaseous Nitrogen Emissions from Grasslands*, 281–5. CAB International, Wallingford.

Jarvis S.C., Wilkins R.J. & Pain B.F. (1996) Opportunities for reducing the environmental impact of dairy farming management: A systems approach. *Grass and Forage Science* **51**, 21–31.

Jones P.W., Smith G.S. & Bew J. (1977) The effects of the microflora in cattle slurry on the survival of *Salmonella* dublin. *British Veterinary Journal* **133**, 1–8.

Kay R.M. & Lee P.A. (1997) Ammonia emissions from pig buildings and characteristics of slurry produced by pigs offered low crude protein diets. In Voermans J.A.M. & Monteny G.J. (eds) *Ammonia and Odour Control from Animal Production Facilities*, 253–60. NVTL, AB Rosmalen, The Netherlands.

Keener H.M., Marugg C., Hansen R.C. & Hoitink H.A.J. (1993) Optimizing the efficiency of the composting process. In Hoitink H.A.J. & Keener H.M. (eds) *Science and Engineering of Composting: Design, Environmental, Microbiological and Utilization Aspects*, 59–94. Renaissance Publishing, Worthington, OH.

Kiely P.V. (1987) The effect of spreading method on slurry N utilization by grassland. In *Proceedings of the 12th General Meeting of the European Grassland Federation, Dublin*, 353–7.

Laws J.A. (1996) *Effects of Surface Applied Slurry on Herbage Yield, Dietary Preference and Grazing Behaviour of Cattle*. M. Phil Thesis, University of Plymouth.

Linton A.H. & Hinton M.H. (1984) The ecology of antibiotic-resistant bacteria in animals and their environment. In Woodbine M. (ed) *Antimicrobials and Agriculture*, 533–49. Butterworth, London.

Lorenz F. & Steffans G. (1997) Effect of application technique on ammonia losses and herbage yield following slurry application to grassland. In Jarvis S.C. & Pain B.F. (eds) *Gaseous Nitrogen Emissions from Grasslands*, 287–92. CAB International, Wallingford.

MAFF (Ministry of Agriculture, Fisheries and Food) (1994) *Fertilizer Recommendations for Agricultural and Horticultural Crops*. Reference Book 209, 6th edn. HMSO, London.

Mannetje L.'t (1994) Towards sustainable grassland management in The Netherlands. In Mannetje L.'t & Frame J. (eds) *Grassland and Society. Proceedings of the 15th General Meeting of the European Grassland Federation, Wageningen*, 3–18.

Martin D. & Dewes T. (1992) Loss of nitrogenous compounds during composting of animal wastes. *Bioresource Technology* **42**, 103–11.

Martin G.C. & Donker J.D. (1966) Animal excrement as a factor influencing acceptability of grazed forage. In Hill A.G.G. (ed) *Proceedings of the 10th International Grassland Congress, Helsinki*, 359–61.

Martinez J. (1997) Solepur: A soil treatment process for pig slurry with subsequent denitrification of drainage water. *Journal of Agricultural Engineering Research* **66**, 51–62.

Mawdsley J.L., Bargett R.D., Merry R.J., Pain B.F. & Theodorou M.K. (1995) Pathogens in livestock waste, their potential for movement through the soil and environmental pollution. *Applied Soil Ecology* **2**, 1–15.

McAllister J.V.S. (1971) Nutrient balance on livestock farms. *First Colloquium of the Potassium Institute Ltd*, 113–21.

Meer H.G. van der, Thompson R.B., Snijders P.J.M. & Geurink J.H. (1987) Utilization of nitrogen from injected and surface-spread slurry applied to grassland. In Meer H.G. van der, Unwin R.J., Van Dijk T.A. & Ennik G.C. (eds) *Animal Manure on Grassland and Fodder Crops. Fertilizer or Waste?*, 47–71. Martinus Nijhoff, Dordrecht.

Minson D.J. (1990) *Forage in Ruminant Nutrition. Magnesium*, 265–90. Academic Press, London.

Misselbrook T.H., Pain B.F., Stone A.C. & Scholefield D. (1995) Nutrient run-off following application of livestock wastes to grassland. *Environmental Pollution* **88**, 51–6.

Misselbrook T.H., Laws J.A. & Pain B.F. (1996) Surface application and shallow injection of cattle slurry to grassland: Nitrogen losses, herbage yields and nitrogen recovery. *Grass and Forage Science* **51**, 270–7.

Misselbrook T.H., Chadwick D.R., Pain B.F. & Headon D.M. (1998) Dietary manipulation as a means of reducing N losses and methane emissions and improving N uptake following application of pig slurry to grassland. *Journal of Agricultural Science, Cambridge* **130**, 183–91.

Monteny G.J. & Kant P.P.H. (1997) Ammonia emission and possibilities for its reduction in dairy cow houses: A review of Dutch developments. In Voermans J.A.M. & Monteny G.J. (eds) *Ammonia and Odour Control from Animal Production Facilities*, 355–64. NVTL, AB Rosmalen, The Netherlands.

Nesheim L., Boller B.C., Lehmann J. & Walther U. (1990) The effect of nitrogen in cattle slurry and mineral fertilizers on nitrogen fixation by white clover. *Grass and Forage Science* **45**, 91–7.

Pain B.F. & Broom D.M. (1978) The effects of injected and surface-spread slurry on the intake and grazing behaviour of dairy cows. *Animal Production* **26**, 75–83.

Pain B.F., Leaver J.D. & Broom D.M. (1974) Effects of cow slurry on herbage production, intake by grazing cattle and grazing behaviour. *Journal of the British Grassland Society* **29**, 85–91.

Pain B.F., Hepherd R.Q. & Pittman R.J. (1978) Factors affecting the performance of four slurry-separating machines. *Journal of Agricultural Engineering Research* **23**, 231–42.

Pain B.F., Smith K.A. & Dyer C.J. (1986) Factors affecting the response of cut grass to the nitrogen content of dairy cow slurry. *Agricultural Wastes* **17**, 189–202.

Pain B.F., Phillips V.R., Clarkson C.R. & Klarenbeek J.V. (1989) Loss of nitrogen through ammonia volatization during and following the application of pig or cattle slurry to grassland. *Journal of the Science of Food and Agriculture* **47**, 1–12.

Pain B.F., Thompson R.B., Rees Y.J. & Skinner J.H. (1990) Reducing gaseous losses of nitrogen from cattle slurry applied to pasture by the use of additives. *Journal of the Science of Food and Agriculture* **50**, 141–53.

Pain B.F., Phillips V.R., Huijsmans J.F.M. & Klarenbeek J.V. (1991) Anglo-Dutch experiments on odour and ammonia emissions following the spreading of piggery wastes on arable land. 28 pp. Rapport 91-9 IMAG-DLO, Wageningen, The Netherlands.

Pain B.F., Misselbrook T.H. & Rees Y.J. (1994) Effects of nitrification inhibitor and acid additions to cattle slurry on nitrogen losses and herbage yields. *Grass and Forage Science* **49**, 209–15.

Pain B.F., van der Weerden T.J., Chambers B.J., Phillips V.R. & Jarvis S.C. (1998) A new inventory for ammonia emissions from UK agriculture. *Atmospheric Environment* **32**, 309–13.

Peel S., Chambers B.J., Harrison R. & Jarvis S.C. (1997) Reducing nitrogen emissions from complete dairy farm systems. In Jarvis S.C. & Pain B.F. (eds) *Gaseous Nitrogen Emissions from Grasslands*, 383–90. CAB International, Wallingford.

Prins W.H. & Snijders P.J.M. (1987) Negative effects of animal manure on grassland due to surface spreading and injection. In Meer H.G. van der, Unwin R.J., Van Dijk J.A. & Ennik G.C. (eds) *Animal Manure on Grassland and Fodder Crops. Fertilizer or Waste?*, 119–35. Martinus Nijhoff, Dordrecht.

Rees Y.J., Pain B.F., Phillips V.R. & Misselbrook T.H. (1993) The influence of surface and sub-surface methods for pig slurry application on herbage yields and nitrogen recovery. *Grass and Forage Science* **48**, 38–44.

Roelofs J.G.M. & Houdijk A.L.F.M. (1990) Ecological effects of ammonia. In Nielsen V.C., Voorburg J.H. & L'Hermite P. (eds) *Odour and Ammonia Emissions from Livestock Farming*, 10–16. Elsevier Applied Science, Barking.

Rose J.B., Darbin H. & Gerba C.P. (1988) Correlations of the protozoa *Cryptosporidium* and *Giardia* with water quality variables in a watershed. *Water Science Technology* **20**, 271–6.

Schröder J., De Groot W.J.M. & van Dijk W. (1992) Nitrogen losses from continuous maize as affected by cover crops. In Archer J.R. (ed) *Nitrates and Farming Systems. Aspects of Applied Biology* **30**, 317–26.

Smith K.A. & Chambers B.J. (1993) Utilizing the nitrogen content of organic manures on farms – problems and practical solutions. *Soil Use and Management* **9**, 105–12.

Smits M.L.J., Valk G.J., Monteny G.J. & Vuuren van A.M. (1997) Effects of protein nutrition on ammonia emissions from cow houses. In Jarvis S.C. & Pain B.F. (eds) *Gaseous Nitrogen Emissions from Grasslands*, 101–107. CAB International, Wallingford.

Sommer S.G. & Olesen J.E. (1991) Effects of dry matter content and temperature on ammonia loss from surface-applied cattle slurry. *Journal of Environmental Quality* **20**, 679–83.

Sommer S.G., Christensen B.T., Nielsen N.E. & Schjorring J.K. (1993) Ammonia volatilization during storage of pig and cattle slurry: Effect of surface cover. *Journal of Agricultural Science, Cambridge* **121**, 63–71.

Stentiford E.I. (1996) Composting control: principles and practice. In de Bertoldi M (ed) *The Science of Composting. EC International Symposium*, 49–59. Blackie, Glasgow.

Stevens R.J., Laughlin R.J. & Frost J.P. (1992) Effects of separation, dilution, washing and acidification on ammonia volatilization from cow and pig slurries. *Journal of Agricultural Science, Cambridge* **119**, 383–9.

Sutton M.A., Place C.J., Eager M., Fowler D. & Smith R.I. (1995) Assessment of the magnitude of ammonia emissions in the United Kingdom. *Atmospheric Environment* **29**, 1393–411.

Thompson R.B., Pain B.F. & Rees Y.J. (1990) Ammonia volatilization from cattle slurry following surface application to grassland. 11. Influence of application rate, wind speed and applying in narrow bands. *Plant and Soil* **125**, 119–28.

Tunney H., Molloy S. & Codd F. (1980) Effects of cattle slurry, pig slurry and fertilizer on yield and quality of grass silage. In: Gasser J.K.R. (ed) *Effluents from Livestock*, 327–43. Applied Science Publishers, London.

Unwin R.J., Pain B.F. & Whinham W.N. (1986) The effect of rate and time of application of nitrogen in cow slurry on grass cut for silage. *Agricultural Wastes* **15**, 253–68.

Chapter 14
The Role and Management of Grassland in Organic Farming

D. Younie

14.1 Introduction

The decoupling of the relationship between agricultural production and statutory support for agriculture in the European Union, which commenced with the CAP reforms of 1992, reflects concern about the production surpluses and environmental impacts of intensive agricultural systems. Growing concerns about animal welfare, food quality and safety, including pesticide residues and disease problems associated with food, have also focused public interest on the methods employed in food production. In the UK this has led to a number of quality assurance schemes introduced variously by producer groups, processors and retailers. Organic farming ('ecological' or 'biological' agriculture in continental Europe), is in effect a food quality assurance scheme with the most clearly defined standards of production, which has been operating since the mid-1970s, and which is recognized at an international level (CEC 1991, 1999, IFOAM 1996).

Since the mid-1980s, there has been a very rapid global expansion in organic farming. In the 15 countries of the European Union, for example, the total number of organic farms increased from 7000 to 80 000 between 1986 and 1997 (Lampkin 1997). The total area of land farmed organically increased by 17 times over the same period, although a wide disparity exists between countries in terms of the proportion of land farmed organically. Demand for organic food also increased by an estimated 30% per annum during the 1990s. It seems certain that organic farming will continue to constitute a major component of the agricultural industry.

In organic systems, the supply of N is derived ultimately from N-fixation by legumes – in cool temperate climates almost universally forage legumes. For most arable crops in such systems the main sources of N are mineralized soil N and organic manures. Hence grass–clover swards, and the ruminant livestock enterprises required for their utilization, comprise the major source of soil-N enhancement and are essential features of most successful organic systems. Grassland also maintains and improves soil structure and provides a green winter cover (thus minimizing leaching losses and soil erosion risk), whilst the associated ruminant enterprises fulfil an additional important role through their utilization of arable crop residues.

This chapter discusses the principles of organic farming and the framework for achieving organic certification of production, as well as the role of grassland in organic systems, the environmental implications, and organic grassland systems in practice.

14.2 Principles of organic farming

The detailed husbandry standards of organic farming are primarily based on the principles of enhancement and exploitation of the natural biological cycles in *soil* (e.g. N fixation, nutrient cycling in the soil), in *crops* (e.g. manipulation of the competitive ability of crops and populations of natural predators of crop pests) and in *livestock* (e.g. rumen digestion in ruminants, development of natural immunity in young animals, interruption of host/pathogen relationships). In addition, there is strong emphasis on optimizing animal welfare, avoidance of pollution and improvement of the environmental infrastructure of the farm.

The aim, therefore, is to work *with* natural processes rather than seek to dominate them, as is often the case in 'conventional' systems using soluble fertilizers and pesticides, particularly the more intensive systems, and to minimize the use of non-renewable natural resources such as the fossil fuel used for manufacture of fertilizers and pesticides (IFOAM 1996, UKROFS 1997).

14.3 Certification of organic farming

14.3.1 Legal framework

Within the EU, the sale of food as 'organic' is controlled under EU Regulation 2092/91 (CEC 1991), which became operational for plant products in January 1993 and was supplemented by Regulation 1804/1999 to cover livestock production (CEC 1999). This regulation in effect defines organic farming. It sets out the minimum standards of production and defines how certification procedures must operate. In addition to organic production and processing within the EU, the regulation also covers certification of produce imported into the EU.

Under the regulation, each member state is required to establish a National Certifying Authority to ensure adherence to the law (in the UK, the UK Register of Organic Food Standards (UKROFS)). Organic farmers must be registered with a certification body which must be approved by the National Certifying Authority. There are six approved certification schemes in the UK (Soil Association (SA), Organic Farmers and Growers (OFG), Scottish Organic Producers Association (SOPA), Irish Organic Farmers and Growers Association (IOFGA), Organic Food Federation (which is primarily concerned with processors) and the Biodynamic Agricultural Association (BDAA) (biodynamic agriculture is a specific form of organic agriculture, first defined by the Austrian philosopher Rudolf Steiner in the 1920s (Steiner 1974)).

14.3.2 Standards for organic food production

The standards of the OFG, SOPA and SA schemes are fairly similar and, on the whole, mirror those of UKROFS. The BDAA has more wide-ranging standards. The main features of UKROFS standards are given below (UKROFS 1997).

- *It is not necessary to convert the whole of a holding, although the unit to be converted must be large enough to impose a valid crop rotation.*
- *It is not necessary to convert the whole unit at once – a field-by-field conversion is possible (and may even be desirable in order that a grass–clover ley can be established in each field in turn to ensure that soil fertility during the conversion phase is satisfactory).*
- *A 2-year conversion period is required between the last application of a prohibited substance and the sowing of the first full organic crop.*
- *The crop rotation must have a balance between fertility-building crops (e.g. grass–clover ley) and exploitative crops (e.g. cereals, potatoes), e.g. a cycle of 3 years of ley, followed by a succession of 1 year each of cereal, roots, then cereal.*
- *Permanent grassland is permitted.*
- *Regular inputs of organic matter, e.g. farmyard manure (FYM), must be made.*
- *Conventionally produced FYM (from ethically acceptable livestock systems) may be brought in but must be composted prior to application.*
- *Fertilizers such as lime and rock phosphate, which are slowly soluble in the soil, are permitted, but soluble mineral fertilizers are prohibited.*
- *Most manufactured agrochemicals are prohibited, but some natural biocides are permitted.*
- *Ruminant livestock must be fed a diet which is at least 60% green forage, on a daily dry matter basis (i.e. maximum 40% concentrates).*
- *Livestock diets must be based principally on organically produced feedstuffs, but a small proportion can be of conventional origin (e.g. 10% of daily dry matter intake for beef animals).*
- *Feeds derived from genetically modified organisms, and solvent-extracted feeds, are prohibited. Fishmeal is prohibited in ruminant diets.*
- *Housed animals must be provided with bedding – totally slatted systems are prohibited.*
- *Livestock must have access to pasture during the growing season.*
- *Livestock health policy must be based on preventative management strategies; no routine treatment of healthy animals with drugs, except in the case of a known farm problem. However, chemotherapy of individual sick animals is permitted, although withdrawal periods are extended. In the case of animals treated with organophosphate-based medicines, organic status is lost.*
- *Origin of animals: animals intended for breeding and/or milk production may be brought in from a conventional source but must undergo a conversion*

period; animals intended for meat production must have been born and reared under full organic management.

The inspection and certification scheme is operated on an annual basis. Detailed records of inputs must be kept, and the farmer has to submit an annual return describing the inputs to each field/livestock enterprise.

14.4 Contribution of grassland and forage legumes to organic farming systems

Any analysis of the components of organic farming must take account of its holistic nature. The components of the whole farm system interact closely, and grassland (in particular, forage legumes) plays the central role in this intricate web, including the arable cropping phase, especially in relation to nitrogen supply via its influence on N-fixation, soil organic matter, structure and biological activity, but also on negative aspects of the farming system, such as N loss to the wider environment. In organic crop rotations, the grass–clover ley, or the forage legume green manure crop, also have major roles to play in restricting the build-up of arable weeds and of soil-borne crop diseases.

Ruminant livestock share this central role with grassland on most successful organic farms, whether they be all-grass farms based on permanent grassland, or on mixed ley/arable farms. A housed cattle enterprise, for the generation of manure, is also an important element on any organic farm where significant off-take of nutrients takes place on specific fields (e.g. arable fields or grass fields cut for conservation crops) and the return of nutrients is required. Ruminant enterprises also fulfil an additional important role through their utilization of arable crop residues.

There are relatively few true stockless organic farms, i.e. systems in which the nutrient supply is largely internally generated, with no significant external nutrient inputs and no reliance on livestock wastes, but even on these farms the N supply is primarily based on N-fixation by forage legumes. There is at present a lack of information on the long-term reliability of stockless organic systems, although positive short-to-medium term results have been reported from a number of sources (Bulson *et al.* 1996, Drinkwater *et al.* 1996, Holle 1996, Cormack 1997). Certainly there is a need for such systems to be developed, since the financial constraints involved in converting to the standard mixed organic farming model may be prohibitive for existing conventional all-arable farms, given the capital investment involved in purchase of stock, livestock accommodation, fencing, etc., and the loss of statutory arable area support payments. As stockless systems develop further, their reliance on forage legumes will continue, and perhaps even expand, exploiting techniques such as bi-cropping (Clements *et al.* 1996) and undersowing (Younie 1998).

14.4.1 The role of grassland and forage legumes in the nitrogen economy of organic farming systems

Organic farming systems and nitrogen fixation

Nitrogen is the single most important nutrient required for herbage growth. In organic systems, nitrogen is also supplied from mineralized soil organic N and from applications of organic manures, but the ultimate source of N input to the system is atmospherically derived N fixed by legumes such as white clover (*Trifolium repens*), red clover (*T. pratense*) and lucerne (*Medicago sativa*). The N resource in organic farming is therefore a renewable resource in comparison with fossil-fuel-derived artificial N fertilizers. Whitehead (1995), quoting seven series of trials in the UK, reported amounts of fixed N ranging from 0 to 445 kg N ha^{-1} per annum, with an overall average of 152 kg N ha^{-1} per annum. Other workers have reported N fixation input to grassland as a fertilizer N equivalent value, ranging from 100 and 250 kg N ha^{-1} per annum for temperate European conditions. The lower end of this range is associated with soils of high potential for N mineralization, continuous grazing management and smaller leaved white clovers, whilst the upper end of the range is associated with soils of low potential for N mineralization, cutting or rotational grazing management, large-leaved white clovers and red clover or lucerne (Reid 1970, Morrison 1981, Frame 1990, Davies & Hopkins 1996). The upright, higher yielding species red clover and lucerne would normally be expected to fix more N than white clover under a cutting management. Stopes (1993), in a comparison of legumes cut and mulched for 1-year green manuring in the context of an organic stockless rotation, reported N accumulation values of 371, 328 and 211 kg N ha^{-1} per annum for red clover, white clover and trefoil (*Medicago lupulina*), respectively. Cormack (1997) has also reported mean annual N accumulations of over 300 kg N ha^{-1} for 2-year red clover green manures in an organic stockless rotation.

The level of N fixation is notoriously difficult to measure accurately (Wood 1996), but is accepted to be primarily dependent on the clover content of the sward, particularly in the early years of a ley (Frame & Newbould 1986, Van der Meer & Baan Hofman 1989). At the farm level, clover ground-cover is more easily estimated than clover content in the dry matter. Kristensen *et al.* (1995) have used clover ground cover as the basis for estimating the amount of atmospheric N fixed in grass–clover systems (Table 14.1).

Table 14.1 Effect of clover content and age of ley on estimated N-fixation in organic grass–clover leys (kg N ha^{-1} per annum) (from Kristensen *et al.* 1995).

Clover content (% ground cover)		10–29	30–49	Above 49
Clover content (% in dry matter)		3–16	17–29	Above 29
Age of ley (years):	1st, 2nd	80	157	248
	3rd, 4th and 5th	47	84	128

Using the same estimation technique, Halberg *et al.* (1995) estimated atmospherically derived N inputs on six organic mixed farms in Denmark. Their estimates, which include a figure of 21 kg N ha^{-1} from precipitation, range from 39 to 56 kg N ha^{-1} per annum in permanent grass and from 120 to 408 kg N ha^{-1} per annum in grass–clover leys (the latter figure of 408 kg N ha^{-1} being derived from lucerne).

In contrast to herbage legumes, the N-fixation of grain legumes can often be rather modest, partly because of their shorter growing season, but their contribution to soil N status is likely to be much less than herbage legumes because much of the fixed N is removed in the grain crop itself (Fisher 1996).

In addition to contributing to herbage growth, a proportion of the atmospherically fixed N is stored in roots and stubble or is immobilized in soil organic matter, i.e. it contributes to the longer-term build-up of soil organic matter and soil N status. Simultaneously, a proportion of the N in the soil organic pool is mineralized and becomes available for uptake by pasture plants. As the supply of N from mineralized soil organic matter increases (e.g. in the later years of a ley), the relative contribution from N-fixation to the overall supply of N to the sward declines (Younie 1992, Davies & Hopkins 1996), as is also implied in the data of Kristensen *et al.* (1995) above.

The increase in soil organic matter and N content is asymptotic, and eventually an equilibrium is reached between immobilization and mineralization of N. The time taken to reach this equilibrium will vary with climate, soil type and management, but is likely to be between 50 and 200 years (Jenkinson 1988). There is no information on the rate of build-up of N or soil organic matter in organic leys specifically, but Clement & Williams (1967) reported annual increases of 70 kg N ha^{-1} and 100–110 kg N ha^{-1} for perennial ryegrass only and ryegrass–clover swards, respectively, over a 3-year period. The latter figure represents an annual increase of about 0.005% N.

Halberg *et al.* (1995) estimated that on Danish organic farms the amount of surplus N available annually for immobilization in the soil and for loss to the environment was between 18 and 231 kg N ha^{-1} on grass–clover leys and between 3 and 59 kg N ha^{-1} on permanent grass. The equivalent figure calculated by Watson and Younie (1995a) for an organic beef farmlet was 103 kg N ha^{-1}, compared with 216 kg N ha^{-1} for a conventional high-N fertilizer system. On a whole-farm basis, the average N surplus for 14 commercial organic farms calculated by Halberg *et al.* (1995) was 124 kg N ha^{-1} per annum, whilst for a comparable 16 conventional farms the N surplus was 240 kg N ha^{-1} per annum.

Thus, herbage legumes, normally in the form of grass–clover leys, play a vitally important role in organic systems through the capture of the atmospheric N resource, enabling its subsequent exploitation by ruminant livestock and arable crops. In a self-contained organic farming system the proportion of grass–clover ley or herbage legume crop in the rotation will determine the total N input, which will in turn determine arable crop yields (Younie *et al.* 1995). The ideal proportion of fertility-building crop (i.e. legume) in the rotation has never been definitively determined. A minimum of 50% is often suggested, but good crop yields have been obtained in good soils, at least over the short term, with a stockless rotation containing only one year in five of red clover green manure, plus a grain legume (Cormack 1997). The data of Kristensen *et al.* (1995) would suggest that short-term (2- to 3-year) leys are a more efficient way, compared with longer-term leys, of utilizing the atmospheric N resource within the overall context of fertility building and exploitation. Johnson *et al.* (1994) have also suggested that the optimum length of ley in terms of its residual value to following crops is 3 years.

Soil structure

A well-aerated soil is an essential aim in organic farming because of the beneficial effect this has on soil N mineralization and the reduced risk of N loss through denitrification. The positive influence of grassland on soil structure is well established (Tisdall & Oades 1980, Eder & Harrod 1996), and it has been suggested that legumes in particular enhance the soil aggregation process (Angers & Carter 1996). Reganold *et al.* (1993), in a paired-farm comparison between biodynamic and conventional farms in New Zealand, found that pasture soils on biodynamic farms had significantly lower bulk density and lower penetration resistance than pastures on conventional farms, indicating better soil structure and less compaction. These authors suggested no reasons for these differences, but a higher clover content in the swards may have contributed.

Although a grass–clover ley will undoubtedly improve soil structure relative to the arable cropping phase, soil compaction does occur in grassland situations, including organic farming, and indeed is of particular importance in the organic context, given the heavy reliance on mineralized N in organic farming. Heavy silage-making equipment or high stocking densities, particularly when soil conditions are wet (e.g. around feeding troughs in winter), will all cause compaction and reduce soil N availability, both to the grass–clover ley itself and possibly to the following arable crop. The effect of compaction on N off-take in unfertilized grass (reflecting N mineralization) caused by a typical sequence of silage harvesting operations is shown in Table 14.2. Outdoor pig production also

Table 14.2 Effect of compaction on annual N off-take ($kg\,ha^{-1}$) from three silage cuts (zero N fertilizer input) (Douglas & Crawford 1993).

Soil condition	1988	1989	1990	Average
Non-compacted	41.3	34.7	39.5	38.5
Typically compacted	25.8	27.1	25.7	26.2

causes serious soil compaction, which can have an adverse residual effect on the subsequent arable crop (Edwards & Watson 1997).

There is a need for further research, within the context of organic farming, to quantify these effects and to examine practical methods of ameliorating soil compaction in organic grassland.

Soil biological activity

A high degree of soil biological activity is associated with good soil structure and soil N mineralization. Manipulation and enhancement of this biological resource is a major aim of organic farming. As a result of the permanent green cover and vigorous root growth afforded by grassland, by the absence of soil disturbance, and by the consequent increase in soil organic matter, soil biological activity is greater under grassland than under arable cropping. This is one of the primary reasons for the inclusion of grass–clover leys in organic rotations. Earthworm populations increase during the ley phase and decline under cropping as a result of increased frequency of cultivation, as shown in Fig. 14.1 for organic crop rotation experiments in NE Scotland (G. Armstrong, unpublished information, 1998).

From the same set of experiments, Watson *et al.* (1996) also showed that soil microbial biomass and activity were greater under grass–clover than under arable crops. These results are not unique and have been mirrored elsewhere (e.g. Follett & Schimel 1989, Weil *et al.* 1993).

Because of the higher investment costs and lower profitability of livestock systems relative to cropping systems, stockless organic rotations are an attractive proposition economically. However, an unanswered question remains about

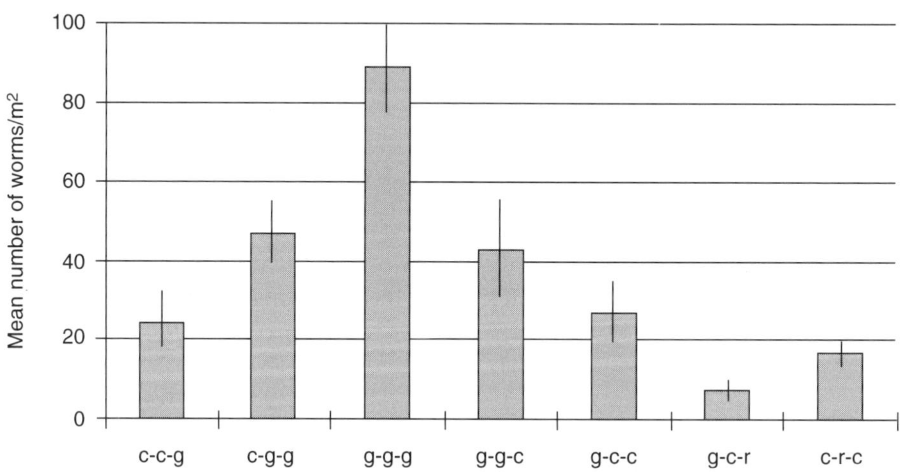

Fig. 14.1 Effect of cropping sequence on populations of *Lumbricus terrestris*. Cropping sequence: c, cereal; g, grass; r, roots.

the level of nitrogen supply (and therefore crop yield) which such systems can sustain with a reduced proportion of herbage legume/grass–clover ley in the rotation. Part of this question relates to the maintenance of soil biological activity in a system with high cultivation and cropping frequency. It would seem logical that the nearer to a continuous cropping system we approach, and the shorter the ley or green manure phase, the lower the level of soil biological activity, the lower the soil organic matter content and the lower the potential for soil N mineralization. Indeed, the long-term Rodale Farming Systems Trial in the USA has shown this to be the case; soil biological activity was greater in the livestock-based organic system than in the stockless system (Wander & Traina 1996). It is therefore important that research on stockless rotations should consider soil biology responses.

Higher soil biological activity (e.g. earthworm populations) has sometimes been reported in organic grassland compared with conventional grassland (Reganold & Palmer 1995), but this is most likely to be simply a result of the addition of more organic matter in the form of animal manures, rather than because of an intrinsic advantage of organic management. Whilst biocides may indeed reduce earthworm populations, the intensive use of mineral fertilizer will actually improve the food supply for micro- and macrofauna (Younie & Armstrong 1996).

14.4.2 The impact of organic grassland on the environment

Care for the environment, including wildlife habitats, and minimal use of non-renewable resources are enshrined as principles of organic farming. One of the strategic objectives at farm level is to minimize leaching of nutrients into surface and ground waters. For the organic farmer in particular, this is important from a production, as well as from an environmental, point of view. The extent to which organic farming can be regarded as a sustainable system can also be assessed by the potential for reduced use of fossil fuel resources and for enhanced biodiversity. Grassland plays an important role in these elements of organic philosophy.

Nutrient emissions to the aquatic environment

It is now accepted that the risk of nitrate-N leaching from grassland is influenced more by the total quantity of N circulating in the system than by the source of the N. Swards with similar stock-carrying capacity will result in similar environmental losses of N, whether based on N derived from N-fixation by legumes or supplied as fertilizer N (Titchen & Philipps 1996). However, whilst clover-based systems do have the potential to leak N into the environment, leaching losses from organic grass–clover swards will be lower than from the most intensive fertilizer-based systems (i.e. those receiving more than 200 kg N ha^{-1} per annum). A number of comparative studies have shown that the N surplus, and hence the potential for nitrate-N leaching, is less from clover-based systems, including organic grassland, than from intensively fertilized grassland (Halberg *et al.* 1995, Watson & Younie 1995a, Tyson *et al.* 1996).

Maximizing green cover of soil over winter is also encouraged in organic farming, both to ensure uptake of surplus mineral N from the soil and so minimize leaching losses, and also to minimize the risk of soil erosion. The inclusion of a ley in the crop rotation obviously goes some way towards achieving this objective, but other forage species are often also sown specifically as winter covers during the arable cropping phase of the rotation. Species such as mustard (*Sinapis alba*) and *Phacelia* are effective scavengers of soil N and are commonly used as cover crops, but graminaceous species such as grazing rye (*Secale cereale*) and Italian ryegrass (*Lolium multiflorum*) can also be used. At farm level, however, the farmer must balance the perceived benefits in terms of the amount of N saved, or prevented from leaking to the environment, against the establishment costs, including seed costs. The extent of any benefit will depend not only on the species used as a cover crop, but also on how well it establishes, and on how much mineral N is present in the soil initially. In this latter regard, the winter following the first arable crop in the rotation is normally when soil N is at its highest level in the cropping phase of the rotation, and therefore when winter cover crops offer the greatest potential benefit. A further consideration in relation to establishment of winter covers is the length of the growing season available for establishment, following the harvest of the main crop. Effective establishment of winter covers can be achieved from direct sowing following early harvested crops such as early potatoes or vegetables, or in locations with a long growing season; however, in areas with a short growing season, or following a late harvested cereal, for example, direct sowing of winter cover crops can often give poor establishment (Watson & Younie 1995b). In these situations, undersowing the winter cover with the previous cereal crop will give better establishment.

In the context of organic mixed ley–arable systems, the management of the transition phase from ley to arable cropping, the most critical point for N leaching in organic rotations, is of crucial importance. It is clear that autumn ploughing (usually for establishment of a winter cereal) leads to more leaching of N than spring ploughing. Cobb *et al.* (1997) measured leaching losses of between 119 and 132 $kg\,N\,ha^{-1}$ in organic winter wheat in England following the ploughing of leys. A switch to the production of spring crops would reduce losses significantly (Philipps & Stopes 1995), but winter wheat has considerable economic advantages over spring cereals, and for heavy clay soils autumn ploughing is essential for satisfactory seedbed preparation. In any case, this transition phase from ley to arable cropping represents only 15–20% of the rotation, and the lower losses from the other phases of the organic rotation reduce the average losses over the whole rotation to 10–20 $kg\,N\,ha^{-1}$ per annum (Philipps & Stopes 1995).

Date of ploughing of the ley (or indeed of a graminaceous winter cover crop) relative to the sowing date of the following crop is important not only in relation to minimizing leaching loss, but also in controlling N immobilization by the incorporated ley vegetation. If a ley (or any other vegetation type with a high carbon : nitrogen ratio) is ploughed too close to the sowing date of the crop, N can be immobilized, and this can significantly reduce crop N uptake and yield. Younie and Watson (1995) measured a $1.3\,t\,ha^{-1}$ reduction in grain yield of spring barley following a March ploughing of an organic grass–clover ley, compared with a January ploughing date. A delay of at least 1 month is necessary between the ploughing of a ley or a graminaceous winter cover crop and the sowing of the subsequent arable crop.

Nutrient emissions to the atmosphere

Ammonia volatilization from livestock systems makes a major contribution to acid rainfall. Ruminant animals play a major role in organic systems, and so the potential leakage of N into the atmosphere as ammonia from associated livestock housing, manure storage and land spreading cannot be dismissed as insignificant. Halberg *et al.* (1995) estimated ammonia losses from organic dairy farms in Denmark to vary over the same range (22–52 kg N per livestock unit per annum) as comparable conventional dairy farms, although stocking rates on organic farms tend to be lower than on conventional farms.

In addition to ammonia volatilization, grassland and ruminant livestock systems represent a major source of the greenhouse gases methane and nitrous oxide, and carbon dioxide is released in the manufacture of inputs such as fertilizers and the burning of fossil fuels on farms. Bakken *et al.* (1994) compared the emissions of greenhouse gases from high-intensity and low-intensity (organic) dairy farms in Denmark (Table 14.3).

They calculated that, on a per hectare basis, the organic system produced less CO_2 equivalents, but because of the lower output per hectare of the organic system, the amount of CO_2 equivalent per unit of milk produced was almost identical for the two types of farm. However, the energy consumption on the organic farm used by Bakken *et al.* (1994) for their study may be unusually high because thermal weed control was used in the fodder beet crop. Nitrous oxide emissions from grassland on the organic farm were only one-third that from the

Table 14.3 Effect of grassland farming intensity on global warming gas emissions (expressed as t CO_2 equivalent per ha per annum) (from Bakken *et al.* 1994).

	High intensity farm	Low intensity farm (organic)
Gas type		
N_2O	1.8	1.0
CH_4	5.0	4.1
Energy	2.3	1.4
Total	9.1	6.5

conventional farm, as a result of lower fluxes from the soil (due to lower N levels) and the absence of emissions from the industrial production of nitrate. The largest contributor of greenhouse gases was methane, originating primarily from rumen digestion (contributing approximately 60% of CO_2 equivalents), and the organic farm produced only 17% less per hectare than the conventional farm, simply mirroring the lower stocking rate.

Thus, organic grassland systems do appear, on a per hectare basis, to produce less pollution of the aquatic and atmospheric environments, although to some extent this is simply a reflection of lower stocking rates. Since the main sources of pollution from organic livestock farming are derived from integral components within ruminant livestock systems (e.g. volatilization of ammonia, methane from rumen digestion) rather than from excessive nutrient or energy inputs, the main approach towards reducing emission of nutrients and greenhouse gases must be the further improvement of internal nutrient cycling (e.g. slurry injection in early spring rather than surface-spreading in early winter; see Chapter 13). Nutrient loss is particularly undesirable for organic crop production.

Energy use in organic grassland

Energy consumption is discussed in the previous section in the context of CO_2 release and the contribution of organic grassland to global warming. It is also pertinent to discuss the influence of organic grassland systems in terms of consumption of non-renewable fossil fuel resources. Refsgaard *et al.* (1998), in a modelling exercise based on case studies, compared the energy utilization of the crop enterprises on 14 organic and 17 conventional dairy farms in Denmark. Using as a basis for comparison the energy productivity of a given crop (quantity of yield per unit of energy required), they found that grass–clover leys and lucerne on organic farms were approximately three times more productive than those on conventional farms (Table 14.4).

Silage harvesting had the biggest demand for direct energy (i.e. diesel fuel) in both organic and conventional systems (51–53%), but organic farms consumed more direct energy for manure handling, particularly for dry manure (17% and 6% for organic and conventional systems, respectively). However, the main difference in energy consumption between the two farm types was in indirect energy use, in which conventional grassland exceeded organic grassland by a factor of eight, primarily because of the energy required in the manufacture of fertilizer N. Grass–clover was more energy-productive than fodder beet or whole-crop silage, and increasing the proportion of grazed rather than ensiled forage improved the productivity still further.

Organic grassland appears to be relatively energy-efficient, principally as a result of the non-use of N fertilizer. Scope for further improvements in energy utilization appears to lie principally in modifications to the silage-making process and to manure handling and application. Adoption of high dry-matter silage and big balers instead of forage harvesting may offer opportunities for

Table 14.4 Model of energy productivity in grass–clover swards in organic and conventional dairy farms; non-irrigated sand, Denmark (Refsgaard *et al.* 1998).

	Conventional	Organic
51% of grass–clover harvested for silage		
Sum, direct energy (MJ/ha)	2745	2454
Sum, indirect energy (MJ/ha)	10922	1378
Total energy requirement (MJ/ha)	13667	3832
Yield (SFU/ha)	6000	5200
Energy productivity (SFU/MJ)	0.44	1.36
100% harvested for silage		
Energy productivity (SFU/MJ)	0.37	0.84
100% grazed		
Energy productivity (SFU/MJ)	0.57	7.03

SFU, Scandinavian Feed Units (= feeding value of 1 kg barley).

reducing the energy costs associated with silage making. There is a need to model the energy costs of the manure handling process; composting will increase costs, but subsequent spreading may be easier and incur lower energy costs.

Floristic diversity of organic grassland

A consideration of the floristic diversity of organic grassland needs to recognize three different situations: (1) short-term leys; (2) permanent swards from which the main objective is a high level of herbage production and livestock output; (3) permanent swards managed for their nature conservation value, e.g. species richness. Grassland in this third category is usually associated with low nutrient status and therefore low levels of output, and there may be further opportunities, e.g. in Environmentally Sensitive Areas, for organic management to be linked to management schemes which attract payments for environmental objectives (see Chapter 11).

Short-term organic leys can be excluded from any discussion of species-rich swards. A division of permanent grassland into the other two categories listed above (production and nature conservation) is evidenced in a survey of 91 organic grassland sites in Schleswig–Holstein by Worner & Taube (1995), who found that the index of stand value (for agriculture) was closely related to the presence of perennial ryegrass and white clover, the percentage of legume in the sward increased with intensity of utilization, and the content of white clover tended to be negatively correlated with species diversity. Similarly, Younie & Armstrong (1996) observed relatively little species diversity in an intensively managed and highly productive organic sward over a period of 9 years from sowing. Over 95% of ground cover in this sward comprised perennial ryegrass and white clover.

However, a number of organic/conventional comparisons of long-established permanent grassland have shown that organic swards generally contain more plant species than conventional swards. Frieben & Köpke (1996) found, in the Lower Rhineland, that organic permanent pastures had 14% more species than comparable conventional permanent pastures and 45% more than conventional sown pastures (but admitted that there may be little or no difference in species diversity between organic pastures and extensively managed conventional grassland). At the Isle of Terschelling, The Netherlands, Jansen *et al.* (1985) found almost 40 species per parcel in six organic farms compared with 30 on nine conventional intensive farms. Baars *et al.* (1982) found smaller differences between organic and conventional dairy farms, but found much higher species numbers in the field margins of the organic farm on clay soil. On a well-established organic dairy farm in Freisland, The Netherlands, Ennik *et al.* (1982) found 28–35 species per field in old permanent pastures, which is typical for this type of extensive organic grassland.

14.5 Organic grassland management in practice

14.5.1 Management objectives

Grassland management objectives on organic farms are not necessarily prioritized in the same order as on conventional farms. The restrictions of the organic standards in relation to plant nutrient inputs and routine veterinary treatment of livestock mean that grassland management strategies (a) to maintain high animal health status, and (b) to maintain soil nutrient status are equally, if not more, important than (c) maximizing herbage output per hectare in the short-term, which is normally recognized as the primary objective in conventional situations.

14.5.2 Maintaining livestock health at grass

Stomach worm infection in lambs is the most common health problem in organic lambs (Halliday *et al.* 1991, Roderick *et al.* 1996). Grazing management is the main component of worm-control strategy and must be designed primarily to minimize parasite infection. Essentially this involves restricting the stocking rate of susceptible animals, particularly young stock, and minimizing the contamination of pasture with parasite larvae and the parasite challenge to these animals, for example, by:

(1) having an enterprise mix on the farm which includes both cattle and sheep (since the parasites which affect one species do not, in general, affect the other);

(2) having a mix of adult stock (which are largely immune to parasite attack) as well as young stock on the farm;
(3) alternating cattle, sheep, and/or forage conservation from year to year on any one field (which will minimize the over-winter contamination of pasture with worm larvae);
(4) avoiding high stocking rates *per se*;
(5) regular reseeding, where possible, which will provide uncontaminated pasture in the first year.

Parasite challenge to susceptible animals is likely to be least on low-ground, mixed farms with a mix of cattle and sheep enterprises, where opportunities exist to utilize crop stubbles, crop residues and annual forage crops as well as grassland, where young, larvae-free reseeds are available on an annual basis, where all grass fields can be mown for conservation and included in alternating clean grazing systems, and where cattle can provide an effective dilution of the parasite challenge to lambs. Conversely, parasite and disease problems are likely to be greatest on the sheep-only upland farm, where opportunities to take forage conservation cuts are limited and a clean grazing system is difficult to establish.

Even where it is possible to establish an alternating clean grazing system, it is not always effective in controlling parasitic gastroenteritis in lambs, primarily because of the pasture contamination caused by the ewes themselves at lambing (the periparturient egg rise) (Halliday *et al.* 1994, Gray *et al.* 1997). There is a need to consider other approaches to reducing the problem, e.g. breeding for resistance to parasites in the sheep flock, exploiting the differential parasite burdens of different pasture species (Niezen *et al.* 1996), and developing integrated strategies of worm control, with clean grazing as one of the main components.

Grassland and ruminant livestock management on organic farms should be planned to make the maximum use of grazing. Apart from considerations of the cost of conserving grass (and the concomitant nutrient removal from the soil, see below) and the cost of organic concentrates, health problems tend to be fewer in animals kept outdoors, particularly where calving and lambing are undertaken outdoors (provided weather and soil conditions permit). Stress, which can contribute to metabolic disorders (e.g. hypomagnesaemia) or breakdown in natural immunity, should be minimized at all times in the grazing situation, for example by appropriate construction of gateways and feeding stances, so as to minimize poaching (and hence potential foot problems), by provision of access to shelter (e.g. by hedging along field boundaries), and by careful grouping, feeding and handling of animals.

Bloat is often mentioned as a potential risk in legume-based grassland systems. Whilst the risk exists (in the grazing situation but not with silage, and with cattle but not with sheep), it is a relatively infrequent occurrence and can be predicted (and therefore avoided) to a large extent. The risk is greatest in the second half of the year, when animals are offered relatively rank herbage with a high clover content and low dry matter content (e.g. silage aftermaths or around dung pats, on wet or dewy mornings). Hungry animals should never be introduced to such herbage in these conditions. Animals should be introduced to aftermaths soon after the crop has been removed, when the regrowth is very short, so that the clover grows up to the animals rather than vice versa. In high-risk situations, a straw buffer feed reduces the risk, but poloxalene bloat preventative, fed with a carrier feed such as barley or brewer's grains, is the most effective insurance.

14.5.3 Maintaining soil nutrient status

In relation to the maintenance of soil nutrient status, the main grassland management priority concerns the removal of nutrients from fields cut for hay or silage. With a herbage yield of 10 t DM ha^{-1} per annum, nutrient off-take in the herbage will be in the region of 80 kg P_2O_5 ha^{-1} per annum and 260 kg K_2O ha^{-1} per annum (Fowler *et al.* 1993, Younie *et al.* 1998). This can obviously have a major effect on soil K and P content and consequently on herbage and crop yield.

In a systems comparison in Germany, soil P content decreased by 39% over 5 years in the organic system (Weber *et al.* 1993). Similarly, in two out of three organic and conventional paired-farm comparisons in New Zealand, soil P content in pasture soils was significantly lower on the organic farms, reflecting negative P budgets on these farms (Nguyen *et al.* 1995). Soil P content in the top 5 cm in a sandy soil in The Netherlands, managed organically, fell over 5 years by 25% under cutting management and by 18% under grazing/cutting management, even where 50 kg P_2O_5 ha^{-1} per annum as farmyard manure was applied (Younie & Baars 1997).

Figure 14.2 shows the effect of first-cut silage yield on exchangeable soil K content 9 months later in grass–white clover swards in an organic crop rotations trial on a sandy loam soil (10% clay) in northeast Scotland. The relationship is significant at the $P < 0.001$ level.

Similarly, on a sandy soil in The Netherlands, Younie & Baars (1997) reported reductions of 62% and 39%, respectively, in soil K content over 3 years in cut and cut/grazed organic grass–clover leys yielding 9–11 t DM ha^{-1}. In contrast, on similarly managed swards on clay soils, no decline in soil K content was detected.

Clovers are poor competitors for soil P and K compared with grass. The reductions in K content in the experiments of Baars described above caused a significant loss of clover from the swards. Fothergill *et al.* (1995) have also shown the significant effect of P and K on white clover content in upland pasture. Low clover content resulting from restricted K input has also been observed in swards in a long-term trial on sandy loam soil in northeast Scotland (D. Younie, unpublished information, 1998). This adverse effect of low soil P and K content

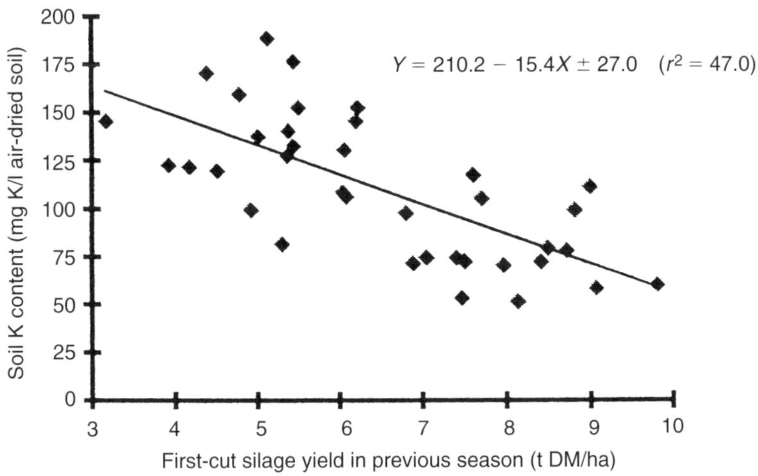

Fig. 14.2 Effect of first-cut silage yield in previous season on K status of a sandy loam soil.

on clover content will restrict the input of atmospherically fixed N in the system and have a knock-on effect on herbage yield. Newton & Stopes (1995) reported that soil P status was the best predictor of herbage production in a survey of organic farms.

It is essential, therefore, that organic grass–clover swards intended for conservation be given high priority in the distribution of manures, particularly on sandy soils. In a field trial in which a range of manure and silage effluent treatments were applied to an organic grass–clover sward on loamy sand soil (5% clay) for two cuts of silage, it was estimated that, in order to maintain a zero change in soil K content, an annual application of 233 kg K ha^{-1} (282 kg K_2O ha^{-1}) was necessary (Younie *et al.* 1998). The same trial indicated that where two cuts were being harvested, a single spring application of 20 t FYM ha^{-1} was less effective in maintaining soil K content than two applications of 10 t ha^{-1}, one for each cut. Luxury uptake of K can result from large spring applications of manure, and therefore manure and effluent applications should be spread evenly throughout the season, perhaps even with an emphasis towards late-season application, although this increases the risk of leaching loss of N and K.

Because of the labile nature of K_2O in manure and in soils, the risk of loss during the nutrient cycling process itself is substantial, in contrast to the situation with P_2O_5. Berner (1986) estimated losses of 50–60% of total K from farmyard manure heaps on commercial organic farms. Nolte & Werner (1994) estimated that 78% of whole-farm K losses were lost during the internal nutrient cycling process, but only 25% of P losses, and suggested that the K balance could be improved substantially by improving manure handling and storage procedures.

Inevitably there will be transfer of nutrients, of K in particular, from grassland to arable cropping land, given that FYM is one of the major sources of nutrient for arable and vegetable crops in organic systems, although in a mixed ley/arable system the rotation of crops means that these nutrients will ultimately be uniformly distributed around the farm. However, even if 100% efficiency in nutrient cycling were possible, there would still be a considerable loss of P and K in crop and livestock products sold off the farm and in leaching losses in the field (on biodynamic farms, Nolte & Werner (1994) estimated 14% of K losses and 73% of P losses were in products sold off the farm). The inference is therefore that soil P and K in grassland will be depleted over time, or that supplies of P and K must be brought into the system. The rate and extent of depletion, and therefore the need for inputs, will depend on the level of crop yield and nutrient off-take and the level of soil nutrient reserves. Some clay soils have large reserves of potassium ions and have the capacity to release K into the exchangeable fraction, and hence support good crop growth, over a long period of time. Sandy or coarse-textured soils with a low clay content, on the other hand, have a low reserve of non-exchangeable K and will be unable to sustain high production levels without K inputs.

Rock phosphate is permitted as a nutrient input in organic farming and is a relatively effective source of P, but unfortunately there are few effective sources of K input which are acceptable in organic systems. Simpson & Stopes (1991) assessed a range of potassium fertilizers for organic farming and found that rock potash was a relatively ineffective source of potash, whereas lime kiln dust gave useful responses. Potassium sulphate, Kainit, Sylvanite and Cummulus K (Kali Vinesse) are effective, but their use is restricted by UK organic certification authorities to soils where clay content is low (below 20%) and soil K is deficient. A further potential source of potash is in straw bought in for bedding but, assuming a stocking rate of one livestock unit per forage hectare, this will supply only about 13 kg K ha^{-1} if 1 tonne of barley straw is purchased annually per livestock unit.

14.5.4 Optimizing herbage production and utilization per hectare

Choice of species

White clover is by far the most appropriate and widely used forage legume for organic farming systems in temperate maritime climates, because of its adaptability to a range of management and soil fertility conditions. It is persistent, tolerant of a reasonably wide range of soil pH and drainage conditions, and can be used for management regimes ranging from continuous sheep grazing (for which small-leaved varieties are most suitable) to lax defoliation, including cutting (for which larger-leaved varieties are most suitable). Red clover is also undemanding in terms of soil conditions and is highly productive, but is not persistent and should only be sown in mixtures intended for short-term leys (up to 3 years), primarily for cutting. It is probably the best species for use as a 1- or 2-year green manure. Lucerne and sainfoin (*Onobrychis sativa*) also have considerable potential as crops for conservation, but are suitable only for soils with a high pH and, in the case of lucerne, good drainage.

Although perennial ryegrass has been criticized as being more appropriate for intensive high-N systems than for organic or low-input systems, it is undoubtedly the most suitable species for ley farming in temperate maritime conditions, given its ease of establishment, yield potential, persistence and quality characteristics. Tetraploid varieties are known to promote a higher clover content in the sward than diploid varieties because of their more open growth habit, and should constitute at least 75% of the perennial ryegrass component in organic seed mixtures.

A number of herb species, often included in grass-based seed mixtures in the past, have deep tap root systems and have the potential to explore deeper soil layers, exploiting a greater volume of soil than conventional herbage species, and potentially making available an enhanced supply of nutrients for consumption by grazing livestock and consequent recycling to surface soil layers. Herbs tend to be more drought-resistant than grasses, and their tap root systems may also lead to a more open soil structure.

Chemical analysis indicates that a number of herb species consistently have higher concentrations of some important minerals than perennial ryegrass (Jones 1990, Swift *et al.* 1990, Fisher & Baker 1996). Wilman & Derrick (1994) reported higher calcium and magnesium status in lambs fed a range of herbs, including ribwort (*Plantago lanceolata*) and dandelion (*Taraxacum officinale*), compared with perennial ryegrass. Similarly, again in comparison with perennial ryegrass, Younie *et al.* (1997) demonstrated significant improvements in calcium, sodium, magnesium, copper and vitamin B_{12} (cobalt) status of lambs from feeding a 100% chicory diet (*Cichorium intybus*). The extent of any benefit in mineral status of animals from feeding herbs is likely to depend on the site (the inherent mineral status of the soil, and hence the potential mineral content of the herbage), and the proportion of herbs in the diet of the animals.

Despite these potential nutritional and agronomic advantages, forage herbs have been largely overlooked in practice, even in organic farming (Foster 1988). This is changing to some extent, at least in New Zealand and Australia, where breeding effort has led to the commercial release of varieties of chicory and ribwort (Rumball 1986, Moloney & Milne 1993). Perhaps the main reason for the low level of use of forage herbs is their generally low contribution to sward biomass, largely resulting from low seed rates, poor establishment and persistence. This may be due, at least in part, to their inability to persist under intensive management regimes which involve frequent utilization by both cutting and grazing – regimes to which vigorously tillered grass species such as perennial ryegrass are ideally suited. Research in New Zealand has already indicated that chicory should be managed according to a rotational grazing regime, and that, compared with 8-week regrowth intervals, 4-week intervals (Clark *et al.* 1990) and a hard/lax grazing sequence (Li *et al.* 1997) increase the ratio of leaves to stems, thus enhancing the palatability and digestibility of the material. New thinking is also required in relation to the most appropriate companion species and mixtures for forage herbs. Umrani (1998) has shown that perennial ryegrass, with its vigorous, densely tillered growth habit, is a major competitor. It significantly reduced root length, root weight and shoot weight of individual herb plants. The upright and less well tillered growth habit of timothy is less competitive and, coupled with the N-fixing ability of white clover, makes for a more suitable mixture. Such timothy/herb/clover combinations will be best sown in strips or as separate swards, with the main bulk of the field sown to a ryegrass-based mixture, given that perennial ryegrass, with all its other advantages, will probably remain as the mainstay of organic grassland.

Control of perennial weeds in organic grassland

Control of perennial weeds such as dock and creeping thistle presents a major agronomic challenge in long-term organic grassland, since herbicide use is prohibited. Where the site is ploughable, adoption of a regular programme of grass renewal (i.e. a ley system, using 3–5-year leys) will prevent the problem developing in the first place. Hopkins *et al.* (1997) have suggested that high grass and white clover seed rates can reduce dock infestation at sward establishment. This ley policy also has other major advantages in the organic farming context, such as the annual provision of swards uncontaminated with worm larvae, and the opportunity to exploit fertility built up in the soil by growing arable cash crops or annual forage crops for finishing lambs. However, there is an expense involved in reseeding, and there will be situations where site conditions prevent regular ploughing and reseeding.

Ragwort is effectively controlled in long-term grassland by sheep grazing in winter and early spring, but this is not effective for dock and creeping thistle. With these weeds, other approaches are necessary. A dock infestation becomes worse much more rapidly under a cutting than under a grazing management (Courtney 1985), and so continuous cutting for hay or silage should be avoided in permanent grassland, if possible. Regular topping of docks in grazed swards, and soil aeration to improve grass growth and cause physical damage to dock roots, may also limit the increase in dock density (Hopkins *et al.* 1997). When dock-infested grassland is destroyed prior to establishment of a winter cereal crop, rotovating the turf to a depth of 8–10 cm, followed by desiccation for 8–10 weeks before ploughing, has been successful in farm practice (Welsh 1995). Topping is also the only method advised for control of creeping thistle in organic grassland.

Livestock systems for organic grassland

Factors relating to livestock nutrient supply (from grassland and supplementary feeds) have a major influence on the choice of livestock system on organic farms (although a number of other factors, such as marketing and farm infrastructure constraints, are also crucially important). Organic standards specify that ruminant animals should obtain most of their nutrient supply from forage. In any case, organic cereals and concentrates are generally very expensive. Thus, livestock systems which rely primarily on forage, particularly grazed grass, are most appropriate; for example, calving in spring or early summer rather than in autumn so that the nutrient requirements of the animals matches the grass growing season. In the case of sheep, early lambing in January is not an appropriate system, given the level of concentrate feeding involved. In contrast, lambing later than normal, in late April or May, to coincide with the later spring growth of clover-based organic swards, is well suited to organic farming and reduces the input of concentrates and the overall cost of production. During the winter feeding period, productive classes of livestock such as lactating dairy cows have a high requirement for protein. Sources of supplementary protein which meet the organic standards are scarce and expensive, and there is therefore a need to maximize home-produced protein from grain legumes and silage, e.g. from red clover or lucerne.

14.5.5 Physical and financial output from organic grassland

Given satisfactory levels of soil P and K and pH status (and soil moisture supply), the level of herbage production and livestock output from organic grassland is largely dependent on its legume content. In this regard it equates closely with conventional, low external input grass–clover swards. A considerable body of research literature now exists on legume-based grassland systems, and most of this is relevant to organic farming (Thomson 1984, Frame & Newbould 1986, Plancquaert & Haggar 1989, Hopkins *et al.* 1994, Sheldrick *et al.* 1995, Frame *et al.* 1998).

In order to illustrate the potential levels of production possible in organic systems, published data are presented in Table 14.5. These generally reflect the conditions under which the swards were monitored, namely site class, legume species, cutting date or regrowth interval.

Financial analysis of organic farming should be carried out on a whole farm basis, given the holistic nature of the system, but enterprise analysis is appropriate when undertaking an organic/conventional comparison restricted to the grassland and livestock components of the farm system.

Profitability of livestock enterprises is a function of physical output times unit sale price, minus input costs. Physical output per head from organic beef and sheep enterprises compares well with equivalent conventional grass-based enterprises (Younie & Mackie 1996, Younie 1997), with beef animals reaching slaughter condition in 17–21 months. On the other hand, average milk yield per cow from organic dairy herds, at around 5000 kg per annum, is normally lower than from conventional units (Houghton & Poole 1990, Redman 1991, Lampkin 1994, Krutzinna *et al.* 1996, Weller *et al.* 1996), reflecting the generally lower levels of concentrate input (limited by the organic standards to 40% of daily

Table 14.5 Production per hectare from organic grassland.

Reference	Sward type	Farm and stock type	Herbage production ($t\,DM\,ha^{-1}$)			Stocking rate ($LU\,ha^{-1}$)
			Annual (grazing)	Annual (mainly cut)	First-cut silage	
Fowler *et al.* (1993)	G/WC[1]	Lowground, dairy	10.1	10.1		1.6
Houghton & Poole (1990)	Unspecified	Lowground, dairy				1.6
Jones *et al.* (1996) (i)	G/WC	Lowground, dairy	7.2		4.2	1.7
Jones *et al.* (1996) (ii)	G/WC	Lowground, dairy	7.1	8.8		1.6
Jones *et al.* (1996) (iii)	G/RC[2]	Lowground, dairy		14		
Halberg *et al.* (1995)	Varied	Lowground, dairy				1.1
Newton & Stopes (1995)	G/WC	Varied, mixed		13.0[4]		
Redman (1991)	Unspecified	Lowground, dairy				1.8
Van der Meer & Baan Hofman (1989) (i)	G/WC	Lowground, dairy	6.4			1.9
Van der Meer & Baan Hofman (1989) (ii)	G/RC, L[3]	Lowground, dairy		10.9		
Younie & Wightman (1992)	G/WC	Lowground, beef	8.7	8.9	5.9	2.1
Younie (1997)	G/WC	Upland, beef, sheep		7.6	5.4	1.2
Younie (1996b)	G/WC	Lowground, beef, sheep		8.0	5.5	

[1] Grass–white clover.
[2] Grass–red clover.
[3] Lucerne.
[4] Cut to ground level.

DM intake), and the greater use by organic dairy farmers of lower-yielding breeds such as Jerseys or Ayrshires. However, the high cost of organic concentrates can result in relatively high feed costs per head, particularly in dairy and finishing beef enterprises, although this is offset by reduced forage costs. Total variable costs per head are therefore generally only slightly lower than in conventional enterprises.

Large and frequent fluctuations in unit sale price of both organic and conventional products make detailed financial comparisons difficult, and valid only in the short term. The demand for organic livestock products has outstripped supply, but lack of supply and a poorly developed market infrastructure have limited the development of the market and thus premium prices have not always been widely available. However, where organic premiums can be achieved, gross margin per head for organic livestock enterprises is normally substantially higher than the conventional equivalent. On a per hectare basis, organic systems, which are clover-based, will normally support a stocking rate and livestock output equivalent to a conventional system receiving up to about 180–200 kg N ha^{-1} per annum, and should therefore achieve similar or better gross margins compared with such conventional enterprises, even without the benefit of organic premium prices. With the added bonus of higher prices, organic systems have the potential to be substantially more profitable on a per hectare basis than such conventional enterprises (e.g. Keatinge 1997, Lampkin 1997, Younie 1997). On the other hand, compared with intensively fertilized conventional enterprises receiving more than 200 kg N ha^{-1} per annum, organic premiums are probably essential if organic systems are to match such conventional systems on gross margin per hectare.

The availability of organic price premiums is a matter of concern for farmers considering conversion, since the relative profitability depends so much on these premiums. However, continuing downward pressure on conventional prices towards world market levels, and increasing consumer concerns over food safety and quality, have focused farmers' attentions increasingly on the potential for product differentiation (and hence the ability to command higher prices). Organic produce has a very strong brand image in the marketplace, and therefore would appear to have the potential to maintain, in the long term, a price differential over conventional products.

14.6 References

Angers D.A. & Carter M.R. (1996) Aggregation and organic matter storage in cool, humid agricultural soils. In Carter M.R. & Stewart B.A. (eds) *Structure and Organic Matter Storage in Agricultural Soils*, 193–211. CRC Press, Boca Raton, FL.

Baars T., van der Klundert A.F. & Reijer A.W.J. de (1982) Vegetatie, bodem, mineralenbalans van enkele gangbare en biologische bedrijven in Noord-Holland. *Doctoraalvak Lanschapsecologie en Natuurbeheer*, Universiteit Utrecht.

Bakken L., Refsgaard K., Christensen S. & Vatn A. (1994) Energy use and emission of greenhouse gases from grassland agricultural systems. In 't Mannetje L. & Frame J. (eds) *Grassland and Society. Proceedings 15th General Meeting European Grassland Federation, Wageningen, June 1994*, 361–76.

Berner A. (1986) Einfluß des Sickersaftes von Kompostmieten auf die Umwelt. *Forschungsinstitut für Biologischen Landbau*. Oberwil, Switzerland.

Bulson H.A.J., Welsh J.P., Stopes C.E. & Woodward L. (1996) Agronomic viability and potential economic performance of three organic four-year rotations without livestock, 1988–1995. In Clarke J.H., Davies D.H.K. & Dampney P.M.R. (eds) *Rotations and Cropping Systems. Aspects of Applied Biology* **47**, 277–86.

CEC (1991) *Council Regulation No 2092/91 on Organic Production*. Official Journal of the European Communities **L198**, 1–15.

CEC (1999) *Council Regulation No 1804/1999 Supplementing Regulation No 2092/91 on organic production*. Official Journal of the European Communities L222, 1–28.

Clark D.A., Anderson C.B. & Berquist T. (1990) Growth rates of Grasslands Puna chicory (*Cichorium intybus* L.) at various cutting intervals and heights and rates of nitrogen. *New Zealand Journal of Agricultural Research* **33**, 213–17.

Clement C.R. & Williams T.E. (1967) Leys and soil organic matter. II. The accumulation of nitrogen in soils under different leys. *Journal of Agricultural Science, Cambridge* **69**, 133–8.

Clements R.O., George S., Martyn T., Donaldson G. & Balsdon S. (1996) Clover: cereal bi-cropping – a low-input cost-effective method of growing whole-crop winter wheat. In Younie D. (ed) *Legumes in Sustainable Farming Systems*. British Grassland Society Occasional Symposium No. 30, 234–46. BGS, Reading.

Cobb R., Feber R., Hopkins A. & Stockdale E. (1997) The organic farming study. *Elm Farm Research Centre Bulletin*, No. 33, 8–10.

Cormack W. (1997) Testing the sustainability of a stockless arable rotation on a fertile soil in eastern England. In Isart J. & Llerena J.J. (eds) *Resource Use in Organic Farming. Proceedings of Third Workshop of European Network for Scientific Coordination in Organic Farming (ENOF), University of Ancona, Italy*, June 1997, 127–35.

Courtney A.D. (1985) Impact and control of docks in grassland. In Brockman J.S. (ed) *Weeds, Pests and Diseases of Grassland and Herbage Legumes*. British Grassland Society Occasional Symposium No. 18, (British Crop Protection Council Monograph No. 29), 120–7.

Davies D.A. & Hopkins A. (1996) Production benefits of legumes in grassland. In Younie D. (ed) *Legumes in Sustainable Farming Systems*. British Grassland Society Occasional Symposium No. 30, 234–46. BGS, Reading.

Douglas J.T. & Crawford C.E. (1993) The response of a ryegrass sward to wheel traffic and applied nitrogen. *Grass and Forage Science* **48**, 91–100.

Drinkwater L.E., Fiorina L.J. & Wagoner P. (1996) *Rodale's Farming Systems Trial: First 15 years. Proceedings 11th IFOAM International Scientific Conference, Copenhagen*, 62.

Eder G. & Harrod T.R. (1996) Influence of grassland on soil erosion, aggregate stability and water quality. In Parente G., Frame J. & Orsi S. (eds) *Grassland and Land Use Systems. Proceedings 16th General Meeting of European Grassland Federation, Grado, Italy, September 1996*, 683–93.

Edwards S.A. & Watson C. (1997) An approach to investigating the environmental impact of an outdoor pig production system. In Sorensen J.T. (ed) *Livestock Farming Systems – More than Food Production*. EAAP Publication No. 89, 335–40. Wageningen Pers, Wageningen.

Ennik G.C., Baan Hofman T., Wieling H. & Altena H.J. (1982) Grasproduktie zonder kunstmeststikstof op het bedrijf van de familie Cuperus te Boksum (F) *CABO-verslag Nr. 42*.

Fisher G. & Baker L. (1996) The chemical composition of forb species in grassland. In Parente G., Frame J. & Orsi S. (eds) *Grassland and Land Use Systems. Proceedings of*

16th General Meeting of European Grassland Federation, Grado, Italy, September 1996, 429–31.
Fisher N.M. (1996) The potential of grain and forage legumes in mixed farming systems. In Younie D. (ed) *Legumes in Sustainable Farming Systems*. British Grassland Society Occasional Symposium No. 30, 290–9. BGS, Reading.
Follett R.F. & Schimel D.S. (1989) Effect of tillage practices on microbial biomass dynamics. *Soil Science Society of America Journal* **53**, 1091–6.
Foster L. (1988) Herbs in pastures. Development and research in Britain, 1850–1984. *Biological Agriculture and Horticulture* **5**, 97–133.
Fothergill M., Davies D.A. & Morgan C.T. (1995) The effect of extensification of upland pasture on white clover. In Pollott G.E. (ed) *Grassland into the 21st Century*. British Grassland Society Occasional Symposium No. 29, 171–2. BGS, Reading.
Fowler S.M., Watson C.A. & Wilman D. (1993) N, P and K on organic farms: Herbage and cereal production, purchases and sales. *Journal of Agricultural Science, Cambridge* **120**, 353–60.
Frame J. (1990) The role of red clover in United Kingdom pastures. *Outlook on Agriculture* **19**, 24–55.
Frame J. & Newbould P. (1986) Agronomy of white clover. *Advances in Agronomy* **40**, 1–88.
Frame J., Charlton J.F.L. & Laidlaw A.S. (1998) *Temperate Forage Legumes*, 336 pp. CAB International, Wallingford.
Frieben B. & Köpke U. (1996) Effects of farming systems on biodiversity. In Isart J. & Llerena J.J. (eds) *Organic Farming in Land Use Systems. Proceedings of First Workshop of European Network for Scientific Coordination in Organic Farming (ENOF), University of Bonn, December 1995*, 11–21.
Gray D., Cowie R., Tierney J. & Younie D. (1997) Epidemiology of PGE in a May lambing organic sheep flock. *Proceedings of Scottish Agricultural College Veterinary Conference, 1997*, Appendix 1, poster 7.
Halberg N., Kristensen E.S. & Kristensen I.S. (1995) Nitrogen turnover on organic and conventional mixed farms. *Journal of Agricultural and Environmental Ethics* **8**, 30–51.
Halliday G., Ramsay D.A., Scanlan S. & Younie D. (1991) *A Survey of Organic Livestock Health and Treatment*. Kintail Land Research Foundation, Glasgow, in association with Scottish Agricultural College, 45 pp.
Halliday G.J., Gray D. & Younie D. (1994) Some animal health problems on a small organic farm in Scotland. *Actes du Congrés de l'Élevage en Agriculture Biologique, Brioude, France*, 134–6. Institut Technique de l'Agriculture Biologique, Paris.
Holle R. (1996) Stockless farming in Northern Germany. *Proceedings 11th IFOAM International Scientific Conference, Copenhagen*, 53.
Hopkins A., Davies D.A. & Doyle C. (1994) *Clovers and other Grazed Legumes in UK Pasture Land*. Institute of Grassland and Environmental Research Technical Review No. 1, 61 pp. IGER, Aberystwyth.
Hopkins A., Jones E.L., Bowling P.J. & Johnson R.H. (1997) Cultural methods of dock control in permanent pasture. *Proceedings, British Grassland Society Fifth Research Conference, Newton Abbot, September 1997*, 39–40.
Houghton M. & Poole A.H. (1990) *Organic Milk Production*. Genus Information Unit Report No. 70, 31 pp.
IFOAM (1996) *Basic Standards of Organic Agriculture*. International Federation of Organic Agriculture Movements, Tholey-Theley, Germany.
Jansen S.P.J., van Hooff W. & Smeets P.J.A.M. (1985) Processen en structuren in het agrarisch landschap van Terschelling en Schiermonnikoog. *Dorschkamprapport nr. 405*, 163 pp.
Jenkinson D.S. (1988) Soil organic matter and its dynamics. In Wild A. (ed) *Russell's Soil Conditions and Plant Growth*, 11th edn, 564–607. Longman, Harlow.
Johnson A.E., McEwen J., Lane P.W., Hewitt M.V., Poulton P.R. & Yeoman D.P. (1994)

Effects of one- to six-year-old ryegrass–clover leys on soil nitrogen and on the subsequent yields and fertilizer nitrogen requirements of the arable sequence winter wheat, potatoes, winter wheat, winter beans (*Vicia faba*) grown on sandy loam soil. *Journal of Agricultural Science, Cambridge* **122**, 73–89.

Jones E.L., Bowling P.J. & Haggar R.J. (1996) Forage yields and quality during conversion. In Haggar R.J. & Padel S. (eds) *Conversion to Organic Milk Production*. Institute of Grassland and Environmental Research Technical Review No. 4, 57–65. IGER, Aberystwyth.

Jones L. (1990) Forage chicory at a site in south-east England. In Pollott G. (ed) *Milk and Meat from Forage Crops*. British Grassland Society Occasional Symposium No. 24, 67–9. BGS, Hurley.

Keatinge R. (1997) Organic hill farming: Does it pay? *New Farmer and Grower* No. 55, 24–5.

Kristensen E.S., Høgh-Jensen H. & Kristensen I.S. (1995) A simple model for estimation of atmospherically derived nitrogen in grass–clover systems. *Biological Agriculture and Horticulture* **12**, 263–76.

Krutzinna C., Boehncke E. & Herrmann H.-J. (1996) Organic milk production in Germany. *Biological Agriculture and Horticulture* **13**, 351–8.

Lampkin N.H. (1994) Economics of organic farming in Britain. In Lampkin N.H. & Padel S. (eds) *The Economics of Organic Farming: An International Perspective*, 71–89. CAB International, Wallingford.

Lampkin N. (1997) Opportunities for profit from organic farming. In *Organic Farming – Science into Practice*, Royal Agricultural Society of England Conference, November 1997 (unpaginated). RASE, Stoneleigh.

Li G.D., Kemp P.D. & Hodgson J. (1997) Regrowth, morphology and persistence of Grasslands Puna chicory (*Cichorium intybus* L.) in response to grazing frequency and intensity. *Grass and Forage Science* **52**, 33–41.

Moloney S.C. & Milne G.D. (1993) Establishment and managment of Grasslands Puna chicory used as a specialist, high quality forage herb. *Proceedings of the New Zealand Grassland Association* **55**, 113–18.

Morrison J. (1981) The potential of legumes for forage production. In *Legumes and Fertilizers in Grassland Systems*. British Grassland Society Winter Meeting 1981, 1.1–1.9. BGS, Hurley.

Newton J. & Stopes C. (1995) Grassland productivity on organic farms, 1992–1994. *Elm Farm Research Centre Bulletin*, No. 18, 2–6.

Nguyen M.L., Haynes R.J. & Goh K.M. (1995) Nutrient budgets and status in three pairs of conventional and alternative mixed cropping farms in Canterbury, New Zealand. *Agriculture, Ecosystems and Environment* **52**, 149–62.

Niezen J.H., Charleston W.A.G., Hodgson J., Mackay A.D. & Leathwick D.M. (1996) Controlling internal parasites in grazing ruminants without recourse to anthelmintics: Approaches, experiences and prospects. *International Journal of Parasitology* **26**, 983–92.

Nolte C. & Werner W. (1994) Investigations on the nutrient cycle and its components of a biodynamically managed farm. *Biological Agriculture and Horticulture* **10**, 235–54.

Philipps L. & Stopes C. (1995) The impact of rational practice on nitrate leaching losses in organic farming systems in the United Kingdom. In Kristensen L. (ed) *Nitrogen Leaching in Ecological Agriculture*, 123–34. AB Academic Publishers, Bicester.

Plancquaert P. & Haggar R.J. (eds) (1989) *Legumes in Farming Systems*, Proceedings of a Workshop. Kluwer, Dordrecht, 181 pp, as *Developments in Plant and Soil Science*, vol. **37**.

Redman M. (1991) Organic dairy costings: Update. *New Farmer and Grower* No. 33, 20–21.

Refsgaard K., Halberg N. & Kristensen E.S. (1998) Energy utilization in crop and dairy production in organic and conventional livestock production systems. *Agricultural Systems* **57**, 599–630.

Reganold J.P. & Palmer A.S. (1995) Significance of gravimetric versus volumetric measurements of soil quality under biodynamic, conventional, and continuous grass management. *Journal of Soil and Water Conservation* **50**, 298–305.

Reganold J.P., Palmer A.S., Lockhart J.C. & Macgregor A.N. (1993) Soil quality and financial performance of biodynamic and conventional farms in New Zealand. *Science, Washington* **260**, 344–9.

Reid D. (1970) The effects of a wide range of nitrogen application rates on the yields from a perennial ryegrass sward with and without white clover. *Journal of Agricultural Science, Cambridge* **74**, 227–40.

Roderick S., Short N. & Hovi M. (1996) Organic livestock production: Animal health and welfare research priorities. Veterinary Epidemiology and Economics Research Unit, Department of Agriculture, University of Reading.

Rumball W. (1986) 'Grasslands Puna' chicory (*Cichorium intybus* L.) *New Zealand Journal of Experimental Agriculture* **14**, 165–71.

Sheldrick R.D., Newman G. & Roberts D.J. (1995) *Legumes for Milk and Meat*. 2nd edn, 109 pp. Chalcombe, Canterbury.

Simpson D. & Stopes C. (1991) The requirement and availability of potassium fertilizers in organic agriculture. *Ecology and Farming* **3**, 14–16.

Steiner R. (1974) *Agriculture: A Course of Eight Lectures*. 3rd edn. Rudolf Steiner, London.

Stopes C. (1993) Green manure trials. *Elm Farm Research Centre Bulletin*, No. 8, 6–7.

Swift G., Davies D.H.K., Tiley G.E.D. & Younie D. (1990) The nutritive value of broad-leaved weeds and forage herbs in grassland. *Scottish Agricultural College, Technical Note No. 223.*

Thomson D.J. (ed) (1984) *Forage Legumes*. British Grassland Society Occasional Symposium No. 16, 236 pp. BGS, Hurley.

Tisdall J.M. & Oades J.M. (1980) The effect of crop rotation on aggregation in a red-brown earth. *Australian Journal of Soil Resources* **18**, 423–33.

Titchen N.M. & Philipps L. (1996) Environmental effects of legume-based grassland systems. In Younie D. (ed) *Legumes in Sustainable Farming Systems*. British Grassland Society Occasional Symposium No. 30, 257–61. BGS, Reading.

Tyson K., Stone A.C., Scholefield D. & Jarvis S.C. (1996) A comparison of animal output and nitrogen leaching losses from grassland receiving 200 kgN/ha/annum and grass/clover receiving no fertilizer nitrogen. In Younie D. (ed) *Legumes in Sustainable Farming Systems*. British Grassland Society Occasional Symposium No. 30, 273–4.

UKROFS (1997) *UKROFS Standards for Organic Food Production*. United Kingdom Register of Organic Food Standards, London.

Umrani A.P. (1998) *Sustainable Approaches for Rangeland Management and Livestock Production in Pakistan*. PhD Thesis, University of Aberdeen.

Van der Meer H.G. & Baan Hofman T. (1989) Contribution of legumes to yield and nitrogen economy of leys on a biodynamic farm. In Plancquaert P. & Haggar R.J. (eds) *Legumes in Farming Systems. Developments in Plant and Soil Science* **37**, 25–36.

Wander M.M. & Traina S.J. (1996) Organic matter fractions from organically and conventionally managed soils. 1. Carbon and nitrogen distribution. *Soil Science Society of America Journal* **60**, 1081–7.

Watson C.A. & Younie D. (1995a) Nitrogen balances in organically and conventionally managed beef production systems. In Pollott G.E. (ed) *Grassland into the 21st Century*. British Grassland Society Occasional Symposium No. 29, 197–9. BGS, Reading.

Watson C.A. & Younie D. (1995b) The contribution of winter cover crops to the sustainability of agricultural systems. In Pollott G.E. (ed) *Grassland into the 21st Century*. British Grassland Society Occasional Symposium No. 29, 194–6. BGS, Reading.

Watson C.A., Ritz K., Younie D. & Franklin M. (1996) Nitrogen and soil biomass dynamics in ley/arable crop rotations. In Clarke J.H., Davies D.H.K. & Dampney P.M.R. (eds) *Rotations and Cropping Systems. Aspects of Applied Biology* **47**, 43–50.

Weber S., Pabst K., Schulte-Corne H., Westphal R. & Gravert H.O. (1993) Funfjahrige

Untersuchung zur Umstellung auf okologische Milcherzeugung. 1. Produktionstechnik. [Five years studies on the conversion to ecological milk production. 1. Production techniques.] *Zuchtungskunde* **65**, 325–37.
Weil R.R., Lowell K.A. & Shade H.M. (1993) Effects of intensity of agronomic practices on a soil ecosystem. *American Journal of Alternative Agriculture* **8**, 5–14.
Weller R.F., Cooper A. & Padel S. (1996) Animal production during conversion. In Haggar R. & Padel S. (eds) *Conversion to Organic Milk Production*. Institute of Grassland and Environmental Research Technical Review No. 4, 84–96. IGER, Aberystwyth.
Welsh J.P. (1995) Dock control demonstration trial. *Elm Farm Research Centre Bulletin* No. 18, 6–7.
Whitehead D.C. (1995) *Grassland Nitrogen*, 397 pp. CAB International, Wallingford.
Wilman D. & Derrick R.W. (1994) Concentration and availability to sheep of N, P, K, Ca, Mg and Na in chickweed, dandelion, dock, ribwort and spurrey, compared with perennial ryegrass. *Journal of Agricultural Science, Cambridge* **122**, 217–23.
Wood M. (1996) Nitrogen fixation: How much and at what cost? In Younie D. (ed) *Legumes in Sustainable Farming Systems*. British Grassland Society Occasional Symposium No. 30, 26–35. BGS, Reading.
Worner M. & Taube F. (1995) Artenzusammensetzung des Dauergrunlandes im okologischen Landbau – eine Erhebung auf norddeutschen Praxisflachen. [Species composition of permanent grassland in ecological agriculture – a survey on North German farms.] *Kongressband 1992 Gottingen. Vortrage zum Generalthema des 104.* VDLUFA-Kongresses vom 14–19.9.92 in Gottingen: Okologische Aspekte extensiver Landbewirtschaftung, 1992, 623–6.
Younie D. (1992) Potential output from forage legumes in organic systems. In Peeters A. & Van Bol V. (eds) *Potential and Limits of Organic Farming, Proceedings of an EC Workshop, Louvain-la-Neuve, Belgium, September 1992*, 116–29.
Younie D. (ed) (1996a) *Legumes in Sustainable Farming Systems*. British Grassland Society Occasional Symposium No. 30, 332 pp. BGS, Reading.
Younie D. (1996b) A comparison of crop rotations in organic farming: Agronomic performance. In Clarke J.H. *et al.* (eds) *Rotations and Cropping Systems. Aspects of Applied Biology* **47**, 379–82.
Younie D. (1997) Organic beef and sheep systems. In *Grass is Greener*. British Grassland Society Winter Meeting 1997 (unpaginated). BGS, Reading.
Younie D. (1998) Establishment of herbage legumes and grass–clover leys in mixed farming systems. In van Keulen H., Lantinga E.A. & van Laar H.H. (eds) *Mixed Farming Systems in Europe. Proceedings of an International Workshop, Wageningen Agricultural University*, APMinderhoudhoeve-reeks No. 2, 173–8.
Younie D. & Armstrong G. (1996) Botanical and invertebrate diversity in organic and intensively fertilized grassland. In Isart J. & Llerena J. (eds) *Biodiversity and Land Use: The Role of Organic Farming. Proceedings of first Workshop of European Network for Scientific Coordination in Organic Farming (ENOF), Bonn, December 1995*, 35–44.
Younie D. & Baars T. (1997) Resource use in organic grassland: The Central Bank and the Art Gallery of Organic Farming. In Isart J. & Llerena J. (eds) *Resource Use in Organic Farming, Proceedings of Third Workshop of European Network for Scientific Coordination in Organic Farming (ENOF), University of Ancona, Italy, June 1997*, 43–60.
Younie D. & Mackie C.K. (1996) Factors affecting profitability of organic, low-input and high-input beef systems. In Parente G., Frame J. & Orsi S. (eds) *Grassland and Land Use Systems. Proceedings 16th General Meeting of European Grassland Federation, Grado, Italy, September 1996*, 879–82.
Younie D. & Watson C.A. (1995) Crop growth and nitrogen recovery following ploughing of grass/clover swards. In Pollott G.E. (ed) *Grassland into the 21st Century*. British Grassland Society Occasional Symposium No. 29, 206–208. BGS, Reading.
Younie D. & Wightman P.S. (1992) Herbage production over eight years from

clover-based and intensively fertilized swards under grazing and silage management. *Proceedings, British Grassland Society Third Research Conference, Greenmount, September 1992*, 99–100. BGS, Reading.

Younie D., Watson C.A., Ramsay D.A. & Mackie C.K. (1995) Systems interactions and the role of grassland in sustainability. In Pollott G.E. (ed) *Grassland into the 21st Century*. British Grassland Society Occasional Symposium No. 29, 314–16. BGS, Reading.

Younie D., Umrani A.P., Gray D. & Coutts M. (1997) Influence of chicory or ryegrass diets on trace element status of lambs. In *Organic Farming – Science into Practice, Royal Agricultural Society of England Conference, November 1997* (unpaginated). RASE, Stoneleigh.

Younie D., Wilson J.F., Reid C. & Wightman P.S. (1998) Maintenance of soil potassium in organic farming silage systems. In Nagy G. & Peto K. (eds) *Ecology and Grassland. Proceedings 17th General Meeting of European Grassland Federation, Debrecen, Hungary, May 1998*, 289–92.

Chapter 15
Economic Aspects of Grassland Production and Utilization

J.P. McInerney

15.1 Introduction

A popular misconception amongst those who have not studied the subject is that economics is about money. Thus, book chapters with the title 'Economic Aspects of . . .' typically create an expectation that what follows will be details of the costs and revenues of the particular activity, a discussion of whether it is financially worthwhile, and what factors affect the profit that can be made. That expectation, however, is too narrow an interpretation of economics because it focuses only on how well the individual producer/firm does out of the activity. Private commercial gain (although clearly of interest to the accountant) is not the sole criterion of economic relevance. Economic analysis is concerned just as much with the impacts on a wider group of people – consumers, input suppliers, third parties who might be affected and 'society' in general.

Economics is about creating *value* – i.e. something that human beings perceive as worth having, making them 'better off', satisfying their requirements. Money comes into it because that is the only measurement unit by which to aggregate very diverse components of gain or loss in value. It does not imply that monetized commodities (items that carry a price label) are the only things that have economic value or that should influence decisions about grassland farming, for example. Indeed, much analysis has to deal with components that are clearly economic benefits or costs but do not have money prices attached to them because they are not bought or sold. This is the case for many things associated with agricultural land use – for example, wildlife, biodiversity, visual amenity, access, chemical pollution, animal welfare – and increasingly the economic aspects of farming need to take them into account. The magnitude of profits may indicate success in creating value as seen from the narrow standpoint of a commercial producer, but that person is only one of the interested parties in economic activity. A few obvious examples make this clear. Suppose those profits have been made at the expense of other people's benefit – paying low wages to the workforce, using market power to sell to consumers at high prices, creating serious pollution or other negative effects on third parties, depleting resources or damaging the opportunities of future producers; all of these things are common in commercial practice, and are not necessarily illegal, but they could hardly be called 'good economics'. Alternatively, suppose the activity was highly profitable but left many other people feeling uncomfortable (hence experiencing a loss in value) – such as imposing a burden on the welfare of farm animals, dealing in pornography, excluding vulnerable (poor, disabled) groups in society, or generating benefits only for those judged to be extremely well off already – all things that commonly occur in everyday economic activity and sometimes lead governments to intervene on behalf of a wider economic interest. Finally, the concept of profit has no merit in measuring the success of all those value-creating activities undertaken by public agencies where the output is 'given away' rather than sold – education, health service provision, road building, national defence, environmental protection, etc. These form a substantial part of the economic activity and the economic benefits gained in all societies, and are prominent in modern agriculture where the role of farmers in providing such public goods is becoming increasingly important.

These initial cautionary comments highlight the fact that (a) the creation of value and the incurring of costs have to be viewed in the wider context as well as at the level of the individual firm or industry, (b) there are public, as well as private, economic goods (benefits and costs) that need to be brought into the analysis if economic activity is to be undertaken to best effect, and (c) even though private enterprise and free consumer choice are the driving forces in economic activity, the conditions under which the pursuit of profit is consistent with the maximization of overall economic benefit are commonly not fulfilled. The market will not necessarily produce the most desirable outcomes, prices may not correctly reflect true economic values, and therefore the actions and strategies indicated by conventional financial accounting may be the 'best' in only a very limited context. All this may seem a long way from the problems of grassland production and utilization, but, like every production activity in farming and beyond, it is embedded in this complex system of human activity and, as we shall see, this makes its economic aspects less than straightforward.

15.2 The framework for economic analysis

Economics is about how to use resources to best effect. This simple statement covers the economic problem faced by a farm, a household, a multi-national company, an industry, a region, a country, the European Union or the whole of global society. For each level of aggregation the essential principles are the same.

All economic activity consists of taking resources (inputs) and transforming them into outputs, which may be physical commodities ('goods') or 'services'. Resources have no value in themselves; it is what they produce that has value, and this is gained when the goods and services are consumed (i.e. used, enjoyed) by people. Because no individual, group or society can ever achieve everything it would like to have, the central concern is to gain the maximum value or benefit from the resources available. This statement can be interpreted in the context of the individual firm/farm as how to maximize the profits to the owner(s), or in the context of a national food system as how to provide the level and pattern of food products demanded at the lowest prices to consumers. In all cases it highlights why the focus in economic analyses is on (a) identifying and measuring *benefits*, or the value gained, and how they are generated, (b) identifying and measuring *costs*, which are reflected in the resources used and how they relate to the level of output, and (c) maximizing the *difference between them* (note: neither maximizing benefits nor minimizing costs is in itself an appropriate target).

The general thrust of this approach implies the need for concepts of *efficiency* to measure the performance of the system, and we turn to these in the next section. In order to assess efficiency correctly, of course, the valuation of benefits and costs has to be accurate, and this introduces two complications. First, prices may not reflect real economic values or costs. Government policy may distort prices and make the value to the individual very different from the value to the economy, and so what is efficient for the producer may not be efficient for society. For example, where market price support leads to surplus production, a tonne of wheat may still be worth $120 to the farmer who sells it into intervention, but it is worth nothing to the economy which already has more wheat than it wants. Similarly, milk quota may be traded between dairy farmers at a price of £0.5 per litre but it has no actual value to the economy (it is the *milk*, not the entitlement to sell it, that is the real economic commodity). Also, the use of certain inputs is subsidized (e.g. capital grants, tax concessions on investment) or taxed (nitrogen taxes, fuel taxes), thereby causing their money cost to diverge from their real (economic) cost. Second, there may be no observable prices or any information to indicate economic value. For many products there are no effective markets and so their valuation is problematical because it cannot be directly captured in terms of a price; this is essentially the case for 'grass' as an output, but is true also for a wide range of inputs (climatic inputs, the farmer's own time), and outputs (slurry, nitrogen run off) in farming – not to mention the whole array of outputs from land that people value as 'the environment'.

None of this may be of particular concern to the *individual producer*, who reacts to costs and prices as they are and pursues the course of action that is best for him/her, but it does highlight the difficulty of defining what would be the best pattern of resource use for *the economy* (which is the most important consideration). As a result of these complications, the conventional financial accounting used to portray the economics of the farm business is generally only an incomplete and rather crude approximation to the real flows of costs and benefits involved in the process. However, this is something that has to be accepted in a market economy where resources are privately owned and the individuals concerned are largely free to use them as they choose. Information on the commercial profitability of production options is the only guide available to farmers in choosing how they will use their resources – but very few are driven exclusively by the market prices of their inputs and outputs or the compulsion to exploit all the technical possibilities in production anyway. Maximum nitrogen applications to short-term grass leys sown to modern seeds mixtures and utilized through high-yielding Holstein dairy cows may promise impressive levels of disposable income, yet a choice to retain some old pastures, run an Ayrshire herd and seek a less intense balance in farming operations could still be entirely rational because that was the individual's preferred way. Economics is about the creation of value for people, not a set of immutable rules that have to be followed like the laws of physics. The 'profit maximizing producer' of elementary economic theory, therefore, like the frictionless plane in physics, is not a description of reality but simply an analytical benchmark against which to assess performance.

15.3 Efficiency in grassland farming

The pursuit of 'efficiency', as a shorthand term for the best balance between what is put in (cost) and what is gained (benefit), is a readily acceptable aim in a wide range of contexts. However, the meaning of the term needs to be specified with some care when used as a guide for economic activity because the popular interpretations of efficiency are often dangerously misleading. For example, achieving high output per person, high milk yields per cow, high dry matter production per hectare, or high rates of liveweight gain are all commonly used as indicating efficient production. Such measures are defective for two reasons. First, they are purely technical coefficients, being the ratio of a physical output quantity to a physical input quantity. Without knowing the economic values of the two components, one cannot judge the merits of the achievement. Broadly speaking, only if it reflects high levels of valuable output per unit of the most expensive – i.e. the scarcest – inputs is any such measure likely to be an indication of economic efficiency; otherwise, technical output : input ratios (high slurry yield per cow?) are not guaranteed to have any useful meaning. It is worth noting in this context that Third World agriculture is often regarded as 'inefficient' because its output per unit of labour is relatively low, but there is no merit in utilizing 'efficient' Western techniques requiring very scarce and expensive capital inputs that achieve high output per worker but leave the bulk of the agricultural labour force unemployed. Second, such 'efficiency' measures are entirely partial, effectively associating all the output with just one of the many inputs in a production process, and thereby apparently dismissing the contribution of the others. High output per person ('labour productivity'), for example, could be due entirely to the use of substantial amounts of capital inputs which, by contrast, would therefore show very low measured output per unit (i.e. inefficiency!). High DM production per hectare suggests the land input is utilized effectively, but the amount of fertilizer, chemicals, labour and machinery inputs that were also required to achieve this are left completely out of the reckoning.

It is for these reasons that economic analysis avoids such simple technical efficiency ratios and develops a more informative set of criteria. Economic efficiency concepts are based on two fundamental propositions.

(1) Output is derived by using an array of resource inputs – land, and labour, and machinery, and animals, and fertilizer . . . and so efficiency measures must relate to the way *all* inputs are utilized. (This introduces the notion of input efficiency.)
(2) The resources available to a farm can be used to produce a number of different outputs – milk, beef, sheep, different arable crops, . . . and so efficiency measures must reflect that the inputs are put to their best economic uses. (This highlights the notion of output efficiency.)

The principles of economic efficiency are intuitively obvious. They require that the maximum total value (benefit) is gained from the resources that are used (costs incurred). This in turn requires three distinct criteria to be satisfied.

(1) For any one input, there is a 'most efficient' (margin maximizing) amount that should be used in relation to the level of output – for example, an optimal rate of fertilizer application per hectare of land or of concentrate input per cow (this explains why simply *maximizing* the output : input ratio is meaningless).
(2) For any particular level of output, there is a 'most efficient' (cost minimizing) combination of the various inputs to be used – for example, the optimal mix of fodder and concentrates for a high yielding cow, or the economically appropriate amounts of land, seed, fertilizer and agrochemicals to produce 10t of wheat (this explains why neither high labour productivity, land productivity, nor high capital productivity in itself is necessarily essential).
(3) For any set of resources, there is a 'most efficient' (value maximizing) combination of outputs to produce – for example, the pattern of crop and livestock products that generates greatest revenue from the farm's resources, or the most valued balance of agricultural products and environmental goods from a given land area (this demonstrates why economic performance cannot be validly assessed in terms of any one physical output).

In all those cases the efficiency criteria involve more than just physical data; a primary determinant of what is 'efficient' is the economic parameters – the unit costs of the different inputs and the unit values of the different outputs. Because these price parameters are subject to (often wide) change over time and from one situation to another, it is *never* possible to define an invariant economic recommendation about what it is best to produce or how it should best be done. It may well be feasible to define a specific feeding regime designed to maximize the milk yield of a typical Holstein dairy cow, but there is no comparable possibility of specifying a standard feeding regime that will maximize the profits from dairying. Nor is it possible to declare that dairying is a more economically efficient way of utilizing grassland than is beef or sheep production (even though that has largely been the case in recent decades in the particular market and policy conditions in the UK). For this reason, no attempt is made in this chapter to offer recommendations based on 'typical' financial figures for livestock enterprises or grassland production systems, since such data carry little genuine information outside their specific time and context. Rather, the emphasis is on the principles and considerations that underlie efficient grassland farming.

The search for resource use efficiency in grassland farming is conceptually even more complex because it involves optimizing over two distinct processes – plant growth followed by livestock production. Although sequential, the production–utilization activities are not independent. Grazing livestock production is an integral *system*, both technically and economically, so it is not appropriate to consider the most efficient way of using resources in grass production without simultaneously considering how best to use them in livestock production. The swards, conservation techniques and grazing management that are economically the best in dairying are different from those for beef or sheep production; intensive (i.e. high levels of inputs per hectare) and extensive land use involve very different husbandry techniques in both grassland and livestock management. The grass production and utilization patterns for producing store cattle, or for finishing cattle, or for producing quality-assured beef for specialist market outlets, although similar in principle are all very different in practice. All these modes of grassland farming may be options in a given situation, and the identification of the best to adopt (i.e. the economically most efficient) in practice involves complex modelling and an integrated analytical approach. It is not feasible to do more than set out the underlying principles here.

15.4 Defining optimality

The conceptual framework that underlies the economic analysis of efficient resource use is the *production function*. This inherently simple idea states the self-evident proposition that the level of output (Q) is determined by the level of the various inputs (X_i) used, and that the relationship between them is systematic. Hence, the production process can be represented in the form of a formal mathematical relationship:

$$Q = f(X_1, X_2, X_3 \cdots X_n)$$

The precise nature of the functional form of f is a matter for agricultural science, although empirical estimation of production functions from experimental or farm survey data is a long-established research area in agricultural economics (Heady & Dillon 1961, Dillon 1968). However, the criteria for economic efficiency can be derived logically because all production functions possess the same general characteristic, captured in the well-known (and widely misused) term of 'diminishing returns'. This logic starts from the proposition that, although a large number of different inputs contribute to the level of output, invariably many of them are fixed in quantity at any given time (e.g. the land area of the farm, its capital stock in terms of buildings and machinery, the size of the dairy herd, etc.). In practice, therefore, output can be adjusted by manipulating only some of the inputs in the set $[X_i]$ – fertilizer application, feed inputs, use of contractors' services, etc. The effective production function is thus represented by

$$Q = f(X_1, X_2 \cdots X_j \mid X_k \cdots X_n)$$

or

$$Q = f(X_1, X_2 \cdots X_j \mid Z)$$

where $X_1 \ldots X_j$ are the *variable inputs* and $X_k \ldots X_n$ are the *fixed inputs* represented collectively by the variable Z. Under such circumstances there is no avoiding the 'law of diminishing marginal returns'. This states that, as the use of any of the variable inputs is increased, output will rise but at a diminishing rate. In other words, the response is not linear, but increments in output (the marginal returns) from successive increments in input will progressively decline. This is not an economic theory, but an unavoidable reality arising from the fact that the pattern of inputs gets progressively out of technical balance (the fixed inputs become relatively more scarce and hence restricting) as output expansion is pursued.

We can illustrate this diagrammatically (Fig. 15.1) by assuming the simplest case of a 2-input production process where the use of input X can be varied but the level of input Z is fixed. This can be represented as $Q = x(X \mid \bar{Z})$, meaning Q is a function of X given Z is fixed at level $\bar{Z}$; it might be interpreted as portraying the response of grass output as more fertilizer is applied to an area of land, or the milk yield of a given dairy herd in response to increasing feed inputs. At low levels (x_1, q_1), the output response to additional input is quite high. As the input use increases, although the total output also rises the marginal (additions to) output progressively tails off. At x_3, the output is as high as it is possible to attain, and this would represent a point of maximum *technical* efficiency, as discussed earlier (maximum DM per hectare, maximum milk yield per cow). However, it is not economically efficient because the last unit of input used added no extra output and therefore provides no added value to cover its cost.

The exact point of efficiency depends on the price (benefit) of a unit of output

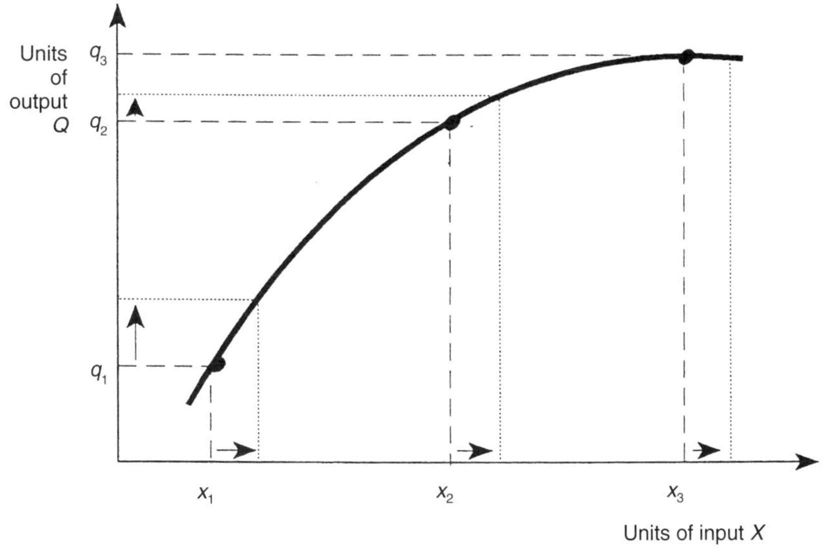

Fig. 15.1 The production function: $Q = x(X \mid \bar{Z})$. Relationship between output (Q) and level of an input (X) with all other inputs (Z) remaining constant.

P_Q and the price (cost) of a unit of input p_x. Each extra unit of input adds $\Delta X.p_X$ to cost and returns $\Delta Q.P_Q$ in additional output value, with ΔQ getting progressively smaller due to the inescapable effect of diminishing marginal returns. As long as $\Delta Q.P_Q > \Delta X.p_x$, then benefit gained at the margin is greater than cost incurred, and so extra input use is adding extra net economic benefit. At the point where $\Delta Q.P_Q = \Delta X.p_x$, the extra output value just covers the extra input cost, but beyond this the cost of further inputs would not be returned. Hence, this is the point where, by definition, the net economic benefits are at a maximum and economic efficiency has been achieved. This process of *marginal analysis* is the core of modern economic technique, and underlies all resource use principles. The condition for optimality in this situation is more normally represented as

$$\frac{\Delta Q}{\Delta X} = \frac{p_x}{P_Q}$$

or the marginal product of input X is equal to the ratio of input:output prices. Note that the marginal product here is defined in terms of discrete increments of output, ΔQ, relative to discrete increments in input ΔX. Mathematically the marginal product is given by $\delta Q/\delta X$, the first partial differential of the production function with respect to X. This allows the whole of economic efficiency analysis to be generalized to n input variables and handled within the context of a straightforward mathematical model.

The above indicates the derivation of the first, and simplest, of the three economic efficiency criteria mentioned earlier. It should give the flavour of how economic analysis seeks to combine technical data relating to the production process (the production function and its parameters, the marginal productivity of inputs) with relevant economic information (prices/values of inputs and outputs) to define the economically best levels of an input and an output to aim for. The other two efficiency criteria, relating to optimal combinations of inputs and optimal combinations of outputs, involve a similar procedure of marginal analysis and can be derived using elementary differential calculus applied to the production functions in the appropriate manner. However, since the intention here is not to offer a treatise on economic theory, the interested reader is referred to any standard elementary textbook in agricultural production economics (e.g. Hill & Ray 1987, Hill 1990) for the details. The important point to note is that economic efficiency is, in every case, defined in terms of *marginal* output : input ratios – i.e. the marginal products of individual inputs, the marginal rates of substitution between inputs. By contrast, the 'popular' indicators of efficiency – such as conventional yields or labour productivity measures – all relate to *average* output : input ratios, and for this reason fail to provide either the guides to resource use or the unambiguous indicators of performance that are required.

15.5 Measuring economic performance in production

The logical aim in all resource use is to achieve the maximum attainable *net* economic benefit, which in turn implies maximizing the difference between the value of the outputs gained and the value of the inputs that are used. Evaluated at the economy level, this involves assessing the value to society of all the outputs produced, whether from private or public goods, and including all benefits that may arise outside of those recorded in market transactions, and setting against them the real costs of all the resources used – plus any damage costs or other so-called 'externality' losses (like pollution or landscape destruction). Not surprisingly, such overall accounting is very rarely undertaken, because although conceptually this is the correct economic approach, the required data are never available. At the individual farm or industry level, where commercial considerations are the guiding framework, the situation is more straightforward, because maximum economic benefit (for farmers) is seen as equivalent to generating maximum profits or net income. However, even in this case the situation is not crystal clear. The accounting conventions produce several quite different measures of farming income depending on their purpose (tax liability or resource-use planning) and on assumptions about the appropriate charge to make for farmers' own labour and use of own versus borrowed capital, landownership earnings, depreciation rates, etc. The several definitions of income commonly in use are explained in the annual publication *Farm Incomes in the United Kingdom* (MAFF 1998).

15.5.1 Gross margin accounting

Going back to the production function, the net economic benefit from production (call it 'profit' π) is the value of the output Q gained, referred to as total revenue and measured as $P_Q.Q$, minus the total cost of all the inputs used, $\Sigma_i p_i X_i$ (where $i = 1,2, \ldots, n$). Hence,

$$\pi = P_Q.Q - \Sigma_i p_i X_i$$

Because the input set is split into two groups, variable inputs ($X_i \ldots X_j$) and fixed inputs ($X_k \ldots X_n$), the total costs similarly consist of two components – the *variable costs* and *fixed costs*, respectively. By definition, fixed costs do not alter as output is varied, since they relate to a set of resources whose availability – and hence use – is fixed. Furthermore, whether those resources are employed fully, or partially, or are lying idle, the costs associated with them have to be paid and so the fixed costs are inescapable; they represent a fixed charge (K) against revenues regardless of the level of production. By contrast, since output is changed by altering the use of variable inputs, then revenue is directly related to the level of variable costs.

The division of cost items into the category of variable or fixed is fairly clear-cut but not entirely mechanistic in practice. Variable costs are identified as those which change with the level of output, and in grazing livestock production would include concentrate feed, veterinary and medicine expenses and marketing costs, plus the fertilizer, seed, sprays and contract costs associated with grass production. The typical elements of fixed cost, which by definition do not alter as output (of the farm or the enterprise) is changed, are regular and farmers' own labour input, land rent and rates, buildings and machinery depreciation and repairs, insurances and other general farm overheads.

$$\pi = P.Q - \sum_{i=1}^{j} p_i X_i - K$$

$$\underbrace{\text{Profit} = \text{Revenue} - \text{Variable costs}}_{\text{Gross margin}} - \text{Fixed costs}$$

The difference between total revenue and total variable costs is called the *gross margin*. For a given production situation (farm or enterprise) which is defined by a particular set of fixed resources and hence fixed costs, it is obvious that maximizing profit is achieved by maximizing the gross margin. The gross margin is therefore used as a major indicator of economic performance at the farm level. It is also a parameter for the planning and selection of enterprises and production systems because, other things being equal, the higher the gross margin the more revenue there is available to pay off the fixed costs and leave a residue for profit. Some representative values for gross margins from livestock production in the UK are shown in Table 15.1.

Table 15.1 Representative gross margins for grazing livestock enterprises.

	£ per animal	£ per hectare
Dairy cows		
Low yield, low stocking rate	872	1525
Low yield, high stocking rate	827	2065
High yield, low stocking rate	1207	2110
High yield, high stocking rate	1162	2905
Beef cows		
Lowland*	248	645
Upland*	335	400
Beef cows (multiple suckling)*	450	790
Beef (traditional finishing)	49	196
Beef (18-month finishing)*	226	735
Sheep		
Lowland lambing	42	503
Hill flocks*	51	425
Red deer (finishing)	27	405

* If headage payments (i.e. subsidies – financial transfers from the public purse which top up the 'economic' market receipts) are ignored, these figures are reduced significantly. This highlights the point about the difference between the value to the farmer and the value to the economy (source: Nix 1997).

The strength of the gross margin measure is its accounting simplicity and its conceptual validity as an economic efficiency indicator. Maximizing the gross margin from a farm or an enterprise is equivalent to fulfilling the criterion for optimal resource use shown earlier for the production function $Q = f(X \mid \bar{Z})$. The weakness of the gross margin measure, particularly in the context of grazing livestock production, is that it has to be used carefully and knowingly as a comparative indicator of economic merit. First, as shown in Table 15.1, the resource input to which it relates is important. This links to the earlier discussion about which resource (in this case the animal or the land) to use in calculating an efficiency measure. Since land is generally a farm's most restrictive resource, it is high gross margin per hectare that is the appropriate target; a high gross margin per animal but associated with a low stocking density can result in an overall lower economic performance. Second, gross margins are only comparable, and hence usable as valid economic performance indicators, within the context of a given structure of fixed costs – and these, of course, vary from one livestock production system to another. The gross margins achieved per hectare can be usefully compared to assess the relative efficiency of two farms practising effectively the same dairying system, but cannot be used to determine whether beef production is more profitable than sheep production. Similarly, the gross margins from upland and lowland sheep production are not strictly comparable, nor are the gross margins from dairy herds of 70 cows and 170 cows because their fixed cost structures are very different. However, with a little thought none of this should be surprising. Referring back to our original definition of the central economic problem, the aim is to gain maximum benefit from the resources available – which means, in principle, the fixed resources (since the 'availability' of the variable resources is, by definition, not the primary constraint). Different livestock enterprises and farming situations represent different sets (and qualities) of available resources, and the gross margins that are achievable – even under the best management conditions – will obviously vary.

15.5.2 Net margin accounting

A further operational weakness of gross margin as a measure of profitability is that it places no focus on fixed costs – which are just as much a drain on profits as are variable costs. It thus ignores the possibility that production systems or enterprises might be adjusted to have lower fixed costs in such a way that, even if gross margin was also lower, a higher overall profit resulted. This introduces the idea of identifying fixed costs that are allocatable to a specific enterprise, and calculating a *net margin*.

In the case of a farm business that has only one revenue-generating enterprise – a specialist dairy producer or a hill sheep farm – the economic performance of the enterprise and of the business are identical. However, many farms keep beef cattle, sheep and dairy cows on their grassland area, or produce arable crops as well as livestock. In such circumstances the economic performance of each enterprise needs to be assessed separately to ensure it is using its share of the farm's resources efficiently. The different enterprise revenues are readily identifiable because the outputs (sales) are distinct, as are the variable costs attributable to each – but how can one account for the fixed resources they use? Some fixed costs cannot be associated with any particular enterprise but are a charge on the overall farm business, regardless of what or how much it produces. This includes depreciation of buildings and non-specific machinery such as tractors, the farmer's labour and management time, hedge maintenance, and a series of 'general farm overheads' representing the cost of the basic resources upon which all the different income-earning activities of the farm business are built. Other fixed costs, although equally not directly related to output levels, can nevertheless be attributed to specific enterprises because if they were not practised those costs could be eliminated. The fixed costs associated with the housing cubicles, milking parlour, dairy equipment and cowman, for example, are clearly attributable exclusively to the dairy enterprise. Similarly, the incurring of certain other fixed costs – lambing shed, pens and races, shepherd, etc. – can usually be linked directly with a sheep enterprise.

Where such allocations of fixed resources can be made, the relevant measure of enterprise performance is the net margin, calculated as the gross margin less enterprise-specific fixed costs. Clearly, the maximization of net margin is a more valid economic goal than is the maximization of gross margin, and is the closest one can get using farm accounts data to a comparative measure of the economic efficiency of the enterprises using grassland. Table 15.2 gives an indication of both how widely fixed costs can vary from one enterprise to another, and the representative net margins from using grassland in different types of livestock.

Unfortunately, as every scientist knows, without data the most convincing of principles cannot be put into practice. The allocation of fixed costs is easily accomplished only where operations and outputs are largely separate and few resources are shared – such as between arable and livestock enterprises – and in many cases the validity of allocating fixed costs to particular activities is somewhat suspect. Anyway, it is usually a backward-looking accounting exercise rather than providing useful decision-making information. Most farmers find themselves with a largely unalterable pattern of fixed costs to pay (arising from their land rent, labour force, buildings and machinery stock etc.) which are associated with their particular farming system, and need to maximize the total gross margin they can gain from production. If total fixed costs can be lowered in some way without affecting output (e.g. by using contract services to replace the costs of under-utilized grass conservation machinery, or reducing the regular labour force and employing more casual workers), then this is one route to greater profitability and efficiency in using (the remaining) resources. Otherwise, the time-honoured way of increasing efficiency is to expand production and total gross margin, thereby 'spreading the fixed costs over a larger output' (and so achieving a higher net margin). As is evident from Table 15.3, fixed costs per hectare are almost always lower the larger the scale of the farming operations, and this leads to the general conclusion that although small farms are often more *technically* efficient (in terms of achieving higher yields per animal or per hectare) they are rarely more *economically* efficient.

In the grazing livestock complex, it is usually impossible to logically attribute the fixed costs of the grassland production and conservation side of the operations. In the case of a specialist dairy, beef or sheep farm there is no great

Table 15.2 Enterprise-specific fixed costs and net margins.

	Fixed costs		Net margin (£/ha)
	£/animal	£/ha	
Dairying[1]	438	736	743
Lowland sheep[2]	38	508	52
Beef[3] – smaller herds	129	435	70
– fattening herds	219	351	61

Sources: [1] Farrar & Franks (1998); [2] Fogerty & Turner (1996); [3] Jenkins *et al.* (1998).

Table 15.3 Fixed costs (£ per hectare) in some different livestock farming systems.

	Small-scale		Large-scale
Mainly dairy	1400		1045
Lowland sheep and cattle	760		500
Livestock rearing – upland		410	
– hill		270	

Source: Nix 1997.

problem, since all the resources used in grass production are simply part of the wider structure of (fixed and variable) costs in producing the specific livestock revenue, and can be included in the conventional margin calculations. However, since many grassland holdings run beef, sheep and dairy enterprises, it is not possible to logically attribute either the fixed or the variable costs of grassland husbandry among them. It is possible to contrive a partitioning of total grassland costs amongst the utilizing enterprises on the basis of some secondary criterion, or by collecting additional data. For example, one could attribute costs in proportion to the value of output from each livestock type – although this then forces the arbitrary assumption that £100 of dairy output requires the same contribution from grass as £100 of sheep output. Alternatively (and ostensibly more accurately), for each class of livestock one could account for the conserved forage consumed and estimate, using standard scientific/experimental parameters, the amount of grass directly grazed, and then allocate grassland costs accordingly. But what practical farmer can plausibly deal with such detailed recording and data analysis? And what would be the purpose anyway? In both the above instances, the allocation of grassland costs is essentially an arbitrary accounting manoeuvre to produce a specific figure; its conceptual validity in guiding decisions about producing more or less of something (whether grass or a specific livestock output) in the search for greater efficiency is totally suspect.

15.5.3 Assessing grassland productivity

Any attempt to assess grassland productivity raises the question – how can the economic efficiency of the production of grass and forage as an input into the livestock production process be assessed? As mentioned earlier, efficiency measurement depends on the accurate measurement and valuation of inputs and outputs, and the major difficulty is the lack of any clear monetary value to place on the grass output, whether grazing DM or in the form of hay/silage. There are some market indicators, because conserved fodder is traded at specific prices and the sale of grass keep on an annual basis is a kind of valuation on the grass itself. However, except where a figure is required for pure accounting purposes, the use of these prices to value the output from a grassland system is questionable. The sale prices of hay or silage off the farm are highly variable from year to year, depending largely on specific climatic conditions during the season, and on how much of the total bulk produced in that year is actually placed on the market. For a farmer considering how best to produce grass to process through a dairy herd or sheep flock, those prices give little indication of what level of input costs it is worth incurring in managing the grassland. Similarly, grass-keep prices are rentals for an area of grassland, and less directly interpretable as the sale price for a specific quantity of grass DM.

All this implies there is no relevant output price for the grass product, in whatever form, and so no practical means of applying to the grassland production process the formula set out earlier defining economic efficiency in resource use. With no way of calculating a gross margin, the choice of a best grass production and management system for a particular farm, from the range of technical options available between the highly intensive and the more extensive approach to land use, is not obvious. It cannot be assumed that the high grassland productivity technically attainable will not be rendered less efficient (less profitable) by the high input cost it requires compared to a lower input–lower output system – and this is becoming increasingly true in the modern era of more 'market oriented' (i.e. less assisted by government price support policy) livestock production conditions. Under such circumstances, the only workable approach to achieving efficiency is to minimize the resource cost in whatever grassland production system (and hence grass output level) is selected. In effect, producing as much grass as required by the livestock enterprise but as cheaply as possible becomes the guiding axiom. This then implies a data framework for estimating the cost (per unit of DM, say) of growing and conserving grass, and to a large extent the resources used in the 'grass enterprise' can be identified and valued. The annual costs of grassland production are fairly straightforward – fertilizer, sprays, specific machinery used in conservation (mowers, forage harvesters, balers), labour directly utilized on grassland operations, contractors costs, field rent, etc. – and an allowance for a share of unattributable fixed costs (e.g. tractor and general machinery use) can be included if required. However, one of the 'overhead' costs of grass production is the original establishment cost of the sward – which is irrelevant in the case of permanent pasture, but in shorter-term reseeded leys this represents an investment whose 'depreciation' cost should be allocated on an annual basis over the life of the crop. The problem, of course, is that both the quantity and quality of the output, whether grazed grass, hay or silage, are susceptible to wide variation from year to year even under the same management system, so trying to quote a typical cost or value per unit is a largely pointless exercise.

Nevertheless, for indicative purposes, the data in Table 15.4 are of some interest. It emphasizes the fact that, as a feed input in the production of ruminant livestock products for the human food chain, grass has almost no economic

Table 15.4 Indicative costs of different types of grazing livestock feed.

	Cost per tonne DM (£)	Cost per MJ of ME (pence)
Grazed grass	38.6	0.33
Kale	42.0	0.38
Fodder turnips	38.1	0.37
Grass silage (clamp)	69.1	0.63
Grass silage (big-bale)	74.1	0.68
Second-cut silage	37.0	0.34
Purchased hay	76.5	0.87
Brewers grains	108.5	1.09
Concentrates	155.2	1.21

Source: ICI, quoted in Nix 1997.

competitors. As grazed fodder it is about the cheapest obtainable, whether measured in cost per unit of DM or of ME, and in conserved form it is substantially cheaper than any non-grass feed input. The reason why it is nevertheless economic to use the considerably more expensive concentrates is all to do with the nature of the feed input production function and the logic of marginal analysis. The output potential of a high-yielding dairy cow cannot be realized through grass alone because it has not the physical capacity to ingest and process the necessary quantity of nutrients in this form, and its milk yield would thereby be constrained. However, the value of the additional milk output that can be gained by filling some of the cow's intake capacity with concentrate feed (even though it is some four times as expensive per unit of ME) is well in excess of this high cost. Producing milk solely from (cheap) grass may well generate a high financial margin per litre, but if many more additional litres can be produced on top of this, even though at a much lower margin per litre, it still adds up to a higher total margin from the cow.

It will be concluded from the foregoing discussion that "the economics of" grassland production and utilization are very difficult to get a handle on, even at the individual farm level. The appropriate procedures for input and output valuation in order to generate workable measures of economic performance and comparative efficiency, although reasonably clear in a textbook conceptual sense, are extremely elusive in practice. The farm recording and data frameworks required to make these concepts workable and to guide the selection of the 'best' systems and input levels, and the subsequent management of resource use in the face of inevitable price changes in livestock outputs and grassland inputs, so as to sustain efficiency in the operation, are generally beyond the means and practicality of everyday farming. Small wonder then that the standard accounting procedures adopted involve fairly arbitrary allocations and assumptions, and result (not really very helpfully) in measurements of what costs or margins *were* in the period that has passed – whereas the useful role of all data and analytical techniques (whether in economics or other areas) is to assist strategy selection, decision making and management of what to do *now*, or in the future. This is hard enough in a world of predictability and systematic relationships; in the real world of uncertainty, incomplete information, unforeseen change and uncontrollable events affecting the system as is typical of most economic activity, and particularly so in farming, it is impossible to achieve the standards of performance that 'science' identifies as possible. In economics, the concept of 'profit' – a payment in excess of the cost of resources used in production – is recognized explicitly to be the reward (or penalty, if the operator doesn't get it right and the profit is a loss!) for the human manager who struggles with these decision-making problems and takes on the role of putting the production process together.

15.6 Determinants of profitability in grassland farming

As is clear from the discussion above there can be no specific formula, no 'standard recommended practice' that declares how grass is best grown and utilized. The rules of economics indicate simply that success lies in achieving the right balance of physical inputs and outputs in relation to their respective prices, and the whole science of economic analysis contains the procedures for identifying what 'right' means in any particular situation. It is impossible to assert that utilizing production methods at the frontier of technology, or pursuing high-yield systems, or practising low-cost methods, or specializing, or diversifying – or any of the specific strategies frequently propagated – will be better than any other. There are, however, a number of factors which are often used to characterize successful grass production systems. These include such obvious things as choice of seeds mixture, level of nitrogen use, stocking rate, yield (milk yield, lambs per ewe, liveweight gain), feed cost, seasonality of production and fodder conservation method – but listing variables does not define what their optimal values should be.

While one cannot specify what *will* result in economic success, the major factors that explain good economic performance are apparent. Some of these are under the decision-making control of the farmer, while others are unchangeable characteristics of the production situation.

15.6.1 Resource productivity

By resource productivity we mean the inherent capacity of a unit of resource to produce valuable output. Other things being equal, the more a given unit of land/labour/capital can produce the higher the revenue it generates for its given cost, and the higher margin to the producer. In the context of grassland farming it is land productivity that generally holds the key. However, there is little the farmer can do to alter this apart from certain investments in land improvement (drainage, clearance, reseeding of old pasture). Otherwise, the capacity of land to produce is governed by such immutable factors as soil type and quality, geographical location, topography and the physical nature of the farm. There is no avoiding the fact that hill farms have poorer, thinner soils, harsher climate, rougher terrain, a shorter growing season and overall less favourable conditions for grass growth and livestock productivity than their lowland counterparts. As a result, although hill areas have a *comparative* advantage in grazing livestock production – i.e. they are better suited to this than to arable or other forms of farming – they have an *absolute disadvantage* in relation to lowland areas and can generally never expect to compete on equal terms with them in the same product markets. This is recognized in their designation as 'less favoured areas' in the context of European agricultural policy, with the consequent transfer of special extra-market payments (Hill Livestock Compensatory Allowances, headage payments) for the largely socio-political purpose of maintaining the farming population in these areas against all the signals of economic efficiency.

15.6.2 Technology

The term technology relates to the productive quality of the capital inputs used in production. 'Modern' technology consists essentially of a set of capital inputs, some relatively novel, others improved versions of conventional inputs, which have the capacity to contribute more to output per unit than earlier versions. The technology of grass production is embodied in the seed varieties, fertilizers and weed-control chemicals, fodder conservation techniques and machinery, etc. and the pattern of operations that their use implies. The technology of capital in the livestock production side is captured in the animals' genetic make-up, disease resistance, fecundity, metabolic capability to process feed inputs, etc. and in the associated inputs like feed, drugs and housing facilities which characterize the production system. The advantage conferred by the newer technology inputs, where they can be appropriately applied, comes from their greater productivity in terms of net value added. Thus, although modern technology inputs may be more expensive to purchase per unit than their traditional counterparts, they confer additional output benefits in excess of this, and this again emphasizes that it is not the *average* cost or benefit but the *marginal* effect that matters. Unfortunately, access to, or ability to successfully exploit, new technology is not uniformly available to all but is often conditioned by the two factors considered next, so that many producers have to remain economically disadvantaged by their continued use of 'old technology' systems and methods.

15.6.3 Scale of operations

Generally speaking 'small is beautiful' is not the reality in modern farming methods, because there are almost inevitable advantages in operating on a larger scale. Some of these relate to financial gains from larger volume purchases and sales, or the ability to spread the fixed costs of the farm/enterprise over more units of output and so achieve lower overall unit production costs. Others flow from the obvious inherent technical efficiencies of size – larger fields and machines allow more complex and speedier operations and enhance labour productivity, larger buildings are cheaper per unit of space provided. Furthermore, most modern technology is not scale-neutral in its accessibility, and only the larger units can justify the machines for specialist operations, exploit economically the high horsepower tractors and their associated equipment, or utilize the specialist systems for specific processes such as slurry handling. Scale economies do not continue indefinitely, since the difficulties of integrating intensive field operations over vast land areas or managing the husbandry of huge herds/flocks of livestock obviously start to introduce technical inefficiencies. However, up to quite large and as yet untested limits (300 cow herds? 2000 ewe flocks? 2000 hectares of lowland pasture?), the relationship between average costs per unit of output and scale of enterprise will follow the rather stylised L-shape curve shown in Fig. 15.2, which is quite typical throughout agriculture.

The line AC shows that unit costs of production fall consistently (though at a declining rate) until a scale of operations represented by Q^* is reached,

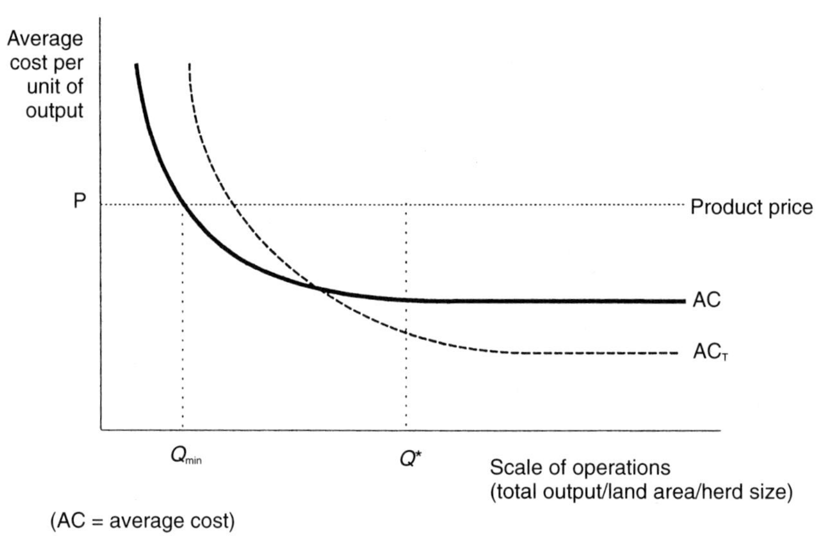

Fig. 15.2 The typical relationship between scale of operations and unit costs of production.

and beyond this there are no discernible economic advantages. If the market price of the product is given by P as shown, then only farms with a scale of operation above $Q_{\min}$ are viable; for units smaller than this the selling price of their output cannot cover the production costs. Profitability per unit of output (the difference between price and average cost) rises as scale increases from $Q_{\min}$ *to* Q^*, and even beyond this point the *aggregate* profits will continue to increase. With the passage of time two things happen to threaten the survival of the small-scale grassland farmer. New technology (buildings, machinery, production systems) offers cost savings through higher resource productivity and causes the overall cost curve to shift downwards to AC_T; but because technology tends to be biased in favour of size, the curve also tends to shift to the right as well, conferring the greatest advantage on larger-scale producers and raising the minimum output level for economically viable operations (Dawson & Hubbard 1987). Farmers who do not, or cannot, adopt the new technologies remain with the old higher-cost structure and become low-profit and increasingly economically vulnerable operators. Hence, there is an unavoidable incentive for everyone to adopt the new technology in order to remain competitive. However, the way markets work is that the higher productivity and extra production at lower cost causes a reduction in product price – excellent for consumers, but tending ultimately to wipe out the profit advantage gained by the innovating producers. This continual process of technological advances and unavoidable pressure for farmers to adopt them has long characterized Western agriculture, and has been labelled 'the treadmill effect' (Cochrane 1958). It explains why consumers have gained greatly from consistent real food price reductions over decades, but agricultural producers have steadily been forced either to leave the industry, increase the size of their farms or seek additional (non-farming) sources of income. The maintenance of agricultural product prices through government support policies has tended to slow up this adjustment effect, but the contemporary policy reforms and withdrawal of market intervention will lead to more rapid price reductions and an acceleration of this process.

15.6.4 Costs versus revenues

For any given form or enterprise, maximum economic performance means achieving the greatest possible revenues with the lowest possible costs. The logic of the production function, however, indicates that this is far from a simple recipe, since higher outputs and revenues come only from using more resources and hence increased costs, while trying to cut costs and resource-use leads to lower outputs. This is the perennial dilemma between high output–high cost (intensive) systems and the alternative of low cost–low output (extensive) production. There is no universal answer to this question, and it depends on the characteristics – resource productivity, scale, management, etc. – of each situation as to which is the best strategy. Other things being equal, the farms which achieve premium performance are those who manage to get the greatest output from their given resources, and farm management data tend to show that, while many farms have fairly similar levels of costs (per animal or per hectare), their differing performance is distinguished by the value of output they achieve from those inputs. The other major item revealed from farm accounts information is that, as an inescapable charge regardless of the level of revenue earned, it is the level of fixed costs that has the dominant influence on overall economic performance. The need to design production systems so as to carry the lowest possible structure of fixed resources consistent with the output level, or to ensure maximum utilization of those resources that give rise to the fixed costs, is a crucial consideration in planning for profitable grassland farming. Finally, a significant determinant of revenues in livestock production for many years under the regimes of the European Common Agricultural Policy have been the direct subsidy payments (usually disguised under obscurantist terminology such as premiums, supplements, compensatory allowances, marketing payments, processing aids, etc.) which livestock output or breeding animals attract – often amounting to one-third or more of output value. These are clearly of importance to the commercial outcome of farming businesses (indeed, reference is often made to 'effectively farming the subsidies'). However, they are not indicative of any genuine *economic* output, but are simply money someone has paid as taxes being transferred to livestock farmers to supplement their market revenues and incomes without representing any economic value added; furthermore, they can rise, fall or disappear as a result of political decision and administrative action, and in this sense are not 'real'.

15.6.5 Management

The technical performance of physical resources is only realized by the human input that assembles them and puts them to work. There is no doubt that the quality of the farmer's management, both technically in grassland and livestock husbandry, and economically in deciding on the appropriate input and output patterns, is a major determinant of economic success. It is the technical ability in managing herbage production and grass conservation processes that allows the higher stocking rates that underpin high economic performance. That same technical ability in animal husbandry is necessary to gain the livestock performance in terms of breeding, rearing, yield, growth rate, etc. In a sense, this is one determinant of profitable grassland farming that the farmer can do most about, since (allowing for the fact that some people have personal characteristics making them inherently better entrepreneurs/managers than others) management ability is built upon observation, awareness, education, information, skills training, careful preparation prior to decision making, and conscious attention to the monitoring and implementation of actions (Giles & Stansfield 1980).

The final component of good management, about which nothing deliberate can be done, is luck.

The influence of these factors is illustrated by the data on gross margins and net margins in different livestock enterprises shown in Table 15.5.

15.7 Responding to the market

So far in this chapter we have considered economic performance in grassland farming from what might be termed the 'supply side' viewpoint. Focus has been on the principles of efficient resource use, the control of costs relative to revenue, the impact of technological factors, and the resource management role in a successful grazing livestock business. This places productivity and its exploitation at the core of the problem, and seems to emphasize that the important information and decisions are about how to get technical production right. It has to be said, however, that this is rapidly becoming an obsolete emphasis in Western economies. In an affluent society where food availability is both abundant and secure, and where food prices in real terms are low, a diminishing proportion of household expenditure is spent on food; in the UK, for example, 30% of household expenditure was on food purchases in 1960, whereas today the proportion is approximately 11% (MAFF 1997). In these circumstances the traditional pressure to raise agricultural productivity, expand production and lower the unit costs of supply – a concern which has been central to the lives of all previous generations – is no longer dominant. Except among low-income groups, consumers do not buy largely on price, and food purchases are no longer driven by the simple aim of eating as cheaply as possible. (For those who find this denial of the conventional wisdom hard to accept, a moment's reflection on the range of specialist, convenience, branded and quality products offered, and purchased, in food supermarkets nowadays, and the comparison with a low-budget minimum-cost basket, should be sufficient.)

In the UK and Europe, the food system has changed its nature from being concerned with supply-side issues of quantity, availability and price to now being driven by demand-side forces. Consumers are increasingly focusing on the characteristics of the food they purchase and seeking to acquire variously perceived quality standards. This is no more evident than in relation to livestock products, where the range of 'quality issues' now influencing demand is quite wide. Prominent among them is safety, a concern fostered by publicity given to the BSE saga, to fears about hormone and antibiotic residues in meat, and to notable scares in relation to *E. coli* and *Salmonella*. There is a widespread image that modern 'factory farming' (a phrase having no clear interpretation) results in products lacking flavour or nutritive quality compared with those from 'traditional' or 'extensive' production systems. This and other preference changes is leading to a growing interest in products derived from organic

Table 15.5 Factors influencing grassland productivity.

1. Location (inherent land productivity)

	Lowland		**Less favoured areas**	
	£/cow	*£/ha*	£/cow	*£/ha*
Dairying				
Gross margin	947	*1648*	856	*1396*
Net margin	496	*863*	395	*644*

2. Scale of operations

	Small herds (<40 cows)		**Large herds (>150 cows)**	
	£/cow	*£/ha*	£/cow	*£/ha*
Dairying				
Gross margin	761	*1112*	1051	*1967*
Net margin	157	*228*	650	*1216*
	Small flocks (<200 ewes)		**Large flocks (>500 ewes)**	
	£/ewe	*£/ha*	£/ewe	*£/ha*
Lowland sheep				
Gross margin	43	*570*	41	*571*
Net margin	4	*–55*	11	*145*
	Small herds (<50 cattle)		**Large herds (>100 cattle)**	
	£/animal	*£/ha*	£/animal	*£/ha*
Beef fattening				
Gross margin	278	*364*	217	*433*
Net margin	–127	*–166*	–33	*–65*
	Small herds (<25 cows)		**Large herds (>50 cows)**	
	£/animal	*£/ha*	£/animal	*£/ha*
Beef suckler				
Gross margin	221	*290*	257	*425*
Net margin	–107	*–138*	57	*94*

3. Management ability

	Bottom quartile		**Top quartile**	
	£/animal	*£/ha*	£/animal	*£/ha*
Dairying				
Gross margin	640	*816*	1113	*2380*
Net margin	125	*159*	659	*1496*
Lowland sheep				
Gross margin	34	*454*	53	*844*
Net margin	–12	*–156*	15	*252*
Suckler cows				
Gross margin	203	*314*	293	*617*
Net margin	–120	*–183*	125	*264*

farming systems (see Chapter 14) which, while more expensive, are perceived to be 'safer' as well as being associated with a desirable responsiveness to environmental considerations. Despite public knowledge of livestock production being minimal, there is nevertheless a growing emphasis on the welfare conditions under which animals have been kept, and this is treated as an innate quality characteristic of the product. Overall, these demand changes demonstrate an increasing interest in the *origins* of food products and not just their price or presentation. In response, food retailers (and producer groups) are constructing a variety of quality assurance schemes, some of which impose quite specific requirements on the supply conditions and identifiable technical characteristics of the livestock product, while others are often little more than labels conferring a particular image ('Aberdeen Angus beef', 'traditional lamb', 'West Country milk') which differentiates it from the standard (unlabelled) commodity without indicating anything about its distinctive characteristics other than those which arise in the consumer's mind. This is not cynical, nor solely a characteristic of the food market. It is simply a part of the wider framework of a consumer society in which, for the large majority of things we purchase and about which we have little detailed technical knowledge, we choose products which seem to meet preferences formed on the basis of our personal perceptions and the images we have gained from association, advertising, experience, hearsay or intuition. (Again, readers who find this hard to accept might wish to ponder how fully and correctly informed, in a scientific sense, their own personal decisions are with respect to the purchase of particular brands of car, electrical goods, clothes, leisure and recreational activities, and financial services, their choice of plumber or house decorator, what charities they support, etc.)

The message for many grazing livestock producers seeking to maximize the economic performance of their business is clear. In the modern demand-determined food system the focus turns increasingly towards selling price not production cost, on securing market rather than expanding production. Opportunities must increasingly be sought for capturing higher selling prices and maximizing the revenues gained from production, more than searching to reduce costs in producing a standard product. In competitive markets, the returns to quality and association with a high-value product image are significant. Linkage into what are (often somewhat disparagingly) referred to as 'niche markets' – through production under a quality assurance label, or offering identifiable 'welfare-friendly' products, or exploiting the distinction conferred by adjectives such as traditional/natural/farm-fresh/home-reared, or exploiting associations of location or origin – all offer product features that many consumers prefer and are increasingly prepared to pay higher prices for. In economic terms, increasing the value of the output is just as valid a way of raising efficiency as is lowering the costs of provision, and if this is what the final consumer seeks it is the producer's role to respond to these market preferences. This is a message that farmers and agricultural scientists have yet to fully accept, and many of our perceptions are still locked into the old fashioned supply-side context of the desirability of raising productivity and lowering supply costs. We should remember the dictum of the early economist Adam Smith, who noted that 'the object of economic activity is not production; it is consumption' (Smith 1776).

15.8 A cautionary note

Discussions of agricultural production are one area where the differences in viewpoint of economics and science emerge. Economic analysis deals with people's behaviour, the responses to markets, the collective actions, interests and requirements of the individuals making up a particular society, the costs and benefits to groups such as consumers, producers and taxpayers. Science, by contrast, explores the variables and relationships of a less anthropocentric world, discerning operating conditions and technical possibilities independently of how they might be valued by any particular group of humans (except, perhaps, by the scientists themselves!). This has relevance to the foregoing discussion of the contemporary food situation, which has been entirely in the context of the affluent and well-fed societies of the UK/EU/industrialized economies. In such societies, what are considered desirable or appropriate technological developments, systems and methods in grassland production and utilization are determined by the needs and preferences associated with an already diverse consumption pattern and high standard of living – and, as argued, pursuit of conventional agricultural productivity growth is no longer high on the list. However, for another and much larger society – that 75% of the world's population whose incomes and consumption are decidedly lower, and in many cases closer to outright poverty – the benefits from developments in agricultural science, improved agricultural technology, better production techniques, more efficient use of available resources, and generally higher levels and greater predictability in the output from farming still represent a substantial force for economic and social gain. European food consumers may validly feel that they themselves require little more in terms of traditional agricultural progress, but the aggregate world food supply situation, both currently and in prospect, implies the need for continual emphasis by agricultural scientists (Anderson 1995). Numerous studies have shown that the returns from agricultural research have made it one of the best investments that a society could make, with rates of return estimated variously at 50–70% (Evenson 1977). In many respects, therefore, the application of the principles and practices set out in the technical chapters of this book are important for the *sustainability* of our (Europe's) supply of affordable, high quality livestock products, but (with due adjustment for geographic location) they are absolutely critical to the *expansion* of food supply to meet the needs and demands of the rest of the world. There is an unfortunate paradox in the fact that economic demands in the rich societies, which can afford it, are moving away from wishing to support conventional productivity-enhancing agricultural research – witness the widespread opposition to genetic modification of agriculture's biological resources – while the poor societies, for whom the benefits would be substantial, have a social and economic need for such research but do not possess the income or the capacity to finance it. In an economic system increasingly driven by the globalization of markets, competitive sourcing and the purchasing power of the individual, the agricultural scientist confronts an uncomfortable dilemma. He can see the *need* for continued technological developments on the moral grounds of 'feeding the world', but the essentially self-centred ethic of the market economy and its associated presumption against public sector expenditures creates a bias against funding R&D as an international public good. In such circumstances, as experience has repeatedly shown, the major technical developments are driven by the research budgets and commercial advantage of multinational corporations and the agricultural supply industries. The likely outcome is then not necessarily the kind of agricultural developments that, on any objective basis, would be considered relevant to small-scale farmers in poor countries; it could equally be the distortion of food markets and the transfer of inappropriate technologies to the developing nations whose populations and agricultures require developments responsive to their own internal conditions.

15.9 Grassland and the non-food system

As they enter the twenty-first century, the Western economies seem to have reached a position of overall (i.e. except for specific groups) food *sufficiency*. Furthermore, their average income levels are such that, should there be a reduction in food supplies on a global basis, they would be well able to purchase the supplies from world markets, leaving the poor nations to suffer the shortages. Hence in this sense, as well as in the dependability of its technology, the West can consider itself to be economically food *secure* as well. When society no longer seeks a continued growth in food output from its agricultural resources, it reasonably begins to consider what other demands those resources could satisfy.

In economic terms, therefore, if efficiency means using available resources to create the greatest value to society, this implies a progressive diversion of land to other uses. In the past this has been referred to in terms of the 'loss' of agricultural land, causing housing development and new roads to be viewed with concern as a threat to food production capacity (Edwards & Wibberley 1971). Yet in reality, with the prospect of a constant stream of new technology based in biological, chemical, mechanical and managerial improvements (which seem to be driven by a force of science-push rather than demand pull), it is clear now that the relatively static quantity of agricultural output required to serve as raw materials for the food system can be met from progressively less and less of the agricultural land area.

There are a number of other uses for rural land which are just as valid in economic terms as producing agricultural products, but at the margin may well yield far greater benefit to society than their continued use in conventional farming. It is necessary to distinguish these uses into two types, depending on whether or not they can be handled within the normal context of commercial forces.

15.9.1 Commercial non-food grassland use

This category includes an array of different commodities that farmers can produce as an alternative to their traditional crop and livestock enterprises. Arable land can be used to supply 'industrial' crops – for oil or pharmaceutical purposes, for bio-fuels – and though they may not be substantial, the markets for these products may offer more profitable opportunities for some landowners. In hill areas, forestry becomes a more genuinely economic form of land use than continuing livestock production, particularly in the face of declining support and market prices and the greater competitive advantage of lowland areas.

However, the greatest economic possibilities under this heading probably lie in the alternative commercial uses of grassland. One of the major areas of expenditure growth in an affluent society is in leisure and recreation, and once food needs have been met, the agricultural land area can make its greatest contribution to society's benefit by supplying such demands. (Recreation expenditure in the UK now exceeds expenditure by households on food purchases, and greatly exceeds the value of output from the agricultural sector (Office for National Statistics 1998).) In an increasing number of rural and suburban households, for example, the ownership of a horse is a significant element in their lifestyle, and it is estimated that in the UK over 3.5% of the grassland area is used to support the horse population (Peat, Marwick, McClintock 1988). Another highly valued way of utilizing grassland is via facilities such as golf courses and sports fields; such land areas may not necessarily remain in the ownership of agricultural producers but, as stated at the beginning of this chapter, it is not solely the *farmers'* interests that define economic value in grassland production and utilization. By contrast, many farmers can operate a commercial recreation enterprise within their main agricultural business in the form of a campsite, caravan park or other grassland area on which people pay to acquire the services provided – whether a picnic area, banger-racing track, hot-air balloon launching site or a tourist attraction such as the farm country parks that are springing up in many rural areas. Finally, the continued development of new housing estates and the expanding interest in gardens means that turf production is both an economically valued form of grassland utilization and a nice little earner for some farmers.

Because our presumptions have grown up from the association of farming and food supply, the automatic reaction of many is to think of all of these things as not 'proper' grassland use, or to dismiss them as 'niche markets'. Although true, this does not make them unimportant. From the standpoint of the economy, *every* activity is a niche market because it occupies only a minority of people or resources; however, although dependable statistics are not available in many areas, it is quite possible that as many people gain an income from 'horseyculture' as from horticulture. Clearly production of commodities for the food system will remain the dominant role of grassland. However, the important areas for study and new information are not necessarily those where roles are established and continuing; they are those where change is taking place, new activities in prospect, with resource-use adjustments and new modes of operation needed. If land use is driven by the developing demands of society, those demands are increasingly focused on the benefits to be derived from the non-food uses – and grassland is more central to this focus than arable land. In the context of the rest of this book, of course, it is not clear that the output value of recreational grassland is very sensitive to detailed consideration of grass species, sward establishment and management, or weeds, and there may not be much scope to apply the conventional concepts of productivity and efficiency, but this will not actually be known until the issues are examined. If these are the grassland uses where the greatest demand growth is likely to take place then, conceptually at least, the same principles of seeking the best way to use resources that have served agriculture so well should be applied.

15.9.2 Non-commercial grassland use

The many non-commercial ways in which grassland can generate economic value is treated separately because its 'outputs' are not of the type that can be exchanged for money between those landowners (producers) who can supply and those consumers who want to acquire. They are in the nature of public goods, to which individuals cannot have property rights, as compared with the private goods (and services) produced and traded on the market. Although genuinely *economic* commodities, in that they are valued by society and require resources to ensure their availability, they are not *commercial*; they are different in their nature, and hence in the appropriate frameworks for their efficient production and utilization. The dominant non-market commodity associated with agricultural land is the array of elements and features contained in the collective term 'the environment' as commonly used. Grassland, in particular, is a major source of these features (see Chapter 11). The list includes not only the diversity of 'natural' species of grass, wild flowers and other flora, but the varied bird, animal and insect life found within it. In addition, it embraces features considered to have landscape or visual merit – traditional pastures, hillsides, rolling downland, riverside meadows – often intangible things that people recognize and value but tend not to think of as 'commodities', but of course they are commodities in terms of economic analysis and resource use. Although all these environmental components exist naturally, their continued existence is frequently in competition with the more profitable use of the land in response to commercial (food and non-food) demands. The market for agricultural products and the technology of land improvement, mechanization, agrochemicals, fertilizers, high-yielding grass species etc. leads to the transformation of grassland and its utilization into what has been viewed as 'more productive' forms, where productivity is measured directly or indirectly in terms of the quantity of saleable produce for the food system. Now, a well-fed, food-secure, mobile and environmentally more aware society increasingly places an intrinsic value on the countryside generally, with particular value associated with specific locations, features and characteristics. It is prepared to pay for the provision (i.e. the retention) of wetlands/buttercup meadows/species-rich grassland/biodiversity/unreclaimed hillsides etc., not necessarily in the allocation of resources to actually *produce* them (which is rarely a technical option), but in the form of foregone agricultural output and marginally cheaper food that might have been available had land use followed totally the dictates of the market. It is this recognition of a value, and the willingness to incur a cost, that makes the environmental characteristics of grassland an economic commodity in society.

There are three important aspects to note in passing here. First, the major economic forces on rural land use in the affluent developed economies will increasingly focus on meeting environmental objectives. Interest in the environment (like in leisure, personal transport, eating out, pension provision and 'lifestyle') is something which arrives and grows with increasing income; food demand, by contrast, after a certain point grows very little with income. Second, this increasing demand for environmental goods does not imply that *all* land should now be used to provide them, any more than all grassland should be used to produce milk or sheepmeat or golf courses; as with all economic choices, it is simply seeking more than there otherwise would be. Third, this demand cannot be expressed through the market, because the elements concerned cannot be paid for by those individuals who want more of them (and not everyone does). Hence, although they have a perceived value, they have no recorded price, nor any easy mechanism for establishing one. In order that the payments necessary to ensure their supply can be made (i.e. landowners are compensated for not destroying them and taking an income from the extra agricultural products they could produce), it is necessary to institute a public payment scheme whereby society collectively (the taxpayer) pays the price and society collectively, rather than just a group of willing buyers, receives the benefit. A major area of current theoretical and empirical research in environmental economics is focused on methods for estimating the 'shadow' prices that a market would generate if it existed (see Hanley *et al.* 1997), and which therefore represent the appropriate payment levels for environmental policies.

15.10 Grassland production and the environment

The changing balance in society's concerns about food production and the countryside environment is initiating new conflicts about the use of rural land. Some conservation interest groups accuse modern 'intensive' farming of engaging in unsustainable monoculture and pursuing commercial production at the expense of major environmental side-effects – pollution, loss of biological diversity, destruction of landscape features, etc. By contrast, farmers assert that the countryside, both physically and visually, is created and maintained by farming activity, that without agriculture there would be no such environment, and that the accusations of environmental damage are overstated. Both sides are correct, of course. Much of what is valued in the landscape is there because of farming and, while there have been countryside changes, they are a continuation of historical agricultural development. On the other hand, the growing commercialization of farming and the particular nature of modern mechanical and chemical technology has had distinctive technical side effects, which involve genuine economic costs not accounted for in the market processes that drive agricultural resource use. In effect this points up a weakness, not just in the framework of the agricultural technologist, but also in the conventional economist's model for analysing resource-use efficiency. More realistically, the production function should be represented as

$$[Q, B] = f(X_1, X_2, \cdots, X_n, \quad X_m, \cdots, X_r)$$

In this formulation, the output from agricultural resource use is recognized as a composite of agricultural 'goods' Q, the conventional crop and livestock prod ucts, and agricultural 'bads' B, the various outputs of zero or negative value such as slurry, silage effluent, nitrogen run-off and other chemical pollutants. Furthermore, in addition to the resources $X_1, \ldots, X_n$ recognized as the conventional farming inputs, $X_m, \ldots, X_r$ represent other elements which are 'used up' in production. These encompass such things as hedgerows, species-rich grassland, wetlands and other habitats that may be lost, wildlife and biodiversity reductions, and even the welfare of farm animals which are exploited; they are not 'resources' in the conventional sense, but nevertheless they represent in the same way elements of value that are given up in gaining the output Q. Both the 'bads' and the 'environmental inputs' are authentic economic commodities because society attaches value to them and their availability is determined by economic action; however, because they have no money prices and are therefore omitted from the accounting frameworks used to guide or assess agricultural resource use, they are ignored in commercial decisions about agricultural production. They are effectively treated as free goods, and as a result there is an inevitable tendency to over-produce the one and over-use the other. This is how the current agriculture/environment conflict has come about, and the more well-fed, food-secure, affluent and environmentally aware society becomes, the higher the valuations it places on these elements. Ultimately they have to come into the reckoning of what an 'efficient' resource use in agriculture means, but their lack of market valuations or money prices means their values have to be reflected via public policy measures. Such measures involve regulating, taxing, or imposing financial penalties on the production of bads, the loss of environmental inputs and reductions in animal welfare, or alternatively making payments to leave unused (conserve) specified environmental resources that society does not wish to be sacrificed in the pursuit of further agricultural output. This has resulted in the definition of wider concepts of value that economic analysis has to deal with (Pearce & Turner 1990). Resources that contribute via their technical productivity to conventional farm output have what is termed a 'use value'; by contrast, countryside resources such as wildlife, visual amenity and species-rich pasture have an 'existence value', yielding their benefit not through their ability to contribute to consumed agricultural products, but simply because of their very nature. The fact that people are prepared to pay to preserve them and prevent their loss in farming processes (either directly via membership of conservation bodies or indirectly via taxation) is indicative of their comparability in economic terms. This is the basis for the agri-environmental schemes that are developing in the contemporary frameworks of agricultural policy.

The unpriced, non-commercial economic values are arguably greater and more varied in grassland production than in arable production. Yellow fields of oilseed rape or blue fields of linseed add a certain variety to the visual scene, but the broad sweep of a green countryside with grazing livestock is probably closer to most people's image of a pleasant landscape. The biodiversity and distinctive habitats associated with grassland are extremely high, in terms of both flora and fauna (implying, paradoxically, that in many instances – upland pastures, wetlands, chalk downland, traditional meadows, for example – grassland is assigned a high value because of its *low* productivity!), and purely from a recreational point of view, the pleasures of walking through, driving through or simply lying in the countryside are undoubtedly greater in a grassland environment. It is an erroneously narrow viewpoint to consider the contribution of grassland to society's economic benefit only in relation to its technical productivity in agricultural production or its ability to generate farm incomes.

The recognition that grassland can be managed to generate either agricultural value or environmental value points up what many see as a conflict of interest between farmers and environmentalists. From the standpoint of the economy, however, it is simply a typical economic problem of the choice between alternative activities in gaining the greatest value from rural land resources. In this sense it is no different in principle from having to decide between using grassland to produce milk or sheepmeat; the only substantial difference being that milk and sheepmeat have market price indicators of their relative value, whereas environmental outputs carry no such monetary weighting – so it is an information problem that prevents the definition of efficiency. The situation can be represented as in Fig. 15.3, which portrays the possibilities for using a stock of grassland to generate either environmental and recreation benefits, or agricultural output benefits. Note there are diverse components within each of these generalized categories which are largely in competition with one another. We are accustomed to the idea of choosing to use grassland

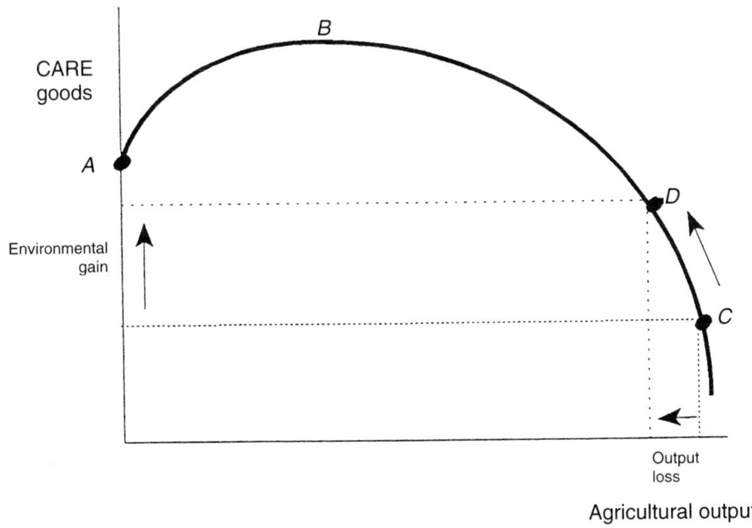

Fig. 15.3 Agriculture/environment conflicts in land use.

for sheep or cattle or dairy cows. It is equally the case that recreational access to the countryside frequently conflicts with the conservation of wildlife and rare plant species, and a 'natural' wilderness habitat is inconsistent with the image of a tidy and visually pleasing landscape.

The concept of 'CARE goods' (standing for conservation, amenity and the rural environment) has been developed as a generic label for the diverse elements of the non-agricultural output that rural land can provide (McInerney 1986). If no effort were made to farm the land (indicated as point *A*), the resulting natural biological environment would provide society with a particular level of CARE goods in the form of a natural wilderness landscape. As agricultural activity increases it results in the provision of more CARE goods (although of a different composition) because the farmed environment is perceived by society as being more visually pleasant, varied and accessible. Over this range up to point *B*, farming and the environment are complementary, but this cannot continue indefinitely and there is an unavoidable logic which changes the relationship to one of competition.

Expansion of agricultural production and the continual pursuit of higher productivity through intensification of land use, aided by sequential developments in agricultural technology, gradually change the nature of the countryside, and its output gains come progressively at the expense of the CARE goods that farmland provides. The inevitable influence of the law of diminishing marginal returns also means that increments in agricultural production are achieved at increasingly greater cost in terms of CARE-good provision. Ultimately a point could be reached (perhaps less likely in grassland than in arable farming) where the land area becomes primarily a factory for agricultural products with very few valuable characteristics as a countryside environment.

Not only does technological development make possible this progression; the economic incentives to landowners actively encourage it. There are no financial returns for CARE-good provision, despite the fact that society widely demands them, because there are no markets where those demands can be reflected in prices and enable farmers to receive payment for supplying them. By contrast the markets for agricultural products provide very clear signals as to the financial gains to be made from exploiting the land's technical productivity in farming – and agricultural policies have for decades magnified those signals even further through market intervention and price support. Not surprisingly, land use is therefore steadily drawn towards a position such as point *C*, where people realize that an imbalance has come about with a perceived excess of agricultural output and a relative deficiency of CARE goods. Hence, the current public clamourings for the reform of agricultural support policy, constraints on land 'improvement' and further agricultural innovation, restrictions on fertilizer and chemical use, and the introduction of incentives for 'extensification', stewardship schemes and the encouragement of environmentally friendly farming. This amounts to a desire to return to some previous land-use position equivalent to point *D*, it being judged that the benefits from a more preferred rural environment (although not measurable in comparable terms) outweigh the small sacrifice in agricultural land productivity. It remains a moot point as to whether this readjustment is in all cases possible, since some agricultural land improvements have irreversibly destroyed the landscape features whose value is now being emphasized.

It is equally arguable that the preferences leading this realignment belong primarily to people whose interests and income allow them to consume CARE goods, whereas there are many low-income (or simply totally disinterested) people in society for whom cheaper food is still of major concern. However true this may be – and there is always a diversity of valid interests in a society – there is no denying the widespread and growing contemporary demands for a realignment of objectives in rural land use that will affect every type of farming. This is the context of the new policy frameworks being sought for European agriculture in the twenty-first century, as outlined in the Agenda 2000 document of the Commission. It also represents a central problem to be resolved in specifying the wider economic aspects of grassland production and utilization. There are no simple recipes to follow or standard recommendations that can be offered to livestock farmers either. It is commonly asserted, for example, that low input/low output grassland systems are needed to deliver environmental benefits (Allen 1995). If this is true, it can only apply to particular types of environmental benefits, and it does not imply that *all* grassland has to provide CARE goods anyway, any more than that all grassland should be devoted to milk production. Like every other economic commodity, the demand for environmental goods is not limitless and has to be satisfied in balance with other economic goods that the land provides. It would be neither necessary nor efficient for every farmer and every hectare of grassland to be emphasizing CARE-goods provision. Food production will remain the dominant focus in livestock farming, and intensive land use will remain the best route for that purpose. It is simply that as demand for CARE goods grows while food demand remains static, some land will find a more valued use in providing, at some cost to agricultural output, the environmental services that grassland can offer.

If the problem is resolved rationally, then livestock farmers will not all find themselves (as many fear) hampered by arbitrary land-use constraints that prevent them from practising efficient and profitable grassland farming. Rather, the meaning of 'efficient and profitable' will change to reflect better the range of benefits society seeks from its land area. Many farmers will increasingly expect to derive their incomes from the dual sources of market revenues from the sale of (not necessarily price-supported) agricultural products to individual buyers, plus specific receipts (not subsidies) for the provision or conservation of various non-food and environmental goods and services to the extent demanded and paid for collectively through public expenditures. The balance of these different income streams will differ from farm to farm and across regions, depending significantly on the productivity of the land in producing valued CARE goods relative to agricultural products.

As stated at the outset, the *real* economics of grassland production and utilization ultimately revolve around such issues. They determine the best way a nation's grassland should be utilized and, consistent with this, the most profitable production methods that individual grassland farmers should adopt. Although the principles are clear, unfortunately the accounting frameworks and economic information needed for specifying the best way of using farming's resources do not at present provide a complete guide for this purpose.

15.11 References

Allen T.D. (1995) Environmental benefits from grassland farming. In Pollott G.E. (ed) *Grassland into the 21st Century: Challenges and Opportunities*. British Grassland Society Occasional Symposium No. 29, 135–42. BGS, Reading.

Anderson J.R. (1995) Food and agriculture – a global perspective. In Marshall B.J. & Miller F.A. (eds) *Priorities for a New Century – Agriculture, Food and Rural Policies in the European Union*. CAS Paper 31, Centre for Agricultural Strategy, University of Reading.

Cochrane W.W. (1958) *Farm Prices, Myth and Reality*. University of Minnesota Press, Minneapolis, MN.

Dawson P.J. & Hubbard L.J. (1987) Management and size economies in the England and Wales dairy sector. *Journal of Agricultural Economics* **38**, 27–38.

Dillon J.L. (1968) *The Analysis of Response in Crop and Livestock Production*. Pergamon, Oxford.

Edwards A. & Wibberley G.P. (1971) *A Land-Use Budget for the UK*. Wye College, University of London.

Evenson R.E. (1977) Comparative evidence on returns to investment in national and international research institutions. In Arndt T.M., Dalrymple D.G. & Ruttan V.W. (eds) *Resource Allocation and Productivity in National and International Agricultural Research.* University of Minnesota Press, Minneapolis, MN.

Farrar J. & Franks J. (1998) *Economics of Milk Production in England & Wales, 1996/97.* Special Studies in Agricultural Economics Report No. 41, University of Manchester.

Fogerty M.W. & Turner M.M. (1996) *Lowland Sheep 1994: Production Economics and Management.* Special Studies in Agricultural Economics Report No. 31, Agricultural Economics Unit, University of Exeter.

Giles A.K. & Stansfield J.M. (1980) *The Farmer as Manager.* George Allen & Unwin, London.

Hanley N., Shogren J.F. & White B. (1997) *Environmental Economics in Theory and Practice.* Macmillan, London.

Heady E.O. & Dillon J.L. (1961) *Agricultural Production Functions.* Iowa State University Press, Ames, IA.

Hill B. (1990) *An Introduction to Agriculture for Students of Agriculture.* Pergamon, Oxford.

Hill B. & Ray D. (1987) *Economics of Agriculture.* Macmillan, London.

Jenkins T., Jones E., McDougall I. & Williams H. (1998) *The Economics of Lowland Beef Production, 1995 and 1996.* University of Wales Aberystwyth, Special Studies in Agricultural Economics, Report No. 36.

MAFF (Ministry of Agriculture, Fisheries & Food) (1997) *National Food Survey 1996.* The Stationery Office, London.

MAFF (Ministry of Agriculture, Fisheries & Food) (1998) *Farm Incomes in the United Kingdom, 1996/97.* The Stationery Office, London.

McInerney J.P. (1986) Agricultural policy at the crossroads. In Gilg A. (ed) *Countryside Planning Yearbook.* Geo Books, Norwich.

Nix J.S. (1997) *Farm Management Pocketbook*, 28th edn. Wye College Press, Wye, Kent.

Office for National Statistics (1998) *Annual Abstract of Statistics.* The Stationery Office, London.

Pearce D.W. & Turner R.K. (1990) *Economics of Natural Resources and the Environment.* Harvester Wheatsheaf, Hemel Hempstead.

Peat, Marwick, McClintock (1988) *The Economic Contribution of the British Equine Industry*, 54 pp. British Horse Society, Stoneleigh.

Smith A. (1776) *The Wealth of Nations.* Reproduced in Everymans Library, Dent, London.

Index